Texts in Computer Science

Editors
David Gries
Fred B. Schneider

For further volumes:
www.springer.com/series/3191

Rudolf Kruse · Christian Borgelt ·
Frank Klawonn · Christian Moewes ·
Matthias Steinbrecher · Pascal Held

Computational Intelligence

A Methodological Introduction

Rudolf Kruse
Faculty of Computer Science
Otto-von-Guericke University Magdeburg
Magdeburg, Germany

Christian Borgelt
Intelligent Data Analysis & Graphical Models Research Unit
European Centre for Soft Computing
Mieres, Spain

Frank Klawonn
FB Informatik
Ostfalia University of Applied Sciences
Wolfenbüttel, Germany

Christian Moewes
Faculty of Computer Science
Otto-von-Guericke University Magdeburg
Magdeburg, Germany

Matthias Steinbrecher
SAP Innovation Center
Potsdam, Germany

Pascal Held
Faculty of Computer Science
Otto-von-Guericke University Magdeburg
Magdeburg, Germany

Series Editors
David Gries
Department of Computer Science
Cornell University
Ithaca, NY, USA

Fred B. Schneider
Department of Computer Science
Cornell University
Ithaca, NY, USA

ISSN 1868-0941 ISSN 1868-095X (electronic)
Texts in Computer Science
ISBN 978-1-4471-5849-3 ISBN 978-1-4471-5013-8 (eBook)
DOI 10.1007/978-1-4471-5013-8
Springer London Heidelberg New York Dordrecht

Printed on acid-free paper

Springer is part of Springer Science+Business Media (www.springer.com)

Preface

Computational Intelligence comprises concepts, paradigms, algorithms and implementations of systems that are supposed to exhibit intelligent behavior in complex environments. It relies heavily on sub-symbolic, predominantly nature-analog or at least nature-inspired methods. These methods have the advantage that they tolerate incomplete, imprecise and uncertain knowledge and thus also facilitate finding solutions that are approximative, manageable and robust at the same time.

The choice of topics in this books reflects the most important fields in the area of Computational Intelligence. Classical fields such as *Artificial Neural Networks*, *Fuzzy Systems* and *Evolutionary Algorithms* are described in considerable detail. However, newer methods such as *Ant Colony Optimization* and *Probabilistic Graphical Models* are discussed as well, although a complete coverage of all approaches and developments is clearly impossible to achieve in a single volume.

Rather than to strive for completeness, our goal is to give a methodical introduction to the area of Computational Intelligence. Hence, we try not only to present fundamental concepts and their implementations, but also explain the theoretical background of proposed problem solutions. In addition, we hope to convey to a reader what is necessary in order to apply these methods successfully.

This textbook is primarily meant as a companion book for lectures on the covered topics in the area of Computational Intelligence. However, it may also be used for self-study by students and practitioners from industry and commerce. This book is based on notes of lectures, exercise lessons and seminars that have been given by the authors for many years. On the book's website

http://www.computational-intelligence.eu/

a lot of additional material for lectures on Neural Networks, Evolutionary Algorithms, Fuzzy Systems and Bayesian Networks can be found, including module descriptions, lecture slides, exercises with solutions, hints to software tools etc.

Magdeburg, Germany

Rudolf Kruse
Christian Borgelt
Frank Klawonn
Christian Moewes
Matthias Steinbrecher
Pascal Held

Preface

Contents

Chapter 1
Introduction

1.1 Intelligent Systems

Complex problem settings in widely differing technical, commercial and financial fields evoke an increasing need for computer applications that must show "intelligent behavior." These applications are desired to support decision making, to control processes, to recognize and interpret patterns or to maneuver vehicles or robots autonomously in unknown environments. Novel approaches, methods, tools and programming environments have been developed to accomplish such tasks.

Seen from a higher level of abstraction, the general requirements for developing such an "intelligent system" are ultimately always the same, namely simulating intelligent thinking and actions in a certain field of application (Russell and Norvig 2009). For this purpose, the knowledge about this field must be represented and processed. The quality of the resulting system mainly depends on how well the knowledge representation problem is solved in the development process. There is no such thing as the "best" method. Rather one has to sift through many available approaches to find those that fit the application area of the intelligent system best.

The mechanisms and processes that underlie intelligent behavior are examined in the research area of artificial intelligence. Similar to most other areas of computer science (or science in general), computational intelligence (CI) comprises both theoretical aspects (*how and why do these system work?*) and application-oriented aspects (*where and when can these systems be used?*) (Luger 2005).

At the beginning of the development of intelligent systems, researchers often focused the idea of seeing "a human being as a machine"—an idea that stems from the age of enlightenment. They ventured to create an (artificial) intelligence that can both think creatively and solve problems in the way a human can. This intelligence was also meant to exhibit forms of both consciousness and emotions. In the infancy of the field, the typical way to design an artificial intelligence was to describe a symbolic basis of the relevant mechanisms. This includes the top-down perspective of problem solving, which mainly addresses the question why these systems work (Minsky 1991). The answer to this question is usually given with the help of a symbolic representation and a logic-based inference mechanism. Techniques

R. Kruse et al., *Computational Intelligence*, Texts in Computer Science,
DOI 10.1007/978-1-4471-5013-8_1, © Springer-Verlag London 2013

that are examples of these approaches include rule-based expert systems, automatic theorem provers, and many operations research techniques which underlie modern planning and scheduling software. Although these traditional approaches have been very successful in some cases, they do have limitations, especially when it comes to scalability. A tiny complication of the problem to solve often causes an increase of complexity that cannot be handled feasibly. As a consequence, although these approaches usually guarantee an optimal, precise or correct solution, they are rarely applicable to practical problems (Hüllermeier et al. 2010).

Consequently, efficient method to represent and process knowledge are still a research topic. For certain types of problems, techniques that are inspired by natural or biological processes proved successful (Brownlee 2011). These approaches signify a paradigm change away from symbolic representations and towards inference strategies for adaptation and learning. Among such methods we find artificial neural networks, evolutionary algorithms and fuzzy systems (Engelbrecht 2007; Mumford and Jain 2009). These novel methods have demonstrated their usefulness in many application areas, often in combination with traditional problem-solving techniques.

1.2 Computational Intelligence

The research area of computational intelligence (CI) comprises concepts, paradigms, algorithms and implementations to develop systems that exhibit intelligent behavior in complex environments. Typically, sub-symbolic and nature-analogous methods are adopted that tolerate incomplete, imprecise and uncertain knowledge. As a consequence, the resulting approaches allow for approximate, manageable, robust and resource-efficient solutions.

The general strategy that is adopted in the area of computational intelligence is to apply approximation techniques and methods that can find coarse, incomplete or only partially valid solutions to given problems. As a reward for dispensing with guaranteed correctness and completeness, solutions are found in a tolerable time frame and within a bearable budget. Such solutions often consist of relatively simple sub-functions, which, through interaction, lead to complex and self-organized behavior. As a consequence, these heuristic approaches can usually not be analyzed in a classical fashion, but, in exchange, they offer the possibility to quickly find approximate solutions to problems that are difficult to solve in other ways.

It is obvious that an area as diverse as computational intelligence cannot be covered exhaustively in about 500 pages. Therefore, we confine ourselves to four core techniques that are frequently used in practice. In the first two parts of this book, so-called *nature-analogous* or *natured-inspired methods* are discussed. Here the governing idea is to analyze problem-solving strategies as they occur in nature. Certain aspects of such solution strategies are then mimicked or simulated in a computer, usually without striving to model the original systems correctly and exhaustively, nor even ensuring the biological plausibility of the simulation. Particularly successful and thus practically relevant representatives of this kind are artificial neural networks (Haykin 2008), evolutionary algorithms (Rozenberg et al. 2012) ant colony optimization (Dorigo and Stützle 2004).

Many ideas and principles in the area of artificial neural networks are inspired by neuroscience. Artificial neural networks are information-processing systems whose structure and functionality simulates the nervous systems and particularly the brain of animals and human beings. They consists of a large number of fairly simple processing units, the so-called neurons, which work in parallel. These neurons send information in the form of action potentials via directed links to other neurons. Based on knowledge about the functionality of biological neural networks, one tries to model and mimic them, especially to achieve learning capability.

Evolutionary algorithms draw on ideas from biological evolution, in which organisms, over many generations, get adapted to environmental conditions. They address certain classes of optimization problems and belong to the family of metaheuristics, which offers algorithms to approximately solve may types of optimization problems. Metaheuristics are defined by an abstract sequence of steps that are applicable to more or less arbitrary problem descriptions. However, every single step must be implemented in a problem-specific fashion. Metaheuristics are often applied to problems for which no efficient solution algorithm is known. Although finding an optimal solution is usually not guaranteed and thus a found solution can, in principle, be arbitrarily bad compared to the optimal solution, metaheuristics offer the possibility to obtain (sufficiently) good solutions in a reasonable time frame.

The latter two parts of the book focus on integrating uncertain, vague and incomplete knowledge into the problem-solving strategy. The governing idea in these parts is that for human beings even imperfect knowledge can be very valuable. Therefore, it is desirable to enable computers to work with such knowledge as well and not just precise and certain knowledge. Particularly successful approaches that can handle vague and uncertain knowledge are fuzzy systems (Kruse et al. 1994; Michels et al. 2006) and Bayesian networks (Borgelt et al. 2009).

In fuzzy systems, vague knowledge, which may be provided by a human expert or formulated intuitively by a system developer, is formalized with the help of fuzzy logic and fuzzy set theory. Fuzzy approaches can also be used to derive inference mechanisms, thus giving rise to approximate reasoning methods. Fuzzy systems are routinely applied in control engineering because in many application scenarios a precise and complete modeling of the system is impractical or even impossible.

Bayesian networks are means to efficiently store and reason with uncertain knowledge in complex application areas. Formally, a Bayesian network is a probabilistic graphical model that represents a set of random variables and their conditional dependences by a directed acyclic graph. Due to the probabilistic representation, one can easily draw inferences based on new information. In addition, Bayesian networks are well-suited for dependence analysis and learning from data.

In many applications, hybrid computational intelligence systems, for instance, neuro-fuzzy systems, have proven to be highly useful. Sometimes these techniques are also combined with related methods, for example, from the area of machine learning or case-based reasoning (Hüllermeier et al. 2010).

1.3 About This Book

Our main objective with this textbook is to give a methodical introduction to the field of computational intelligence. Therefore, we focus on fundamental concepts and their implementation and strive to explain the theoretical background of proposed solutions to certain problems. We hope to convey to a reader all that is necessary for a profound application of the discussed methods. This book requires only fairly basic knowledge of mathematics, as we tried to introduce all necessary concepts and tools in this book in order to make it as self-contained as possible. Furthermore, the four parts about artificial neural networks, evolutionary algorithms, fuzzy systems and Bayesian networks can be studied independently of each other as we tried to avoid dependences between the parts or requiring prerequisites from earlier parts.

This book is intended as an accompanying book for lectures about the field of computational intelligence and is fundamentally based on written notes about lectures that the first author has given periodically for students of different areas for almost 20 years. On the website

http://www.computational-intelligence.eu

module descriptions, lecture slides matching the book, exercise sheets with solutions, sample exams, software demos, literature references, references to organizations, journals, software tools and additional material can be found for all four parts, that is, artificial neural networks, evolutionary algorithms, fuzzy systems and Bayesian networks. This additional material covers a total of four modules (i.e., lectures with corresponding exercise lessons).

References

C. Borgelt, M. Steinbrecher and R. Kruse. *Graphical Models: Representations for Learning, Reasoning and Data Mining*, 2nd edition. J. Wiley & Sons, Chichester, United Kingdom, 2009

J. Brownlee. *Clever Algorithms: Nature-Inspired Programming Recipes*. Lulu Press, Raleigh, NC, USA, 2011

M. Dorigo and T. Stützle. *Ant Colony Optimization*. MIT Press, Cambridge, MA, USA, 2004

A.P. Engelbrecht. *Computational Intelligence—An Introduction*, 2nd edition. J. Wiley & Sons, Chichester, United Kingdom, 2007

S.O. Haykin. *Neural Networks and Learning Machines*, 3rd edition. Prentice Hall, Upper Saddle River, NJ, USA, 2008

E. Hüllermeier, R. Kruse and F. Hoffmann (eds.) *Computational Intelligence for Knowledge-Based Systems Design*. Springer-Verlag, Berlin/Heidelberg, Germany, 2010

R. Kruse, J. Gebhardt and F. Klawonn. *Foundations of Fuzzy Systems*. J. Wiley & Sons, Chichester, United Kingdom, 1994

G.F. Luger. *Artificial Intelligence: Structures and Strategies for Complex Problem Solving*, 5th edition. Pearson Education, Essex, United Kingdom, 2005

K. Michels, F. Klawonn, R. Kruse and A. Nürnberger. *Fuzzy Control: Fundamentals, Stability and Design of Fuzzy Controllers*. Springer-Verlag, Berlin/Heidelberg, Germany, 2006

M. Minsky. Logical Versus Analogical or Symbolic Versus Connectionist or Neat Versus Scruffy. *AI Magazine*, 12(2):647–674. MIT Press, Cambridge, MA, USA, 1991

C.L. Mumford and L.C. Jain (eds.) *Computational Intelligence: Collaboration, Fusion and Emergence*. Springer-Verlag, Berlin/Heidelberg, Germany, 2009

G. Rozenberg, T. Bäck and J.N. Kok (eds.) *Handbook of Natural Computing*. Section III: Evolutionary Computation. Springer-Verlag, Berlin/Heidelberg, Germany, 2012

S.J. Russell and P. Norvig. *Artificial Intelligence—A Modern Approach*, 3rd edition. Prentice Hall, Upper Saddle River, NJ, USA, 2009

Part I
Neural Networks

Chapter 2
Introduction

(*Artificial*) *neural networks* are information processing systems, whose structure and operation principles are inspired by the nervous system and the brain of animals and humans. They consist of a large number of fairly simple units, the so-called *neurons*, which are working in parallel. These neurons communicate by sending information in the form of activation signals, along directed connections, to each other.

A commonly used synonym for "neural network" is the term "connectionist model." The research area that is devoted to the study of connectionist models is called "connectionism." Furthermore, the expression "parallel distributed processing" can often be found in relation to (artificial) neural networks.

2.1 Motivation

(Artificial) neural networks are studied for various reasons: in (neuro-)biology and (neuro-)physiology, but also in psychology, one is mainly interested in their similarity to biological nervous systems. In these areas (artificial), neural networks are used as computational models with which one tries to simulate and thus to understand the mechanisms of nerve and brain functions. Especially in computer science, but also in other engineering sciences, one tries to mimic certain cognitive powers of humans (especially learning ability) by using functional elements of the nervous system and the brain. In physics, certain mathematical models that are analogous to (artificial) neural networks are employed to describe specific physical phenomena. An example are models of magnetism, for instance, the Ising model.

As can already be seen from this brief list, the study of (artificial) neural networks is a highly interdisciplinary research area. However, in this book we widely neglect the use of (artificial) neural networks in physics (even though we draw on examples from physics to explain certain network models) and consider their biological basis only very briefly (see the next section). Rather, we focus on the mathematical and engineering aspects, particularly the use of (artificial) neural networks in the area of computer science that is commonly called "artificial intelligence."

R. Kruse et al., *Computational Intelligence*, Texts in Computer Science,
DOI 10.1007/978-1-4471-5013-8_2, © Springer-Verlag London 2013

While the reasons why biologists study (artificial) neural networks are fairly obvious, we may have to justify why neural networks are (or should be) studied in artificial intelligence. The reason is that the paradigm of classical artificial intelligence (sometimes called, in a somewhat pejorative manner, GOFAI—"good old-fashioned artificial intelligence") is based on a very strong hypothesis about how machines can be made to behave "intelligently." This hypothesis says that the essential requirement for intelligent behavior is the ability to manipulate symbols and symbol structures that are represented by physical structures. Here *symbol* means a token that refers to an object or a situation. This relation is interpreted in an operational manner: the system can perceive and/or manipulate the object referred to. This hypothesis was first formulated explicitly by Newell and Simon (1976):

> **Physical Symbol System Hypothesis**: A physical-symbol system has the necessary and sufficient means for general intelligent action.

As a matter of fact, classical artificial intelligence concentrated, based on this hypothesis, on symbolic forms of representing knowledge and in particular on propositional and predicate logic. (Artificial) neural networks, on the other hand, are no physical symbol systems, since they do not process *symbols*, but rather much more elementary *signals*, which, taken individually, rarely have a (clear) meaning. As a consequence, (artificial) neural networks are often called "sub-symbolic." However, if the ability to process symbols is necessary to produce intelligent behavior, then it is unnecessary to study (artificial) neural networks in artificial intelligence.

There is no doubt that classical artificial intelligence has achieved remarkable successes: nowadays computers can automatically solve many types of puzzles and brain-twisters and can play games like chess and Reversi (also known as Othello) on an extremely high level. However, when it comes to mimicking perception (seeing, hearing etc.), computers usually perform fairly poorly compared to humans—at least if symbolic representations are relied upon: here computers are often too slow, too inflexible and too little tolerant to noise and faults. We may conjecture that the problem is that in order to recognize patterns—a core task of perception—symbolic representations are not very well suited, because there are no adequate symbols on this level of processing. Rather "raw" (measurement) data needs to be structured and summarized before symbolic methods can effectively be applied. Hence, it appears to be reasonable to examine the mechanisms of sub-symbolic information processing in natural intelligent systems—that is, animals and humans—in more detail and possibly to exploit these mechanisms to mimic intelligent behavior.

Additional arguments why studying (artificial) neural networks may be beneficial arise from the following observations:

- Expert systems that use symbolic representations usually become slower with a larger knowledge base, because larger sets of rules need to be traversed. Human experts, however, usually become faster. Maybe a non-symbolic representation (as it is used in natural neural networks) is more efficient.
- Despite the fairly long switching time of natural neurons (in the order of several milliseconds) essential cognitive tasks (like recognizing an object) are solved in a

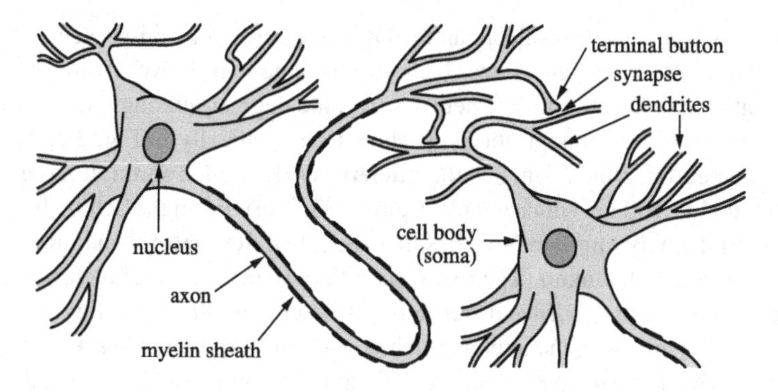

Fig. 2.1 Prototypical structure of biological neurons

fraction of a second. If neural processing were sequential, only about 100 switching operations could be performed ("100-step rule"). Hence, high parallelization must be present, which is easy to achieve with neural networks, but much more difficult to implement with other approaches.

- There is a large number of successful applications of (artificial) neural networks in industry, commerce and finance.

2.2 Biological Background

As already mentioned, (artificial) neural networks are inspired by the structure and the operation principles of the nervous system and particularly the brain of animals and humans. In fact, the neural network models that we study in this book are not very close to their biological original, since they are too simplified to model the characteristics of natural neural networks correctly. Nevertheless, we briefly consider natural neural networks here, because they formed the starting point for investigating artificial neural networks. The description follows (Anderson 1995).

The nervous system of animals consists of the brain (in so-called "lower" life forms often only referred to as the "central nervous system"), the different sensory systems, which collect information from the different body parts (visual, auditory, olfactory, gustatory, thermal, tactile etc. information), and the motor system, which controls movements. The greater part of information processing happens in the brain/central nervous system, although the amount of pre-processing outside the brain can be considerable, for example, in the retina of the eye.

W.r.t. processing information, the neurons are the most important components of the nervous system.[1] According to common estimates, there are about 100 billion (10^{11}) neurons in the human brain, of which a fairly large part is active in parallel. Neurons process information mainly by interacting with each other.

---

[1] The nervous system consists not only of neurons, not even for the largest part. Besides neurons there are various other cells, for instance, the so-called glia cells, which have a supporting function.

A **neuron** is a cell that collects and transmits electrical activity. Neurons exist in many different shapes and sizes. Nevertheless one can derive a "prototypical" neuron that resembles all kinds of neurons to some degree (although this is a fairly severe simplification). This prototype is shown schematically in Fig. 2.1. The **cell body** of the neuron, which contains the **nucleus**, is also called **soma**. It has a diameter of about 5 to 100 μm (micrometer, $1\ \mu\text{m} = 10^{-6}$ m). From the cell body extend several short, heavily ramified branches that are called **dendrites**. In addition, it has a long extension called **axon**. The axon can be between a few millimeters and one meter long. Axon and dendrites differ in the structure and the properties of the **cell membrane**. In particular, the axon is often covered by a **myelin sheath**.

The axons are the fixed paths along which neurons communicate with each other. The axon of a neuron leads to the dendrites of other neurons. At its end, the axon is heavily ramified and possesses at the ends of these branches **terminal buttons**. Each terminal button almost touches a dendrite or the cell body of another neuron. The gap between a terminal button and a dendrite is usually between 10 and 50 nm (nanometer; $1\ \text{nm} = 10^{-9}$ m) wide. Such a place, at which an axon and a dendrite almost touch each other, is called **synapse**.

The most common form of communication between neurons is that a terminal button of the axon releases certain chemicals, the so-called **neurotransmitters**, which act on the membrane of the receiving dendrite and change its polarization (its electrical potential). Usually the inside of the cell membrane, which encloses the whole neuron, is about 70 mV (millivolts; $1\ \text{mV} = 10^{-3}$ V) more negative than its outside, because the concentration of negative ions is greater on the inside, while the concentration of positive ions is greater on the outside. Depending on the type of the released neurotransmitter, the potential difference may be reduced or increased on the side of the dendrite. Synapses that reduce the potential difference are called **excitatory**, those that increase it are called **inhibitory**.

In an adult human, all connections between neurons are completely established and no new connections are created (again this is a severe simplification). An average neuron possesses between 1000 and 10,000 connections to other neurons. The change of the electrical potential that is caused by a single synapse is fairly small, but the individual excitatory and inhibitory effects can accumulate (counting the excitatory influences as positive and the inhibitory ones as negative). If the excitatory net input is large enough, the potential difference in the cell body can be significantly reduced. If the reduction is large enough, the axon's base is depolarized. This depolarization is caused by positive sodium ions entering the cell. As a consequence, the inside of the cell becomes temporarily (for about one millisecond) more positive than its outside. Afterwards the potential difference is rebuilt by positive potassium ions leaving the cell. Finally, the original distribution of sodium and potassium ions is reestablished by special ion pumps in the cell membrane.

The sudden, temporary change of the electrical potential, which is called **action potential**, propagates along the axon. The propagation speed lies between 0.5 and 130 m/s, depending on the properties of the axon. In particular, it depends on how heavily the axon is covered with a myelin sheath (the more myelin, the faster the action potential is propagated). When this nerve impulse reaches the end of the axon, it causes neurotransmitters to be released at the terminal buttons, thus passing the signal on to the next cell, where the process is repeated.

In summary, changes of the electrical potential are accumulated at the cell body of a neuron and, if they reach a certain threshold, are propagated along the axon. This nerve impulse causes that neurotransmitters are released by the terminal buttons at the end of the axon, thus inducing a change of the electrical potential in the receiving neuron. Even though this description is heavily simplified, it captures the essentials of neural information processing on the level of individual neurons.

In the human nervous system, information is encoded by continuously changing quantities, primarily two: the electrical potential of the neuron's membrane and the number of nerve impulses that a neuron transmits per second. The latter is also called the **firing rate** of the neuron. It is commonly assumed that the number of impulses is more important than their shape (in the sense of a change of the electrical potential), although competing theories of **neural coding** exist. A neuron can emit 100 or even more impulses per second. The higher the firing rate, the higher the influence a neuron has on connected neurons. However, in artificial neural networks this frequency coding of information is usually not emulated.

References

J.R. Anderson. *Cognitive Psychology and Its Implications*, 4th edition. Freeman, New York, NY, USA, 1995

M.A. Boden (ed.) *The Philosophy of Artificial Intelligence*. Oxford University Press, Oxford, United Kingdom, 1990

A. Newell and H.A. Simon. Computer Science as Empirical Enquiry: Symbols and Search. *Communications of the Association for Computing Machinery* 19. Association for Computing Machinery, New York, NY, USA, 1976. Reprinted in Boden (1990), 105–132

Chapter 3
Threshold Logic Units

The description of biological neural networks in the preceding chapter makes it natural to model neurons as **threshold logic units**: if a neuron receives enough excitatory input that is not compensated by equally strong inhibitory input, it becomes active and sends a signal to other neurons. Such a model was already examined very early in much detail by McCulloch and Pitts (1943). As a consequence, threshold logic units are also known as **McCulloch–Pitts neurons**. Another name which is commonly used for a threshold logic unit is **perceptron**, even though the processing units that Rosenblatt (1958, 1962) called "perceptrons" are actually somewhat more complex than simple threshold logic units.[1]

3.1 Definition and Examples

Definition 3.1 A **threshold logic unit** is a simple processing unit for real-valued numbers with n inputs $x_1, \ldots, x_n$ and one output y. The unit as a whole possesses a **threshold** θ. To each input x_i a **weight** w_i is assigned. A threshold logic unit computes the function

$$y = \begin{cases} 1 & \text{if } \sum_{i=1}^{n} w_i x_i \geq \theta, \\ 0 & \text{otherwise.} \end{cases}$$

The inputs are often combined into an input vector $\mathbf{x} = (x_1, \ldots, x_n)$ and the weights into a weight vector $\mathbf{w} = (w_1, \ldots, w_n)$. With the help of the scalar product, the condition tested by a threshold logic unit may then also be written as $\mathbf{wx} \geq \theta$.

We depict a threshold logic unit as shown in Fig. 3.1. That is, we draw it as a circle, in which the threshold θ is recorded. Each input is drawn as an arrow that points to the circle and that is labeled with the weight of the input. The output of the threshold logic unit is shown as an arrow that points away from the circle.

[1] In a perceptron there is, besides the actual threshold logic unit, an input layer that executes additional operations on the input signals. However, this input layer consists of immutable functional elements and therefore is often neglected.

R. Kruse et al., *Computational Intelligence*, Texts in Computer Science,
DOI 10.1007/978-1-4471-5013-8_3, © Springer-Verlag London 2013

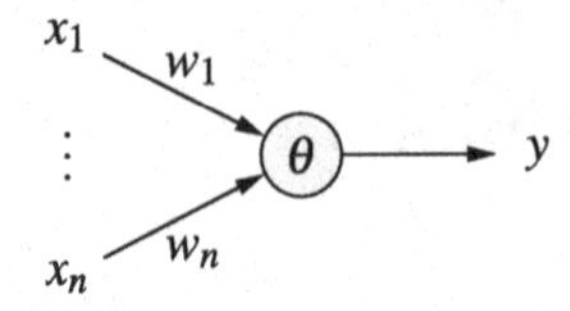

Fig. 3.1 Representation of a threshold logic unit

To illustrate how threshold logic units work and to demonstrate their capabilities, we consider a couple of simple examples. Figure 3.2 shows on the left a threshold logic unit with two inputs x_1 and x_2, which carry the weights $w_1 = 3$ and $w_2 = 2$, respectively. The threshold is $\theta = 4$. If we assume that the input variables can only have values 0 and 1, we obtain the table shown in Fig. 3.2 on the right. Clearly, this threshold logic unit computes the conjunction of its inputs: only if both inputs are active (that is, equal to 1), it becomes active itself and outputs a 1.

Figure 3.3 shows another threshold logic unit with two inputs, which differs from the one shown in Fig. 3.2 by a negative threshold $\theta = -1$ and one negative weight $w_2 = -2$. Due to the negative threshold, it is active (that is, outputs a 1) even if both inputs are inactive (that is, are equal to 0). Intuitively, the negative weight corresponds to an inhibitory synapse: if the corresponding input becomes active (that is, equal to 1), the threshold logic unit is deactivated and its output becomes 0. We can also observe here that positive weights correspond to excitatory synapses: even if the input x_2 inhibits the threshold logic unit, (that is, if $x_2 = 1$), it can become active, namely if it is "excited" by an active input x_1 (that is, by $x_1 = 1$). In summary, this threshold logic unit computes the function show in the table in Fig. 3.3 on the right, that is, the implication $y = x_2 \rightarrow x_1$.

An example for a threshold logic unit with three inputs is shown in Fig. 3.4 on the left. This threshold logic unit already computes a fairly complex function, namely the logical expression $y = (x_1 \wedge \overline{x_2}) \vee (x_1 \wedge x_3) \vee (\overline{x_2} \wedge x_3)$. The truth table of this function and the computations that are carried out by the threshold logic unit for the

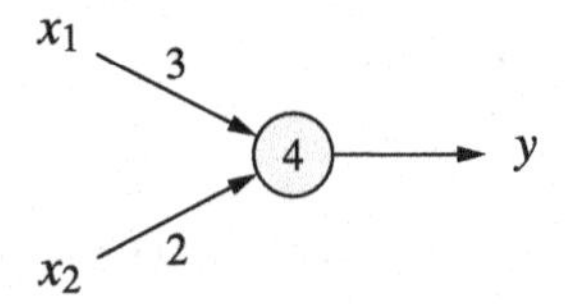

x_1	x_2	$3x_1 + 2x_2$	y
0	0	0	0
1	0	3	0
0	1	2	0
1	1	5	1

Fig. 3.2 A threshold logic unit for the conjunction $x_1 \wedge x_2$

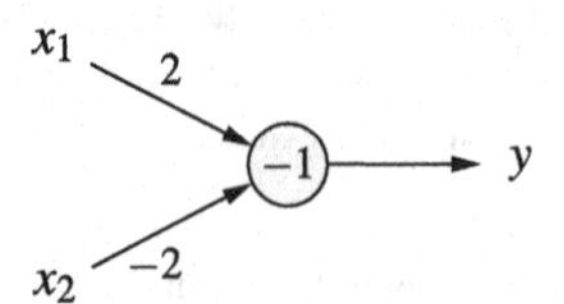

x_1	x_2	$2x_1 - 2x_2$	y
0	0	0	1
1	0	2	1
0	1	−2	0
1	1	0	1

Fig. 3.3 A threshold logic unit for the implication $x_2 \rightarrow x_1$

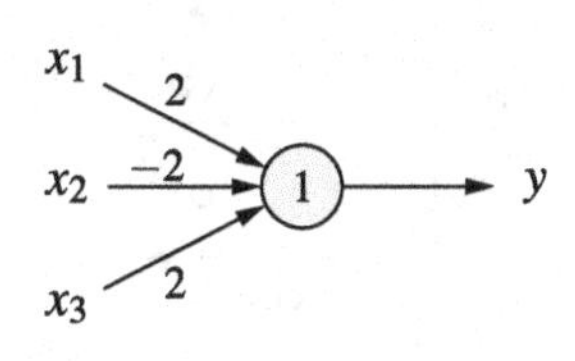

x_1	x_2	x_3	$\sum_i w_i x_i$	y
0	0	0	0	0
1	0	0	2	1
0	1	0	−2	0
1	1	0	0	0
0	0	1	2	1
1	0	1	4	1
0	1	1	0	0
1	1	1	2	1

Fig. 3.4 A threshold logic unit for $(x_1 \wedge \overline{x_2}) \vee (x_1 \wedge x_3) \vee (\overline{x_2} \wedge x_3)$

different input vectors are shown in Fig. 3.4 on the right. This and the preceding threshold logic unit may lead us to presume that negations (in logical expressions) are (often) represented by negative weights.

3.2 Geometric Interpretation

The condition that is tested by a threshold logic unit in order to decide whether it should output a 0 or a 1 is very similar to the equation of a straight line (cf. Sect. 10.1). Indeed, the computation of a threshold logic unit can easily be interpreted geometrically if we turn this condition into a line, plane or hyperplane equation, that is, if we consider the equation

$$\sum_{i=1}^{n} w_i x_i = \theta \quad \text{or} \quad \sum_{i=1}^{n} w_i x_i - \theta = 0.$$

(Note that the line equation differs from the actual condition in that is uses "=" instead of "≥." This is taken care of below, where the inequality is reinstated.) The resulting geometric interpretation is illustrated in Figs. 3.5, 3.6 and 3.8.

Figure 3.5 repeats on the left the threshold logic unit for the conjunction considered above. In the diagram on the right, the input space of this threshold logic unit is shown. The input vectors, which are listed in the table in Fig. 3.2 on the right, are marked according to the output of the threshold logic unit: a filled circle indicates that the threshold logic unit yields output 1 for this point, while an empty circle indicates that it yields output 0. In addition, the diagram shows the straight line described by the equation $3x_1 + 2x_2 = 4$, which corresponds to the **decision border** of the threshold logic unit. It is easy to verify that the threshold logic unit yields output 1 for all points to the right of this line and output 0 for all points to the left of it, even if we allow for other input values than 0 and 1.

On which side the output of the threshold logic unit is 1 and on which it is 0 can also be read from the line equation: it is well known that the coefficients of x_1 and x_2 are the elements of a normal vector of the line (cf. also Sect. 10.1, which

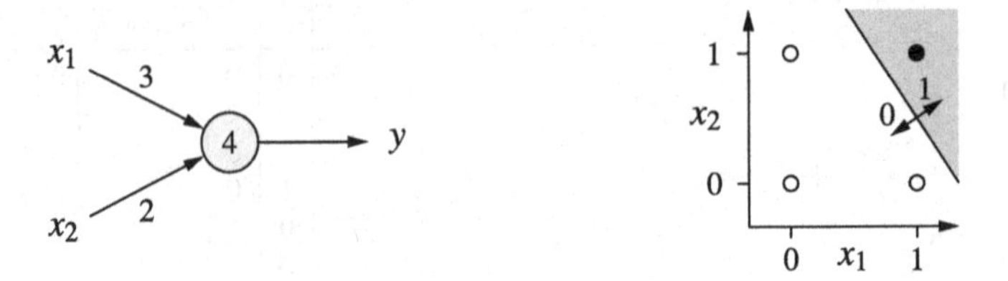

Fig. 3.5 Geometry of the threshold logic unit for $x_1 \wedge x_2$. The *straight line* shown in the right diagram has the equation $3x_1 + 2x_2 = 4$; for the *gray half-plane* it is $3x_1 + 2x_2 \geq 4$

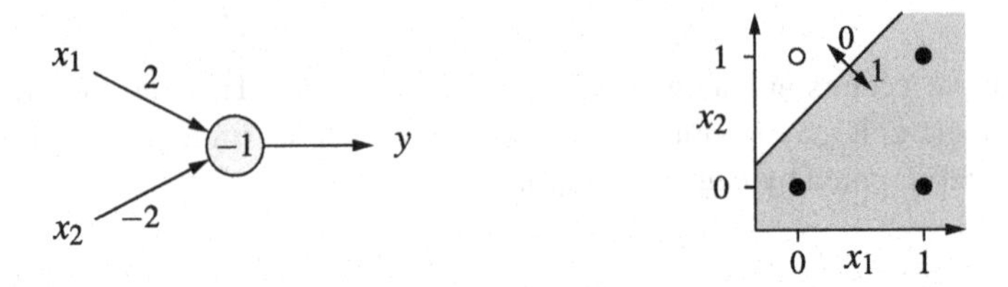

Fig. 3.6 Geometry of the threshold logic unit for $x_2 \rightarrow x_1$: The *straight line* shown in the right diagram has the equation $2x_1 - 2x_2 = -1$; for the *gray half-plane* it is $2x_1 - 2x_2 \geq -1$

collects some important facts about straight lines and their equations.) The side of the line to which this normal vector points if it is attached to a point on the line is the side on which the output is 1. Indeed, the normal vector $\mathbf{n} = (3, 2)$ that can be read from the equation $3x_1 + 2x_2 = 4$ points to the right top and thus to the side on which the point $(1, 1)$ is located. Hence the threshold logic unit yields output 1 in the **half-plane** $3x_1 + 2x_2 \geq 4$, which is shown in gray in Fig. 3.5.

Analogously, Fig. 3.6 shows the threshold logic unit computing the implication $x_2 \rightarrow x_1$ and its inputs space. The straight line drawn into this input space corresponds to its decision border: it separates the points of the input space for which the output is 0 from those for which the output is 1. Since the normal vector $\mathbf{n} = (2, -2)$ that can be read from the equation $2x_1 - 2x_2 = -1$ points to the bottom right, the output is 1 for all points below the line and 0 for all points above it. This coincides with the computations listed in the table in Fig. 3.3, which are represented by filled and empty circles in the diagram in Fig. 3.6. Generally, the threshold logic unit yields output 1 for points in the gray half-plane shown in Fig. 3.6.

Naturally the computations of threshold logic units with more than two inputs can be interpreted geometrically as well. However, due to the limited spatial imagination of humans, we have to confine ourselves to threshold logic units with no more than three inputs. With three inputs, the **separating line** turns into a **separating plane**. We illustrate this by depicting the input space of a threshold logic unit with three inputs as a unit cube as it is shown in Fig. 3.7. With this diagram, let us reconsider the example of a threshold logic unit with three inputs studied in the preceding section, which is repeated in Fig. 3.8. Into the unit cube on the right of this figure the plane with the equation $2x_1 - 2x_2 + 2x_3 = 1$ is drawn in gray, which corresponds to the decision rule of this threshold logic unit. In addition, all input vectors, for which

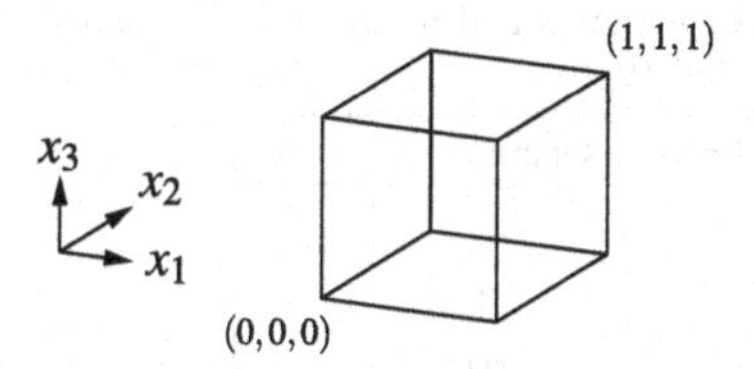

Fig. 3.7 Graphical representation of ternary Boolean functions

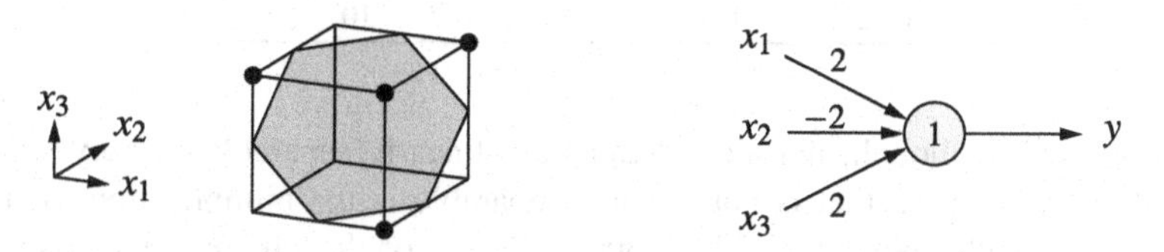

Fig. 3.8 Geometry of the threshold logic unit for the function $(x_1 \wedge \overline{x_2}) \vee (x_1 \wedge x_3) \vee (\overline{x_2} \wedge x_3)$: The plane shown in the left diagram is described by the equation $2x_1 - 2x_2 + 2x_3 = 1$

the table in Fig. 3.4 lists an output of 1, are marked with a filled circle. For all other corners of the unit cube, the output is 0. Like in the two-dimensional case we can again read the side of the plane on which the output is 1 from the normal vector of the plane: from the plane equation we derive the normal vector $\mathbf{n} = (2, -2, 2)$, which points out of the drawing plane to the top right.

3.3 Limitations

The examples studied in the preceding section—especially the threshold logic unit with three inputs—may lead us to presume that threshold logic units are fairly powerful processing units. Unfortunately, though, *single* threshold logic units are severely limited in their expressive and computational power. We know from the geometric interpretation of their computations that threshold logic units can represent only functions that are, as one says, **linearly separable**, that is, functions for which the points with output 1 can be separated from the points with output 0 by a linear function—that is, by a line, plane or hyperplane.

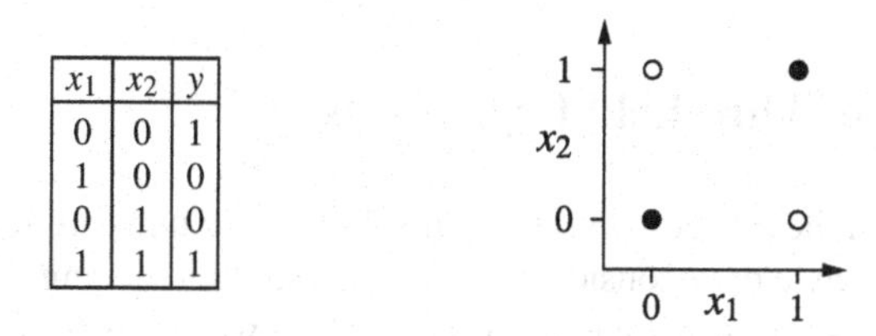

x_1	x_2	y
0	0	1
1	0	0
0	1	0
1	1	1

Fig. 3.9 The biimplication problem: there is no separating straight line

Table 3.1 The number of all Boolean functions of n inputs and the number of them that are linearly separable (Widner 1960 cited according to Zell 1996)

inputs	Boolean functions	linearly separable functions
1	$2^{2^1} = 4$	4
2	$2^{2^2} = 16$	14
3	$2^{2^3} = 256$	104
4	$2^{2^4} = 65536$	1774
5	$2^{2^5} \approx 4.3 \cdot 10^9$	94572
6	$2^{2^6} \approx 1.8 \cdot 10^{19}$	$5.0 \cdot 10^6$

Unfortunately, though, not all functions are linearly separable. A very simple example of a function that is not linearly separable is the biimplication (that is, $x_1 \leftrightarrow x_2$), the truth table of which is shown in Fig. 3.9 on the left. From the graphical representation of this function, which is shown in the same figure on the right, we can already see that there is no separating line. As a consequence, there cannot be any threshold logic unit computing this function.

The formal proof is executed by the common method of *reductio ad absurdum* (proof by contradiction). We assume that there exists a threshold logic unit with weights w_1 and w_2 and threshold θ that computes the biimplication. Then it is

$$\begin{array}{lll} \text{due to } (0,0) \mapsto 1: & 0 \geq \theta, & (1) \\ \text{due to } (1,0) \mapsto 0: & w_1 < \theta, & (2) \\ \text{due to } (0,1) \mapsto 0: & w_2 < \theta, & (3) \\ \text{due to } (1,1) \mapsto 1: & w_1 + w_2 \geq \theta. & (4) \end{array}$$

From (2) and (3) it follows $w_1 + w_2 < 2\theta$, which together with (4) yields $2\theta > \theta$, or $\theta > 0$. However, this contradicts (1). Therefore, there is no threshold logic unit that computes the biimplication.

The fact that only linearly separable functions can be represented may appear to be, at first sight, a small and bearable restriction, since only two of the 16 possible Boolean functions of two variables are *not* linearly separable (namely the biimplication and the exclusive or). However, if the number of inputs is increased, the fraction of all Boolean functions that are linearly separable drops rapidly (see Table 3.1). For a larger number of inputs, (single) threshold logic units can therefore compute "almost no" Boolean functions (relative to all possible ones).

3.4 Networks of Threshold Logic Units

As demonstrated in the preceding section, threshold logic units are severely limited. However, up to now we only considered *single* threshold logic units. The powers of threshold logic units can be increased considerably if we combine several threshold logic units, that is, if we consider networks of threshold logic units.

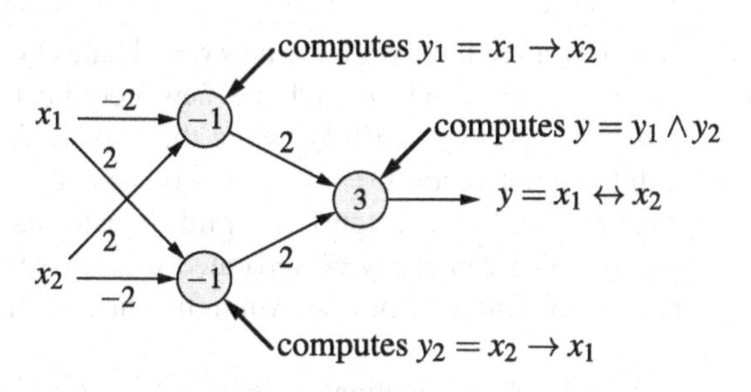

Fig. 3.10 Combining several threshold logic units

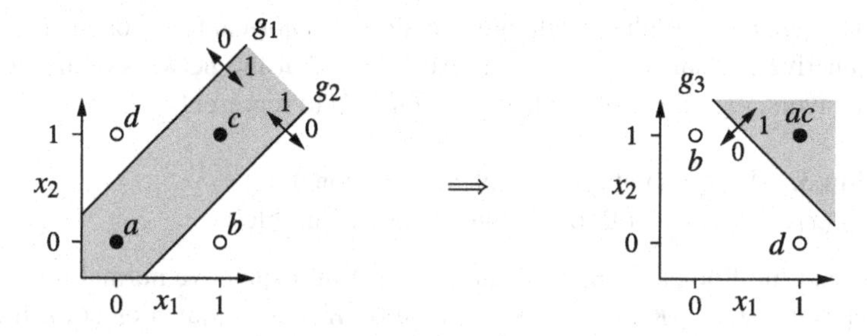

Fig. 3.11 Geometric interpretation of combining multiple threshold logic units into a network to compute the biimplication

As an example, we consider a possible solution for the biimplication problem with the help of three threshold logic units that are organized into two layers. This solution exploits the logical equivalence

$$x_1 \leftrightarrow x_2 \equiv (x_1 \to x_2) \wedge (x_2 \to x_1),$$

which divides the biimplication into three functions. From Figs. 3.3 and 3.6, we already know that the implication $x_2 \to x_1$ is linearly separable. In the implication $x_1 \to x_2$, the variables are merely exchanged, so it is linearly separable as well. Finally, we know from Figs. 3.2 and 3.5 that the conjunction of two Boolean variables is linearly separable. As a consequence, we only have to connect the corresponding threshold logic units, see Fig. 3.10. Thus we obtain a network with two layers, corresponding to the nested structure of the logical expression.

Intuitively, the two threshold logic units on the left (first layer) compute new Boolean coordinates y_1 and y_2 for the input vectors, so that the transformed input vectors in the input space of the threshold logic unit on the right (second layer) become linearly separable. This is illustrated geometrically with the two diagrams shown in Fig. 3.11. The separating line g_1 corresponds to the upper threshold logic unit and describes the implication $y_1 = x_1 \to x_2$: for all points above this line the output is 1, for all points below it the output is 0. The separating line g_2 belongs to the lower threshold logic unit and describes the implication $y_2 = x_2 \to x_1$: for all points above the line the output is 0, for all points below it the output is 1.

The threshold logic units on the left assign the new coordinates $(y_1, y_2) = (0, 1)$ to the input vector $b \mathrel{\hat{=}} (x_1, x_2) = (1, 0)$ and the new coordinates $(y_1, y_2) = (1, 0)$ to the input vector $d \mathrel{\hat{=}} (x_1, x_2) = (0, 1)$, while they assign the coordinates $(y_1, y_2) = (1, 1)$ to both the input vector $a \mathrel{\hat{=}} (x_1, x_2) = (0, 0)$ and the input vector $c \mathrel{\hat{=}} (x_1, x_2) = (1, 1)$ (see Fig. 3.11 on the right). After this transformation, the input vectors for which the output is 1 can easily be separated from those for which the output is 0, for instance, by the line g_3 that is shown in the diagram in Fig. 3.11 on the right.

It can be shown that all Boolean functions with an arbitrary number of inputs can be computed by networks of threshold logic units, simply by exploiting logical equivalences to divide these functions in such a way that all occurring sub-functions are linearly separable. With the help of the disjunctive normal form (or, analogously, the conjunctive normal form), one can even show that the networks only need to have two layers, regardless of the Boolean function to represent:

Algorithm 3.1 (Representation of Boolean Functions)
Let $y = f(x_1, \dots, x_n)$ be a Boolean function of n variables.

1. Represent the Boolean function $f(x_1, \dots, x_n)$ in disjunctive normal form. That is, determine $D_f = K_1 \vee \dots \vee K_m$, where all K_j are conjunctions of n literals, that is, $K_j = l_{j1} \wedge \dots \wedge l_{jn}$ with $l_{ji} = x_i$ (positive literal) or $l_{ji} = \neg x_i$ (negative literal).
2. For each conjunction K_j of the disjunctive normal form create one neuron (with n inputs—one input for each of the variables) where

$$w_{ji} = \begin{cases} 2 & \text{if } l_{ji} = x_i, \\ -2 & \text{if } l_{ji} = \neg x_i, \end{cases} \quad \text{and} \quad \theta_j = n - 1 + \frac{1}{2} \sum_{i=1}^{n} w_{ji}.$$

3. Create one output neuron (with m inputs—one input for each of the neurons created in step 2) where

$$w_{(n+1)k} = 2, \quad k = 1, \dots, m, \quad \text{and} \quad \theta_{n+1} = 1.$$

In the network constructed in this way, every neuron created in step 2 computes a conjunction and the output neuron computes their disjunction.

Intuitively, each neuron in the first layer describes a hyperplane that separates the corner of the hypercube, for which the conjunction is 1, from the rest of the unit hypercube. The equation of this hyperplane is easy to determine: its normal vector points from the center of the unit hypercube to the corner that is cut off and thus is 1 in all components in which the position vector of the corner has value 1, and it is -1 in all components in which the position vector of the corner has the value 0. (As an illustration consider the three-dimensional case.) We multiply this normal vector with 2 in order to obtain an integer threshold. The threshold has to be determined in such a way that it is exceeded only if all inputs that carry a weight of 2 are 1 and all other inputs are 0. The formula stated in step 2 yields such a value.

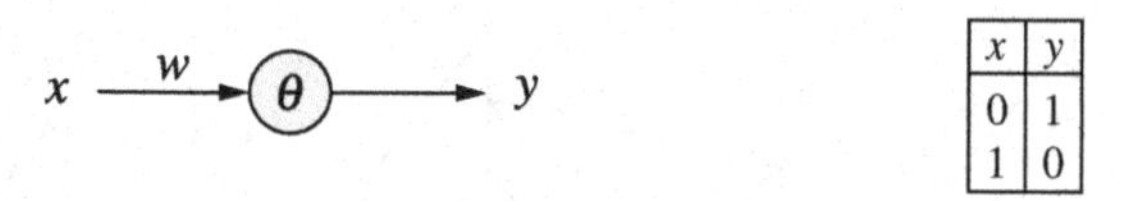

x	y
0	1
1	0

Fig. 3.12 A threshold logic unit with a single input and training examples for the negation

To compute the disjunction of the outputs of the neurons created in step 2, we have to separate, in an m-dimensional unit hypercube of the conjunctions, the corner $(0, \dots, 0)$, for which the output is 0, from all other corners, for which the output is 1. This can be achieved, for instance, with the hyperplane that possesses the normal vector $(1, \dots, 1)$ and the support vector $(\frac{1}{2}, 0, \dots, 0)$. (As an illustration consider the three-dimensional case again.) The parameters of the output neuron stated in step 3 are then simply read from the corresponding equation.

3.5 Training the Parameters

By interpreting the computations of a threshold logic unit geometrically, as shown in Sect. 3.2, we have (at least for functions with 2 and 3 variables) a simple method to find, for a given linearly separable function, a threshold logic unit that computes it: we determine a line, plane or hyperplane that separates the points for which the output is 1 from those for which the output is 0. From the equation describing this line, plane or hyperplane we can then easily read the weights and the threshold.

However, this methods becomes difficult and finally infeasible if the function to compute has more than three arguments, because we cannot imagine the input space of such a function. Furthermore it is impossible to automate this methods, because we find a suitable separating line or plane by "visual inspection" of the point sets to separate. This "visual inspection" cannot be mimicked directly by a computer. In order to be able to determine the parameters of a threshold logic unit with a computer, so that it computes a given function, we need a different approach. The principle consists in starting with randomly chosen values for the weights and the threshold and then changing these values step by step until the desired function is computed. The slow, stepwise adaptation of the weights and the threshold is also called **learning** or—in order to avoid confusions with the much more complex human learning process—the **training** of the threshold logic units.

In order to find a method to adapt the weightsand the threshold, we start from the following consideration: depending on the values of the weights and the threshold, the computation of the threshold logic unit will be more or less correct. Therefore, we can define an error function $e(w_1, \dots, w_n, \theta)$, which states how well, for given weights and threshold, the computed function coincides with the desired one. Our objective is, of course, to determine the weights and the threshold in such a way that the error vanishes, that is, that the error function becomes 0. To achieve this, we try to reduce the value of the error function in every step.

We illustrate this procedure with the help of a very simply example, namely a threshold logic unit with only one input. The parameters of this unit are to be

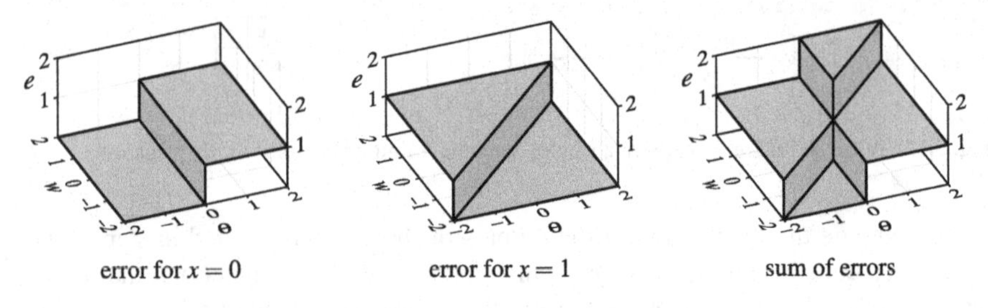

Fig. 3.13 Error of computing the negation w.r.t. the threshold

determined in such a way that it computes the negation. Such a threshold logic unit is shown in Fig. 3.12 together with the two training examples for the negation: if the input is 0, the output should be 1, if the input is 1, the output should be 0.

The error function we define first, as it appears to be natural, as the absolute value of the difference between the desired and the actual output. This function is shown in Fig. 3.13. The left diagram shows the error for the input $x = 0$, for which an output of 1 is desired. Since the threshold logic unit computes a 1 if $xw \geq \theta$, the error is 0 for a negative threshold and 1 for a positive threshold. (Obviously, the weight does not have any influence, because it is multiplied with the input, which is 0.) The middle diagram shows the error for the input $x = 1$, for which an output of 0 is desired. Here both the weight and the threshold have an influence. If the weights are less than the threshold, we have $xw < \theta$ and thus the output and consequently the error is 0. The diagram on the right shows the sum of these individual errors.

From the right diagram, a human can now easily read how the weight and the threshold have to chosen so that the threshold logic unit computes the negation: the values of these parameters must lie in the triangle in the lower left of the w–θ plane, in which the error is 0. However, it is not yet possible to automatically adapt the parameters with this function, because the global "visual inspection" of the error function, which a human relies on to find the solution, cannot be mimicked directly in a computer. Rather we would have to be able to read from the shape of the function at the point, which is given by the current weight and the current threshold, in which directions we have to change the weight and the threshold so that the error is reduced. With this error function, however, this is impossible, because it consists of "plateaus" or "terraces." In "almost all" points (the "edges" of the error function are the only exceptions) the error stays the same in all directions.[2]

In order to circumvent this problem, we modify the error function. Where the threshold logic unit produces the wrong output, we consider how far the threshold is exceeded (for a desired output of 0) or how far it is underrun (for a desired output of 1). Intuitively, we may say that the computation is "the more wrong" the farther the threshold is exceeded for a desired output of 0, or the farther the threshold is underrun for a desired output of 1. The modified error function is shown in Fig. 3.14.

[2]The somewhat imprecise notion "almost all points" can be made mathematically precise by drawing on measure theory: the set of points at which the error function changes has measure 0.

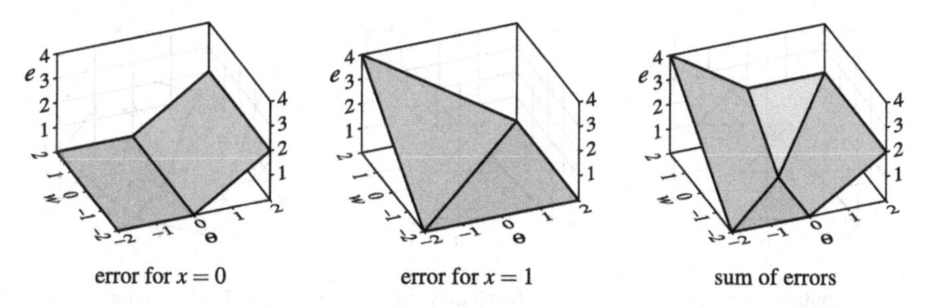

Fig. 3.14 Error of computing the negation w.r.t. how far the threshold is exceeded or underrun

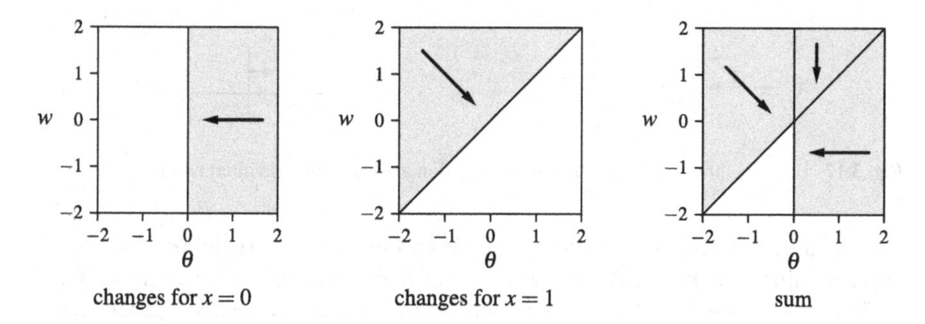

Fig. 3.15 Directions of the weight and threshold changes

Again the left diagram shows the error for the input $x = 0$, the middle diagram the error for the input $x = 1$ and the right diagram the sum of these individual errors.

If a threshold logic unit now produces a wrong output, we adapt the weight and the threshold in such a way that the error is reduced. That is, we try to "descent in the error landscape". With the modified error function this is possible, because we can read from it "locally" (that is, without a visual inspection of the whole error function, but merely by looking at the shape of the error function at the point that is given by the current values of the weight and the threshold) in which directions we have to change the weight and the threshold: we simply move in the direction in which the error function has the strongest downward slope. Intuitively, we follow the common scout advice how to find water: always go downhill. The directions that result from this rule are shown schematically in Fig. 3.15. The arrows indicate how the weight and the threshold should be adapted in different regions of the parameter space. In those regions, in which no arrows are drawn, weight and threshold are left unchanged, because there is no error.

The adaptation rules that are shown in Fig. 3.15 can be applied in two different ways. In the first place, we may consider the inputs $x = 0$ and $x = 1$ alternatingly and may adapt the weight and the threshold according to the corresponding rules. That is, first we adapt the weight and the threshold according to the left diagram, then we adapt them according to the middle diagram, then we adapt them again according to the left diagram and so forth until the error vanishes. This way of

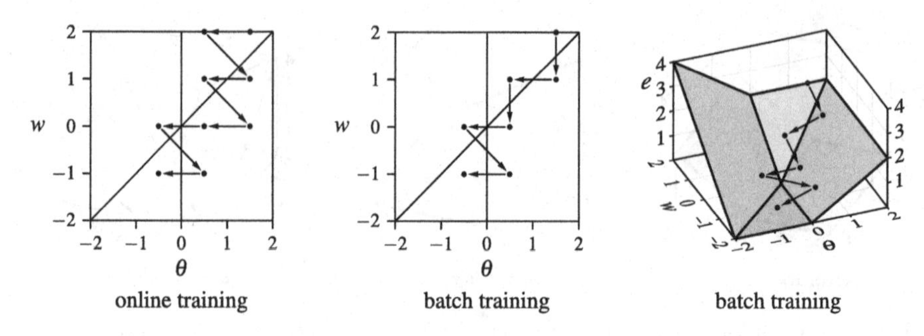

Fig. 3.16 Training processes with initial values $\theta = \frac{3}{2}$, $w = 2$ and learning rate 1

Fig. 3.17 Learned threshold logic unit for the negation and its geometric interpretation

training a neural network is called **online learning** or **online training**, since with every training example that becomes available, a training step can be carried out.

The second option consists in not applying the changes immediately after every training example, but aggregating them over all training examples. Only at the end of a **(learning/training) epoch**, that is, after all training examples have been traversed, the aggregated changes are applied. Then the training examples are traversed again and at the end the weight and the threshold are adapted and so forth until the error vanishes. This way of training is called **batch learning** or **batch training**, since all training examples have to be available together (in a *batch*), and corresponds to adapting weight and threshold according to the diagram on the right.

Figure 3.16 shows the training processes for the initial values $\theta = \frac{3}{2}$ and $w = 2$. Both online training (left) and batch training (middle) use a **learning rate** of 1. The learning rate states by how much the weight and the threshold are changed, and thus how "fast" the training is. (However, the learning rate should also not be chosen arbitrarily large, see Chap. 5.) If the learning rate is 1, the weight and the threshold are increased or reduced by 1. In order to illustrate the "descent in the error landscape", the batch training is repeated in a three-dimensional diagram in Fig. 3.16 on the right. The final, fully trained threshold logic unit (with $\theta = -\frac{1}{2}$ and $w = -1$) is shown, together with its geometric interpretation, in Fig. 3.17.

In this simple example, we derived the adaptation rules directly from a visual inspection of the error function. An alternative way of obtaining these adaptation rules are the following considerations: if a threshold logic unit produces an output of 1 instead of a desired 0, then the threshold is too small and/or the weights are too large. Hence, we should increase the threshold a bit and reduce the weights. Of course, the latter is reasonable only if the corresponding input is 1, as otherwise the weight has no influence on the output. Contrariwise, if a threshold logic unit produces an output of 0 instead of a desired 1, then the threshold is too large and/or

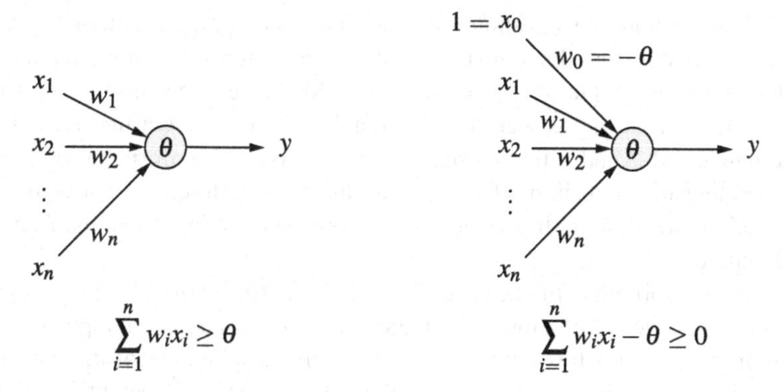

Fig. 3.18 Turning the threshold into a weight

the weights are too small. Hence, the threshold should be reduced and the weights should be increased (provided, of course, that the corresponding input is 1).

For our simple threshold logic unit, the changes shown in Fig. 3.15 have exactly these effects. However, the considerations above have the advantage that they can be applied to threshold logic units with more than one input. Therefore, we can define the following general training method for threshold logic units:

Definition 3.2 Let $\mathbf{x} = (x_1, \ldots, x_n)$ be an input vector of a threshold logic unit, o the desired output for this input vector and y the actual output of the threshold logic unit. If $y \neq o$, then, in order to reduce the error, the threshold θ and the weight vector $\mathbf{w} = (w_1, \ldots, w_n)$ are adapted as follows:

$$\begin{aligned} \theta^{(\text{new})} &= \theta^{(\text{old})} + \Delta\theta \quad &&\text{with } \Delta\theta = -\eta(o - y), \\ \forall i \in \{1, \ldots, n\} : w_i^{(\text{new})} &= w_i^{(\text{old})} + \Delta w_i \quad &&\text{with } \Delta w_i = \eta(o - y)x_i, \end{aligned}$$

where η is a parameter that is called **learning rate**. It determines the severity of the weight and threshold changes. This method is called the **delta rule** or **Widrow–Hoff procedure** (Widrow and Hoff 1960).

In this definition, we have to distinguish between an adaptation of the threshold and adaptations of the weights, because the directions of these changes are opposite to each other (opposite signs for $\eta(t - y)$ and $\eta(t - y)x_i$, respectively). However, we can unify the adaptation rules by turning the threshold into a weight. The rationale of this transformation is illustrated in Fig. 3.18: the threshold is fixed to 0 and to compensate this reduction in the number of parameters, an additional (imaginary) input x_0 is introduced, which has a fixed value of 1. This input is weighted with the negated threshold. The two threshold logic units are clearly equivalent, since the left one tests the condition $\sum_{i=1}^{n} w_i x_i \geq \theta$, the right one tests $\sum_{i=1}^{n} w_i x_i - \theta \geq 0$, in order to determine the output. Note that the same effect is produced if the additional input has the fixed value -1 and is weighted with the threshold directly. This alternative way of turning the threshold into a weight is also very common.

Regardless of how the threshold is turned into a weight, we obtain the same adaptation directions for all parameters. With a negated threshold: if the output is 1 instead of 0, then both the w_i as well as $-\theta$ should be reduced. If the output is 0 instead of 1, both the w_i as well as $-\theta$ should be increased. On the other hand, with an unnegated θ and a fixed input of -1, we obtain a uniform rule, because the needed negative sign is produced by the input. Therefore, we can determine the adaptation direction of all parameters by simply subtracting the actual from the desired output.

Thus, we can formulate the delta rule as follows: Let $\mathbf{x} = (x_0 = 1, x_1, \dots, x_n)$ be an extended input vector of a threshold logic unit (note the additional input $x_0 = 1$), o the desired output for this input vector and y the actual output of the threshold logic unit. If $y \neq o$, then, in order to reduce the error, the extended weight vector $\mathbf{w} = (w_0 = -\theta, w_1, \dots, w_n)$ (note the added weight $w_0 = -\theta$) is adapted as follows:

$$\forall i \in \{0, 1, \dots, n\}: \quad w_i^{(\text{new})} = w_i^{(\text{old})} + \Delta w_i \quad \text{with } \Delta w_i = \eta(o - y)x_i.$$

We point out this possibility here, because it can often be used, for example, to simplify derivations (see, for instance, Sect. 5.4). However, for the sake of clarity, we maintain the distinction of weight and threshold for the remainder of this chapter.

With the help of the delta rule, we can now state two algorithms for training a threshold logic unit: an online version and a batch version. In order to formulate these algorithms, we assume that we are given a set $L = \{(\mathbf{x}_1, o_1), \dots, (\mathbf{x}_m, o_m)\}$ of training examples, each of which consist of an input vector $\mathbf{x}_i \in \mathbb{R}^n$ and the desired output $o_i \in \{0, 1\}$ for this input vector, $i = 1, \dots, m$. Furthermore, let arbitrary initial weights $\mathbf{w}$ and an arbitrary initial threshold θ be given (e.g., chosen randomly). We consider online training first:

Algorithm 3.2 (Online Training of a Threshold Logic Unit)

```
procedure online_training (var w, var θ, L, η);
var y, e;                               (* output, sum of errors *)
begin
  repeat
    e := 0;                             (* initialize the sum of errors *)
    for all (x, o) ∈ L do begin         (* traverse the examples *)
      if (wx ≥ θ) then  y := 1;         (* compute the output of *)
                  else  y := 0;         (* the threshold logic unit *)
      if (y ≠ o) then begin             (* if the output is wrong *)
        θ := θ − η(o − y);              (* adapt the threshold *)
        w := w + η(o − y)x;             (* and the weights *)
        e := e + |o − y|;               (* sum the errors *)
      end;
    end;
  until (e ≤ 0);                        (* repeat the computations *)
end;                                    (* until the error vanishes *)
```

Obviously, this algorithm repeatedly applies the delta rule until the sum of errors over all training examples vanishes. Note that in this algorithm the weight adaptation is written in vector form, which, however, is clearly equivalent to an adaptation of the individual weights. Let us now turn to the batch version:

Algorithm 3.3 (Batch Training of a Threshold Logic Unit)

```
procedure batch_training (var w, var θ, L, η);
var y, e,                                  (* output, sum of errors *)
    θc, wc;                                (* aggregated changes *)
begin
  repeat
    e := 0; θc := 0; wc := 0;              (* initializations *)
    for all (x, o) ∈ L do begin            (* traverse the examples *)
      if (wx ≥ θ) then  y := 1;            (* compute the output of *)
                 else  y := 0;             (* the threshold logic unit *)
      if (y ≠ o) then begin                (* if the output is wrong *)
        θc := θc − η(o − y);               (* sum the threshold and *)
        wc := wc + η(o − y)x;              (* the weight changes *)
        e  := e  + |o − y|;                (* sum the errors *)
      end;
    end;
    θ := θ + θc;                           (* adapt the threshold *)
    w := w + wc;                           (* and the weights *)
  until (e ≤ 0);                           (* repeat the computations *)
end;                                       (* until the error vanishes *)
```

In this algorithm, the delta rule is applied in a modified form, since for each traversal of the training examples the same threshold and the same weights are used. If the output is wrong the computed changes are not applied directly, but summed in the variables θ_c and $\mathbf{w}_c$. Only after all training examples have been visited, the threshold and the weights are adapted with the help of these variables.

To illustrate the operation of these two algorithms, Table 3.2 shows the online training of the simple threshold logic unit considered above, which is to be trained in such a way that it computes the negation. Like in Fig. 3.16 on page 26 the initial values are $\theta = \frac{3}{2}$ and $w = 3$. It is easy to check that the online training shown here in tabular form corresponds exactly to the one shown graphically in Fig. 3.16 on the left. Analogously, Table 3.3 shows batch training. It corresponds exactly to the procedure depicted in Fig. 3.16 in the middle or on the right. Again, the fully trained threshold logic unit with the same parameters is depicted, together with its geometric interpretation, in Fig. 3.19.

As another example, we consider a threshold logic unit with two inputs that is to be trained in such a way that it computes the conjunction of its inputs. Such a threshold logic unit is shown, together with the corresponding training examples, in Fig. 3.20. For this example, we only consider online training. The corresponding training procedure for the initial values $\theta = w_1 = w_2 = 0$ with learning rate 1 is shown in Table 3.4. As for the negation, the training is successful and finally yields

Table 3.2 Online training of a threshold logic unit for the negation with initial values $\theta = \frac{3}{2}$, $w = 2$ and learning rate 1

epoch	x	o	$\mathbf{xw}$	y	e	$\Delta\theta$	Δw	θ	w
								1.5	2
1	0	1	−1.5	0	1	−1	0	0.5	2
	1	0	1.5	1	−1	1	−1	1.5	1
2	0	1	−1.5	0	1	−1	0	0.5	1
	1	0	0.5	1	−1	1	−1	1.5	0
3	0	1	−1.5	0	1	−1	0	0.5	0
	1	0	0.5	0	0	0	0	0.5	0
4	0	1	−0.5	0	1	−1	0	−0.5	0
	1	0	0.5	1	−1	1	−1	0.5	−1
5	0	1	−0.5	0	1	−1	0	−0.5	−1
	1	0	−0.5	0	0	0	0	−0.5	−1
6	0	1	0.5	1	0	0	0	−0.5	−1
	1	0	−0.5	0	0	0	0	−0.5	−1

Table 3.3 Batch training of a threshold logic unit for the negation with initial values $\theta = \frac{3}{2}$, $w = 2$ and learning rate 1

epoch	x	o	$\mathbf{xw}$	y	e	$\Delta\theta$	Δw	θ	w
								1.5	2
1	0	1	−1.5	0	1	−1	0		
	1	0	0.5	1	−1	1	−1	1.5	1
2	0	1	−1.5	0	1	−1	0		
	1	0	−0.5	0	0	0	0	0.5	1
3	0	1	−0.5	0	1	−1	0		
	1	0	0.5	1	−1	1	−1	0.5	0
4	0	1	−0.5	0	1	−1	0		
	1	0	−0.5	0	0	0	0	−0.5	0
5	0	1	0.5	1	0	0	0		
	1	0	0.5	1	−1	1	−1	0.5	−1
6	0	1	−0.5	0	1	−1	0		
	1	0	−1.5	0	0	0	0	−0.5	−1
7	0	1	0.5	1	0	0	0		
	1	0	−0.5	0	0	0	0	−0.5	−1

Fig. 3.19 Learned threshold logic unit for the negation and its geometric interpretation

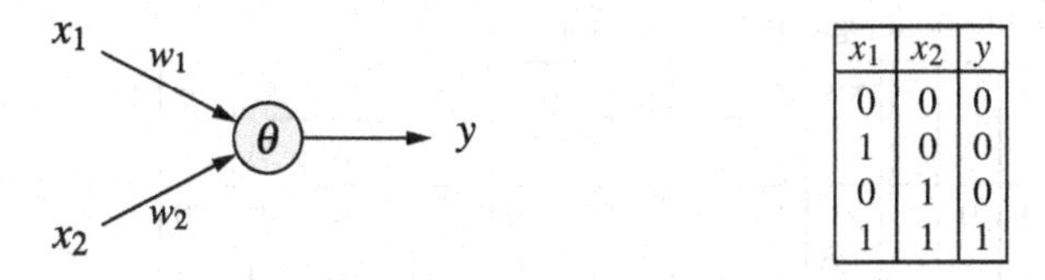

x_1	x_2	y
0	0	0
1	0	0
0	1	0
1	1	1

Fig. 3.20 A threshold logic unit with two inputs and training examples for the conjunction $y = x_1 \wedge x_2$

Table 3.4 Training of a threshold logic unit for the conjunction

Epoch	x_1	x_2	o	$\mathbf{xw}$	y	e	$\Delta\theta$	Δw_1	Δw_2	θ	w_1	w_2
										0	0	0
1	0	0	0	0	1	−1	1	0	0	1	0	0
	0	1	0	−1	0	0	0	0	0	1	0	0
	1	0	0	−1	0	0	0	0	0	1	0	0
	1	1	1	−1	0	1	−1	1	1	0	1	1
2	0	0	0	0	1	−1	1	0	0	1	1	1
	0	1	0	0	1	−1	1	0	−1	2	1	0
	1	0	0	−1	0	0	0	0	0	2	1	0
	1	1	1	−1	0	1	−1	1	1	1	2	1
3	0	0	0	−1	0	0	0	0	0	1	2	1
	0	1	0	0	1	−1	1	0	−1	2	2	0
	1	0	0	0	1	−1	1	−1	0	3	1	0
	1	1	1	−2	0	1	−1	1	1	2	2	1
4	0	0	0	−2	0	0	0	0	0	2	2	1
	0	1	0	−1	0	0	0	0	0	2	2	1
	1	0	0	0	1	−1	1	−1	0	3	1	1
	1	1	1	−1	0	1	−1	1	1	2	2	2
5	0	0	0	−2	0	0	0	0	0	2	2	2
	0	1	0	0	1	−1	1	0	−1	3	2	1
	1	0	0	−1	0	0	0	0	0	3	2	1
	1	1	1	0	1	0	0	0	0	3	2	1
6	0	0	0	−3	0	0	0	0	0	3	2	1
	0	1	0	−2	0	0	0	0	0	3	2	1
	1	0	0	−1	0	0	0	0	0	3	2	1
	1	1	1	0	1	0	0	0	0	3	2	1

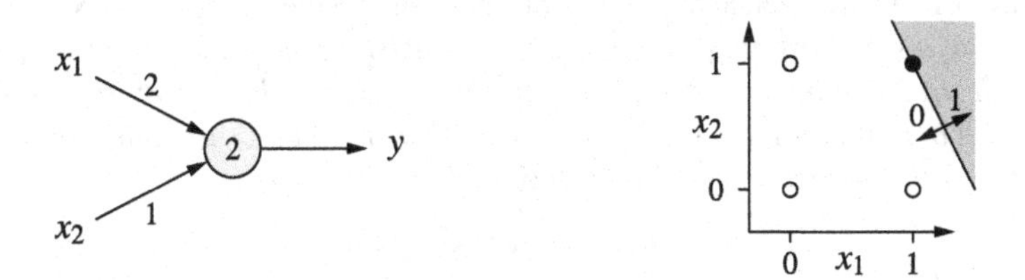

Fig. 3.21 Geometry of the learned threshold logic unit for $x_1 \wedge x_2$. The *straight line* shown on the right is described by the equation $2x_1 + x_2 = 3$

Table 3.5 Training of a threshold logic unit for the biimplication

epoch	x_1	x_2	o	$\mathbf{xw}$	y	e	$\Delta\theta$	Δw_1	Δw_2	θ	w_1	w_2
										0	0	0
1	0	0	1	0	1	0	0	0	0	0	0	0
	0	1	0	0	1	−1	1	0	−1	1	0	−1
	1	0	0	−1	0	0	0	0	0	1	0	−1
	1	1	1	−2	0	1	−1	1	1	0	1	0
2	0	0	1	0	1	0	0	0	0	0	1	0
	0	1	0	0	1	−1	1	0	−1	1	1	−1
	1	0	0	0	1	−1	1	−1	0	2	0	−1
	1	1	1	−3	0	1	−1	1	1	1	1	0
3	0	0	1	0	1	0	0	0	0	0	1	0
	0	1	0	0	1	−1	1	0	−1	1	1	−1
	1	0	0	0	1	−1	1	−1	0	2	0	−1
	1	1	1	−3	0	1	−1	1	1	1	1	0

the threshold $\theta = 3$ and the weights $w_1 = 2$ and $w_2 = 1$. The resulting threshold logic unit is shown, together with its geometric interpretation, in Fig. 3.21. Note that this threshold logic unit indeed computes the conjunction, even though the point (1, 1) lies on the separating line, because it yields output 1 not only for points to the right of the line, but also for all points *on* the line.

After we have seen two examples of successful training, we naturally face the question whether Algorithms 3.2 and 3.3 always achieve their objective. As a first step, we can assert that these algorithms do not terminate if the function to be learned is *not* linearly separable. This is illustrated in Table 3.5 with the help of the online training procedure for the biimplication. Epochs 2 and 3 are clearly identical and will thus be repeated indefinitely, without a solution ever being found. However, this is not surprising, since the training procedure terminates only if the sum of the errors over all training examples vanishes. Since we know from Sect. 3.3 that there is no threshold logic unit that computes the biimplication, the error can never vanish and thus the algorithm cannot terminate.

For linearly separable functions, however, that is, for functions that can actually be computed by a threshold logic unit, it is guaranteed that the algorithms find a solution. That is, the following theorem holds.

Theorem 3.1 (Convergence theorem for the Delta Rule) *Let* $L = \{(\mathbf{x}_1, o_1), \ldots (\mathbf{x}_m, o_m)\}$ *be a set of training examples, each consisting of an input vector* $\mathbf{x}_i \in \mathbb{R}^n$ *and the desired output* $o_i \in \{0, 1\}$ *for this input vector. Furthermore, let* $L_0 = \{(\mathbf{x}, o) \in L \mid o = 0\}$ *and* $L_1 = \{(\mathbf{x}, o) \in L \mid o = 1\}$. *If* L_0 *and* L_1 *are linearly separable, that is, if there exist* $\mathbf{w} \in \mathbb{R}^n$ *and* $\theta \in \mathbb{R}$ *such that*

$$\forall (\mathbf{x}, 0) \in L_0 : \ \mathbf{wx} < \theta \quad \textit{and}$$
$$\forall (\mathbf{x}, 1) \in L_1 : \ \mathbf{wx} \geq \theta,$$

then the Algorithms 3.2 *and* 3.3 *terminate.*

Proof The proof, which we do not want to spell out here, can be found, for example, in Rojas (1996) or in Nauck et al. (1997). □

Since both algorithms terminate only when the error vanishes, the computed values for the threshold and the weights are a solution of the learning problem.

3.6 Variants

All examples that we considered up to now referred to logical functions and we encoded *false* as 0 and *true* as 1. However, this encoding has the disadvantage that with an input of *false* the corresponding weight cannot be changed, because the formula for the weight change contains the input as a factor (see Definition 3.2 on page 27). This disadvantage can, in certain situations, slow down training unnecessarily, since a weight can only be adapted if the corresponding input is *true*.

To avoid this problem, the ADALINE model (ADAptive LINear Element) relies on the encoding *false* $\hat{=} -1$ and *true* $\hat{=} +1$. Thus an input of *false* also leads, provided the output is wrong, to a weight adaptation. Indeed, the delta rule was originally developed for the ADALINE model (Widrow and Hoff 1960), so that strictly we may only speak of the **delta rule** or the **Widrow–Hoff procedure** if the ADALINE model is employed. Although the procedure is equally applicable (and has the same convergence properties) for the encoding *false* $\hat{=} 0$ and *true* $\hat{=} 1$ (see the preceding section), it is sometimes called **error correction procedure** to avoid confusion (Nilsson 1965, 1998). We ignore this distinction here, because it is due to historical rather than conceptual reasons.

3.7 Training Networks

After simple neuro-computers had been used successfully for pattern recognition tasks at the end of the 1950s (for example, Rosenblatt 1958), the simple and fast delta rule training method had been developed by Widrow and Hoff (1960) and the perceptron convergence theorem (corresponds to the convergence theorem for the delta rule) had been proven by Rosenblatt (1962), great expectations were placed in the development of (artificial) neural networks. This started the "first bloom" of neural network research, in which it was believed that one had already discovered the core principles underlying systems that are able to learn.

Only after Minsky and Papert (1969) carried out a careful mathematical analysis of the perceptron and pointed out in all clarity that threshold logic units can compute only linearly separable functions, the limitations of the models and procedures used at the time were properly recognized. Although it was already known from the early works of McCulloch and Pitts (1943), that the limitations of the expressive and computational power can be lifted by using *networks* of threshold logic units (since, for example, such networks can compute arbitrary Boolean functions), training methods were confined to *single* threshold logic units.

Unfortunately, transferring the training procedures to networks of threshold logic units turned out to be a surprisingly difficult problem. For example, the delta rule derives the weight adaptation from the difference between the actual and the desired output (see Definition 3.2 on page 27). However, a desired output is available only for the neuron that yields the output of the network as a whole. For all other threshold logic units, which carry out some kind of preprocessing and transmit their outputs only to other threshold logic units, no such desired output can be given. As an example, consider the biimplication problem and the structure of the network that we proposed as a solution for this problem (Fig. 3.10 on page 21): the training examples do not state desired outputs for the two threshold logic units on the left. One of the main reasons for this is that the necessary coordinate transformation is not uniquely determined: separating lines in the input space may just as well be placed in a completely different way (for example, perpendicular to the bisectrix) or one may direct the normal vectors somewhat differently.

As a consequence (artificial), neural networks were seen as a "research dead end" and the so-called "dark age" of neural network research began. Only when the training procedure of **error backpropagation** was developed, the area was revived. This procedure was described first in Werbos (1974), but did not receive any attention. Only when Rumelhart et al. (1986a, 1986b) independently developed the method again and advertised it in the research community, the modern age ("second bloom") of (artificial) neural network began, which last to the present day.

We consider error backpropagation only in Chap. 5, since it cannot be applied directly to threshold logic units. It requires that the activation of a neuron does not jump at a crisply defined threshold from 0 to 1, but that the activation rises slowly, according to a differentiable function. For networks consisting of pure threshold logic units, still no training method is known.

References

J.A. Anderson and E. Rosenfeld. *Neurocomputing: Foundations of Research.* MIT Press, Cambridge, MA, USA, 1988

M.A. Boden, ed. *The Philosophy of Artificial Intelligence.* Oxford University Press, Oxford, United Kingdom, 1990

W.S. McCulloch. *Embodiments of Mind.* MIT Press, Cambridge, MA, USA, 1965

W.S. McCulloch and W.H. Pitts. A Logical Calculus of the Ideas Immanent in Nervous Activity. *Bulletin of Mathematical Biophysics* 5:115–133, 1943, USA. Reprinted in McCulloch (1965), 19–39, in Anderson and Rosenfeld (1988), 18–28, and in Boden (1990), 22–39

L.M. Minsky and S. Papert. *Perceptrons.* MIT Press, Cambridge, MA, USA, 1969

D. Nauck, F. Klawonn, and R. Kruse. *Foundations of Neuro-Fuzzy Systems.* J. Wiley & Sons, Chichester, United Kingdom, 1997

N.J. Nilsson. *Learning Machines: The Foundations of Trainable Pattern-Classifying Systems.* McGraw-Hill, New York, NY, 1965

N.J. Nilsson. *Artificial Intelligence: A New Synthesis.* Morgan Kaufmann, San Francisco, CA, USA, 1998

R. Rojas. *Theorie der neuronalen Netze—Eine systematische Einführung*. Springer-Verlag, Berlin, Germany, 1996

F. Rosenblatt. The Perceptron: A Probabilistic Model for Information Storage and Organization in the Brain. *Psychological Review* 65:386–408, 1958, USA

F. Rosenblatt. *Principles of Neurodynamics*. Spartan Books, New York, NY, USA, 1962

D.E. Rumelhart and J.L. McClelland, eds. *Parallel Distributed Processing: Explorations in the Microstructures of Cognition, Vol. 1: Foundations*, 1986

D.E. Rumelhart, G.E. Hinton and R.J. Williams. Learning Internal Representations by Error Propagation. In Rumelhart and McClelland (1986), 318–362, 1986a

D.E. Rumelhart, G.E. Hinton and R.J. Williams. Learning Representations by Back-Propagating Errors. *Nature* 323:533–536, 1986b

P.D. Wasserman. *Neural Computing: Theory and Practice*. Van Nostrand Reinhold, New York, NY, USA, 1989

P.J. Werbos. *Beyond Regression: New Tools for Prediction and Analysis in the Behavioral Sciences*. Ph.D. Thesis, Harvard University, Cambridge, MA, USA, 1974

R.O. Widner. Single State Logic. *AIEE Fall General Meeting*, 1960. Reprinted in Wasserman (1989)

B. Widrow and M.E. Hoff. Adaptive Switching Circuits. *IRE WESCON Convention Record*, 96–104. Institute of Radio Engineers, New York, NY, USA, 1960

A. Zell. *Simulation Neuronaler Netze*. Addison-Wesley, Stuttgart, Germany, 1996

Chapter 4
General Neural Networks

In this chapter, we introduce a general model of (artificial) neural networks that captures (more or less) all special forms, which we consider in the following chapters. We start by defining the structure of an (artificial) neural network and then describe generally the operation and finally the training of an (artificial) neural network.

4.1 Structure of Neural Networks

In the preceding chapter, we already considered briefly networks of threshold logic units. The way in which we represented these networks suggests to describe neural networks with the help of a graph (in the sense of graph theory). Therefore, we first define the notion of a graph and a few useful notions, which we draw on in the following definitions and the subsequent chapters.

Definition 4.1 A (directed) **graph** is a pair $G = (V, E)$ consisting of a (finite) set V of **vertices** or **nodes** and a (finite) set $E \subseteq V \times V$ of **edges**. We say that an edge $e = (u, v) \in E$ is **directed** from the vertex u to the vertex v.

It is, of course, also possible to define undirected graphs (for example, by using unordered pairs $\{u, v\}$ for the edges). However, to describe neural networks we only need directed graphs, since the connections between neurons are always directed.

Definition 4.2 Let $G = (V, E)$ be a (directed) graph and $u \in V$ a vertex. The vertices of the set

$$\mathrm{pred}(u) = \{v \in V \mid (v, u) \in E\}$$

are called the **predecessors** of the vertex u and the vertices of the set

$$\mathrm{succ}(u) = \{v \in V \mid (u, v) \in E\}$$

are called the **successors** of the vertex u.

R. Kruse et al., *Computational Intelligence*, Texts in Computer Science,
DOI 10.1007/978-1-4471-5013-8_4, © Springer-Verlag London 2013

Definition 4.3 An (artificial) **neural network** is a (directed) graph $G = (U, C)$ whose vertices $u \in U$ are called **neurons** or **units** and whose edges $c \in C$ are called **connections**. The set U of vertices is divided into the set U_{in} of **input neurons**, the set U_{out} of **output neurons** and the set U_{hidden} of **hidden neurons**. It is

$$U = U_{\text{in}} \cup U_{\text{out}} \cup U_{\text{hidden}},$$
$$U_{\text{in}} \neq \emptyset, \qquad U_{\text{out}} \neq \emptyset, \qquad U_{\text{hidden}} \cap (U_{\text{in}} \cup U_{\text{out}}) = \emptyset.$$

Each connection $(v, u) \in C$ carries a **weight** w_{uv} and to each neuron $u \in U$ three (real-valued) quantities are assigned: the **network input** net_u, the **activation** act_u and the **output** out_u. In addition, each input neuron $u \in U_{\text{in}}$ has a fourth (real-valued) quantity, the **external input** ext_u. Each neuron $u \in U$ also possesses three functions:

the **network input function** $f_{\text{net}}^{(u)} : \mathbb{R}^{2|\text{pred}(u)|+\kappa_1(u)} \to \mathbb{R}$,

the **activation function** $f_{\text{act}}^{(u)} : \mathbb{R}^{\kappa_2(u)} \to \mathbb{R}$, and

the **output function** $f_{\text{out}}^{(u)} : \mathbb{R} \to \mathbb{R}$,

with which the network input net_u, the activation act_u and the output out_u of the neuron u are computed. $\kappa_1(u)$ and $\kappa_2(u)$ depend on the type and the number of arguments of the functions (see below).

The neurons are divided into input, output and hidden neurons in order to specify which neurons receive input from the environment (input neurons) and which emit output to the environment (output neurons). The remaining neurons have no contact with the environment (but only with other neurons) and thus are "hidden."

Note that the set U_{in} of input neurons and the set U_{out} of output neurons need not be disjoint: A neuron can be both input and output neuron. In Chap. 8, we even consider networks (so-called Hopfield networks) in which all neurons are both input and output neurons and there are no hidden neurons.

Note also that in the index of a weight w_{uv} the neuron to which the corresponding connection leads is named first. The reason for this order is that the graph of a neural network is often described by an adjacency matrix which instead of the values 1 (connection) and 0 (no connection) contains the weights of the connections (if a weight is zero, the corresponding connection does not exist). Due to reasons that will be studied in more detail in Chap. 5 it is advantageous to collect the weights of the connections that lead to a specific neuron in the same matrix *row* (and not in the same matrix *column*). Since the elements of a matrix are indexed according to the rule "row first, then column", the neuron is named first to which the connections lead. Thus, we obtain the following scheme (with $r = |U|$):

$$\begin{array}{cccc} u_1 & u_2 & \dots & u_r \end{array}$$
$$\begin{pmatrix} w_{u_1u_1} & w_{u_1u_2} & \dots & w_{u_1u_r} \\ w_{u_2u_1} & w_{u_2u_2} & & w_{u_2u_r} \\ \vdots & & & \vdots \\ w_{u_ru_1} & w_{u_ru_2} & \dots & w_{u_ru_r} \end{pmatrix} \begin{array}{c} u_1 \\ u_2 \\ \vdots \\ u_r \end{array}$$

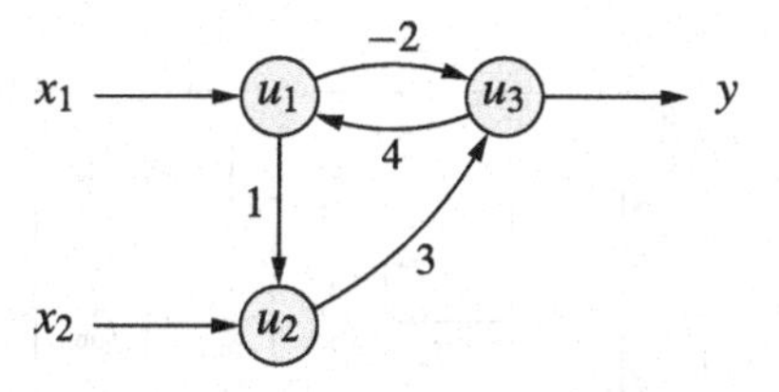

Fig. 4.1 A simple (artificial) neural network

This matrix is to be read from the top to the right: the columns correspond to the neurons, from which the connections emanate, the rows to the neurons to which the connections lead. (Note that neurons may also be connected to themselves—diagonal elements of the above matrix.) This matrix and the corresponding weighted graph (i.e., graph with weighted edges) are called the **network structure**.

According to the network structure, we distinguish two fundamental types of neural networks: if the graph that describes the network structure of a neural network is acyclic—that is, if it does not contain loops[1] and no directed cycles—the network is called a **feed forward network**. However, if the graph contains loops or directed cycles, it is called a **recurrent network**. The reasons for these names are, of course, that in a neural network information can be transmitted only along the (directed) connections. If the graph is acyclic, there is only one direction, namely forward, from the input neurons to the output neurons. However, if there are loops or directed cycles, outputs can be coupled back to inputs. In the subsequent chapters, we first consider different types of feed forward networks, since they are easier to analyze. In Chaps. 8 and 9, we then turn to recurrent networks.

To illustrate the definition of the structure of a neural network, we consider as an example the network with three neurons (that is, $U = \{u_1, u_2, u_3\}$), that is shown in Fig. 4.1. The neurons u_1 and u_2 are input neurons (that is, $U_{\text{in}} = \{u_1, u_2\}$). They receive the external inputs x_1 and x_2, respectively. The neuron u_3 is the only output neuron (that is, $U_{\text{out}} = \{u_3\}$). It produces the output y of the neural network. This network does not contain any hidden neurons (that is, $U_{\text{hidden}} = \emptyset$).

There is a total of four connections between the neurons (that is, the graph has the edges $C = \{(u_1, u_2), (u_1, u_3), (u_2, u_3), (u_3, u_1)\}$), the weights of which are indicated by the numbers with which the arrows are labeled that represent these connections (for example, $w_{u_3u_2} = 3$). This network is recurrent, as there are two directed cycles (for instance, the cycle $(u_1, u_3), (u_3, u_1)$). If we describe the network structure by a matrix (as explained above), we obtain the 3×3 matrix

$$\begin{array}{cc} \begin{matrix} u_1 & u_2 & u_3 \end{matrix} & \\ \begin{pmatrix} 0 & 0 & 4 \\ 1 & 0 & 0 \\ -2 & 3 & 0 \end{pmatrix} & \begin{matrix} u_1 \\ u_2 \\ u_3 \end{matrix} \end{array}$$

[1] A loop is an edge/connection from a vertex to this vertex itself, that is, an edge $e = (v, v)$ with a vertex $v \in V$.

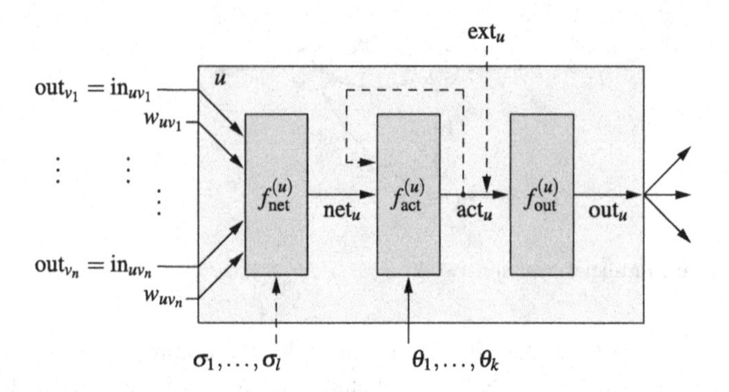

Fig. 4.2 Structure of a generalized neuron

Note that the neuron, from which a connection emanates, selects the column, while the neuron, to which a connection leads, selects the row, in which the corresponding connection weight is entered.

4.2 Operation of Neural Networks

To describe the operation of neural networks, we have to specify (1) how a single neuron computes its output from its inputs (that is, the outputs of its predecessors) and (2) how the computations of the different neurons are organized, in particular, how the external input is processed and in what order the neurons are updated.

Let us first consider a single neuron. Every neuron can be seen as a simple processor, the structure of which is shown in Fig. 4.2. The network input function $f_{\text{net}}^{(u)}$ computes the network input net_u from the inputs $\text{in}_{uv_1}, \ldots, \text{in}_{uv_n}$, which correspond to the outputs $\text{out}_{v_1}, \ldots, \text{out}_{v_n}$ of the predecessors of the neuron u, and the connection weights $w_{uv_1}, \ldots, w_{uv_n}$. This computation can be influenced by additional parameters $\sigma_1, \ldots, \sigma_l$ (see, for instance, Sect. 6.5). From the network input, a certain number of parameters $\theta_1, \ldots, \theta_k$, and possibly a feedback of the current activation of the neuron u (see, for instance, Chap. 9) the activation function $f_{\text{act}}^{(u)}$ computes the new activation act_u of the neuron u. Finally, the output function $f_{\text{out}}^{(u)}$ computes the output of the neuron u from its activation. The external input ext_u sets the (initial) activation of the neuron u, if it is an input neuron (see below).

The number $\kappa_1(u)$ of the additional arguments of the network input function and the number $\kappa_2(u)$ of the arguments of the activation function depend on the type of these functions and the structure of the neuron (for example, whether there is a feedback of the current activation or not). They may differ for each of the neurons of a neural network. Usually the network input function has only $2|\operatorname{pred}(u)|$ arguments (namely the outputs of the predecessor neurons and the corresponding connection weights), since no other parameters enter it. The activation function usually has two arguments: the network input and a parameter, which may be, for instance (as in the preceding chapter), a threshold. The output function, on the other hand, has only the activation as its argument and usually serves the purpose to scale the output to a desired output range, most commonly by a linear mapping.

Note that the network input function is often written with vector arguments:

$$\begin{aligned} f_{\text{net}}^{(u)}(\mathbf{w}_u, \mathbf{in}_u) &= f_{\text{net}}^{(u)}(w_{uv_1}, \ldots, w_{uv_n}, \text{in}_{uv_1}, \ldots, \text{in}_{uv_n}) \\ &= f_{\text{net}}^{(u)}(w_{uv_1}, \ldots, w_{uv_n}, \text{out}_{v_1}, \ldots, \text{out}_{v_n}). \end{aligned}$$

This is analogous to the way in which we have worked with a weight vector $\mathbf{w}$ and an input vector $\mathbf{x}$ in the preceding chapter.

Having clarified the operation of a single neuron, we turn to the neural network as a whole. We divide the computations of a neural network into two phases: the **input phase**, in which the external inputs are fed into the network, and the **work phase**, in which the output of the neural network is computed.

The input phase serves the purpose to initialize the network. In this phase, the activations of the input neurons are set to the values of the corresponding external inputs. The activations of the remaining neurons are initialized arbitrarily, usually by simply setting them to 0. In addition, the output function is applied to the initialized activations, so that all neurons produce initial outputs.

In the work phase, the external inputs are switched off and the activations and outputs of the neurons are recomputed (possibly multiple times). To achieve this, the network input function, the activation function and the output function are applied as described above. If a neuron does not receive any network input, because it does not have any predecessors, we define that it simply maintains its activation (and thus also its output). Essentially this is only important for the input neurons in a feed forward network. For these input neurons, which do not have predecessors, this definition is meant to guarantee that they always possess a well-defined activation (and output), since the external inputs are switched off in the work phase.

The recomputations are terminated either if the network reaches a stable state, that is, if further re-computations do not change the outputs of the neurons anymore, or if a predetermined number of recomputations has been carried out.

The temporal order of the recomputations is not generally fixed (although there are, depending on there network type, certain orders that suggest themselves). For example, all neurons of a network may recompute their outputs at the same time (synchronous update), drawing on the old outputs of their predecessors. Or we may define an order of the neurons in which they compute their new output one after the other (asynchronous update). In this case, the new outputs of other neurons may already be used as inputs for subsequent computations.

For a feed forward network the computations usually follow a **topological ordering**[2] of the neurons, as no redundant computations are carried out in this way.

[2] A topological ordering is a numbering of the vertices of a directed graph, such that all edges are directed from a vertex with a lower number to a vertex with a higher number. A topological ordering exists only for acyclic graphs, that is, for feed forward networks. For feed forward networks, an update following a topological ordering ensures that all inputs of a neuron are already available (have already been computed), before it (re-)computes its own activation and output.

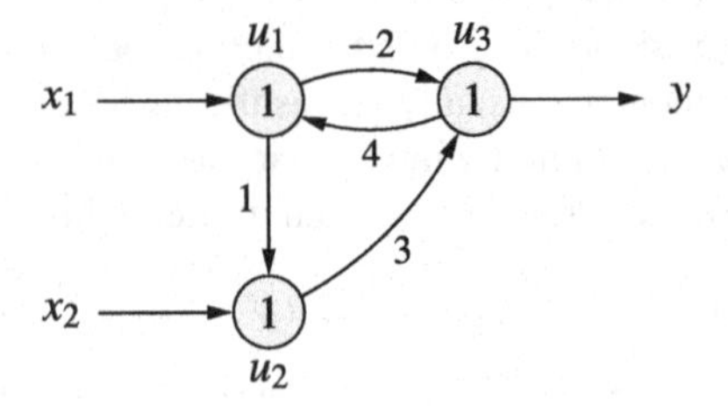

Fig. 4.3 A simple (artificial) neural network. The *numbers* in the neurons state the threshold of the activation function, the *labels at the arrows* the connection weights

Table 4.1 Computations carried out by the simple neural network shown in Fig. 4.1 for the input $(x_1 = 1, x_2 = 0)$ if the activations are updated in the order $u_3, u_1, u_2, u_3, u_1, u_2, u_3, \ldots$

	u_1	u_2	u_3	
input phase	**1**	**0**	**0**	
work phase	1	0	**0**	$\text{net}_{u_3} = -2$
	0	0	0	$\text{net}_{u_1} = 0$
	0	**0**	0	$\text{net}_{u_2} = 0$
	0	0	**0**	$\text{net}_{u_3} = 0$
	0	0	0	$\text{net}_{u_1} = 0$

Note that for recurrent networks the final output may depend on the order in which the neurons recompute their outputs as well as on how many recomputations are carried out.

As an example we reconsider the (artificial) neural network consisting of three neurons that is shown in Fig. 4.1. We assume that all have the weighted sum of the outputs of their predecessors as their network input functions. That is,

$$f_{\text{net}}^{(u)}(\mathbf{w}_u, \mathbf{in}_u) = \sum_{v \in \text{pred}(u)} w_{uv} \text{in}_{uv} = \sum_{v \in \text{pred}(u)} w_{uv} \, \text{out}_v .$$

We also assume that the activation function of all neurons is the threshold function

$$f_{\text{act}}^{(u)}(\text{net}_u, \theta) = \begin{cases} 1, & \text{if } \text{net}_u \geq \theta, \\ 0, & \text{otherwise.} \end{cases}$$

If we write the threshold into the neurons, as we did in the preceding chapter, we can represent the neural network as shown in Fig. 4.3. Finally, we assume that the output function of all neurons is the identity, that is,

$$f_{\text{out}}^{(u)}(\text{act}_u) = \text{act}_u .$$

Therefore, we need not distinguish between activation and output.

We consider first how this network operates if it receives the inputs $x_1 = 1$ and $x_2 = 0$ and updates the outputs of the neurons in the order $u_3, u_1, u_2, u_3, u_1, u_2, u_3, \ldots$. The corresponding computations are shown in Table 4.1.

In the input phase, the activations of the input neurons u_1 and u_2 are initialized with the values of the external inputs $\text{ext}_{u_1} = x_1 = 1$ and $\text{ext}_{u_2} = x_2 = 0$, respectively. The activation of the output neuron u_3 is initialized to the (arbitrarily chosen)

Table 4.2 Computations carried out by the neural network shown in Fig. 4.1 for the input $(x_1 = 1, x_2 = 0)$ if the activations are updated in the order $u_3, u_2, u_1, u_3, u_2, u_1, u_3, \ldots$

	u_1	u_2	u_3	
input phase	**1**	**0**	**0**	
work phase	1	0	**0**	$\text{net}_{u_3} = -2$
	1	**1**	0	$\text{net}_{u_2} = 1$
	0	1	0	$\text{net}_{u_1} = 0$
	0	1	**1**	$\text{net}_{u_3} = 3$
	0	**0**	1	$\text{net}_{u_2} = 0$
	1	0	1	$\text{net}_{u_1} = 4$
	1	0	**0**	$\text{net}_{u_3} = -2$

value 0. Since we assumed that the output function is the identity, we need not carry out any calculations in the input phase, but simply copy the external inputs. The neurons now have the outputs $\text{out}_{u_1} = 1$ and $\text{out}_{u_2} = \text{out}_{u_3} = 0$ (see Table 4.1).

The work phase starts with an update of the neuron u_3. Its network input is the weighted sum of the outputs of neurons u_1 and u_2, which are weighted with -2 and 3, respectively, that is, $\text{net}_{u_3} = -2 \cdot 1 + 3 \cdot 0 = -2$. Since -2 is less than 1, the activation (and thus the output) of the neuron u_3 is set to 0. In the next step of the work phase the output of the neuron u_1 is updated. (Note that its external input is no longer available, but has been switched off.) Since it receives the network input 0, its activation (and thus its output) is set to 0. Likewise, the network input of the neuron u_2 is 0 and thus its activation (and its output as well) is also set to 0 in the third step. After two additional steps, it becomes clear that we have reached a stable state, since after the fifth step of the work phase we have exactly the same situation as after the second step. Therefore, the work phase is terminated and the activation 0 of the output neuron u_3 yields the output $y = 0$ of the neural network.

That a stable state is reached is due to the fact that the neurons were updated in the order $u_3, u_1, u_2, u_3, u_1, u_2, u_3, \ldots$. If we chose, as an alternative, the order $u_3, u_2, u_1, u_3, u_2, u_1, u_3, \ldots$, we observe a completely different behavior that is shown in Table 4.2. In the seventh step of the work phase, it becomes clear that the outputs of all three neurons oscillate and thus that no stable state can be reached: the situation after the seventh step is identical to the one after the first step and thus the computations will repeat indefinitely. Hence we cannot terminate the work phase, because a stable state has been reached, but have to chose a different criterion, for example, that a certain number of update steps have been computed. However, in this case the output of the neural network depends on the step, after which the work phase is terminated. If it is terminated after step k with $(k - 1) \bmod 6 < 3$, the activation of the output neuron u_3 and thus the output of the network is $y = 0$. However, if the work phase is terminated after step k with $(k - 1) \bmod 6 \geq 3$, the activation of the output neuron u_3 and thus the output of the network is $y = 1$.

4.3 Training Neural Networks

One of the most enticing properties of (artificial) neural networks is the possibility to train them for certain tasks with the help of example data. To some degree, we already considered this possibility in the preceding chapter with the help of the delta rule. Although the delta rule is only applicable for single threshold logic units and cannot be transferred to networks directly, it already illustrates the basic principle: training a neural network consists in adapting the connection weights and possibly some other parameters (like thresholds) such that a certain criterion is optimized.

Depending on the type of the training data and the criterion to optimize we can distinguish two fundamental learning tasks: fixed and free.

Definition 4.4 A **fixed learning task** L_{fixed} for a neural network with n input neurons, that is, $U_{\text{in}} = \{u_1, \ldots, u_n\}$, and m output neurons, that is, $U_{\text{out}} = \{v_1, \ldots, v_m\}$, is a set of **training patterns** $l = (\mathbf{i}^{(l)}, \mathbf{o}^{(l)})$, each consisting of an **input vector** $\mathbf{i}^{(l)} = (\text{ext}_{u_1}^{(l)}, \ldots, \text{ext}_{u_n}^{(l)})$ and an **output vector** $\mathbf{o}^{(l)} = (o_{v_1}^{(l)}, \ldots, o_{v_m}^{(l)})$.

If we are given a fixed learning task, we desire to train a neural network in such a way that it produces for all training patterns $l \in L_{\text{fixed}}$ the outputs contained in the output vector $\mathbf{o}^{(l)}$ if the external inputs of the corresponding input vector $\mathbf{i}^{(l)}$ are fed into the network.

In practice, this optimum can rarely be achieved and thus one may have to accept a partial or approximate solution. In order to determine how well a neural network solves a fixed learning task, an error function is employed, which measures how well the actual outputs coincide with the desired outputs in the training patterns. Commonly this error function is defined as the sum of squared deviations of desired and actual output over all training patterns and all output neurons. That is, the error of a neural network w.r.t. a fixed learning task L_{fixed} is defined as

$$e = \sum_{l \in L_{\text{fixed}}} e^{(l)} = \sum_{v \in U_{\text{out}}} e_v = \sum_{l \in L_{\text{fixed}}} \sum_{v \in U_{\text{out}}} e_v^{(l)},$$

where

$$e_v^{(l)} = \left(o_v^{(l)} - \text{out}_v^{(l)}\right)^2$$

is the individual error for a training pattern l and an output neuron v.

The square of the deviations of the desired and the actual output is chosen for various reasons. In the first place, it is clear that we cannot simply sum the deviations directly, since then positive and negative deviations could cancel, thus producing a misleading impression of the actual quality of the network. Therefore, we have to sum (at least) the absolute values of the deviations.

However, the square of the deviation of the actual and the desired output has at least two advantages over the absolute value: in the first place it is continuously differentiable everywhere, while the derivative of the absolute value does not exist/is discontinuous at 0. It is desirable that the error function is continuously differentiable, because this simplifies the derivation of the update rules for the weights (see

Sect. 5.4). Secondly, large deviations from the desired output are weighted more severely, so that there is a tendency that during training individual strong deviations (i.e., for individual training patterns) from the desired value are avoided.

Let us now turn to free learning tasks.

Definition 4.5 A **free learning task** L_{free} for a neural network with n input neurons, that is, $U_{\text{in}} = \{u_1, \ldots, u_n\}$, is a set of **training patterns** $l = (\mathbf{i}^{(l)})$, each of which consists of an **input vector** $\mathbf{i}^{(l)} = (\text{ext}_{u_1}^{(l)}, \ldots, \text{ext}_{u_n}^{(l)})$.

While the training patterns of a fixed learning task contain a desired output, which allows us to compute an error, free learning tasks need a different criterion in order to assess how well a neural network solves the task. In principle, with a free learning task for a neural network we ask for a training result that "produces similar outputs for similar inputs", where the outputs can be chosen by the training method. The objective of the training can be, for example, to group the input vectors into clusters of similar vectors (*clustering* or *cluster analysis*), so that for all vectors in a cluster the same output is produced (see, for instance, Sect. 7.2).

If we are given a free learning task, the most important aspect for training a neural network is how the similarity between the training patterns is measured. This may be defined, for example, with the help of a distance function (details about distance functions can be found in Sect. 6.1). The outputs that are produced for a group of similar input vectors are then often assigned by choosing representatives or by forming prototypes (see Chap. 7 for details).

In the remainder of this section, we consider some general aspects of training neural networks that are relevant in practice. It is, for instance, advisable to normalize the inputs of a neural network in order to avoid certain numerical problems, which can result from an unequal scaling of the different input variables. Most commonly, each input variable is scaled in such a way that it has the arithmetic mean 0 and the variance 1. To achieve this, one computes from the input vectors of the training patterns l of the learning task L for each input neuron u_k

$$\mu_k = \frac{1}{|L|} \sum_{l \in L} \text{ext}_{u_k}^{(l)} \quad \text{and} \quad \sigma_k = \sqrt{\frac{1}{|L|} \sum_{l \in L} \left(\text{ext}_{u_k}^{(l)} - \mu_k\right)^2},$$

that is, the arithmetic mean and the standard deviation of the external inputs.[3] Then the external inputs are transformed according to

$$\text{ext}_{u_k}^{(l)(\text{new})} = \frac{\text{ext}_{u_k}^{(l)(\text{old})} - \mu_k}{\sigma_k}.$$

This normalization can be carried out as a pre-processing step or (in a feed forward network) by the output function of the input neurons.

[3]The second formula is based on the maximum likelihood estimator for the variance of a normal distribution. In statistics often the unbiased estimator is preferred, which differs from the one used above only by using $|L| - 1$ instead of $|L|$. For the normalization this difference is negligible.

Up to now we assumed (sometimes implicitly) that the inputs and outputs of a neural network are real numbers. However, in practice we often face nominal attributes (often also called symbolic), for example, color, vehicle type, marital status etc. If we want to process such attributes with a neural network, we have to transform them into numbers. Although it may seem natural to simply number the different values of such attributes, this can lead to undesired effects if the numbers do not reflect a natural order of the values (and even then it may not be appropriate to choose equal steps between neighboring values). A better option is a so-called 1-in-n encoding, in which each nominal attribute is assigned as many (input or output) neurons as it has values: each neuron corresponds to one attribute value. With the input of a training pattern, the neuron that corresponds to the obtaining value of the nominal attribute is set to 1, while all other neurons that belong to the same attribute are set to 0. That is, only 1 in n neurons (where n is the number of attributes values) is set to one, the others to 0, which explains the name of this encoding.

Chapter 5
Multi-Layer Perceptrons

Having described the structure, the operation and the training of (artificial) neural networks in a general fashion in the preceding chapter, we turn in this and the subsequent chapters to specific forms of (artificial) neural networks. We start with the best-known and most widely used form, the so-called **multi-layer perceptron** (MLP), which is closely related to the networks of threshold logic units we studied in Chap. 3. They exhibit a strictly layered structure (see the definition below) and may employ other activation functions than a step at a crisp threshold.

5.1 Definition and Examples

Definition 5.1 An **r-layer perceptron** is a neural network with a graph $G = (U, C)$ that satisfies the following restrictions:

1. $U_{\text{in}} \cap U_{\text{out}} = \emptyset$,
2. $U_{\text{hidden}} = U_{\text{hidden}}^{(1)} \cup \cdots \cup U_{\text{hidden}}^{(r-2)}$, $\forall 1 \leq i < j \leq r-2 : U_{\text{hidden}}^{(i)} \cap U_{\text{hidden}}^{(j)} = \emptyset$,
3. $C \subseteq (U_{\text{in}} \times U_{\text{hidden}}^{(1)}) \cup (\bigcup_{i=1}^{r-3} U_{\text{hidden}}^{(i)} \times U_{\text{hidden}}^{(i+1)}) \cup (U_{\text{hidden}}^{(r-2)} \times U_{\text{out}})$ or, if there are no hidden neurons ($r = 2$, $U_{\text{hidden}} = \emptyset$), $C \subseteq U_{\text{in}} \times U_{\text{out}}$.

The network input function of each hidden and each output neuron is the weighted sum (weighted with the connection weights) of the inputs, that is,

$$\forall u \in U_{\text{hidden}} \cup U_{\text{out}} : \quad f_{\text{net}}^{(u)}(\mathbf{w}_u, \mathbf{in}_u) = \mathbf{w}_u \mathbf{in}_u = \sum_{v \in \text{pred}(u)} w_{uv} \, \text{out}_v .$$

The activation function of each neuron is a so-called **sigmoid function**, that is, a monotonically non-decreasing function with

$$f : \mathbb{R} \to [0, 1] \quad \text{with} \quad \lim_{x \to -\infty} f(x) = 0 \text{ and } \lim_{x \to \infty} f(x) = 1.$$

The activation function of each output neuron is either also a sigmoid function or a linear function $f_{\text{act}}(\text{net}, \theta) = \alpha \, \text{net} - \theta$.

R. Kruse et al., *Computational Intelligence*, Texts in Computer Science,
DOI 10.1007/978-1-4471-5013-8_5, © Springer-Verlag London 2013

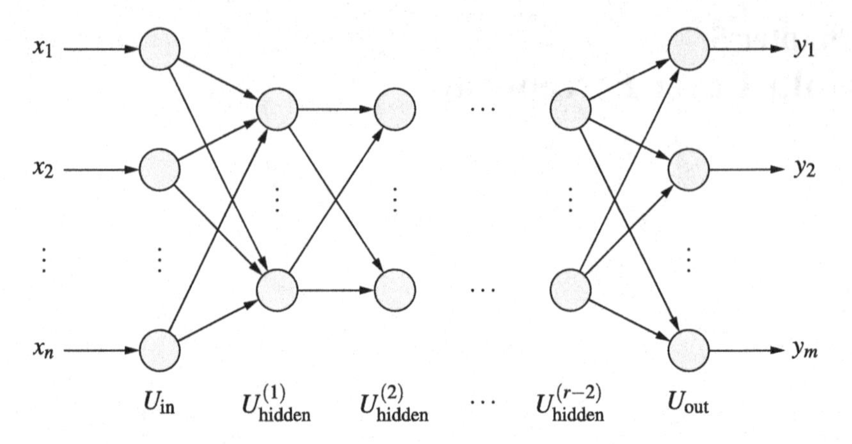

Fig. 5.1 General structure of an r-layered perceptron

Intuitively, the restrictions of the graph in this definition mean that a multi-layer perceptron consists of an input and an output layer (the neurons of the sets U_{in} and U_{out}) and none, one, or several hidden layers (the neurons in the sets $U^{(i)}_{\text{hidden}}$) between them. Connections exist only between the neurons of consecutive layers, that is, between the input layer and the first hidden layer, between consecutive hidden layers and between the last hidden layer and the output layer (see Fig. 5.1). Note that according to this definition a multi-layer perceptron has always at least two layers, namely the input and the output layer.

Examples of sigmoid activation functions, all of which have a parameter, namely a **bias value** θ, are shown in Fig. 5.2. The threshold logic units we studied in Chap. 3 employ exclusively the (Heaviside or unit) step function as their activation function. The advantages of other activation functions are discussed in Sect. 5.2. Here we remark only that instead of the listed **unipolar** sigmoid functions ($\lim_{x\to-\infty} f(x) = 0$) sometimes **bipolar** sigmoid functions ($\lim_{x\to-\infty} f(x) = -1$) are employed. An example of such a function is the hyperbolic tangent (see Fig. 5.3), which is closely related to the logistic function. In addition, it should be clear that any unipolar sigmoid function can be turned into a bipolar one by simply multiplying it by 2 and subtracting 1. Using bipolar sigmoid activation functions does not cause any fundamental differences. In this book, we therefore confine ourselves to unipolar sigmoid activation functions. All considerations and derivations of the subsequent sections can easily be transferred to bipolar functions.

The strictly layered structure of a multi-layer perceptron and the special network input function of the hidden as well as the output neurons suggest to describe the network structure with the help of a weight matrix, as already discussed in Chap. 4. In this way, the computations carried out by a multi-layer perceptron can be written in a simpler way, using vector and matrix notation. However, for this purpose we do not use a weight matrix for the network as a whole (although this would be possible as well), but one matrix for the connections between one layer and the next: let

(Heaviside or unit) step function:

$$f_{\text{act}}(\text{net}, \theta) = \begin{cases} 1 & \text{if net} \geq \theta, \\ 0 & \text{otherwise.} \end{cases}$$

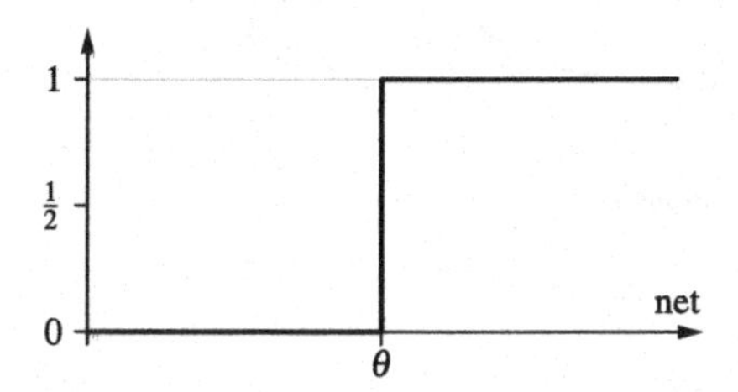

semi-linear function:

$$f_{\text{act}}(\text{net}, \theta) = \begin{cases} 1 & \text{if net} > \theta + \frac{1}{2}, \\ 0 & \text{if net} < \theta - \frac{1}{2}, \\ (\text{net} - \theta) + \frac{1}{2} & \text{otherwise.} \end{cases}$$

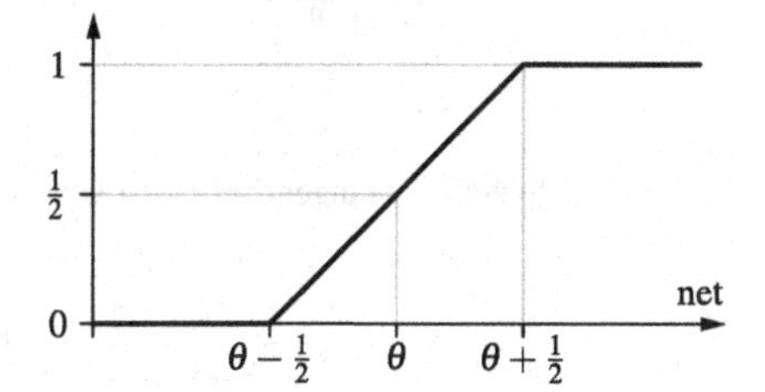

sine up to saturation:

$$f_{\text{act}}(\text{net}, \theta) = \begin{cases} 1 & \text{if net} > \theta + \frac{\pi}{2}, \\ 0 & \text{if net} < \theta - \frac{\pi}{2}, \\ \frac{\sin(\text{net} - \theta) + 1}{2} & \text{otherwise.} \end{cases}$$

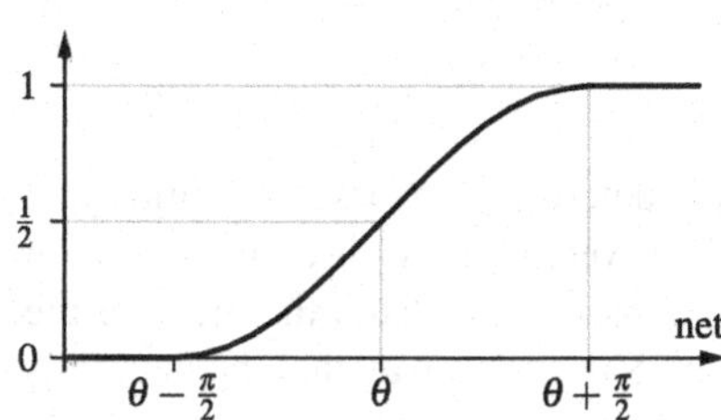

logistic function:

$$f_{\text{act}}(\text{net}, \theta) = \frac{1}{1 + e^{-(\text{net} - \theta)}}$$

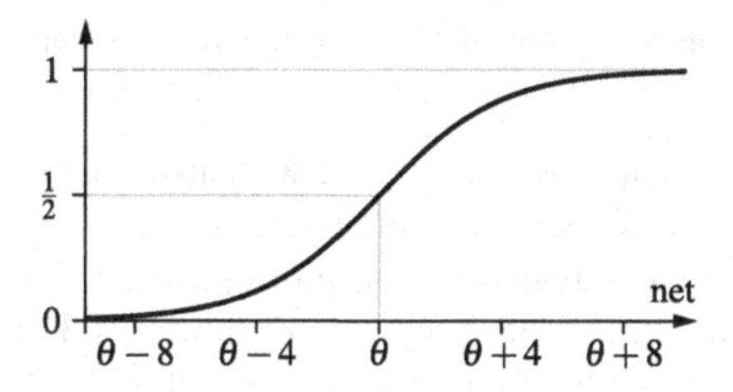

Fig. 5.2 Different unipolar sigmoid activation functions

$U_1 = \{v_1, \ldots, v_m\}$ and $U_2 = \{u_1, \ldots, u_n\}$ be the neurons of two layers of a multi-layer perceptron, where U_2 may follow U_1. We construct an $n \times m$ matrix

$$\mathbf{W} = \begin{pmatrix} w_{u_1 v_1} & w_{u_1 v_2} & \ldots & w_{u_1 v_m} \\ w_{u_2 v_1} & w_{u_2 v_2} & \ldots & w_{u_2 v_m} \\ \vdots & \vdots & & \vdots \\ w_{u_n v_1} & w_{u_n v_2} & \ldots & w_{u_n v_m} \end{pmatrix}$$

of the weights of the connections between these two layers, setting $w_{u_i v_j} = 0$ if there is no connection between neuron v_j and neuron u_i. The advantage of such a matrix is that it allows us to write the network input of the neurons of the layer U_2 as

$$\mathbf{net}_{U_2} = \mathbf{W} \cdot \mathbf{in}_{U_2} = \mathbf{W} \cdot \mathbf{out}_{U_1}$$

where $\mathbf{net}_{U_2} = (\text{net}_{u_1}, \ldots, \text{net}_{u_n})^\top$ and $\mathbf{in}_{U_2} = \mathbf{out}_{U_1} = (\text{out}_{v_1}, \ldots, \text{out}_{v_m})^\top$ (the superscript $\top$ means that the vector is transposed, that is, that it is turned from a row vector into a column vector—as if it were a $n \times 1$ or $m \times 1$ matrix).

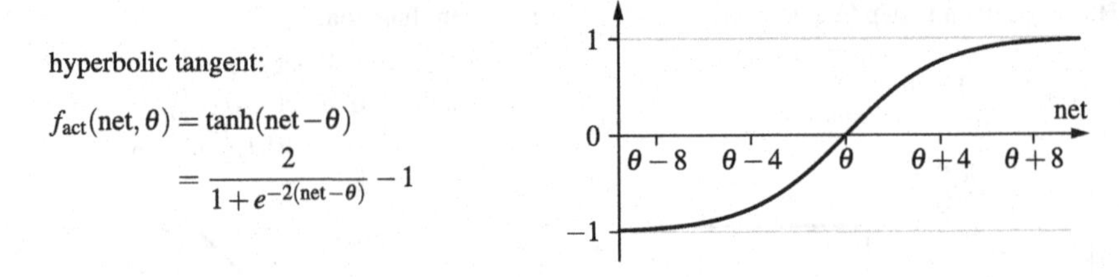

Fig. 5.3 The hyperbolic tangent, a bipolar sigmoid function

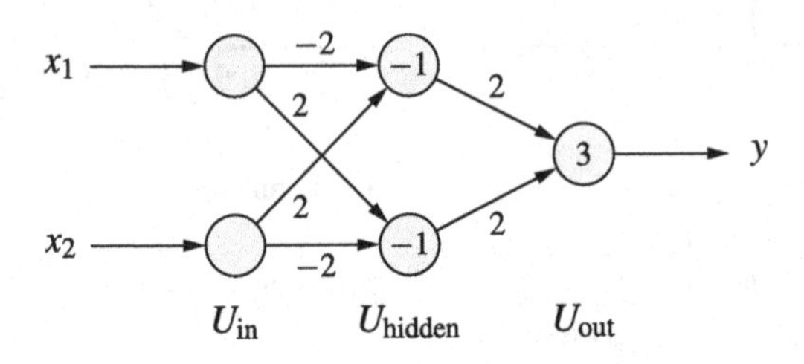

Fig. 5.4 A three-layer perceptron for the biimplication

The placement of the weights in the matrix is determined by the convention that matrix-vector equations are written with column vectors as well as the common rules of matrix-vector multiplication. This explains why we fixed the order of the indices of the weights in Definition 4.3 on page 38 in such a way that the neuron is written first to which the connection leads.

As a first example of a multi-layer perceptron, we reconsider the network of threshold logic units studied in Sect. 3.4 that computes the biimplication. This network is shown in Fig. 5.4 as a three-layer perceptron. Note that compared to Fig. 3.10 on page 21 there are two additional neurons, namely the two input neurons. Formally these two neurons are needed, because our definition of a neural network only allows us to assign weights to the edges of the graph, but not directly to the inputs. Hence, we need the input neuron so that we have edges to the neurons of the hidden layer to which we can assign the input weights. (Note, however, that the input neurons may also transform the input quantities if they possess a suitable output function. For example, if the logarithm of an input is to be used for the computations of a neural network, we simply choose $f_{\text{out}}(\text{act}) \equiv \log(\text{act})$ for the corresponding input neuron.)

To illustrate the matrix notation of the weights, we describe the connection weights of this network by two matrices. We obtain

$$\mathbf{W}_1 = \begin{pmatrix} -2 & 2 \\ 2 & -2 \end{pmatrix} \quad \text{and} \quad \mathbf{W}_2 = \begin{pmatrix} 2 & 2 \end{pmatrix},$$

where the matrix $\mathbf{W}_1$ contains the weights of the connections from the input layer to the hidden layer and the matrix $\mathbf{W}_2$ contains the weights of the connections from the hidden layer to the output layer.

As another example we consider the **Fredkin gate**, which plays an important role in so-called **conservative logic**[1] (Fredkin and Toffoli 1982). This gate has three inputs: s, x_1 and x_2, and three outputs: s, y_1 and y_2 (see Fig. 5.5). The "switch variable" s is always passed through without change. The inputs x_1 and x_2 are connected either parallel or crossed to the two outputs y_1 and y_2, depending on whether the switch variable s has value 0 or value 1. The function that is computed by a Fredkin gate is shown in Fig. 5.5 as a table and geometrically in Fig. 5.6.

Figure 5.7 shows a three-layer perceptron that computes the function of the Fredkin gate (ignoring the switch variable s, which is merely passed through without change). Actually, this network consists of two separate three-layer perceptrons, since there are no connections from any of the neurons in the hidden layer to *both* output neurons. This is, of course, not always the case for multi-layer perceptron with more than one output, but a result of the special function of the Fredkin gate.

To illustrate the matrix notation of the weights, we write the weights of this network in two matrices. We obtain

$$\mathbf{W}_1 = \begin{pmatrix} 2 & -2 & 0 \\ 2 & 2 & 0 \\ 0 & 2 & 2 \\ 0 & -2 & 2 \end{pmatrix} \quad \text{and} \quad \mathbf{W}_2 = \begin{pmatrix} 2 & 0 & 2 & 0 \\ 0 & 2 & 0 & 2 \end{pmatrix},$$

where the matrix $\mathbf{W}_1$ represents the connections from the input layer to the hidden layer and the matrix $\mathbf{W}_2$ the connections from the hidden layer to the output layer. Note that in these matrices zero elements correspond to missing/absent connections.

With the help of the matrix notation of the weights it is easy to show why sigmoid or generally non-linear activation functions are decisive for the computational capabilities of a multi-layer perceptron. Suppose all activation and output functions were linear, that is, functions $f_{\text{act}}(\text{net}, \theta) = \alpha\,\text{net} - \theta$. Then such a multi-layer perceptron can always be reduced to a two-layer perceptron (only input and output layer):

As mentioned above, we have for two consecutive layers U_1 and U_2

$$\mathbf{net}_{U_2} = \mathbf{W} \cdot \mathbf{in}_{U_2} = \mathbf{W} \cdot \mathbf{out}_{U_1}.$$

If all activation function are linear, then the activations of the neurons of the layer U_2 can also be determined by a matrix-vector calculation, namely by

$$\mathbf{act}_{U_2} = \mathbf{D}_{\text{act}} \cdot \mathbf{net}_{U_2} - \boldsymbol{\theta},$$

where $\mathbf{act}_{U_2} = (\text{act}_{u_1}, \ldots, \text{act}_{u_n})^\top$ is the vector of activations of the neurons of layer U_2, $\mathbf{D}_{\text{act}}$ is an $n \times n$ diagonal matrix of the factors α_{u_i}, $i = 1, \ldots, n$, and $\boldsymbol{\theta} = (\theta_{u_1}, \ldots, \theta_{u_n})^\top$ is a bias vector. If the output function is a linear function as

[1] Conservative logic is a mathematical model for computations and computational powers of computers, in which the fundamental physical principles that govern computing machines are explicitly taken into account. Among these principles are, for instance, that the speed with which information can travel as well as the amount of information that can be stored in the state of a finite system are both finite (Fredkin and Toffoli 1982).

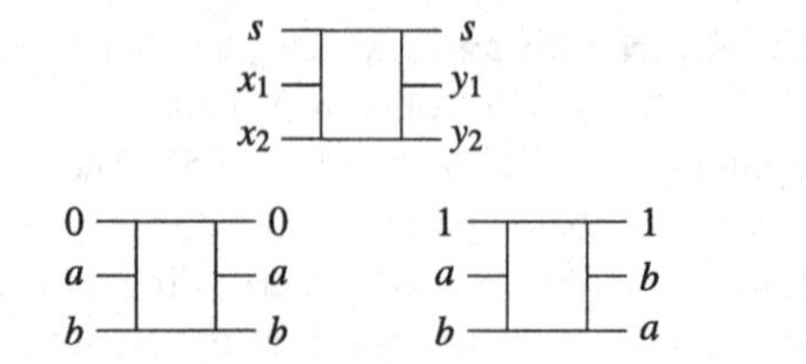

s	0	0	0	0	1	1	1	1
x_1	0	0	1	1	0	0	1	1
x_2	0	1	0	1	0	1	0	1
y_1	0	0	1	1	0	1	0	1
y_2	0	1	0	1	0	0	1	1

Fig. 5.5 The Fredkin gate (Fredkin and Toffoli 1982)

Fig. 5.6 Geometric interpretation of the function that is computed by a Fredkin gate (ignoring the input s which is simply passed through)

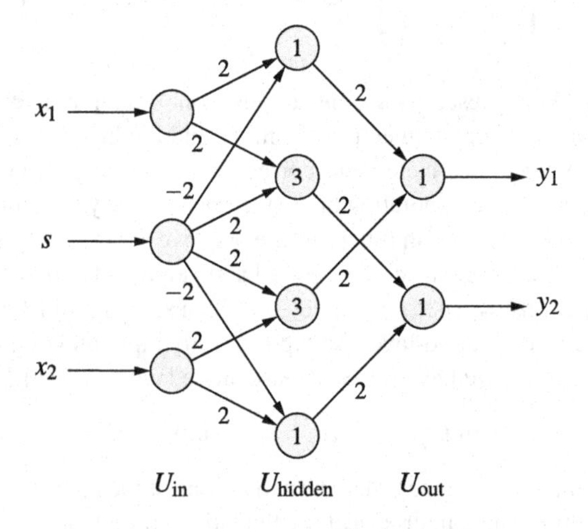

Fig. 5.7 A three-layer perceptron that computes the function of the Fredkin gate (see Fig. 5.5)

well, we have analogously

$$\mathbf{out}_{U_2} = \mathbf{D}_{\text{out}} \cdot \mathbf{act}_{U_2} - \boldsymbol{\xi},$$

where $\mathbf{out}_{U_2} = (\text{out}_{u_1}, \ldots, \text{out}_{u_n})^\top$ is the vector of outputs of the neurons of layer U_2, $\mathbf{D}_{\text{out}}$ is again an $n \times n$ diagonal matrix of factors and finally $\boldsymbol{\xi} = (\xi_{u_1}, \ldots, \xi_{u_n})^\top$ is again a bias vector. Therefore, we can write the computation of the outputs of the neurons of layer U_2 from the outputs of the neurons of the preceding layer U_1 as

$$\mathbf{out}_{U_2} = \mathbf{D}_{\text{out}} \cdot \big(\mathbf{D}_{\text{act}} \cdot (\mathbf{W} \cdot \mathbf{out}_{U_1}) - \boldsymbol{\theta}\big) - \boldsymbol{\xi},$$

which can be simplified to

$$\mathbf{out}_{U_2} = \mathbf{A}_{12} \cdot \mathbf{out}_{U_1} + \mathbf{b}_{12},$$

with an $n \times m$ matrix $\mathbf{A}_{12}$ and an n-dimensional vector $\mathbf{b}_{12}$. Analogously, we obtain for the computations of outputs of the neurons of a layer U_3, which follows layer U_2, from the outputs of the neurons of layer U_2

$$\mathbf{out}_{U_3} = \mathbf{A}_{23} \cdot \mathbf{out}_{U_2} + \mathbf{b}_{23},$$

and therefore for computing the outputs of the neurons of layer U_3 from the outputs of the neurons of layer U_1

$$\mathbf{out}_{U_3} = \mathbf{A}_{13} \cdot \mathbf{out}_{U_1} + \mathbf{b}_{13},$$

where $\mathbf{A}_{13} = \mathbf{A}_{23} \cdot \mathbf{A}_{12}$ and $\mathbf{b}_{13} = \mathbf{A}_{23} \cdot \mathbf{b}_{12} + \mathbf{b}_{23}$. As a consequence, the computations of two consecutive layers can be reduced to a single layer. It should be clear that by iterating this result we can incorporate the computations of arbitrarily many layers. Therefore multi-layer perceptrons can compute only affine transformations if the activation and output functions of all neurons are linear. For more complex tasks, non-linear activation functions are needed.

5.2 Function Approximation

In this section, we study in more detail what we gain compared to threshold logic units (that is, neurons with the (Heaviside or unit) step function as their activation function) if we allow for other activation function.[2] In a first step, we demonstrate that all Riemann-integrable functions can be approximated by four-layer perceptrons with arbitrary accuracy, provided that the output neuron has the identity, instead of a step function, as its activation function.

The principle is illustrated in Figs. 5.8 and 5.9 for a unary function: the function to compute is approximated by a step function (see Fig. 5.8). For each step border x_i we create a neuron in the first hidden layer of a multi-layer perceptron with a total of four layers (see Fig. 5.9). This neuron serves the purpose to determine on which side of the step border an input values lies.

In the second hidden layer we create one neuron for each step, which receives input from the two neurons in the first hidden layer that refer to the values x_i and x_{i+1} marking the borders of this step (see Fig. 5.9). The weights and the threshold are chosen in such a way that the neuron is activated if the input value is no less than x_i, but less than x_{i+1}, that is, if the input values lies in the range of the step. Note that in this way only exactly one neuron on the second hidden layer can be active, namely the one representing the step in which the input value lies.

[2] In the following we assume implicitly that the output function of all neurons is the identity. Only the activation functions are exchanged.

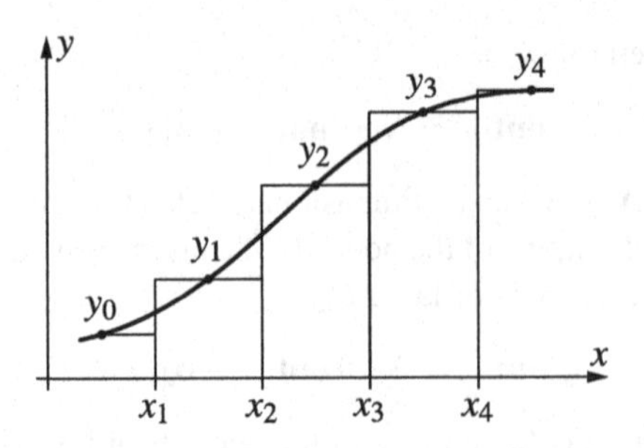

Fig. 5.8 Approximating a continuous function with step functions

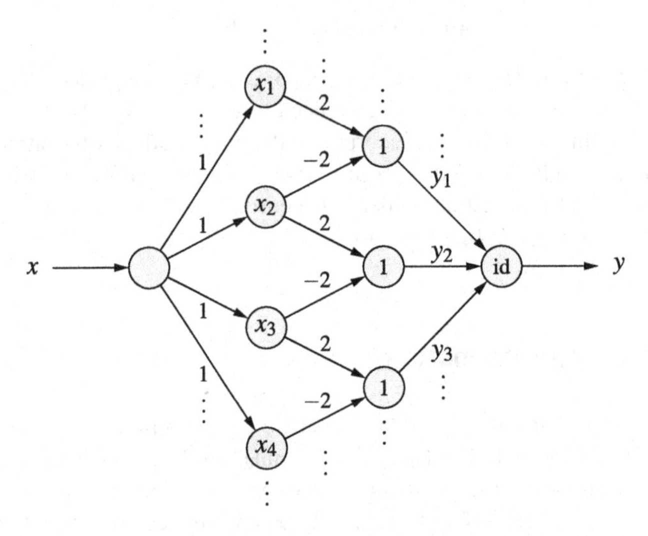

Fig. 5.9 A neural network that computes the step function shown in Fig. 5.8. ("id" instead of a threshold value means that this neuron uses the identity instead of a threshold function)

The connections from the neurons of the second hidden layer to the output neuron are weighted with the function values of the stair steps that are represented by the neurons. Since only one neuron can be active on the second hidden layer, the output neuron receives as input the height of the stair step, in which the input value lies. Since the activation function of the output neuron is the identity, this value is emitted unchanged. As a consequence, the four-layer perceptron shown in Fig. 5.9 computes exactly the step function sketched in Fig. 5.8.

It should be clear that the approximation accuracy can be increased arbitrarily by making the stair steps sufficiently small. It may help to recall the introduction of the notion of an integral in calculus by Riemann upper and lower sums: for any given error limit $\varepsilon > 0$ there exists a step width $\delta(\varepsilon) > 0$, such that the Riemann upper and lower sum differ by less than ε. Therefore we can state the following theorem:

Theorem 5.1 *Any Riemann-integrable function can be approximated with arbitrary accuracy by a multi-layer perceptron.*

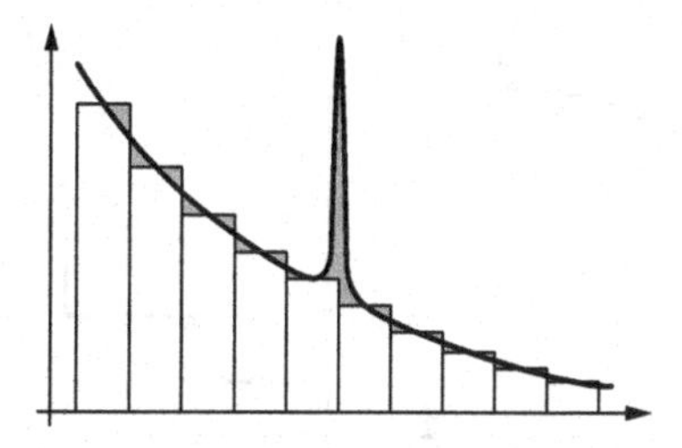

Fig. 5.10 Limits of the theorem about approximating a function by a multi-layer perceptron

Note that this theorem only requires that the function to represent is Riemann-integrable. It need *not* be continuous. That is, the function to represent may have "jumps." However it may have only finitely many "jumps" of finite height in the region in which it is to be approximated by a multi-layer perceptron. In other words, the function must be continuous "almost everywhere."

Note also that in this theorem the approximation error is measured by the *area* between the function to approximate and the output of the multi-layer perceptron. This area can be made arbitrarily small by increasing the number of neurons (i.e., by increasing the number of stair steps). However, this does *not* guarantee that for a given multi-layer perceptron, which achieves a certain approximation accuracy in this sense, the difference between its output and the function to approximate is less than a certain error bound everywhere. The function could, for instance, possess a very thin spike, which is not captured by any stair step (see Fig. 5.10). In such a case, the area between the function to represent and the output of the multi-layer perceptron is small (because the spike is thin and thus encloses only a small area), but at the location of the spike the deviation of the output from the true function value can nevertheless be considerable.

Naturally, the idea to approximate a given function by a step function can directly be transferred to functions with multiple arguments: the input space is divided—depending on the arity of the function—into rectangles, boxes, or generally hyper-boxes, to each of which a function value is assigned. It should be clear that a four-layer perceptron can be constructed again that computes the higher-dimensional "step function." Since we can also increase the approximation accuracy arbitrarily by making the rectangles, boxes or hyperboxes sufficiently small, the above theorem is not limited to unary functions, but holds for functions of arbitrary arity.

Even though the above theorem attests multi-layer perceptrons a high expressive power, one has to concede that it is of little use in practice. The reason is clearly that in order to achieve a sufficiently accurate approximation we have to chose step functions with a very small step width (and thus very many steps). This then forces us to construct multi-layer perceptrons with a possibly huge number of neurons (one neuron for each step and for each step border).

In order to understand how multi-layer perceptrons can approximate functions much better, we consider the case of a unary function in a little more detail. It is easy to see that we can save one layer if we do not use the absolute, but the relative height of a step (that is, the change w.r.t. the preceding step) as the weight of the connection

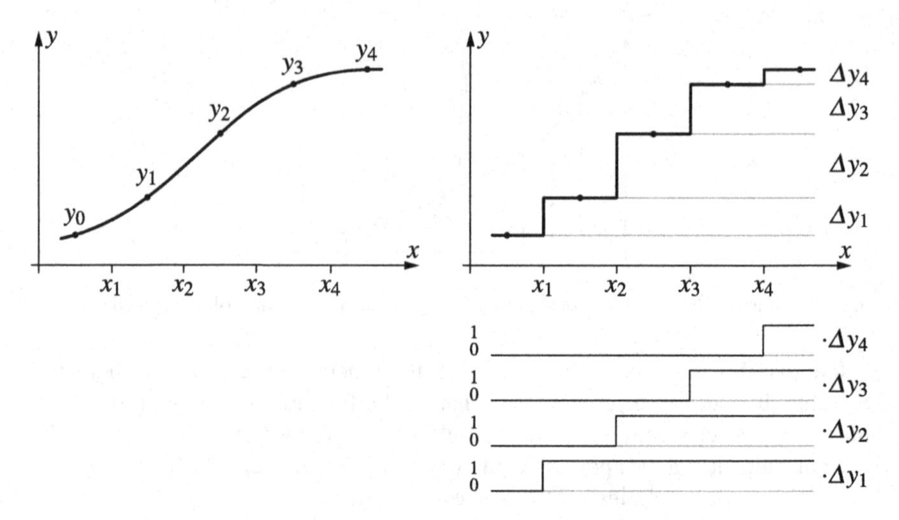

Fig. 5.11 Representing the step function shown in Fig. 5.8 by a weighted sum of (Heaviside) step functions. It is $\Delta y_i = y_i - y_{i-1}$

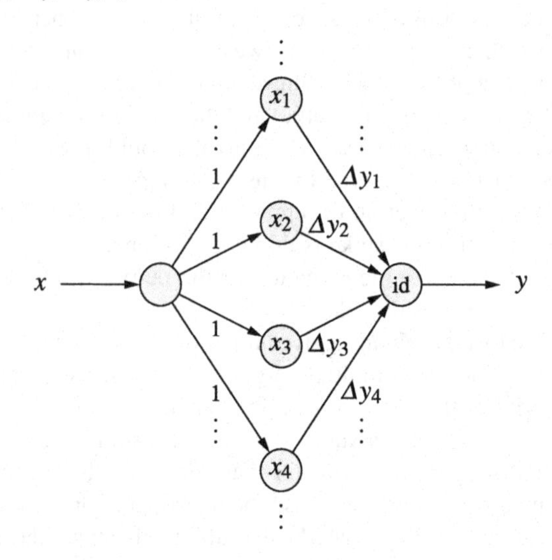

Fig. 5.12 A neural network that computes the step function shown in Fig. 5.8 as a weighted sum of (Heaviside) step functions, cf. Fig. 5.11. ("id" instead of a threshold means that this neuron has the identity instead of a threshold function as its activation function)

to the output neuron. This idea is illustrated in Figs. 5.11 and 5.12. Every neuron of the hidden layer represents a step border and determines whether an input values lies to the left or to the right of this border. If it lies to the right, the neuron becomes active (outputs a 1). If it lies to the left, the neuron remains inactive (outputs a 0). The output neuron then receives as an additional network input the relative height of the stair step (that is, the change compared to the preceding stair step). Since always all those neurons of the hidden layer are active for which the step border lies

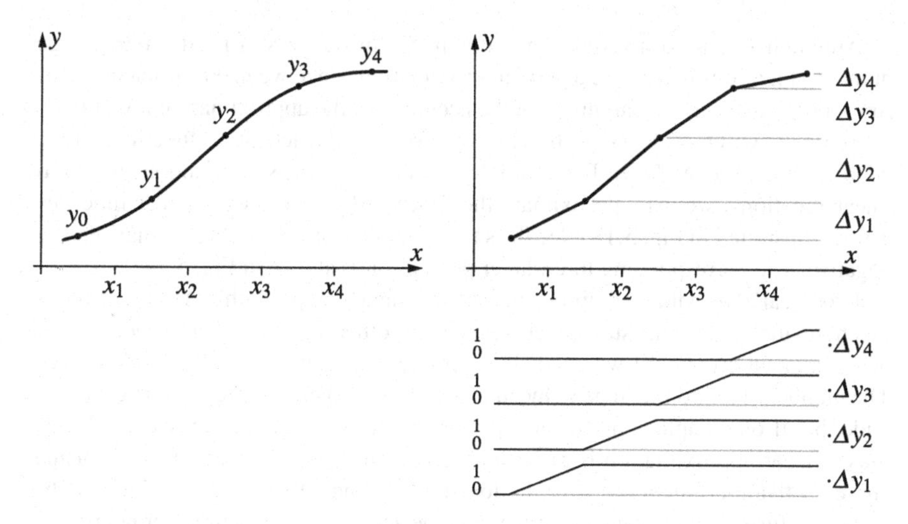

Fig. 5.13 Approximation of a continuous function by a weighted sum of semi-linear functions. It is $\Delta y_i = y_i - y_{i-1}$

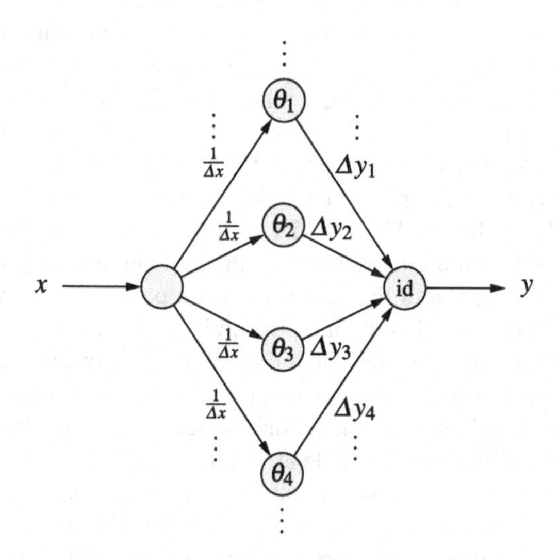

Fig. 5.14 A neural network that computes the piecewise linear function shown in Fig. 5.13 by a weighted sum of semi-linear functions. It is $\Delta x = x_{i+1} - x_i$ and $\theta_i = \frac{x_i}{\Delta x}$. ("id" means again that the activation function of the output neuron is the identity)

to the left of the current input value, the weights add up to the height of the stair step.[3] Note that the (relative) step heights may as well be negative and therefore the function to approximate need not be monotonically non-decreasing.

---

[3]Note that this approach is not easily transferred to functions with multiple arguments. For this to be possible, the influences of the two or more inputs have to be independent in a certain sense.

Although we saved a layer of neurons in this way, we still need a fairly large number of neurons to achieve a good approximation, since we need sufficiently narrow steps. However, we can improve the accuracy of the approximation not only by reducing the width of the steps, but also by changing the activation functions of the neurons in the hidden layer. For example, if we replace the step functions by semi-linear functions, we can approximate the function by a piecewise-linear function. This is illustrated in Fig. 5.13. Needless to say that the step heights Δy_i may also be negative. The corresponding three-layer perceptron is shown in Fig. 5.14.

It is clear that with this approach and the same "step width" Δx, the error is much smaller than with step functions. Or the other way round: in order to stay below a given error limit we need considerably fewer neurons in the hidden layer. The number of neurons can be reduced even further if the steps are not made equally wide, but if one employs narrower ones where the function is heavily curved (and thus a linear approximation is poor) and fewer where it is almost linear. By using curved activation functions—like the logistic function—the approximation may be improved further or the same accuracy may be achieved with even fewer neurons.

The principle that we exploited above to eliminate one hidden layer of the multi-layer perceptron may not be directly transferable to functions with multiple arguments, because in two or more dimensions we certainly have to engird, in two steps, the regions for which the weights of the connections to the output layer state the function values. However, with stronger mathematical tools and a few and fairly weak additional assumptions it can be shown that for functions with multiple arguments also a single hidden layer suffices. To be more specific, it can be shown that a multi-layer perceptron can approximate any continuous function (note that this is a stronger condition than in Theorem 5.1, which only required that the function is Riemann-integrable) on a compact part of the $\mathbb{R}^n$ with arbitrary accuracy, provided the activation function of the neurons is not a polynomial (which, however, is implicitly excluded in our definition by the limit conditions anyway). This statement holds even in the stronger sense that the difference between the output of the multi-layer perceptron and the function to approximate is everywhere smaller than a given error bound ε (while Theorem 5.1 only states that the *area* between the output and the actual function can be made arbitrarily small). An overview of results concerning the approximation powers of multi-layer perceptrons and a proof of the mentioned theorem can be found, for example, in Pinkus (1999).

Note, however, that these results are relevant only in as far as they ensure that it is not the choice of a structure with only one hidden layer that already rules out the possibility to approximate certain (continuous) functions sufficiently well. That is, they ensure that there are no fundamental obstacles. These results do not say anything, though, about how, for a given network structure and particularly a given number of hidden neurons, one can find the parameter values with which the best possible approximation accuracy is achieved.

One should also be careful *not* to read from the mentioned theorem that multi-layer perceptrons with more than one hidden layer are useless, because they do not increase the expressive power of multi-layer perceptrons (even though it is often cited as an argument in this direction). With a second hidden layer, it may sometimes be possible to compute the function to represent in a much simpler fashion. Multi-layer perceptrons with two hidden layers may also have advantages when we have to train them. However, since multi-layer perceptrons with more that one hidden layer are much more difficult to analyze, little is known about such possibilities.

5.3 Logistic Regression

Having convinced ourselves that multi-layer perceptrons with general activation functions have considerable expressive and computational powers, we now turn to the task of determining their parameters with the help of training examples. In Chap. 4, we already mentioned that we need an error function for this and that the most common error function is the sum of the squared errors over the output neurons and the training patterns. This sum of squared errors is to be minimized by suitable adaptations of the weights and the parameters of the activation functions. This approach leads to **method of least squares**, also known as **regression**, which is well-known in calculus and statistics, where it is used to determine best fit lines (regression lines) and generally best fit polynomials for a given set of data points (x_i, y_i). The fundamentals of this methods are recalled in Sect. 10.2.

Although we are not interested in best fit lines or polynomials here, the method is worth studying. The reason is that computing a best fit polynomial can also be used to determine other best fit functions, namely if we succeed in finding an appropriate transformation that reduces the problem to the task of finding a regression polynomial. For instance, best fit functions of the form

$$y = ax^b$$

can be found by determining a regression line: if we take the logarithm of the equation, we obtain

$$\ln y = \ln a + b \cdot \ln x.$$

This equation can be handled by computing a regression line. We merely have to take the logarithms of the data points (x_i, y_i) and work with the transformed values.[4]

For (artificial) neural networks, it is important that for the **logistic function**

$$y = \frac{Y}{1 + e^{a+bx}},$$

where Y, a and b are constants, there also exists a transformation with which the problem can be reduced to the task of computing a regression line (so-called **logistic regression**). The logistic function is very frequently employed as an activation function (see also Sect. 5.4). If we can find a method to determine a logistic regression function, we immediately possess a method to determine the parameters of a

[4]Note, however, that with this approach the sum of squared errors is minimized in the transformed space (coordinates $x' = \ln x$ and $y' = \ln y$), but this does not imply that it is also minimized in the original space (coordinates x and y). Nevertheless this approach usually yields very good results or at least an initial solution that may then be improved by other means.

two-layered perceptron with a single input, since the value of a is the bias value of the output neuron and the value of b is the weight of the input.

However, how can we "linearize" the logistic function, that is, how can we transform it in such a way that the problem is reduced to the task of finding a regression line? We start by forming the reciprocal value of the logistic equation:

$$\frac{1}{y} = \frac{1 + e^{a+bx}}{Y}.$$

Therefore, it is

$$\frac{Y - y}{y} = e^{a+bx}.$$

Taking the logarithm of this equation yields

$$\ln\left(\frac{Y - y}{y}\right) = a + bx.$$

This equation can easily be handled by finding a regression line if we transform the y-values according to the left hand side of this equation. (Note that we need to know the value of Y, which effectively describes a scaling, to compute this transformation.) This transformation is commonly known as **logit transformation**. It corresponds to a kind of inverse of the logistic function. By finding a regression line for the data points that are transformed accordingly, we (indirectly) obtain a regression curve for the original data points.[5]

To illustrate the procedure, we consider a simple example. The table below shows a data set consisting of five points $(x_1, y_1), \dots, (x_5, y_5)$:

x	1	2	3	4	5
y	0.4	1.0	3.0	5.0	5.6

We transform these data points with

$$z = \ln\left(\frac{Y - y}{y}\right), \qquad Y = 6.$$

The transformed data points are

x	1	2	3	4	5
z	2.64	1.61	0.00	−1.61	−2.64

To set up the system of normal equations, we compute

$$\sum_{i=1}^{5} x_i = 15, \qquad \sum_{i=1}^{5} x_i^2 = 55, \qquad \sum_{i=1}^{5} z_i = 0, \qquad \sum_{i=1}^{5} x_i z_i \approx -13.775.$$

[5]Note again that with this procedure the sum of squared errors is minimized in the transformed space (coordinates x and $z = \ln(\frac{Y-y}{y})$), but this does not imply that it is also minimized in the original space (coordinates x and y), cf. the preceding footnote.

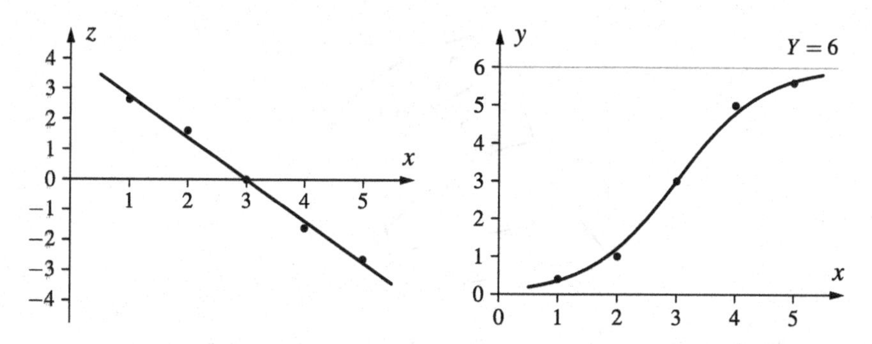

Fig. 5.15 Transformed data (*left*) and original data (*right*) as well as the computed regression line (transformed data) and the corresponding regression curve (original data)

Thus we obtain the (linear) equation system (normal equations)

$$5a + 15b = 0,$$

$$15a + 55b = -13.775,$$

which possesses the solution $a \approx 4.133$ and $b \approx -1.3775$. Hence the regression line for the transformed data is

$$z \approx 4.133 - 1.3775x$$

and the regression curve for the original data consequently

$$y \approx \frac{6}{1 + e^{4.133 - 1.3775x}}.$$

These two regression functions are shown, together with the (transformed and original, respectively) data points, in Fig. 5.15.

The resulting regression curve for the original data can be computed by a neuron with one input x that has the network input function $f_{\text{net}}(x) \equiv wx$ with $w = b \approx -1{,}3775$, the logistic activation function $f_{\text{act}}(\text{net}, \theta) \equiv (1 + e^{-(\text{net} - \theta)})^{-1}$ with the parameter $\theta = a \approx 4.133$ and the output function $f_{\text{out}}(\text{act}) \equiv 6\,\text{act}$.

Note that with the help of logistic regression we can compute not only the parameters of a neuron with a single input, but—in analogy to **multi-linear regression**, see Sect. 10.2—also the parameters of a neuron with multiple inputs. However, since the sum of squared errors can be determined only for output neurons, this method is limited to two-layer perceptrons (that is, with only an input and an output layer, but without any hidden layer). It is not possible to directly transfer it to three- and multi-layer perceptrons. Thus we face essentially the same problem as in Sect. 3.7, where we could not transfer the delta rule. Therefore we consider in the next section a different method, which can be extended to multi-layer perceptrons.

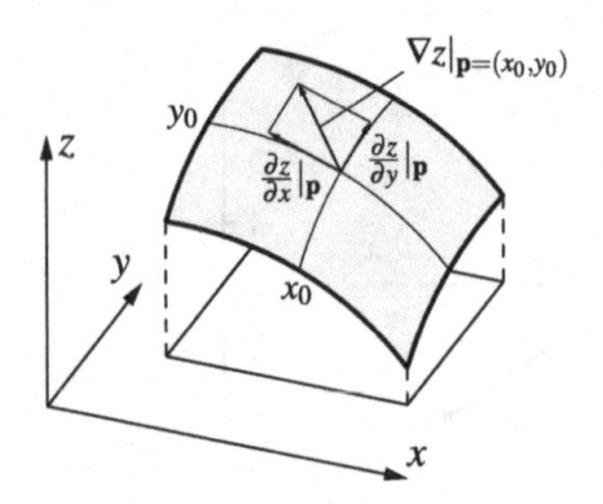

Fig. 5.16 Intuitive interpretation of the gradient of a real-valued function $z = f(x, y)$ at a point $\mathbf{p} = (x_0, y_0)$. It is $\nabla z|_{(x_0,y_0)} = (\frac{\partial z}{\partial x}|_{(x_0,y_0)}, \frac{\partial z}{\partial y}|_{(x_0,y_0)})$

5.4 Gradient Descent

In the following, we consider the method of gradient descent to determine the parameters of a multi-layer perceptron. In principle, this method relies on the same idea as the procedure studied in Sect. 3.5: depending on the values of the weights and the biases, the output of the multi-layer perceptron will be more or less correct. If we can derive from the error function in which directions we have to change the weights and the bias values in order to reduce the error, we obtain a possibility to train the parameters of the network. We simply make a small step into these directions, determine the directions of change again, make another small step and so on—in the same way as we proceeded in Sect. 3.5 (cf. Fig. 3.16 on page 26).

However, in Sect. 3.5 we could not derive the change directions directly from the natural error function (cf. Fig. 3.13 on page 24), but had to invest additional thought in order to modify the error function appropriately. However, this was necessary only, because we used the (Heaviside or unit) step function as the activation function, due to which the error function consisted of "plateaus" or "terraces". In the multi-layer perceptrons, we are studying here, however, we have other choices for the activation functions at our disposal (cf. Fig. 5.2 on page 49). In particular, we may choose a **differentiable activation function**, preferably the logistic function. Such a choice has the following advantage: if the activation function is differentiable, then the error function is differentiable as well.[6] Hence, we can determine the directions, in which the weights and the bias values have to be changed by simply computing the **gradient** of the error function.

Intuitively the gradient describes the slope of a function (see Fig. 5.16). Formally, computing the gradient yields a **vector field**. That is, the gradient assigns to each point of the domain of the function a vector, the elements of which are the **partial derivatives** of the function w.r.t. its arguments (also known as **direction derivatives**). This vector is often simply called the gradient of the function at the given

[6]Unless the output function is not differentiable. However, we usually assume (implicitly) that the output function is the identity and thus does not introduce any problems.

point (see Fig. 5.16). It points into the direction of the steepest slope of the function at this point. Forming a gradient (in a point or for a whole function) is commonly denoted by the differential operator ∇ (pronounced: nabla).

Training a neural network thus becomes very simple: the weights and the bias values are initialized randomly. Then the gradient of the error function is computed at the point that is given by these weights and bias values. Since we want to minimize the error, but the gradient points into the direction of the steepest slope, we make a small step in the opposite direction. At the new point (new weights and bias values), we recompute the gradient etc. until we reach a minimum of the error function.

With these considerations, the general procedure should be clear. Therefore we now turn to a detailed formal derivation of the adaptation rules for the weights and bias values. In order to avoid unnecessary and clumsy distinctions of special cases, we denote in the following the set of neurons of the input layer of an r-layered perceptron with U_0, the sets of neurons of the $r-2$ hidden layers with U_1 to U_{r-2} and the set of neurons of the output layer (sometimes) with U_{r-1}. We start from the total error of a multi-layer perceptron with output neurons U_{out} w.r.t. a fixed learning task L_{fixed}, which is defined as (cf. Sect. 4.3)

$$e = \sum_{l \in L_{\text{fixed}}} e^{(l)} = \sum_{v \in U_{\text{out}}} e_v = \sum_{l \in L_{\text{fixed}}} \sum_{v \in U_{\text{out}}} e_v^{(l)},$$

that is, as the sum of the individual errors over all output neurons v and all training patterns l. Let u be a neuron of the output layer or a hidden layer, that is, $u \in U_k$, $0 < k < r$. Its predecessors are the neurons $\text{pred}(u) = \{p_1, \ldots, p_n\} \subseteq U_{k-1}$; the corresponding (extended) weight vector is $\mathbf{w}_u = (-\theta_u, w_{up_1}, \ldots, w_{up_n})$. Note the additional vector element $-\theta_u$: as shown in Sect. 3.5, a bias value can be turned into a weight, so that all parameters can be treated in a uniform manner (see Fig. 3.18 on page 27). Here we exploit this possibility to simplify the derivations.

We now compute the gradient of the total error w.r.t. these weights, that is,

$$\nabla_{\mathbf{w}_u} e = \frac{\partial e}{\partial \mathbf{w}_u} = \left(-\frac{\partial e}{\partial \theta_u}, \frac{\partial e}{\partial w_{up_1}}, \ldots, \frac{\partial e}{\partial w_{up_n}} \right).$$

As the total error is the sum of the individual errors over the training patterns, we get

$$\nabla_{\mathbf{w}_u} e = \frac{\partial e}{\partial \mathbf{w}_u} = \frac{\partial}{\partial \mathbf{w}_u} \sum_{l \in L_{\text{fixed}}} e^{(l)} = \sum_{l \in L_{\text{fixed}}} \frac{\partial e^{(l)}}{\partial \mathbf{w}_u}.$$

Hence, we can confine ourselves to the error $e^{(l)}$ for a single training pattern l. This error depends on the weights in $\mathbf{w}_u$ only via the network input $\text{net}_u^{(l)} = \mathbf{w}_u \text{in}_u^{(l)}$ with the (extended) network input vector $\text{in}_u^{(l)} = (1, \text{out}_{p_1}^{(l)}, \ldots, \text{out}_{p_n}^{(l)})$. We apply the chain rule and obtain

$$\nabla_{\mathbf{w}_u} e^{(l)} = \frac{\partial e^{(l)}}{\partial \mathbf{w}_u} = \frac{\partial e^{(l)}}{\partial \,\text{net}_u^{(l)}} \frac{\partial \,\text{net}_u^{(l)}}{\partial \mathbf{w}_u}.$$

Since $\text{net}_u^{(l)} = \mathbf{w}_u \mathbf{in}_u^{(l)}$, we get for the second factor immediately

$$\frac{\partial \,\text{net}_u^{(l)}}{\partial \mathbf{w}_u} = \mathbf{in}_u^{(l)}.$$

For the first factor, we consider the error $e^{(l)}$ for the training pattern $l = (\mathbf{i}^{(l)}, \mathbf{o}^{(l)})$. This error is

$$e^{(l)} = \sum_{v \in U_{\text{out}}} e_u^{(l)} = \sum_{v \in U_{\text{out}}} \left(o_v^{(l)} - \text{out}_v^{(l)}\right)^2,$$

that is, the sum of errors over all output neurons. Therefore, we have

$$\frac{\partial e^{(l)}}{\partial \, \text{net}_u^{(l)}} = \frac{\partial \sum_{v \in U_{\text{out}}} (o_v^{(l)} - \text{out}_v^{(l)})^2}{\partial \, \text{net}_u^{(l)}} = \sum_{v \in U_{\text{out}}} \frac{\partial (o_v^{(l)} - \text{out}_v^{(l)})^2}{\partial \, \text{net}_u^{(l)}}.$$

Since the actual output $\text{out}_v^{(l)}$ of an output neuron v depends on the network input $\text{net}_u^{(l)}$ of the considered neuron u, it is

$$\frac{\partial e^{(l)}}{\partial \, \text{net}_u^{(l)}} = -2 \underbrace{\sum_{v \in U_{\text{out}}} \left(o_v^{(l)} - \text{out}_v^{(l)}\right) \frac{\partial \, \text{out}_v^{(l)}}{\partial \, \text{net}_u^{(l)}}}_{\delta_u^{(l)}}.$$

This also introduces the abbreviation $\delta_u^{(l)}$ for the sum over the output neurons that occurs here and that plays an important role below.

To determine the sums $\delta_u^{(l)}$ we have to distinguish two cases: if u is an output neuron, we can simplify the expression for $\delta_u^{(l)}$ considerably, because the output of all other output neurons are clearly independent of the network input of the neuron u. Therefore, all terms of the sum vanish except the one with $v = u$. We obtain

$$\forall u \in U_{\text{out}}: \quad \delta_u^{(l)} = \left(o_u^{(l)} - \text{out}_u^{(l)}\right) \frac{\partial \, \text{out}_u^{(l)}}{\partial \, \text{net}_u^{(l)}}.$$

Therefore, the gradient is

$$\forall u \in U_{\text{out}}: \quad \nabla_{\mathbf{w}_u} e_u^{(l)} = \frac{\partial e_u^{(l)}}{\partial \mathbf{w}_u} = -2\left(o_u^{(l)} - \text{out}_u^{(l)}\right) \frac{\partial \, \text{out}_u^{(l)}}{\partial \, \text{net}_u^{(l)}} \mathbf{in}_u^{(l)},$$

which implies the general weight change

$$\forall u \in U_{\text{out}}: \quad \Delta \mathbf{w}_u^{(l)} = -\frac{\eta}{2} \nabla_{\mathbf{w}_u} e_u^{(l)} = \eta \left(o_u^{(l)} - \text{out}_u^{(l)}\right) \frac{\partial \, \text{out}_u^{(l)}}{\partial \, \text{net}_u^{(l)}} \mathbf{in}_u^{(l)}.$$

The negatives sign disappears, because we have to minimize the error and thus have to move in the direction opposite to the gradient, since the gradient points into the direction of the steepest slope of the error function. The constant factor 2 is incorporated into the **learning rate** η.[7] A typical value for the learning rate is $\eta = 0.2$.

Note, however, that this is only the weight change that results from a single training pattern l, since we neglected the sum over the training patterns at the beginning. In other words, this is the adaptation rule for **online training**, in which the weights are adapted after each training pattern (cf. page 25f and Algorithm 3.2 on page 28).

[7] In order to avoid this factor right from the start, the error of an output neuron is sometimes defined as $e_u^{(l)} = \frac{1}{2}(o_u^{(l)} - \text{out}_u^{(l)})^2$. In this way the factor 2 simply cancels in the derivation.

For **batch training,** we have to sum the changes described by the above formula over all training patterns rather than changing the parameters directly (cf. page 26f and Algorithm 3.3 on page 29f), since the weights are adapted only at the end of a (learning/training) epoch, that is, after all training patterns have been visited.

In the above formula for the weight changes, we cannot determine the derivative of the output $\mathrm{out}_u^{(l)}$ w.r.t. the network input $\mathrm{net}_u^{(l)}$ generally, because the output is computed from the network input with the help of the output function f_{out} and the activation function f_{act} of the neuron u. That is, we have

$$\mathrm{out}_u^{(l)} = f_{\mathrm{out}}\big(\mathrm{act}_u^{(l)}\big) = f_{\mathrm{out}}\big(f_{\mathrm{act}}\big(\mathrm{net}_u^{(l)}\big)\big),$$

and there are several choices for these two functions.

To simplify matters, we assume here that the activation function does not take any extra parameters.[8] It may be, for instance, the logistic function, which is the most common choice. Furthermore, we assume for the sake of simplicity that the output function f_{out} is the identity and thus that we can neglect it. Then we obtain

$$\frac{\partial\, \mathrm{out}_u^{(l)}}{\partial\, \mathrm{net}_u^{(l)}} = \frac{\partial\, \mathrm{act}_u^{(l)}}{\partial\, \mathrm{net}_u^{(l)}} = f'_{\mathrm{act}}\big(\mathrm{net}_u^{(l)}\big),$$

where the prime ($'$) means taking the derivative w.r.t. the argument $\mathrm{net}_u^{(l)}$. For the logistic activation function in particular, that is, for

$$f_{\mathrm{act}}(x) = \frac{1}{1+e^{-x}},$$

the relation

$$\begin{aligned} f'_{\mathrm{act}}(x) &= \frac{\mathrm{d}}{\mathrm{d}x}\big(1+e^{-x}\big)^{-1} = -\big(1+e^{-x}\big)^{-2}\big(-e^{-x}\big) \\ &= \frac{1+e^{-x}-1}{(1+e^{-x})^2} = \frac{1}{1+e^{-x}}\left(1-\frac{1}{1+e^{-x}}\right) \\ &= f_{\mathrm{act}}(x)\cdot\big(1-f_{\mathrm{act}}(x)\big) \end{aligned}$$

holds and therefore (as we assume that the output function is the identity)

$$f'_{\mathrm{act}}\big(\mathrm{net}_u^{(l)}\big) = f_{\mathrm{act}}\big(\mathrm{net}_u^{(l)}\big)\cdot\big(1-f_{\mathrm{act}}\big(\mathrm{net}_u^{(l)}\big)\big) = \mathrm{out}_u^{(l)}\big(1-\mathrm{out}_u^{(l)}\big).$$

We obtain for the weight adaptation

$$\Delta\mathbf{w}_u^{(l)} = \eta\big(o_u^{(l)} - \mathrm{out}_u^{(l)}\big)\,\mathrm{out}_u^{(l)}\big(1-\mathrm{out}_u^{(l)}\big)\mathbf{in}_u^{(l)},$$

which makes the computations particularly simple.

[8] Note that the bias value θ_u is already contained in the extended weight vector.

5.5 Error Backpropagation

In the preceding section we considered the term $\delta_u^{(l)}$ only for output neurons u. That is, the resulting adaptation rule applies only to the weights of the connections from the last hidden layer to the output layer (or, alternatively, only to two-layer perceptrons). In this situation we found ourselves already with the delta rule (see Definition 3.2 on page 27) and faced the problem that the procedure cannot be extended to networks, because we did not have desired outputs for the hidden neurons. The gradient descent approach, however, can be extended to multi-layer perceptrons, since the differentiable activation functions allow us to differentiate the output also w.r.t. the weights of the connections from the input layer to the first hidden layer or w.r.t. the weights of the connections between two consecutive hidden layers.

Let u be a neuron of a hidden layer, that is, let $u \in U_k$, $0 < k < r - 1$. In this case the output $\mathrm{out}_v^{(l)}$ of an output neuron v for a training pattern l depends on the network input $\mathrm{net}_u^{(l)}$ of the neuron u only indirectly via the successors of u, that is, $\mathrm{succ}(u) = \{s \in U \mid (u, s) \in C\} = \{s_1, \ldots, s_m\} \subseteq U_{k+1}$, namely via their network input $\mathrm{net}_s^{(l)}$. By applying the chain rule we obtain

$$\delta_u^{(l)} = \sum_{v \in U_{\mathrm{out}}} \sum_{s \in \mathrm{succ}(u)} \left(o_v^{(l)} - \mathrm{out}_v^{(l)}\right) \frac{\partial\, \mathrm{out}_v^{(l)}}{\partial\, \mathrm{net}_s^{(l)}} \frac{\partial\, \mathrm{net}_s^{(l)}}{\partial\, \mathrm{net}_u^{(l)}}.$$

Since both sums are finite, we can exchange the summations and thus arrive at

$$\delta_u^{(l)} = \sum_{s \in \mathrm{succ}(u)} \left(\sum_{v \in U_{\mathrm{out}}} \left(o_v^{(l)} - \mathrm{out}_v^{(l)}\right) \frac{\partial\, \mathrm{out}_v^{(l)}}{\partial\, \mathrm{net}_s^{(l)}} \right) \frac{\partial\, \mathrm{net}_s^{(l)}}{\partial\, \mathrm{net}_u^{(l)}} = \sum_{s \in \mathrm{succ}(u)} \delta_s^{(l)} \frac{\partial\, \mathrm{net}_s^{(l)}}{\partial\, \mathrm{net}_u^{(l)}}.$$

Finally, we have to determine the partial derivative of the network input. It is

$$\mathrm{net}_s^{(l)} = \mathbf{w}_s \mathbf{in}_s^{(l)} = \left(\sum_{p \in \mathrm{pred}(s)} w_{sp}\, \mathrm{out}_p^{(l)} \right) - \theta_s,$$

where one element of the vector $\mathbf{in}_s^{(l)}$ is the output $\mathrm{out}_u^{(l)}$ of the neuron u. Obviously, $\mathrm{net}_s^{(l)}$ depends on $\mathrm{net}_u^{(l)}$ only via this element $\mathrm{out}_u^{(l)}$. Therefore,

$$\frac{\partial\, \mathrm{net}_s^{(l)}}{\partial\, \mathrm{net}_u^{(l)}} = \left(\sum_{p \in \mathrm{pred}(s)} w_{sp} \frac{\partial\, \mathrm{out}_p^{(l)}}{\partial\, \mathrm{net}_u^{(l)}} \right) - \frac{\partial \theta_s}{\partial\, \mathrm{net}_u^{(l)}} = w_{su} \frac{\partial\, \mathrm{out}_u^{(l)}}{\partial\, \mathrm{net}_u^{(l)}},$$

since all terms vanish except the one with $p = u$. As a consequence, we have

$$\delta_u^{(l)} = \left(\sum_{s \in \mathrm{succ}(u)} \delta_s^{(l)} w_{su} \right) \frac{\partial\, \mathrm{out}_u^{(l)}}{\partial\, \mathrm{net}_u^{(l)}}.$$

Thus, we arrived at a layer-wise recursion formula for computing the δ-values of the neurons of the hidden layers.

If we compare this result to the one obtained in the preceding section for the output neurons, we see that the sum

$$\sum_{s \in \text{succ}(u)} \delta_s^{(l)} w_{su}$$

plays the role of the difference $o_u^{(l)} - \text{out}_u^{(l)}$ of the desired and the actual output of the neuron u for the training pattern l. Therefore, we may see it as an error value for a neuron in a hidden layer, like $o_u^{(l)} - \text{out}_u^{(l)}$ is the error value of an output neuron. As a consequence, the error values of any (hidden) layer of a multi-layer perceptron can be computed from the error values of is successor layer. We may also say that an error signal is transmitted from the output layer backwards through the hidden layers. Therefore, this method is also called **error backpropagation**.

The weight adaptation is

$$\Delta \mathbf{w}_u^{(l)} = -\frac{\eta}{2} \nabla_{\mathbf{w}_u} e^{(l)} = \eta \delta_u^{(l)} \mathbf{in}_u^{(l)} = \eta \left(\sum_{s \in \text{succ}(u)} \delta_s^{(l)} w_{su} \right) \frac{\partial \, \text{out}_u^{(l)}}{\partial \, \text{net}_u^{(l)}} \mathbf{in}_u^{(l)}.$$

Note, however, that this is only the change resulting from a single training pattern l. For batch training, these changes have to be summed over all training patterns.

For the rest of the derivation, we again make the simplifying assumption, as in the preceding section, that the output function is the identity. Furthermore, we consider the special case of a logistic activation function. This yields the particularly simple weight adaptation rule (cf. the derivations for output neurons on page 65).

$$\Delta \mathbf{w}_u^{(l)} = \eta \left(\sum_{s \in \text{succ}(u)} \delta_s^{(l)} w_{su} \right) \text{out}_u^{(l)} \left(1 - \text{out}_u^{(l)}\right) \mathbf{in}_u^{(l)}.$$

Figure 5.17 combines all formulas that we need to execute and to train a multi-layer perceptron, which employs logistic activation functions in the hidden and the output neurons. Where the formulas are applied is marked by encircled numbers.

5.6 Gradient Descent Examples

To illustrate the gradient descent method, we consider training a two-layer perceptron for the negation, as we used it already in Sect. 3.5. This perceptron and the corresponding training examples are shown in Fig. 5.18. In analogy to Fig. 3.13 on page 24, Fig. 5.19 shows the (sum of) squared errors of computing the negation depending on the values of the weights and the bias value. We assumed a logistic activation function, as can clearly be seen from the shape of the error function. Note that the fact that the activation function is now (meaningful) differentiable, the error function is as well and does no longer consist of "plateaus" or "terraces". Thus, we can now execute a gradient descent on the (unmodified) error function.

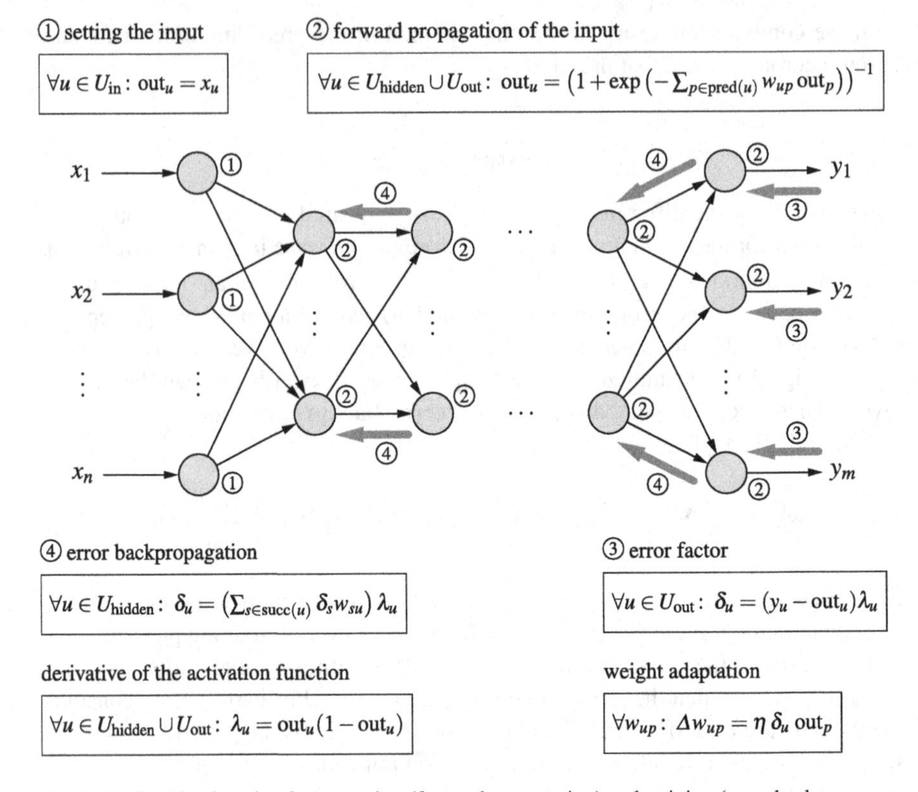

Fig. 5.17 Cookbook recipe for executing (forward propagation) and training (error backpropagation) a multi-layer perceptron with logistic activation functions in the hidden and output neurons

The course of this gradient descent, starting with the initial values $\theta = 3$ and $w = \frac{7}{2}$ and proceeding with learning rate 1 is shown in Table 5.1 (online training on the left, batch training on the right). The two courses are very similar, which is due to the low number of training examples and the smoothness of the error function. Figure 5.20 shows a graphical representation of the course of the training, with the left and the middle diagram displaying, for the sake of comparison, the regions we used in Sect. 3.5 (cf. Fig. 3.15 on page 25 and Fig. 3.16 on page 26). The dots indicate the state of the network every 20 epochs. In the three-dimensional diagram on the right it can be seen particularly well how the error is slowly reduced and how finally a region is reached in which the error almost vanishes.

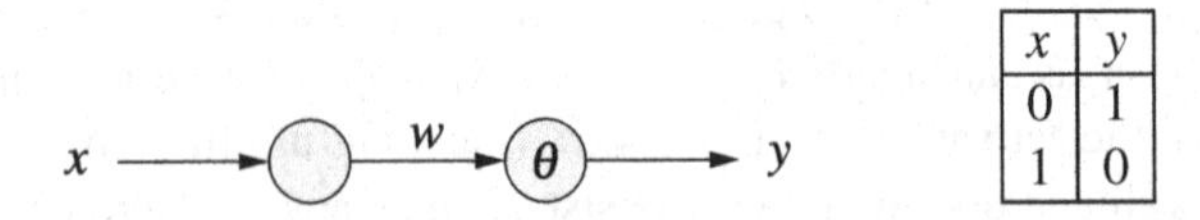

x	y
0	1
1	0

Fig. 5.18 A two-layer perceptron with a single input and training examples for the negation

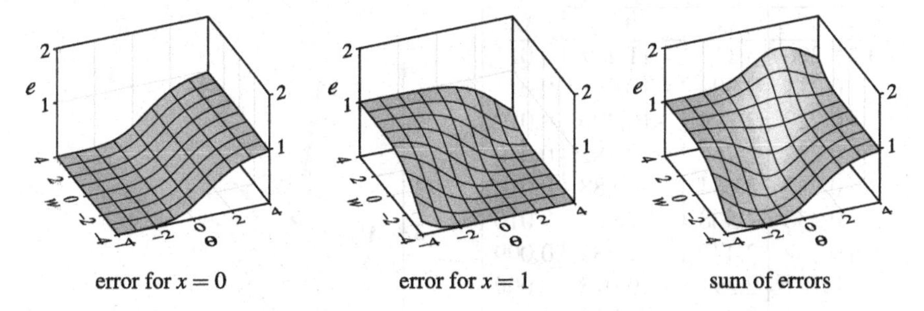

Fig. 5.19 (Sum of) squared errors for computing the negation with a logistic activation function

Table 5.1 Training processes with initial values $\theta = 3$, $w = \frac{7}{2}$ and learning rate 1

online training

Epoch	θ	w	error
0	3.00	3.50	1.307
20	3.77	2.19	0.986
40	3.71	1.81	0.970
60	3.50	1.53	0.958
80	3.15	1.24	0.937
100	2.57	0.88	0.890
120	1.48	0.25	0.725
140	−0.06	−0.98	0.331
160	−0.80	−2.07	0.149
180	−1.19	−2.74	0.087
200	−1.44	−3.20	0.059
220	−1.62	−3.54	0.044

batch training

Epoch	θ	w	error
0	3.00	3.50	1.295
20	3.76	2.20	0.985
40	3.70	1.82	0.970
60	3.48	1.53	0.957
80	3.11	1.25	0.934
100	2.49	0.88	0.880
120	1.27	0.22	0.676
140	−0.21	−1.04	0.292
160	−0.86	−2.08	0.140
180	−1.21	−2.74	0.084
200	−1.45	−3.19	0.058
220	−1.63	−3.53	0.044

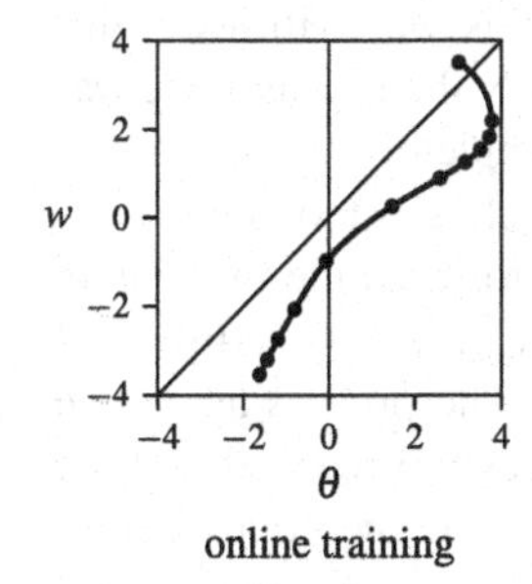

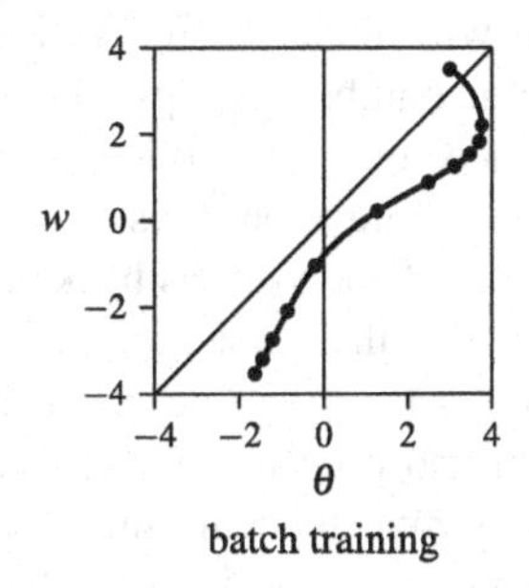

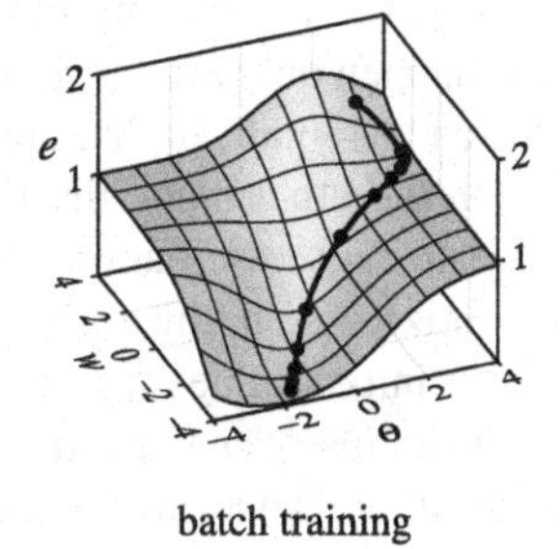

Fig. 5.20 Training processes with initial values $\theta = 3$, $w = \frac{7}{2}$ and learning rate 1

i	x_i	$f(x_i)$	$f'(x_i)$	Δx_i
0	0.200	3.112	−11.147	0.011
1	0.211	2.990	−10.811	0.011
2	0.222	2.874	−10.490	0.010
3	0.232	2.766	−10.182	0.010
4	0.243	2.664	−9.888	0.010
5	0.253	2.568	−9.606	0.010
6	0.262	2.477	−9.335	0.009
7	0.271	2.391	−9.075	0.009
8	0.281	2.309	−8.825	0.009
9	0.289	2.233	−8.585	0.009
10	0.298	2.160		

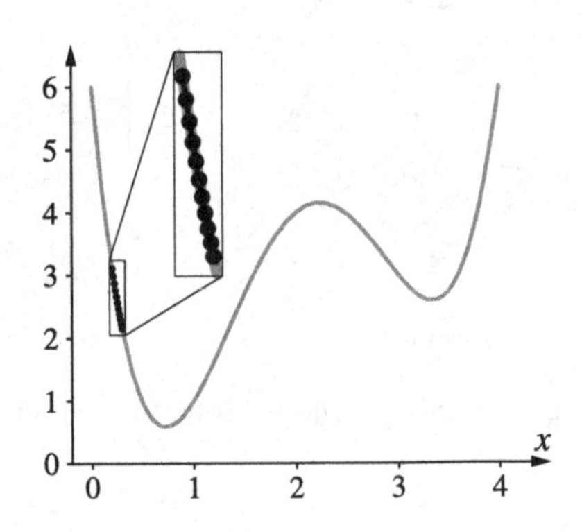

Fig. 5.21 Gradient descent with initial value 0.2 and learning rate 0.001

As a second example, we study how one can use gradient descent to find the minimum of a function, here specifically

$$f(x) = \frac{5}{6}x^4 - 7x^3 + \frac{115}{6}x^2 - 18x + 6.$$

Although this function has no direct relationship to the error function of a multi-layer perceptron, it is well suited to demonstrate certain problems that gradient descent can run into. We start be finding the derivative of the function, that is,

$$f'(x) = \frac{10}{3}x^3 - 21x^2 + \frac{115}{3}x - 18,$$

which corresponds to the gradient (the sign indicates the direction of the steepest slope). The computations then proceed according to the scheme

$$x_{i+1} = x_i + \Delta x_i \quad \text{with } \Delta x_i = -\eta f'(x_i),$$

where x_0 is a chosen initial value and η is the learning rate.

We first consider the course of the gradient descent for an initial value $x_0 = 0.2$ and a learning rate $\eta = 0.001$ as it is shown in Fig. 5.21. Starting at an initial point on the left branch of the function, small steps are carried out towards the minimum. Clearly, the (global) minimum will finally be reached in this way. However, this will happen only after a very large number of steps. Obviously the learning rate was chosen too small, so that the procedure takes too long to complete.

On the other hand, the learning rate should also not be chosen too large, since this can cause oscillations or even chaotic jumps back and forth on the function to minimize. As an illustration consider the gradient descent that is shown in Fig. 5.22 for an initial value $x_0 = 1.5$ and a learning rate $\eta = 0.25$. The process again and again jumps over the minimum and after a few steps values are reached that are even farther away from the minimum than the initial value. If the computations were continued for a few steps more, we could even observe a jump over the local minimum in the middle, thus resulting in values on the right branch of the function.

i	x_i	$f(x_i)$	$f'(x_i)$	Δx_i
0	1.500	2.719	3.500	−0.875
1	0.625	0.655	−1.431	0.358
2	0.983	0.955	2.554	−0.639
3	0.344	1.801	−7.157	1.789
4	2.134	4.127	0.567	−0.142
5	1.992	3.989	1.380	−0.345
6	1.647	3.203	3.063	−0.766
7	0.881	0.734	1.753	−0.438
8	0.443	1.211	−4.851	1.213
9	1.656	3.231	3.029	−0.757
10	0.898	0.766		

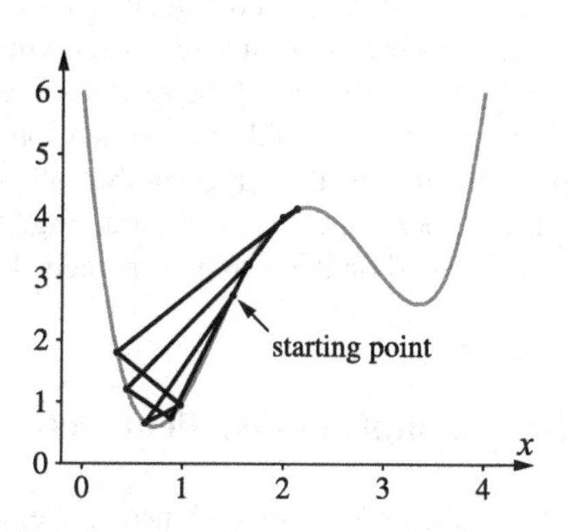

Fig. 5.22 Gradient descent with initial value 1.5 and learning rate 0.25

i	x_i	$f(x_i)$	$f'(x_i)$	Δx_i
0	2.600	3.816	−1.707	0.085
1	2.685	3.660	−1.947	0.097
2	2.783	3.461	−2.116	0.106
3	2.888	3.233	−2.153	0.108
4	2.996	3.008	−2.009	0.100
5	3.097	2.820	−1.688	0.084
6	3.181	2.695	−1.263	0.063
7	3.244	2.628	−0.845	0.042
8	3.286	2.599	−0.515	0.026
9	3.312	2.589	−0.293	0.015
10	3.327	2.585		

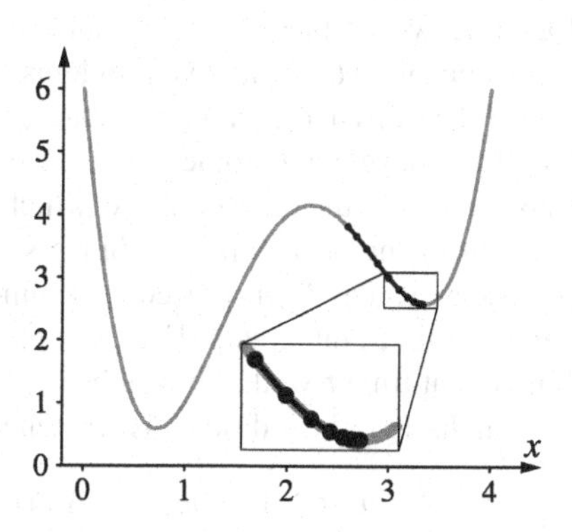

Fig. 5.23 Gradient descent with initial value 2.6 and learning rate 0.05

In general, with a large learning rate the behavior is often entirely erratic and thus cannot be expected to ever approach a local, let alone the global minimum.

However, even if a suitable value is chosen for the learning rate, we cannot guarantee that the procedure will be successful. As can be seen in Fig. 5.23, which shows the course of the gradient descent for an initial value $x_0 = 2.6$ and a learning rate $\eta = 0.05$, the closest minimum is approached quickly. However, this minimum is only a local one; the global minimum is missed. This problem is caused mainly by the initial value and thus cannot be solved by changing the learning rate.

For an illustration of the error backpropagation, we recommend trying the visualization programs `wmlp` (for Microsoft Windows™) and `xmlp` (for Unix/Linux), that are available on the web page

http://www.borgelt.net/mlpd.html

With these programs a three-layer perceptron can be trained in such a way that the biimplication, the exclusive or, or (an approximation of) one of two different real-valued functions are computed. After each training step the computations of the neural network are visualized by drawing the current position of the separating lines or the current output of the neurons over their input space. Although this does not allow to follow directly the descent on the error function (which is impossible in principle due to the large number of parameters), one gets a good impression of what happens in the course of training. An extensive explanation of these programs can be found on the web page named above.

5.7 Variants of Gradient Descent

One of the problems mentioned above, namely "getting stuck" in a local minimum, cannot be avoided in principle. It can only be mitigated by training the network multiple times, each time with different initial values, and choosing the best result. This increases that chances that the global minimum (or at least a good local minimum) is found. However, there is no guarantee that the global minimum will be discovered.

To cope with the other two problems, which concern the learning rate and thus the size of the steps in the parameter space, however, several variants of gradient descent have been developed. Some of these we discuss in the following. We describe them by stating the rules according to which a weight has to be changed based on the gradient of the error function. Since some of these methods draw on gradients or parameter values from preceding training steps, we introduce a parameter t that denotes this training step. For example, $\nabla_w e(t)$ denotes the gradient of the error function at time t w.r.t. the weight w. As a comparison, the weight adaptation rule for standard gradient descent is written as

$$w(t+1) = w(t) + \Delta w(t) \quad \text{with } \Delta w(t) = -\frac{\eta}{2}\nabla_w e(t)$$

(cf. the derivations on page 64 and page 67). We do not distinguish explicitly between batch and online training, since the difference only consists in whether the total error $e(t)$ or the single pattern error $e^{(l)}(t)$ is used.

5.7.1 *Manhattan Training*

In the preceding section, we saw that training can take very long if the learning rate is too small. However, even if the learning rate is chosen well, training can be slow, namely if it takes place in a region of the parameter space in which the error function is "flat", that is, in which the gradient is fairly small. In order to eliminate this dependence on the magnitude of the gradient, one may employ so-called Manhattan training, which considers only the sign of the gradient. Thus, the weight adaptation rule becomes

$$\Delta w(t) = -\eta \operatorname{sgn}\bigl(\nabla_w e(t)\bigr).$$

Note that this adaptation rule is also obtained if one uses the sum of the absolute deviations of the actual from the desired output as the error function and completes the derivative at 0 (at which it does not exist/is discontinuous) in a suitable fashion.

The advantage of this approach is that training proceeds with constant speed (in the sense of a fixed step width), irrespective of the shape of the error function. A disadvantage is, though, that the weights can only assume certain discrete values (taken from a grid with the grid width η), which can make it impossible in principle to get arbitrarily close to the minimum of the error function. In addition, the problem of how to chose the learning rate appropriately still prevails.

5.7.2 Lifting the Derivative of the Activation Function

Often the error function is fairly flat in some region of the parameter space, because the activation functions are in their saturation region (that is, are evaluated far away from the bias value θ, cf. Fig. 5.2 on page 49), where the gradient is very small or even vanishes completely. In order to speed up training in such a case, the derivative f'_{act} of the activation function may be lifted artificially by a fixed value α, so that sufficiently large training steps are carried out even in these saturation regions (Fahlman 1988). Choosing $\alpha = 0.1$ often leads to good results. This modification of gradient descent is also known as **flat spot elimination**.

Lifting the derivative of the activation function has the additional advantage that it counteracts a weakening of the error signal in the error backpropagation procedure. Such a weakening is due to the fact that, for example, the derivative of the logistic function is bounded by 0.25 (which results for the function value 0.5, that is, for the location of the bias value). As a consequence the error value has a tendency to become smaller and smaller with every layer through which it is propagated, so that training is slower in the front layers of the network.

5.7.3 Momentum Term

The momentum term procedure (Rumelhart et al. 1986) adds a fraction of the preceding weight change to a normal gradient descent step. With this additional term, the weight adaptation rule reads

$$\Delta w(t) = -\frac{\eta}{2}\nabla_w e(t) + \beta \Delta w(t-1),$$

where β is a parameter that has to be less than 1 in order to render the procedure stable. Typically β is chosen between 0.5 and 0.95. Larger values can lead to increasingly larger weight changes and thus unstable behavior.

The additional term $\beta \Delta w(t-1)$ is called **momentum term**, since its effect is similar to the momentum that is gained by a ball rolling down a slope. The longer the ball rolls into the same direction, the faster it gets. Therefore, it tends to continue to roll in the old direction (momentum term), but nevertheless follows (though delayed) the shape of the surface (gradient term).

By introducing a momentum term, the training can be accelerated in regions of the parameter space in which the error function is fairly flat, but has a uniform slope in one direction. In addition, the problem of how to choose the learning rate is mitigated somewhat, since the momentum term increases or reduces the step width according to the shape of the error function. However, a momentum term cannot fully compensate a learning rate that was chosen much too small, since the step width $|\Delta w|$ in case of a constant gradient $\nabla_w e$ is bounded by $s = |\frac{\eta \nabla_w e}{2(1-\beta)}|$. Furthermore, a learning rate that is too large can still lead to oscillations and chaotic jumps.

5.7.4 Self-adaptive Error Backpropagation

The method of super self-adaptive error backpropagation (SuperSAB) (Jakobs 1988; Tollenaere 1990) introduces a separate learning rate η_w for each parameter of a neural network, that is, for each weight and each bias value. In addition, these learning rates are not constant, but are adapted, prior to their use in a training step, depending on the current and the previous gradient according to the following rule:

$$\eta_w(t) = \begin{cases} c^- \cdot \eta_w(t-1) & \text{if } \nabla_w e(t) \cdot \nabla_w e(t-1) < 0, \\ c^+ \cdot \eta_w(t-1) & \text{if } \nabla_w e(t) \cdot \nabla_w e(t-1) > 0 \\ & \wedge \nabla_w e(t-1) \cdot \nabla_w e(t-2) \geq 0, \\ \eta_w(t-1) & \text{otherwise} \end{cases}$$

c^- is a shrinkage factor ($0 < c^- < 1$), with which the learning rate is decreased if the current and the previous gradient have opposite signs. The intuitive reason is that in such a case the training step must have leaped over the minimum (as the gradient now indicates the opposite direction to reach it), and thus smaller steps are necessary to actually approach it. Typically c^- is chosen between 0.5 and 0.7.

c^+ is a growth factor ($c^+ > 1$), with which the learning rate is increased if the current and the previous gradient have the same sign. In this case, two training steps were made in the same direction and thus it is plausible to assume that a longer slope of the error function is currently traversed. Hence, the learning rate should be increased in order to traverse it more quickly. (This is similar in spirit to the idea of the momentum term, see above.) Typically c^+ is chosen between 1.05 and 1.2, so that the learning rate grows only slowly.

The second condition for applying the growth factor c^+ is meant to prevent that the learning rate is increased again immediately after it has been reduced. This is usually implemented in such a way that the old gradient is set to zero after the learning rate has been reduced, in order to signal the executed reduction. Although this also suppresses a repeated reduction, this procedure eliminates the need to store the gradient $\nabla_w e(t-2)$ or to create a corresponding marker.

In order to avoid large jumps as well as slow training, it is common to bound the learning rate from above and from below. Furthermore, (super) self-adaptive error backpropagation should only be applied with batch training, since online training is often unstable with it.

5.7.5 Resilient Error Backpropagation

Resilient backpropagation (Rprop) can be seen as combining the ideas of Manhattan training and self-adaptive error backpropagation (Riedmiller and Braun 1992, 1993). It introduces a *step width* Δw for each parameter of the neural network, that is, for each weight and each bias value, which is adapted according to the following rule, depending on the current and the previous gradient:

$$\Delta w(t) = \begin{cases} c^- \cdot \Delta w(t-1) & \text{if } \nabla_w e(t) \cdot \nabla_w e(t-1) < 0, \\ c^+ \cdot \Delta w(t-1) & \text{if } \nabla_w e(t) \cdot \nabla_w e(t-1) > 0 \\ & \wedge \nabla_w e(t-1) \cdot \nabla_w e(t-2) \geq 0, \\ \Delta w(t-1) & \text{otherwise.} \end{cases}$$

Like for self-adaptive error backpropagation c^- is a shrinkage factor ($0 < c^- < 1$) and c^+ a growth factor ($c^+ > 1$), with which the step width is reduced or increased, respectively. Applying these factors is justified in essentially the same way as for self-adaptive error backpropagation. The ranges of typical values for these parameters coincide as well, namely $c^- \in [0.5, 0.7]$ and $c^+ \in [1.05, 1.2]$.

Like the value of the learning rate in self-adaptive error backpropagation, the value of the step width is bounded from above and from below in order to avoid large jumps as well as slow training. Furthermore, resilient error backpropagation should only be applied with batch training, since online training is even less stable as for self-adaptive error backpropagation.

Resilient error backpropagation has proven to be significantly better and faster than other methods (momentum term, self-adaptive error backpropagation, but also the quick-propagation method discussed below) in many applications. It belongs to the most highly recommended training methods for multi-layer perceptrons.

5.7.6 Quick-Propagation

The quick-propagation method (Fahlman 1988) approximates the error function at the location of the current weights by a parabola (see Fig. 5.24) and computes from the current and the previous gradient the location of the apex of this parabola. Training then "jumps" directly to this apex, that is, the weight is set to the computed location of the apex. If the error function is "benevolent", training may thus get very close to the actual minimum in a single training step.

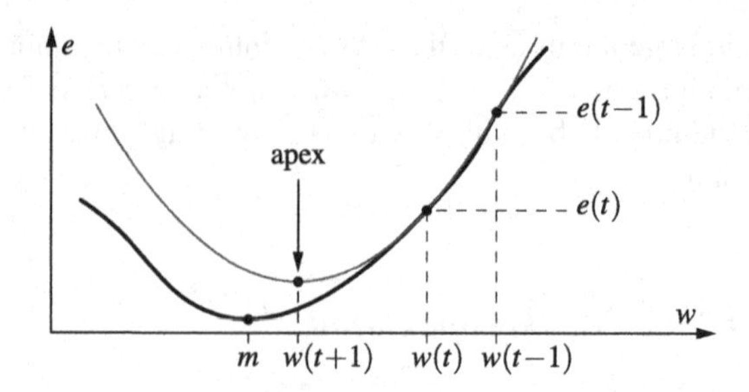

Fig. 5.24 Quick-propagation relies on a local approximation of the error function by a parabola. m is the true minimum

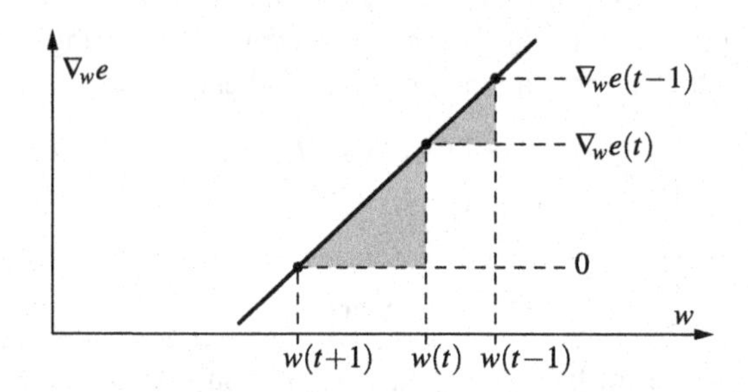

Fig. 5.25 The weight adaptation formula can be computed from the slope triangles of the derivative of the approximation parabola

The weight adaptation rule can be obtained, for example, from two slope triangles of the derivative of the parabola (see the gray triangles in Fig. 5.25). We have

$$\frac{\nabla_w e(t-1) - \nabla_w e(t)}{w(t-1) - w(t)} = \frac{\nabla_w e(t)}{w(t) - w(t+1)}.$$

By solving for $\Delta w(t) = w(t+1) - w(t)$ and exploiting $\Delta w(t-1) = w(t) - w(t-1)$, we arrive at

$$\Delta w(t) = \frac{\nabla_w e(t)}{\nabla_w e(t-1) - \nabla_w e(t)} \cdot \Delta w(t-1).$$

Note, however, that the above equation does not distinguish between a parabola that opens upwards and one that opens downwards, so that it may happen that a (local) maximum of the error function is approached. Although this may be avoided by a test whether

$$\frac{\nabla_w e(t-1) - \nabla_w e(t)}{\Delta w(t-1)} < 0$$

(this indicates a parabola that opens upwards), implementations often skip this test. Instead a parameter is introduced that limits the increase of the weight change relative to the preceding step. That is, it is made sure that

$$|\Delta w(t)| \leq c \cdot |\Delta w(t-1)|$$

holds, where c is a parameter typically chosen between 1.75 and 2.25. This improves the behavior, but does *not* ensure that the weight is adapted in the correct direction.

Furthermore, implementations often add a standard gradient step to the weight adaptation rule stated above, provided the gradients $\nabla_w e(t)$ and $\nabla_w e(t-1)$ have the same sign, that is, provided the minimum does not lie between the current and the preceding weight value. In addition it is advisable to bound the weight change from above in order to avoid large jumps and unstable behavior.

If the assumptions of the quick-propagation method hold, namely that the error function can be locally approximated by a parabola that opens upwards and that the parameters are largely independent, and if batch training is employed, quick-propagation is one of the fastest training algorithms for multi-layer perceptrons, thus justifying its name. Otherwise it exhibits a tendency towards unstable behavior.

5.7.7 Weight Decay

It is generally unfavorable if training leads to very large values for the connection weights of a neural network. The reasons are, in the first place, that with large weights one easily reaches the saturation region of a logistic activation function, in which the gradients almost vanishing and thus can make training very slow or can even bring it to a halt. Secondly, large weights increase the risk of overfitting the network to accidental properties of the training data, so that the performance of the network on new data falls short of what can be achieved otherwise.

The weight decay method (Werbos 1974) serves the purpose to avoid a heavy growth of the weights. To prevent an excessive growth, each weight is reduced in each step by a small factor, for example, with

$$\Delta w(t) = -\frac{\eta}{2}\nabla_w(t) - \xi\, w(t)$$

for standard gradient descent. Alternatively, each weight may be multiplied before its adaptation by the factor $(1-\xi)$, which is often easier to implement. The value of ξ should be chosen very small, so that the weights are not kept permanently at low values. Typical values for ξ are in the range between 0.005 to 0.03.

Note that we may obtain the weight decay rule by using an extended error function that penalizes large weights:

$$e^* = e + \frac{\xi}{2} \sum_{u \in U_{\text{out}} \cup U_{\text{hidden}}} \left(\theta_u^2 + \sum_{p \in \text{pred}(u)} w_{up}^2 \right).$$

The derivative of this modified error leads to the weight adaption stated above.

5.8 Examples for Some Variants

To illustrate gradient descent with a momentum term we consider, in analogy to Sect. 5.6, training a two-layer perceptron for the negation, again starting from the initial values $\theta = 3$ and $w = \frac{7}{2}$. The courses of training without a momentum term and with such a term with the factor $\beta = 0.9$ are shown in Table 5.2 and in Fig. 5.26. Clearly, the training process advances in almost the same way, only that with a

Table 5.2 Training with and without a momentum term ($\beta = 0.9$)

epoch	θ	w	error
0	3.00	3.50	1.295
20	3.76	2.20	0.985
40	3.70	1.82	0.970
60	3.48	1.53	0.957
80	3.11	1.25	0.934
100	2.49	0.88	0.880
120	1.27	0.22	0.676
140	−0.21	−1.04	0.292
160	−0.86	−2.08	0.140
180	−1.21	−2.74	0.084
200	−1.45	−3.19	0.058
220	−1.63	−3.53	0.044

without momentum term

Epoch	θ	w	error
0	3.00	3.50	1.295
10	3.80	2.19	0.984
20	3.75	1.84	0.971
30	3.56	1.58	0.960
40	3.26	1.33	0.943
50	2.79	1.04	0.910
60	1.99	0.60	0.814
70	0.54	−0.25	0.497
80	−0.53	−1.51	0.211
90	−1.02	−2.36	0.113
100	−1.31	−2.92	0.073
110	−1.52	−3.31	0.053
120	−1.67	−3.61	0.041

with momentum term

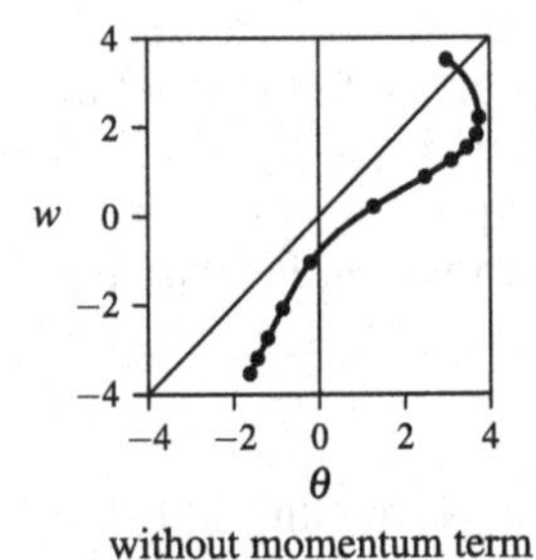

without momentum term

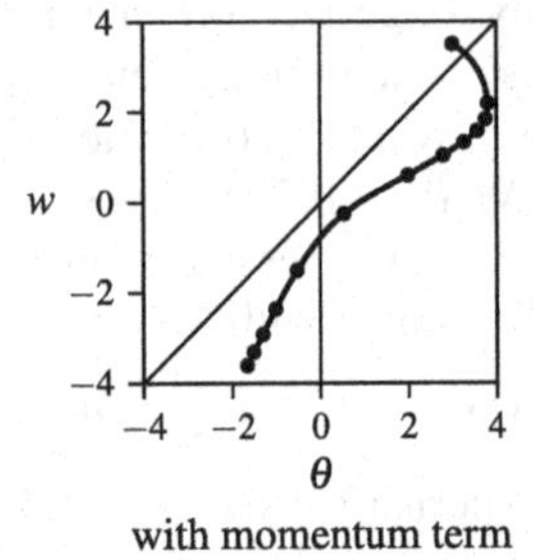

with momentum term

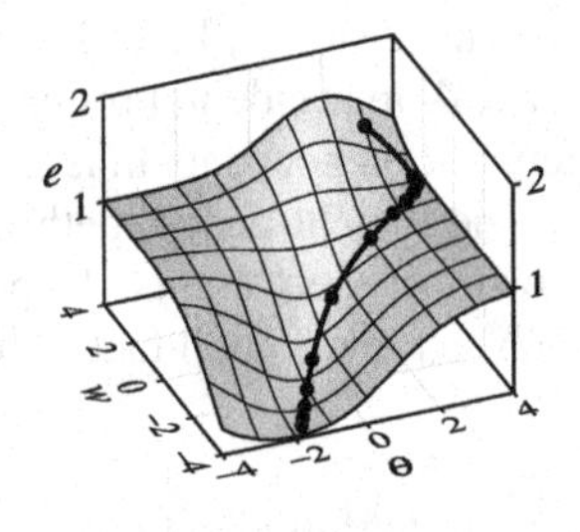

with momentum term

Fig. 5.26 Training with and without a momentum term ($\beta = 0.9$); the points show the values of the weight w and the bias value θ every 20 or 10 epochs, respectively

momentum term merely about half the number of epochs are needed in order to obtain the same error value. That is, the momentum term about doubled the training speed. If bigger networks are trained with a larger number of training patterns, the difference in training speed can be even much larger.

As another example we reconsider the minimization of the function studied in Sect. 5.6 (see page 70f). With the help of a momentum term the very slow descent of Fig. 5.21 can be accelerated considerably, as shown in Fig. 5.27. However, the very small learning rate cannot be compensated completely. The reason is, as already mentioned in the preceding section, that even for a constant gradient $f'(x)$ the step width is bounded by $s = |\frac{\eta f'(x)}{1-\beta}|$, as already mentioned.

By using an adaptive learning rate even the chaotic back and forth jumps can be avoided, which we observed in Fig. 5.22 on page 71. This is demonstrated by

i	x_i	$f(x_i)$	$f'(x_i)$	Δx_i
0	0.200	3.112	−11.147	0.011
1	0.211	2.990	−10.811	0.021
2	0.232	2.771	−10.196	0.029
3	0.261	2.488	−9.368	0.035
4	0.296	2.173	−8.397	0.040
5	0.337	1.856	−7.348	0.044
6	0.380	1.559	−6.277	0.046
7	0.426	1.298	−5.228	0.046
8	0.472	1.079	−4.235	0.046
9	0.518	0.907	−3.319	0.045
10	0.562	0.777		

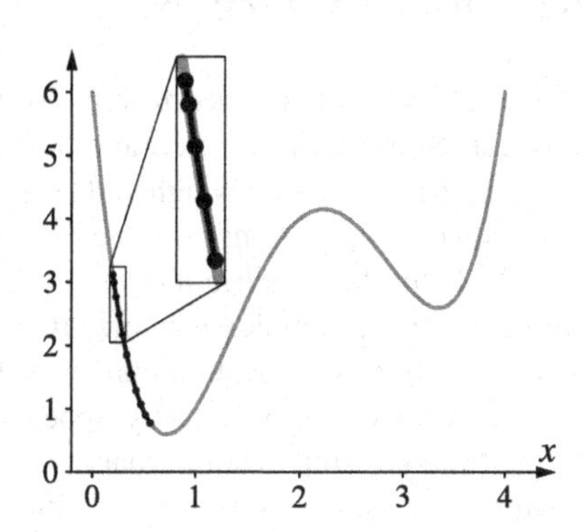

Fig. 5.27 Gradient descent with a momentum term ($\beta = 0.9$) starting from the initial value 0.2 and with a learning rate 0.001

i	x_i	$f(x_i)$	$f'(x_i)$	Δx_i
0	1.500	2.719	3.500	−1.050
1	0.450	1.178	−4.699	0.705
2	1.155	1.476	3.396	−0.509
3	0.645	0.629	−1.110	0.083
4	0.729	0.587	0.072	−0.005
5	0.723	0.587	0.001	0.000
6	0.723	0.587	0.000	0.000
7	0.723	0.587	0.000	0.000
8	0.723	0.587	0.000	0.000
9	0.723	0.587	0.000	0.000
10	0.723	0.587		

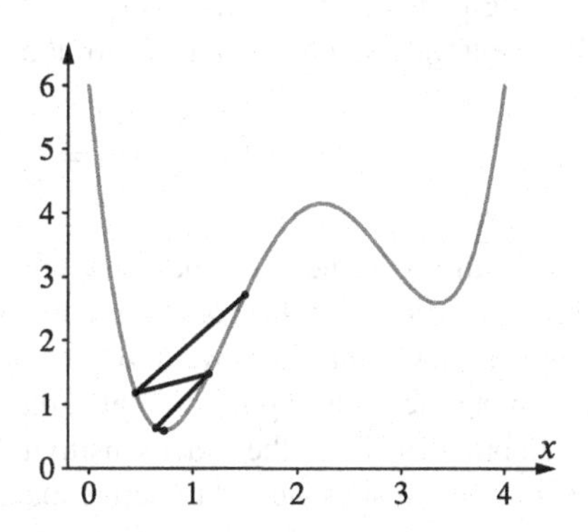

Fig. 5.28 Gradient descent with adaptive learning rate (with $\eta_0 = 0.3$, $c^+ = 1.2$, $c^- = 0.5$) starting from the initial value 1.5

Fig. 5.28, which shows the training procedure for the even larger learning rate $\eta = 0.3$ (compared to $\eta = 0.25$ in Fig. 5.22). Although the initial value is too large, it is quickly corrected by the shrinkage factor, so that the minimum of the function is reached in astonishingly few steps.

To illustrate error backpropagation with a momentum term, we refer again to the programs `wmlp` and `xmlp`, which we already mentioned on page 71. They allow to introduce a momentum term, which accelerates the training considerably. It is instructive to observe how much faster a solution for the biimplication or the exclusive or is found and how much more swiftly the network output approximates the given training examples for the two function fitting tasks.

Finally, we remark that with the command line programs that can be found at

http://www.borgelt.net/mlp.html

arbitrary multi-layer perceptrons can be trained and then executed on new data. They contain all of the variants of gradient descent that we discussed here.

5.9 Sensitivity Analysis

(Artificial) neural networks have the serious disadvantage that the knowledge that they learned from training examples is often difficult to understand, because it is encoded in the connection weights, that is, stored in a matrix of real-valued numbers. In the preceding sections, we tried to visualize the operation of neural networks with the help of geometric interpretations, but such an approach encounters severe problems for the complex networks that we meet in practice. Especially if the input space is high-dimensional, human imagination is bound to fail completely. A complex neural network thus easily appears to be a "black box", which produces its output from its input in somewhat mysterious ways.

However, we can improve this situation to some degree if we carry out a so-called **sensitivity analysis**, which determines what influence the different inputs have on the output of the network. To execute a sensitivity analysis, we sum the derivatives of the output w.r.t. the external inputs over all output neurons and all training patterns. This sum is divided by the number of training patterns, to make the result independent of the size of the training data set. That is, we compute

$$\forall u \in U_{\text{in}}: \quad s(u) = \frac{1}{|L_{\text{fixed}}|} \sum_{l \in L_{\text{fixed}}} \sum_{v \in U_{\text{out}}} \frac{\partial \operatorname{out}_v^{(l)}}{\partial \operatorname{ext}_u^{(l)}}.$$

The resulting value $s(u)$ indicates how "important" the input that is assigned to the neuron u is for the computations and eventually the output of the multi-layer perceptron. On this basis we may then, for example, simplify the neural network by removing the least important inputs, that is, those with the lowest values $s(u)$.

In order to obtain the exact sensitivity computation rule, we start by applying the chain rule—just as we did to derive the gradient descent rules:

$$\frac{\partial \operatorname{out}_v}{\partial \operatorname{ext}_u} = \frac{\partial \operatorname{out}_v}{\partial \operatorname{out}_u} \frac{\partial \operatorname{out}_u}{\partial \operatorname{ext}_u} = \frac{\partial \operatorname{out}_v}{\partial \operatorname{net}_v} \frac{\partial \operatorname{net}_v}{\partial \operatorname{out}_u} \frac{\partial \operatorname{out}_u}{\partial \operatorname{ext}_u}.$$

If the output function of the input neurons is the identity, as we assume here, the last factor can be neglected, because then

$$\frac{\partial \operatorname{out}_u}{\partial \operatorname{ext}_u} = 1.$$

For the second factor, we obtain in the general case

$$\frac{\partial \operatorname{net}_v}{\partial \operatorname{out}_u} = \frac{\partial}{\partial \operatorname{out}_u} \sum_{p \in \operatorname{pred}(v)} w_{vp} \operatorname{out}_p = \sum_{p \in \operatorname{pred}(v)} w_{vp} \frac{\partial \operatorname{out}_p}{\partial \operatorname{out}_u}.$$

On the right-hand side we see a derivative of the output of a neuron p w.r.t. the output of the input neuron u, so that we arrive at the layer-wise recursion formula

$$\frac{\partial \operatorname{out}_v}{\partial \operatorname{out}_u} = \frac{\partial \operatorname{out}_v}{\partial \operatorname{net}_v} \frac{\partial \operatorname{net}_v}{\partial \operatorname{out}_u} = \frac{\partial \operatorname{out}_v}{\partial \operatorname{net}_v} \sum_{p \in \operatorname{pred}(v)} w_{vp} \frac{\partial \operatorname{out}_p}{\partial \operatorname{out}_u}.$$

However, in the first hidden layer (or for a two-layer perceptron) we obtain

$$\frac{\partial \operatorname{net}_v}{\partial \operatorname{out}_u} = w_{vu}, \quad \text{and thus} \quad \frac{\partial \operatorname{out}_v}{\partial \operatorname{out}_u} = \frac{\partial \operatorname{out}_v}{\partial \operatorname{net}_v} w_{vu},$$

since all terms vanish except the one having $p = u$. This formula defines the starting point of the recursion. Then we apply the recursion formula to it until we reach the output layer, where we can finally compute the term of the value $s(u)$ that corresponds to the training pattern l by summing over the output neurons.

Like when we derived error backpropagation we also consider here the special case of a logistic activation function and the identity as the output function. In this case, we obtain a particularly simple recursion formula

$$\frac{\partial \operatorname{out}_v}{\partial \operatorname{out}_u} = \operatorname{out}_v(1 - \operatorname{out}_v) \sum_{p \in \operatorname{pred}(v)} w_{vp} \frac{\partial \operatorname{out}_p}{\partial \operatorname{out}_u}$$

and the recursion start (v is a neuron in the first hidden layer)

$$\frac{\partial \operatorname{out}_v}{\partial \operatorname{out}_u} = \operatorname{out}_v(1 - \operatorname{out}_v) w_{vu}.$$

The command line programs mentioned at the end of the preceding section allow to carry out a sensitivity analysis of a multi-layer perceptron based on these formulas.

It should be noted, though, that a sensitivity analysis produces reliable and reproducible results only if the training is carried out with weight decay (see page 77f). Otherwise the initial conditions (initial weights and bias values) can have a strong effect on the assessment of the relative importance of the different inputs.

References

S.E. Fahlman. An Empirical Study of Learning Speed in Backpropagation Networks. In: Touretzky et al. (1988)

E. Fredkin and T. Toffoli. Conservative Logic. *International Journal of Theoretical Physics* 21(3/4):219–253. Plenum Press, New York, NY, USA, 1982

R.A. Jakobs. Increased Rates of Convergence Through Learning Rate Adaption. *Neural Networks* 1:295–307. Pergamon Press, Oxford, United Kingdom, 1988

A. Pinkus. Approximation Theory of the MLP Model in Neural Networks. *Acta Numerica* 8:143–196. Cambridge University Press, Cambridge, United Kingdom, 1999

M. Riedmiller and H. Braun. Rprop—A Fast Adaptive Learning Algorithm. Technical Report, University of Karlsruhe, Karlsruhe, Germany, 1992

M. Riedmiller and H. Braun. A Direct Adaptive Method for Faster Backpropagation Learning: The RPROP Algorithm. *Int. Conf. on Neural Networks (ICNN-93, San Francisco, CA)*, 586–591. IEEE Press, Piscataway, NJ, USA, 1993

D.E. Rumelhart, G.E. Hinton and R.J. Williams. Learning Representations by Back-Propagating Errors. *Nature* 323:533–536, 1986

T. Tollenaere. SuperSAB: Fast Adaptive Backpropagation with Good Scaling Properties. *Neural Networks* 3:561–573, 1990

D. Touretzky, G. Hinton and T. Sejnowski (eds.) *Proc. of the Connectionist Models Summer School (Carnegie Mellon University)*. Morgan Kaufman, San Mateo, CA, USA, 1988

P.J. Werbos. *Beyond Regression: New Tools for Prediction and Analysis in the Behavioral Sciences*. Ph.D. Thesis, Harvard University, Cambridge, MA, USA, 1974

Chapter 6
Radial Basis Function Networks

Like multi-layer perceptrons, radial basis function networks are feed-forward neural networks with a strictly layered structure. However, the number of layers is always three, that is, there is exactly one hidden layer. In addition, radial basis function networks differ from multi-layer perceptrons in the network input and activation functions, especially in the hidden layer. In this hidden layer **radial basis functions** are employed, which are responsible for the name of this type of neural network. With these functions a kind of "catchment region" is assigned to each neuron, in which it mainly influences the output of the neural network.

6.1 Definition and Examples

Definition 6.1 A **radial basis function network** (RBF network) is a neural network with a graph $G = (U, C)$ that satisfies the following conditions:

1. $U_{\text{in}} \cap U_{\text{out}} = \emptyset$,
2. $C = (U_{\text{in}} \times U_{\text{hidden}}) \cup C', C' \subseteq (U_{\text{hidden}} \times U_{\text{out}})$

The network input function of each hidden neuron is a **distance function** of the input vector and the weight vector, that is,

$$\forall u \in U_{\text{hidden}}: \quad f_{\text{net}}^{(u)}(\mathbf{w}_u, \mathbf{in}_u) = d(\mathbf{w}_u, \mathbf{in}_u),$$

where $d : \mathbb{R}^n \times \mathbb{R}^n \to \mathbb{R}_0^+$ is a function that satisfies $\forall \mathbf{x}, \mathbf{y}, \mathbf{z} \in \mathbb{R}^n$:

$$\begin{aligned}
&\text{(i)} && d(\mathbf{x}, \mathbf{y}) = 0 \quad \Leftrightarrow \quad \mathbf{x} = \mathbf{y}, \\
&\text{(ii)} && d(\mathbf{x}, \mathbf{y}) = d(\mathbf{y}, \mathbf{x}) && \text{(symmetry)}, \\
&\text{(iii)} && d(\mathbf{x}, \mathbf{z}) \leq d(\mathbf{x}, \mathbf{y}) + d(\mathbf{y}, \mathbf{z}) && \text{(triangle inequality)},
\end{aligned}$$

and thus fulfills the definition of a *distance* or a *metric*.

R. Kruse et al., *Computational Intelligence*, Texts in Computer Science,
DOI 10.1007/978-1-4471-5013-8_6, © Springer-Verlag London 2013

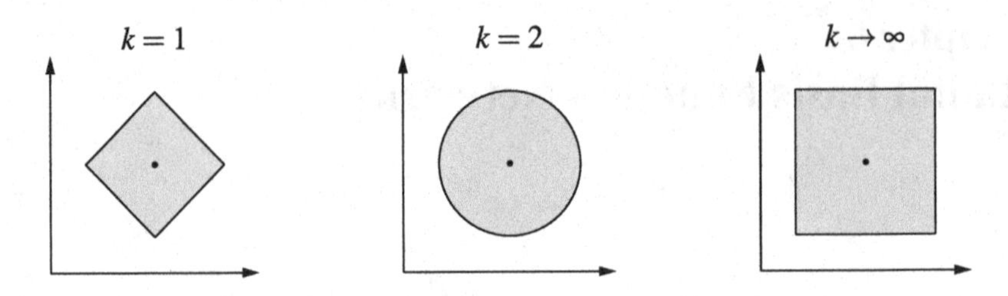

Fig. 6.1 Circles for different distance functions. All circles have the same radius

The network input function of the output neurons is the weighted sum of the inputs (weighted with the connection weights), that is,

$$\forall u \in U_{\text{out}}: \quad f_{\text{net}}^{(u)}(\mathbf{w}_u, \mathbf{in}_u) = \mathbf{w}_u \mathbf{in}_u = \sum_{v \in \text{pred}(u)} w_{uv} \,\text{out}_v .$$

The activation function of each hidden neuron is a **radial function** (as we call it here), that is, a monotone non-increasing function

$$f : \mathbb{R}_0^+ \to [0, 1] \quad \text{with } f(0) = 1 \text{ and } \lim_{x\to\infty} f(x) = 0.$$

The activation function of each output neuron is a linear function, namely

$$f_{\text{act}}^{(u)}(\text{net}_u, \theta_u) = \text{net}_u - \theta_u .$$

Note that a radial basis function network always has exactly three layers and that the input layer and the hidden layer are always fully connected because of the distance computation (that is, all coordinates are used to determine the distance).

The network input function and the activation functions of a hidden neuron describe a kind of "catchment region" of this neuron. The weights of the connections from the input layer to a neuron of the hidden layer state the **center** of this region, since the distance (network input function) is measured between the weight vector and the input vector. The type of distance function determines the shape of the catchment region. To illustrate this fact, we consider the well-known family of distance functions (the so-called *Minkowski family*) that is defined as

$$d_k(\mathbf{x}, \mathbf{y}) = \left(\sum_{i=1}^{n} (x_i - y_i)^k \right)^{\frac{1}{k}} .$$

Well-known special members of this family are:

$k=1$: Manhattan or city block distance,
$k=2$: Euclidean distance,
$k \to \infty$: Maximum distance, that is, $d_\infty(\mathbf{x}, \mathbf{y}) = \max_{i=1}^{n} |x_i - y_i|$.

Distance functions like these can easily be illustrated by considering how a circle looks with them. The reason is that a circle is defined as the set of points that have

rectangular function:

$$f_{\text{act}}(\text{net}, \sigma) = \begin{cases} 0 & \text{if net} > \sigma, \\ 1 & \text{otherwise.} \end{cases}$$

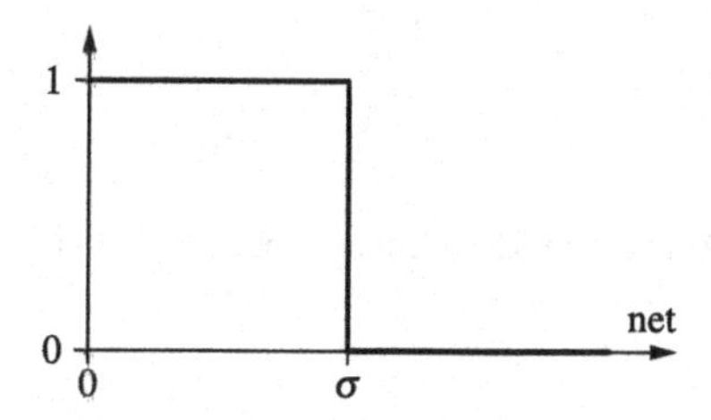

triangular function:

$$f_{\text{act}}(\text{net}, \sigma) = \begin{cases} 0 & \text{if net} > \sigma, \\ 1 - \frac{\text{net}}{\sigma} & \text{otherwise.} \end{cases}$$

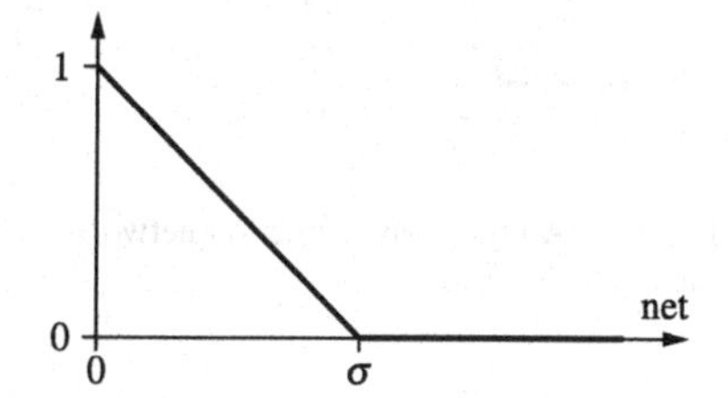

cosine down to zero:

$$f_{\text{act}}(\text{net}, \sigma) = \begin{cases} 0 & \text{if net} > 2\sigma, \\ \frac{\cos\left(\frac{\pi}{2\sigma}\text{net}\right)+1}{2} & \text{otherwise.} \end{cases}$$

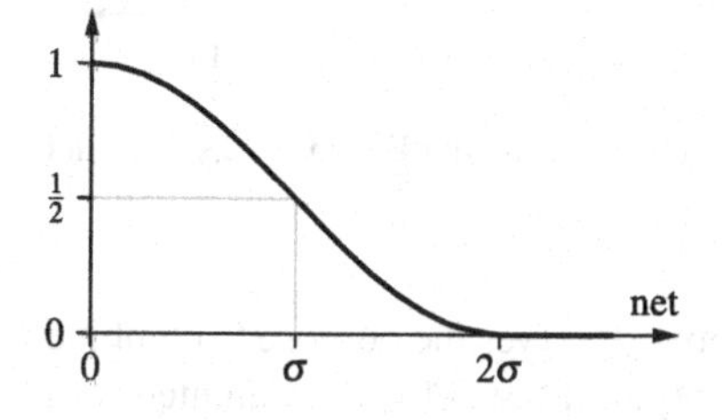

Gaussian function:

$$f_{\text{act}}(\text{net}, \sigma) = e^{-\frac{\text{net}^2}{2\sigma^2}}$$

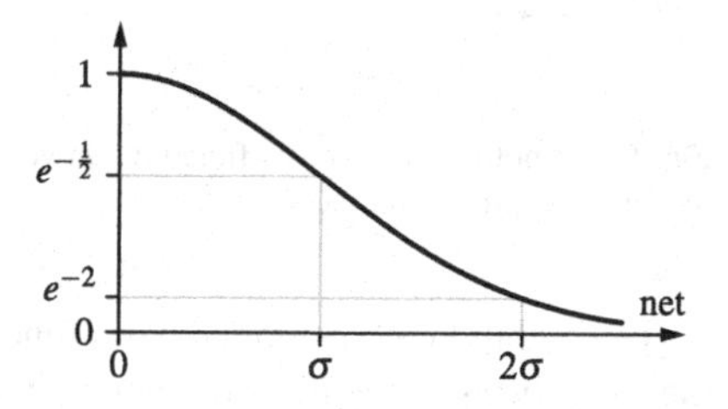

Fig. 6.2 Different radial activation functions

the same fixed distance from a given point. This fixed distance is called the *radius* of the circle. For the three special cases listed above, circles are shown in Fig. 6.1. All three circles have the same radius. With these examples we have an immediate impression of the possible shapes of the catchment region of a hidden neuron.

Intuitively, the activation function of a hidden neuron and its parameters determine the "size" of the catchment region of the neuron by specifying how strong the influence of an input vector is depending on its distance from the weight vector. We call this activation function a **radial function**, because it is defined along a ray (lat. *radius*: ray) from a center, which is described by the weight vector and thus assigns to each radius (that is, to each distance from the center) an activation. Examples of radial activation functions, all of which possess a parameter, namely a **(reference) radius** σ, are shown in Fig. 6.2 (cf. also Fig. 5.2 on page 49).

Note that not all of these radial activation functions limit the catchment region crisply. That is, not for all of these functions there exists a radius, beyond which the activation is 0. For instance, the Gaussian function yields a positive activation regardless of how far an input vector is from the center, even though this activation may be, due to the exponential decay of the Gaussian function, extremely small.

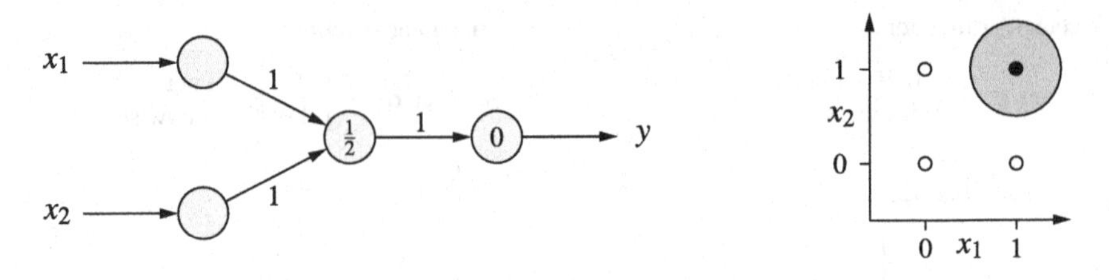

Fig. 6.3 A radial basis function network for the conjunction with Euclidean distance and rectangular activation function

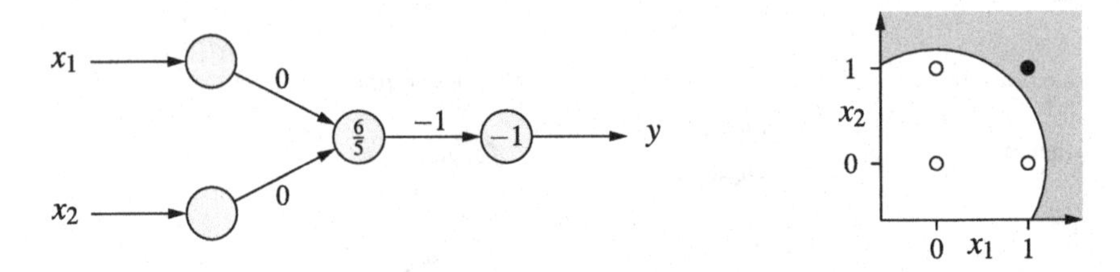

Fig. 6.4 Another radial basis function network for the conjunction with Euclidean distance and rectangular activation functions

The output layer of a radial basis function network serves the purpose to combine the activations of the hidden neurons into the output of the network (weighted sum as the network input function), similar to the operation of a multi-layer perceptron. Note, however, that the activation function of the output neurons in a radial basis function network is a linear function. The reason for this choice, which is important for initializing the parameters, is explained in Sect. 6.3.

As a first example we consider, in analogy to Sect. 3.1, how the conjunction of two Boolean variables x_1 and x_2 is computed. A radial basis function network that solves this task is shown in Fig. 6.3 on the left. It possesses only a single hidden neuron, whose weight vector (center of the radial basis function) is the input vector for which an output of 1 is desired, that is, the point $(1, 1)$. The (reference) radius of the activation function is $\frac{1}{2}$. Like the parameter θ (bias value) of a neuron in a multi-layer perceptron it is written into the circle that represents the hidden neuron. The diagram does not express explicitly that we use a Euclidean distance and a rectangular activation function. Due to the fact that the connection to the output neuron has weight 1 and the fact that the output neuron has the bias value 0, the output of the network coincides with the output of the hidden neuron.

In the same manner as the computations of threshold logic units (cf. Sect. 3.2), the computations of radial basis function can be interpreted geometrically, especially, if rectangular activation functions are used, see Fig. 6.3 on the right. The radial function describes a circle with radius $\frac{1}{2}$ around the point $(1, 1)$. Inside this circle the activation of the hidden neuron (and thus the output of the network) is 1, outside it is 0. In this way, it is easy to see that the network shown in Fig. 6.3 on the left indeed computes the conjunction of its inputs.

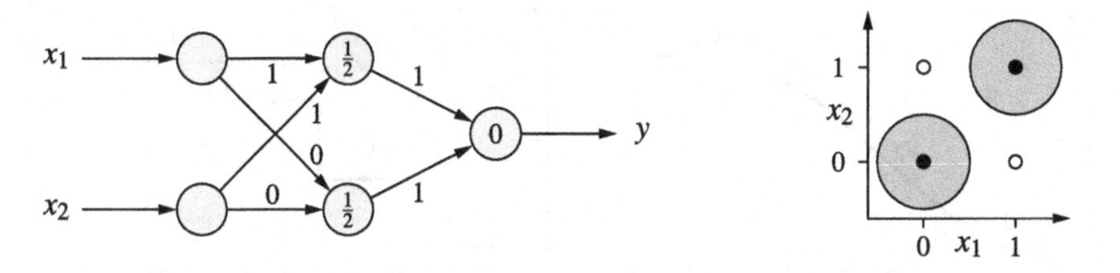

Fig. 6.5 A radial basis function network for the biimplication with Euclidean distance and rectangular activation functions

Of course, the network shown in Fig. 6.3 is not the only possible one for computing the conjunction. For example, we may use a different radius, as long as it is less than 1, or we may shift the center a little, as long as the point $(1, 1)$ stays inside the circle and none of the other points enters the circle. We may also change the distance function or the activation function. However, we may also find a solution that exploits an entirely different principle, as it is shown, for instance, in Fig. 6.4. By using a bias value of -1 in the output neuron a base output of 1 is produced, which is reduced to zero inside a circle with radius $\frac{6}{5}$ around the point $(0, 0)$ (note the negative weight of the connection to the output neuron). Intuitively, we may say that we punched a circular disk from a "carpet" of thickness 1 in such a way that all points, for which an output of 0 is desired, lie inside this disk.

As another example we consider a radial basis function network that computes the biimplication, as shown in Fig. 6.5 on the left. It contains two hidden neurons, which are assigned to the two points for which an output of 1 is desired (namely $(0, 0)$ and $(1, 1)$). Inside circles of radius $\frac{1}{2}$ around these two points the corresponding hidden neuron is activated (that is, outputs a 1), see Fig. 6.5 on the right. The output neuron merely combines these outputs: the output of the network is 1 if the input vector lies inside one of the two circles.

From a logical point of view, the top hidden neuron computes the conjunction of the inputs, the bottom hidden neuron their negated disjunction. The output neuron combines the outputs of the hidden neurons disjunctively (where at most one of the two hidden neurons can be active). That is, the biimplication is represented by exploiting the logical equivalence

$$x_1 \leftrightarrow x_2 \equiv (x_1 \wedge x_2) \vee \neg(x_1 \vee x_2).$$

(Compare the related decomposition used in Sect. 3.4.)

Note that here we may as well draw on the possibility to create a base output of 1 (by using a bias value of -1 in the output neuron), which is reduced to zero by circles of radius $\frac{1}{2}$ (or some other radius less than 1) around the points $(1, 0)$ and $(0, 1)$. Note also that, like for threshold logic units, there is no way to compute the biimplication with only a single (hidden) neuron, unless we employ the *Mahalanobis distance* as the distance function. However, this extension of the computing power of radial basis function networks is discussed only later, namely in Sect. 6.5.

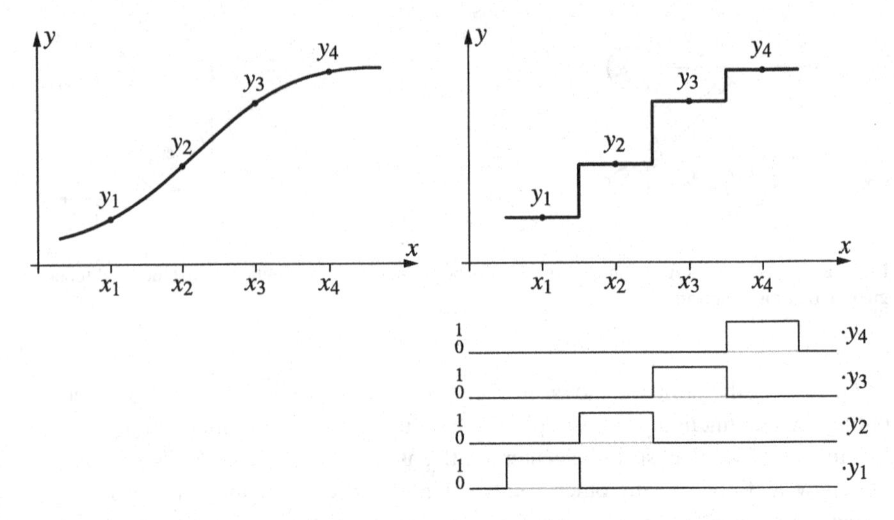

Fig. 6.6 Representing a step function by a weighted sum of rectangular functions (with centers x_i). Naturally, the step heights y_i may also be negative. However, at the step borders incorrect function values are computed (sum of the step heights)

6.2 Function Approximation

After the examples of the preceding section, in which we examined only simple logical functions, we consider now, in analogy to Sect. 5.2, how we can approximate real-valued functions with the help of radial basis function networks. The principle is the same as in Sect. 5.2: the function to represent is approximated by a step function, which can easily be computed by a radial basis function network if we model it as a weighted sum of rectangular functions. We illustrate this principle with the help of the same example function as in Sect. 5.2, see Fig. 6.6.

For each step a radial basis function is employed, whose center lies in the middle of the step and whose radius is half the step width. In this way rectangular pulses are described (see Fig. 6.6 on the bottom right), which are weighted with the corresponding step height and then summed. Thus, we obtain the step function shown in Fig. 6.6 on the top right. The corresponding radial basis function network, which possesses one neuron for each rectangular pulse, is shown in Fig. 6.7.

Note that we actually compute the sum of the step heights at the borders (not shown in Fig. 6.6), since the rectangular pulses overlap at these points. This does not affect the approximation quality, though, because the error is measured, as in Sect. 5.2, as the area between the step function and the function to approximate. The deviations occur at finitely many points and thus do not contribute to the error.

Since we used the same principle as in Sect. 5,2, we can immediately transfer the approximation theorem derived there:

Theorem 6.1 *Every Riemann-integrable function can be approximated with arbitrary accuracy by a radial basis function network.*

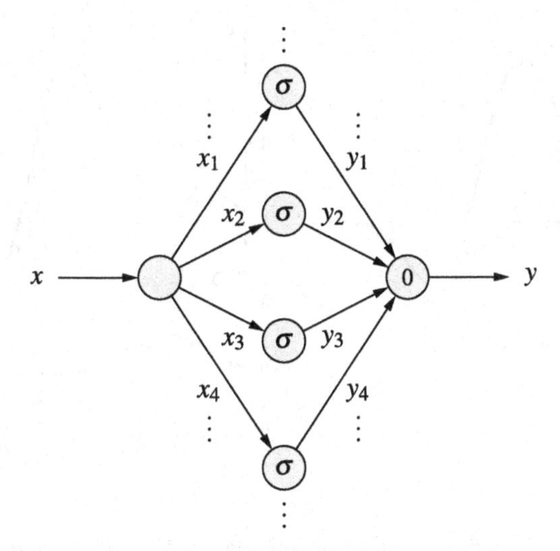

Fig. 6.7 A radial basis function network that computes the step function shown in Fig. 6.6 or the piecewise linear function shown in Fig. 6.8 (depending in the activation functions of the hidden neurons). It is $\sigma = \frac{1}{2}\Delta x = \frac{1}{2}(x_{i+1} - x_i)$ or $\sigma = \Delta x = x_{i+1} - x_i$, respectively

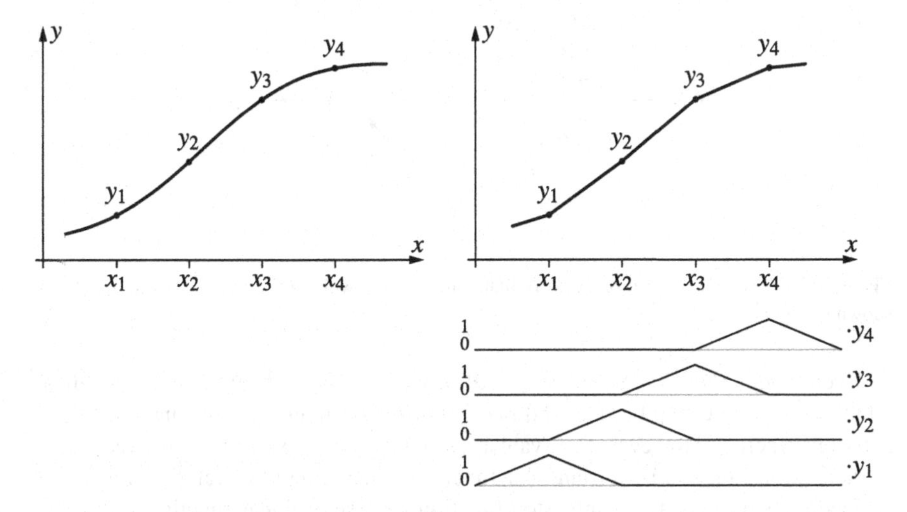

Fig. 6.8 Representing a piecewise linear function by a weighted sum of triangular functions (with centers x_i)

Note here again that, like for the approximation theorem about multi-layer perceptrons, this theorem only requires that the function to represent is Riemann-integrable. It need *not* be continuous. That is, the function to represent may have "jumps." However, it may have only finitely many "jumps" of finite height in the region in which it is to be approximated by a radial basis function network. In other words, the function must be continuous "almost everywhere."

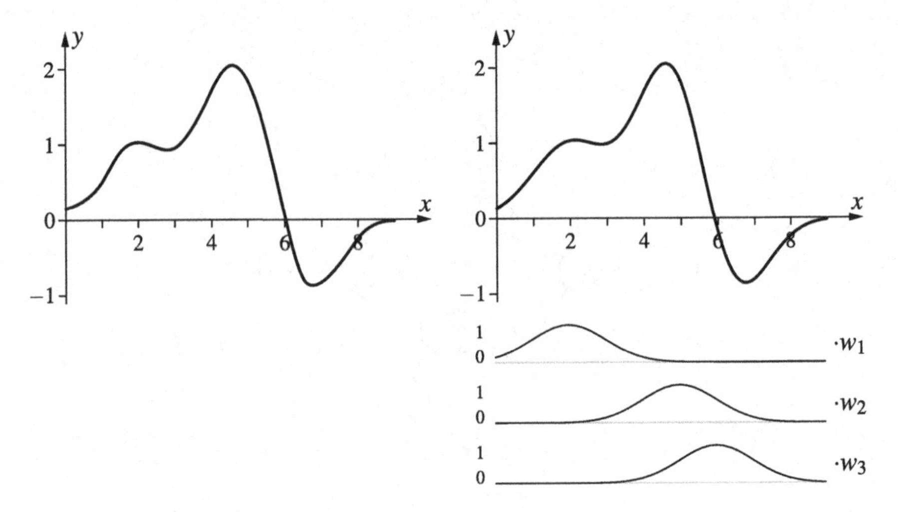

Fig. 6.9 Approximating a function by a sum of Gaussian functions with radius $\sigma = 1$. It is $w_1 = 1$, $w_2 = 3$ and $w_3 = -2$

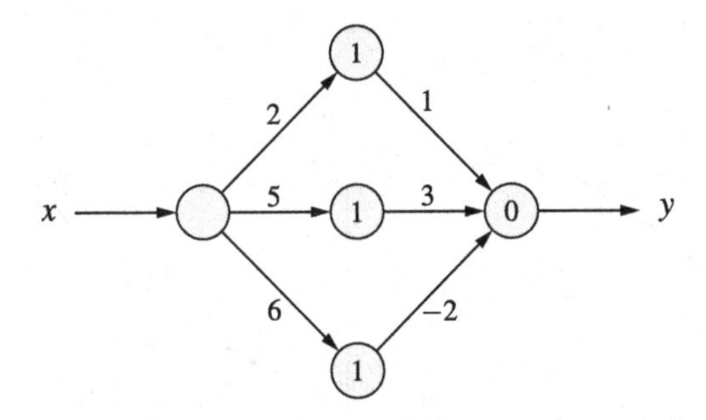

Fig. 6.10 A radial basis function network that computes the weighted sum of Gaussian functions shown in Fig. 6.9

Even though it does not impede the validity of the above theorem, the anomalies, which occur at the step borders and make it differ from a pure step function, are at least unattractive. However, they vanish automatically if we replace the rectangular activation function by a triangular function—in analogy to Sect. 5.2, where we replaced the (Heaviside or unit) step function by a semi-linear function. With this modification the step function is turned into a piecewise linear function, which is computed by a radial basis function network as the weighted sum of overlapping triangular functions, see Fig. 6.8. Thus the approximation is considerably improved (or the same approximation quality can be achieved with fewer neurons).

The approximation quality can be improved even further if one increases the number of support points, especially in regions where the function is strongly curved (as already mentioned in the analogous discussion in Sect. 5.2). In addition, we can eliminate the kinks of the piecewise linear function if we employ an activation function like the Gaussian function, which can produce "smooth" transitions.

To illustrate how a function can be approximated by Gaussian functions, we consider how the function shown in Fig. 6.9 on the left can be approximated by a weighted sum of three Gaussian bell curves (see Fig. 6.9 on the right). The corresponding radial basis function network is shown in Fig. 6.10.

It should be clear that the principle of approximating a real-valued function, which we exploited in this section, is not limited to unary functions, but may as well be transferred to functions with multiple arguments. In contrast to multi-layer perceptrons, we see immediately here that three layers are always sufficient, because the basis functions influence the output of the network only locally.

6.3 Initializing the Parameters

When we discussed multi-layer perceptrons, we treated initializing the parameters (that is, the connection weights and bias values) only in passing, since it is trivial: simple choose random values, the absolute values of which are not too large (≤ 0.5). In principle, we can apply the same method to radial basis function networks, but it usually leads to fairly poor results. In addition, the hidden layer and the output layer of a radial basis function network differ considerably, since they possess different network input and activation functions—in contrast to a multi-layer perceptron, in which the hidden layers and the output layer are homogeneous. As a consequence, these two layers should be treated separately and thus we devote a separate section to how to initialize the parameters of a radial basis function network.

In order to simplify the presentation as much as possible, we start with the special case of a so-called **simple radial basis function network**, in which each training example is covered by its own radial basis function. That is, the hidden layer contains exactly as many neurons as there are training examples. The weights of the connections from the input neurons to the neurons of the hidden layer are determined by the training examples: each hidden neuron receives a training example and the weights of the connections to the hidden neurons are simply initialized with the elements of the input vector of this training example.

Formally: let $L_{\text{fixed}} = \{l_1, \ldots, l_m\}$ be a fixed learning task with m training patterns $l = (\mathbf{i}^{(l)}, \mathbf{o}^{(l)})$. Since each pattern is used as the center of its own radial function, there are m neurons in the hidden layer. Let these neurons be $v_1, \ldots, v_m$. We set

$$\forall k \in \{1, \ldots, m\}: \quad \mathbf{w}_{v_k} = \mathbf{i}^{(l_k)}.$$

If the most commonly employed Gaussian activation function is chosen, the radii σ_k are often initialized to equal values according to the heuristics

$$\forall k \in \{1, \ldots, m\}: \quad \sigma_k = \frac{d_{\max}}{\sqrt{2m}},$$

where $d_{\max}$ is the maximal distance between the input vectors of two training patterns (computed with the network input function chosen for the hidden neurons, which is a distance function d), that is,

$$d_{\max} = \max_{l_j, l_k \in L_{\text{fixed}}} d\big(\mathbf{i}^{(l_j)}, \mathbf{i}^{(l_k)}\big).$$

This choice has the advantages that the Gaussian bell curves are not too narrow (that is, they are not isolated peaks in the input space), but also not too wide (that is, they are also not overlapping too much, at least if the data set is "benevolent" or "well behaved", that is, there are no individual training examples that are located far away from all other training examples—no (extreme) outliers).

The weights from the hidden layer to the output layer and the bias values of the output neurons are computed with the following idea: since the parameters of the hidden layer (centers and radii) are known, we can compute the output of the hidden neurons for every training example. We now have to determine the connection weights and the bias values in such a way that from these outputs the desired outputs of the network are computed. Since the network input function of the output neurons is a weighted sum of its inputs and its activation and output functions are both linear, each training example l yields for each output neuron u one linear equation

$$\sum_{k=1}^{m} w_{uv_m} \operatorname{out}_{v_m}^{(l)} - \theta_u = o_u^{(l)}.$$

(This is the main reason for choosing linear activation and output functions for the output neurons.) Thus, we obtain for each output neuron a linear equation system with m equations (one equation for each training example) and $m + 1$ unknowns (m weights and one bias value). That this equation system is under-determined (more unknowns than equations), we can easily fix by simply setting the surplus parameter $\theta_u = 0$. In matrix and vector notation, the equation system to solve reads

$$\mathbf{A} \cdot \mathbf{w}_u = \mathbf{o}_u,$$

where $\mathbf{w}_u = (w_{uv_1}, \ldots, w_{uv_m})^\top$ is the weight vector of the output neuron u and $\mathbf{o}_u = (o_u^{(l_1)}, \ldots, o_u^{(l_m)})^\top$ is the vector of the desired outputs of the output neuron u for the different training examples. $\mathbf{A}$ is an $m \times m$ matrix with the outputs of the neurons of the hidden layer for the different training examples, namely

$$\mathbf{A} = \begin{pmatrix} \operatorname{out}_{v_1}^{(l_1)} & \operatorname{out}_{v_2}^{(l_1)} & \ldots & \operatorname{out}_{v_m}^{(l_1)} \\ \operatorname{out}_{v_1}^{(l_2)} & \operatorname{out}_{v_2}^{(l_2)} & \ldots & \operatorname{out}_{v_m}^{(l_2)} \\ \vdots & \vdots & & \vdots \\ \operatorname{out}_{v_1}^{(l_m)} & \operatorname{out}_{v_2}^{(l_m)} & \ldots & \operatorname{out}_{v_m}^{(l_m)} \end{pmatrix}.$$

That is, each matrix row contains the outputs of the different neurons for one training example, each column contains the outputs of one hidden neuron for the different training examples. Since the elements of this matrix can be computed from the training examples and the desired outputs are known as well, the weights can be found by simply solving this equation system with the usual methods of linear algebra.

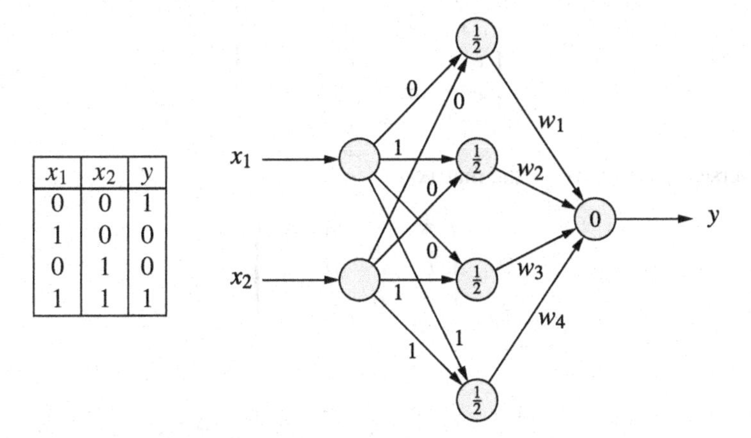

x_1	x_2	y
0	0	1
1	0	0
0	1	0
1	1	1

Fig. 6.11 Training examples for the biimplication and the simple radial basis function network as it is already partially determined by them

For the subsequent considerations, it is advantageous to compute the solution of the linear equation system by inverting the matrix **A**, that is, by

$$\mathbf{w}_u = \mathbf{A}^{-1} \cdot \mathbf{o}_u,$$

even though this method requires that the matrix **A** has full rank. In practice, this is usually, but not always the case. If **A** does not have full rank, weights have to be chosen randomly until the remaining equation system is uniquely solvable.

Note that this initialization method already guarantees that the error of a simple radial basis function network vanishes on the training data. Since the equation system, we have to solve is at most under-determined, we can always find connection weights with which the desired outputs are computed exactly. Hence, it is not necessary to train a simple radial basis function network.

To illustrate the procedure, we consider a radial basis function network for the biimplication $x_1 \leftrightarrow x_2$, in which the neurons of the hidden layer possess a Gaussian activation function. The training examples as well as the radial basis function network as it is already determined by them is shown in Fig. 6.11. The radii σ were chosen according to the heuristics mentioned above: clearly, we have $d_{\max} = \sqrt{2}$ (diagonal of the unit square) and $m = 4$, therefore $\sigma = \frac{\sqrt{2}}{\sqrt{2 \cdot 4}} = \frac{1}{2}$.

We are left with determining the four weights $w_1, \dots, w_4$. (Note that the bias value of the output neuron is set to 0, since otherwise the equation system to solve is under-determined.) In order to compute these weights, we set up the matrix

$$\mathbf{A} = (a_{jk}) \quad \text{with } a_{jk} = e^{-\frac{|\mathbf{i}_j - \mathbf{i}_k|^2}{2\sigma^2}} = e^{-2|\mathbf{i}_j - \mathbf{i}_k|^2},$$

where $\mathbf{i}_j$ and $\mathbf{i}_k$ are the input vectors of the jth and kth training example (numbered according to the table shown in Fig. 6.11). Therefore, it is

$$\mathbf{A} = \begin{pmatrix} 1 & e^{-2} & e^{-2} & e^{-4} \\ e^{-2} & 1 & e^{-4} & e^{-2} \\ e^{-2} & e^{-4} & 1 & e^{-2} \\ e^{-4} & e^{-2} & e^{-2} & 1 \end{pmatrix}.$$

The inverse of this matrix is the matrix

$$\mathbf{A}^{-1} = \begin{pmatrix} \frac{a}{D} & \frac{b}{D} & \frac{b}{D} & \frac{c}{D} \\ \frac{b}{D} & \frac{a}{D} & \frac{c}{D} & \frac{b}{D} \\ \frac{b}{D} & \frac{c}{D} & \frac{a}{D} & \frac{b}{D} \\ \frac{c}{D} & \frac{b}{D} & \frac{b}{D} & \frac{a}{D} \end{pmatrix},$$

where

$$D = 1 - 4e^{-4} + 6e^{-8} - 4e^{-12} + e^{-16} \approx 0.9287$$

is the determinant of the matrix $\mathbf{A}$ and

$$\begin{aligned} a &= 1 - 2e^{-4} + e^{-8} \approx 0.9637, \\ b &= -e^{-2} + 2e^{-6} - e^{-10} \approx -0.1304, \\ c &= e^{-4} - 2e^{-8} + e^{-12} \approx 0.0177. \end{aligned}$$

From this matrix and the output vector $\mathbf{o}_u = (1, 0, 0, 1)^\top$, we can now easily compute the connection weights. We obtain

$$\mathbf{w}_u = \mathbf{A}^{-1} \cdot \mathbf{o}_u = \frac{1}{D} \begin{pmatrix} a + c \\ 2b \\ 2b \\ a + c \end{pmatrix} \approx \begin{pmatrix} 1.0567 \\ -0.2809 \\ -0.2809 \\ 1.0567 \end{pmatrix}.$$

The computations of the radial basis function network that is initialized in this way are shown in Fig. 6.12. The left diagram shows a single basis function, namely the one with the center $(0, 0)$. The middle diagram shows all four basis functions (overlain, no sum). The output of the whole network is shown in the right diagram. It is clearly visible how the radial basis functions of the two centers, for which an output of 1 is desired, are weighted positively, while the two others are weighted negatively, so that in total exactly the desired outputs are computed.

Obviously, simple radial basis function networks are very easy to initialize, because the training examples already fix the parameters of the hidden layer. The weights from the hidden layer to the output layer can be computed, as we saw, by solving a simple linear equation system. In practice, however, simple radial basis function networks are of little use. In the first place, the number of training example is generally too large to create a separate neuron for each of them: the resulting network would become too large to be feasible. In addition one desires, for obvious reasons, that several training examples are covered by the same radial basis function.

Therefore, **radial basis function networks** (without the qualifier "simple") possess fewer neurons in the hidden layer than there are training examples. To initialize

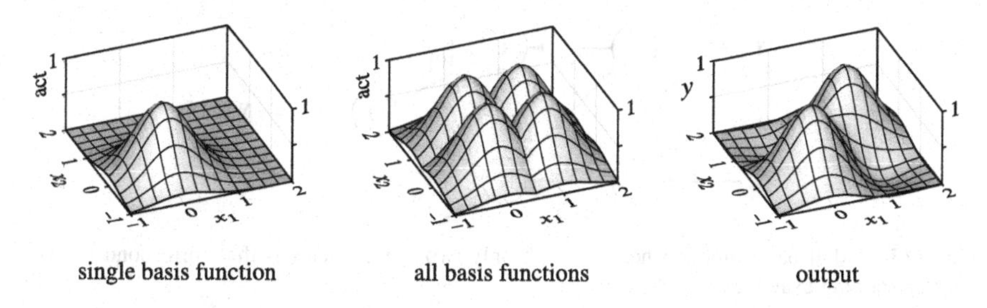

Fig. 6.12 Radial basis functions and output of a radial basis function network with four hidden neurons for the biimplication

them one often selects a random (though hopefully representative) subset of the training examples as the centers of the radial basis functions, namely one training example for each hidden neuron. (However, this is only one possible method for choosing the center coordinates; another is discussed below.) The weights from the input layer to the hidden layer are again fixed by these selected training examples: the coordinates of the input vectors are simply copied into the weight vectors. Likewise, the radii are chosen heuristically, using the same values as above (although we should rather refer to the selected subset of training examples instead of the whole set) in order to avoid having to deal with square roots.

In the step in which the weights of the connections from the hidden layer to the output neurons are determined, we now face the problem that the equation system to solve is over-determined. Since we selected a subset of the training examples, say k examples, we have m equations for each output neuron (one for each training example in the full set), but only $k+1$ unknowns (k weights and one bias value) with $k < m$. Thus we do not choose the bias value to be simply zero, but treat it by converting it into a weight (cf. Fig. 3.18 on page 27).

In analogy to simple radial basis function networks, we set up a $m \times (k+1)$ matrix

$$\mathbf{A} = \begin{pmatrix} 1 & \mathrm{out}_{v_1}^{(l_1)} & \mathrm{out}_{v_2}^{(l_1)} & \dots & \mathrm{out}_{v_k}^{(l_1)} \\ 1 & \mathrm{out}_{v_1}^{(l_2)} & \mathrm{out}_{v_2}^{(l_2)} & \dots & \mathrm{out}_{v_k}^{(l_2)} \\ \vdots & \vdots & \vdots & & \vdots \\ 1 & \mathrm{out}_{v_1}^{(l_m)} & \mathrm{out}_{v_2}^{(l_m)} & \dots & \mathrm{out}_{v_k}^{(l_m)} \end{pmatrix}$$

of the activations of the hidden neurons. (Note the ones in the first column, which refer to the fixed input 1 for the bias value.) We now have to determine for each output neuron u an (extended) weight vector $\mathbf{w}_u = (-\theta_u, w_{uv_1}, \dots, w_{uv_k})^\top$, so that

$$\mathbf{A} \cdot \mathbf{w}_u = \mathbf{o}_u,$$

where again $\mathbf{o}_u = (o_u^{(l_1)}, \dots, o_u^{(l_m)})^\top$. However, since the equation system is over-determined (at least if the matrix has a rank $> k+1$), this equation does not gen-

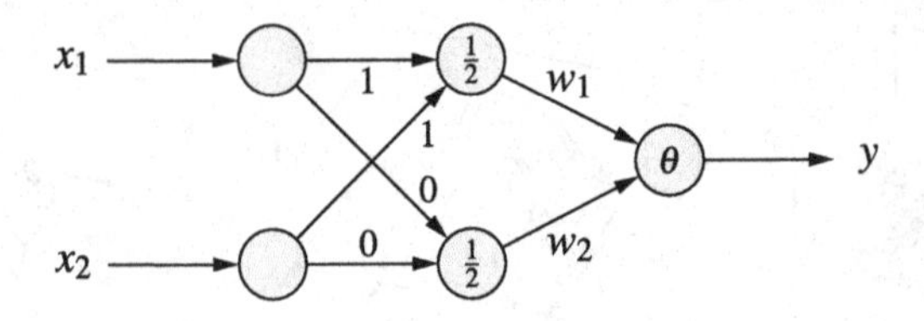

Fig. 6.13 Radial basis function network with only two hidden neurons that correspond to two selected training examples

erally possess a solution. In other words: the matrix $\mathbf{A}$ is not square and thus not invertible. Fortunately, a good approximate solution (so-called minimum norm solution) can be found with the help of the so-called (Moore–Penrose) **pseudo inverse** $\mathbf{A}^+$ of the matrix $\mathbf{A}$ (Albert 1972). This pseudo inverse is computed as

$$\mathbf{A}^+ = \left(\mathbf{A}^\top \mathbf{A}\right)^{-1} \mathbf{A}^\top.$$

The weights can finally be determined with the equation

$$\mathbf{w}_u = \mathbf{A}^+ \cdot \mathbf{o}_u = \left(\mathbf{A}^\top \mathbf{A}\right)^{-1} \mathbf{A}^\top \cdot \mathbf{o}_u$$

(cf. the computations on page 94).

We illustrate this procedure again with the help of the biimplication. From the training examples, we choose the first, that is, $l_1 = (\mathbf{i}^{(l_1)}, \mathbf{o}^{(l_1)}) = ((0,0),(1))$, and the last, that is, $l_4 = (\mathbf{i}^{(l_4)}, \mathbf{o}^{(l_4)}) = ((1,1),(1))$. That is, we start from the partially determined radial basis function network shown in Fig. 6.13 and now have to compute the weights w_1 and w_2 and the bias value θ. To do so, we set up the 4×3 matrix

$$\mathbf{A} = \begin{pmatrix} 1 & 1 & e^{-4} \\ 1 & e^{-2} & e^{-2} \\ 1 & e^{-2} & e^{-2} \\ 1 & e^{-4} & 1 \end{pmatrix}.$$

The pseudo inverse of this matrix is the matrix

$$\mathbf{A}^+ = \left(\mathbf{A}^\top \mathbf{A}\right)^{-1} \mathbf{A}^\top = \begin{pmatrix} a & b & b & a \\ c & d & d & e \\ e & d & d & c \end{pmatrix},$$

where

$$a \approx -0.1810, \qquad b \approx 0.6810,$$
$$c \approx 1.1781, \qquad d \approx -0.6688, \qquad e \approx 0.1594.$$

From this matrix and the output vector $\mathbf{o}_u = (1,0,0,1)^\top$, we can compute the connection weights easily. We obtain

$$\mathbf{w}_u = \begin{pmatrix} -\theta \\ w_1 \\ w_2 \end{pmatrix} = \mathbf{A}^+ \cdot \mathbf{o}_u \approx \begin{pmatrix} -0.3620 \\ 1.3375 \\ 1.3375 \end{pmatrix}.$$

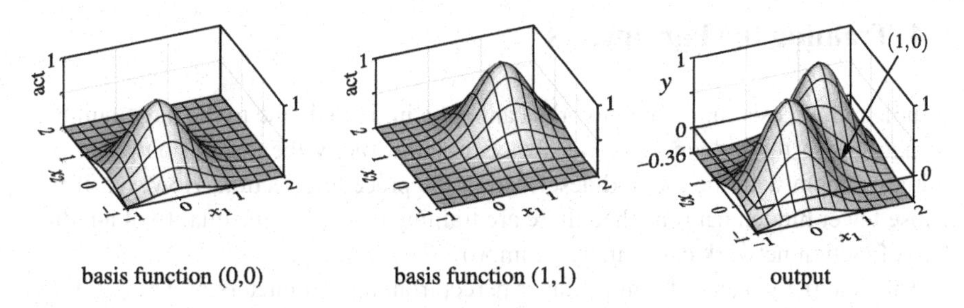

Fig. 6.14 Radial basis functions and output of a radial basis function network with two hidden neurons for the biimplication

The computations of the network that is initialized in this way are shown in Fig. 6.14. The left and the middle diagram show the two radial basis functions, the right diagram the output of the network. Note the non-vanishing bias value and that (somewhat surprisingly) exactly the desired outputs are computed. In practice, this is usually not the case, because the equation system to solve is over-determined. Thus, in general, one has to be satisfied with an approximate solution. However, since the equation system is not really over-determined here (due to reasons of symmetry: from the matrix $\mathbf{A}$ and the output vector $\mathbf{o}_u$ it is easy to see that the second and the third equation are actually identical), we obtain an exact solution nevertheless.

Up to now we chose the centers of the radial basis functions (randomly) from the training examples. It would be better, though, if we could determine proper centers with a different method, because a random selection does not guarantee that the centers fit the training examples sufficiently well to allow for a good approximation of the desired outputs. In order to find the centers, basically any prototype-based clustering method can be applied. Among these is, for instance, learning vector quantization, which is treated in the next chapter. Here we consider a method of classical statistics instead, which is known as **c-means clustering** (also known as k-means clustering) (Hartigan and Wong 1979). The letter c (or k) in the name of this method stands for a parameter, namely the number of clusters to be found.

This method is very simple and straightforward. At the beginning c cluster centers are chosen randomly, usually from the training examples. Then the training examples are divided into c groups (clusters) by assigning to each cluster center all training examples that are closer to it than to any other cluster center. In a second step new cluster centers are computed by finding the "centers of gravity" of the formed groups of data points. That is, one computes the vector sum of the training examples of a group and divides by the number of these training example. The result is the new center. Next the first step, that is, forming the groups is executed again and so forth, until the cluster centers do not change anymore. The cluster centers found in this way can immediately be used to initialize the centers of a radial basis function network. Even the radii can be chosen with this method from the data: for example, one chooses the average distance of the training examples that are assigned to a cluster center from this cluster center.

6.4 Training the Parameters

Simple radial basis function networks cannot be improved: due to the large number of neurons in the hidden layer we always obtain exactly the desired output if we initialize them with the method described in the preceding section. However, if we chose fewer hidden neurons than there are training examples, the quality of a radial basis function network can usually be improved by training.

Like the parameters of a multi-layer perceptron, the parameters of a radial basis function network are trained by **gradient descent**. Therefore, in order to find the adaptation rules for the weights, we proceed in basically the same manner as in Sect. 5.4. For the parameters of the output neurons, that is, for the weights of the connections from the hidden neurons and the bias values of the output neurons, we even obtain the same result as for a multi-layer perceptron: the gradient for a single output neuron u and a single training pattern l is (see page 64)

$$\nabla_{\mathbf{w}_u} e_u^{(l)} = \frac{\partial e_u^{(l)}}{\partial \mathbf{w}_u} = -2\left(o_u^{(l)} - \text{out}_u^{(l)}\right) \frac{\partial\, \text{out}_u^{(l)}}{\partial\, \text{net}_u^{(l)}} \mathbf{in}_u^{(l)},$$

where $\mathbf{in}_u$ is the vector of outputs of the predecessors of the neuron u, $o_u^{(l)}$ is the desired output of the neuron u, $\text{net}_u^{(l)}$ is its network input and $\text{out}_u^{(l)}$ its actual output if the input vector $\mathbf{i}^{(l)}$ of the training pattern l is fed into the network. Recall that the actual output $\text{out}_u^{(l)}$ of the neuron u for the training pattern l is computed from its network input via its output function f_{out} and its activation function f_{act}, that is,

$$\text{out}_u^{(l)} = f_{\text{out}}\left(f_{\text{act}}\left(\text{net}_u^{(l)}\right)\right).$$

Like in Sect. 5.4 we assume, for reasons of simplicity, that the output function is the identity. In addition, we insert the linear activation function of the output neuron of a radial basis function network. Thus, we obtain

$$\frac{\partial\, \text{out}_u^{(l)}}{\partial\, \text{net}_u^{(l)}} = \frac{\partial\, \text{net}_u^{(l)}}{\partial\, \text{net}_u^{(l)}} = 1.$$

Note that the bias value θ_u of the output neuron u is already contained in the network input $\text{net}_u^{(l)}$, since we work with extended input and weight vectors in order to avoid inconvenient distinctions (just as we did in Sect. 5.4). Therefore, we have

$$\nabla_{\mathbf{w}_u} e_u^{(l)} = \frac{\partial e_u^{(l)}}{\partial \mathbf{w}_u} = -2\left(o_u^{(l)} - \text{out}_u^{(l)}\right) \mathbf{in}_u^{(l)},$$

from which we derive the online adaptation rule

$$\Delta \mathbf{w}_u^{(l)} = -\frac{\eta_3}{2} \nabla_{\mathbf{w}_u} e_u^{(l)} = \eta_3\left(o_u^{(l)} - \text{out}_u^{(l)}\right) \mathbf{in}_u^{(l)}$$

for the weights (and thus implicitly also the bias value θ_u). Note that the negative sign of the gradient vanishes, because we have to “descend in the error landscape” and thus have to move against the direction of the gradient. The factor 2 is incorporated into the learning rate η_3. (The index 3 of this learning rate already indicates that we will meet two more learning rates.) As usual, for batch training the weight changes $\Delta \mathbf{w}_u$ have to be summed over all training patterns and are applied to the weights only after all training patterns have been processed.

The adaptation rules for the weights of the connections from the input neurons to the hidden neurons and for the radii of the radial basis functions are derived in a similar way as error backpropagation was derived in Sect. 5.5. We merely have to take the special network input and activation functions into account. This implies, though, that we can no longer work with extended weight and input vectors, but have to treat the weights (the centers of the radial basis functions) and the radius separately. Hence, for reasons of clarity, we go through the whole derivation here.

We start from the total error of a radial basis function network with output neurons U_{out} w.r.t. a fixed learning task L_{fixed}:

$$e = \sum_{l \in L_{\text{fixed}}} e^{(l)} = \sum_{u \in U_{\text{out}}} e_u = \sum_{l \in L_{\text{fixed}}} \sum_{u \in U_{\text{out}}} e_u^{(l)}.$$

Let v be a neuron in the hidden layer. Let its predecessors (input neurons) be the neurons $\text{pred}(v) = \{p \in U_{\text{in}} \mid (p, v) \in C\} = \{p_1, \dots, p_n\}$. Furthermore, let $\mathbf{w}_v = (w_{vp_1}, \dots, w_{vp_n})$ be the corresponding weight vector and σ_v the corresponding radius. We start by computing the gradient of the total error w.r.t. the connection weights (center coordinates):

$$\nabla_{\mathbf{w}_v} e = \frac{\partial e}{\partial \mathbf{w}_v} = \left(\frac{\partial e}{\partial w_{vp_1}}, \dots, \frac{\partial e}{\partial w_{vp_n}} \right).$$

Since the total error is a sum over all training patterns, we have

$$\frac{\partial e}{\partial \mathbf{w}_v} = \frac{\partial}{\partial \mathbf{w}_v} \sum_{l \in L_{\text{fixed}}} e^{(l)} = \sum_{l \in L_{\text{fixed}}} \frac{\partial e^{(l)}}{\partial \mathbf{w}_v}.$$

Hence we can confine ourselves, in analogy to Sect. 5.5, to the error $e^{(l)}$ for a single training pattern l. This error depends on the weights in $\mathbf{w}_v$ only via the network input $\text{net}_v^{(l)} = d(\mathbf{w}_v, \mathbf{in}_v^{(l)})$ with the network input vector $\mathbf{in}_v^{(l)} = (\text{out}_{p_1}^{(l)}, \dots, \text{out}_{p_n}^{(l)})$. Therefore we can apply the chain rule and obtain in analogy to Sect. 5.5:

$$\nabla_{\mathbf{w}_v} e^{(l)} = \frac{\partial e^{(l)}}{\partial \mathbf{w}_v} = \frac{\partial e^{(l)}}{\partial \text{net}_v^{(l)}} \frac{\partial \text{net}_v^{(l)}}{\partial \mathbf{w}_v}.$$

To compute the first factor, we examine the error $e^{(l)}$ for the pattern $l = (\mathbf{i}^{(l)}, \mathbf{o}^{(l)})$:

$$e^{(l)} = \sum_{u \in U_{\text{out}}} e_u^{(l)} = \sum_{u \in U_{\text{out}}} \left(o_u^{(l)} - \text{out}_u^{(l)}\right)^2,$$

that is, the error sum over all output neurons. Thus, we obtain

$$\frac{\partial e^{(l)}}{\partial \text{net}_v^{(l)}} = \frac{\partial \sum_{u \in U_{\text{out}}} (o_u^{(l)} - \text{out}_u^{(l)})^2}{\partial \text{net}_v^{(l)}} = \sum_{u \in U_{\text{out}}} \frac{\partial (o_u^{(l)} - \text{out}_u^{(l)})^2}{\partial \text{net}_v^{(l)}}.$$

Since only the actual output $\text{out}_u^{(l)}$ of an output neuron u depends on the network input $\text{net}_v^{(l)}$ of the neuron v under consideration, it is

$$\frac{\partial e^{(l)}}{\partial \text{net}_v^{(l)}} = -2 \sum_{u \in U_{\text{out}}} \left(o_u^{(l)} - \text{out}_u^{(l)}\right) \frac{\partial \text{out}_u^{(l)}}{\partial \text{net}_v^{(l)}}.$$

Let the neurons $\text{succ}(v) = \{s \in U_{\text{out}} \mid (v, s) \in C\}$ be the successors (output neurons) of the neuron v we consider. The output $\text{out}_u^{(l)}$ of an output neuron u is affected by the network input $\text{net}_v^{(l)}$ of the considered neuron v only if there exists a connection from v to u, that is, if u is among the successors $\text{succ}(v)$ of v. Therefore, we can confine the sum over the output neurons to the successors of v. Furthermore, the output $\text{out}_s^{(l)}$ of a successor s of neuron v depends on the network input $\text{net}_v^{(l)}$ of the considered successor only via the network input $\text{net}_s^{(l)}$ of the considered neuron v. Thus, we obtain with the chain rule

$$\frac{\partial e^{(l)}}{\partial \text{net}_v^{(l)}} = -2 \sum_{s \in \text{succ}(v)} \left(o_s^{(l)} - \text{out}_s^{(l)}\right) \frac{\partial \text{out}_s^{(l)}}{\partial \text{net}_s^{(l)}} \frac{\partial \text{net}_s^{(l)}}{\partial \text{net}_v^{(l)}}.$$

Since the successors $s \in \text{succ}(v)$ are output neurons, we can insert (like when we consider the output neurons)

$$\frac{\partial \text{out}_s^{(l)}}{\partial \text{net}_s^{(l)}} = 1.$$

We are left with computing the partial derivative of the network input. Since the neurons s are output neurons, it is

$$\text{net}_s^{(l)} = \mathbf{w}_s \mathbf{in}_s^{(l)} - \theta_s = \left(\sum_{p \in \text{pred}(s)} w_{sp} \, \text{out}_p^{(l)} \right) - \theta_s,$$

where one element of $\mathbf{in}_s^{(l)}$ is the output $\text{out}_v^{(l)}$ of the neuron v under consideration. Clearly, $\text{net}_s^{(l)}$ depends on $\text{net}_v^{(l)}$ only via this element $\text{out}_v^{(l)}$. Therefore, it is

$$\frac{\partial \text{net}_s^{(l)}}{\partial \text{net}_v^{(l)}} = \left(\sum_{p \in \text{pred}(s)} w_{sp} \frac{\partial \text{out}_p^{(l)}}{\partial \text{net}_v^{(l)}} \right) - \frac{\partial \theta_s}{\partial \text{net}_v^{(l)}} = w_{sv} \frac{\partial \text{out}_v^{(l)}}{\partial \text{net}_v^{(l)}},$$

since all terms vanish except the one having $p = v$. In total, we have for the gradient

$$\nabla_{\mathbf{w}_v} e^{(l)} = \frac{\partial e^{(l)}}{\partial \mathbf{w}_v} = -2 \sum_{s \in \text{succ}(v)} \left(o_s^{(l)} - \text{out}_s^{(l)}\right) w_{su} \frac{\partial \text{out}_v^{(l)}}{\partial \text{net}_v^{(l)}} \frac{\partial \text{net}_v^{(l)}}{\partial \mathbf{w}_v},$$

from which we obtain the online adaptation rule

$$\Delta \mathbf{w}_v^{(l)} = -\frac{\eta_1}{2} \nabla_{\mathbf{w}_v} e^{(l)} = \eta_1 \sum_{s \in \text{succ}(v)} \left(o_s^{(l)} - \text{out}_s^{(l)}\right) w_{sv} \frac{\partial \text{out}_v^{(l)}}{\partial \text{net}_v^{(l)}} \frac{\partial \text{net}_v^{(l)}}{\partial \mathbf{w}_v}.$$

Note again that the minus sign vanishes, since we have to move against the direction of the gradient and that the factor 2 is incorporated into the learning rate η_1. Note also the index 1 of the learning rate, which indicates that it differs from the learning rate η_3 for the connection weights from the hidden to the output layer. For batch training, these weight changes have to be summed over all training patterns and are applied to the weights only after all training patterns have been processed.

Unfortunately, it is not possible to generally compute the derivative of the output w.r.t. the network input or the derivative of the network input w.r.t. the weights, which are still contained in the weight adaptation rule, since radial basis function networks can employ different distance and different radial functions. We consider here the special case of a Euclidean distance and a Gaussian activation function, which are the most common choices. In this case (Euclidean distance), we have

$$d\big(\mathbf{w}_v, \mathbf{in}_v^{(l)}\big) = \sqrt{\sum_{i=1}^{n} \big(w_{vp_i} - \mathrm{out}_{p_i}^{(l)}\big)^2}.$$

Therefore, the second factor becomes

$$\frac{\partial\, \mathrm{net}_v^{(l)}}{\partial \mathbf{w}_v} = \left(\sum_{i=1}^{n} \big(w_{vp_i} - \mathrm{out}_{p_i}^{(l)}\big)^2\right)^{-\frac{1}{2}} \big(\mathbf{w}_v - \mathbf{in}_v^{(l)}\big).$$

For the first factor (Gaussian function), we obtain (under the simplifying assumption that the output function is the identity)

$$\frac{\partial\, \mathrm{out}_v^{(l)}}{\partial\, \mathrm{net}_v^{(l)}} = \frac{\partial f_{\mathrm{act}}(\mathrm{net}_v^{(l)}, \sigma_v)}{\partial\, \mathrm{net}_v^{(l)}} = \frac{\partial}{\partial\, \mathrm{net}_v^{(l)}} e^{-\frac{(\mathrm{net}_v^{(l)})^2}{2\sigma_v^2}} = -\frac{\mathrm{net}_v^{(l)}}{\sigma_v^2} e^{-\frac{(\mathrm{net}_v^{(l)})^2}{2\sigma_v^2}}.$$

Finally, we have to compute the gradient for the radius parameter σ_v of the hidden neurons. In principle, computing this gradient follows the same paths as computing the gradient for the weights. Indeed, it is even somewhat simpler, since we need not take the network input function into account. Hence, we only state the result here:

$$\frac{\partial e^{(l)}}{\partial \sigma_v} = -2 \sum_{s \in \mathrm{succ}(v)} \big(o_s^{(l)} - \mathrm{out}_s^{(l)}\big) w_{su} \frac{\partial\, \mathrm{out}_v^{(l)}}{\partial \sigma_v}.$$

As the online weight adaptation, we thus obtain

$$\Delta\sigma_v^{(l)} = -\frac{\eta_2}{2} \frac{\partial e^{(l)}}{\partial \sigma_v} = \eta_2 \sum_{s \in \mathrm{succ}(v)} \big(o_s^{(l)} - \mathrm{out}_s^{(l)}\big) w_{sv} \frac{\partial\, \mathrm{out}_v^{(l)}}{\partial \sigma_v}.$$

As usual, the negative sign disappears because we have to move against the gradient direction and the factor 2 is incorporated into the learning rate. Naturally, for batch training the radius changes have to summed over all training patterns and are applied to the radius σ_v only after all training patterns have been processed.

The derivative of the output of the neuron v w.r.t. the radius σ_v cannot be determined in a general form, since the neurons of the hidden layer may employ different radial functions. Therefore we consider again, as an example, the special case of a Gaussian activation function (and, for reasons of simplicity, the identity as the output function). Then it is

$$\frac{\partial \operatorname{out}_v^{(l)}}{\partial \sigma_v} = \frac{\partial}{\partial \sigma_v} e^{-\frac{(\operatorname{net}_v^{(l)})^2}{2\sigma_v^2}} = \frac{(\operatorname{net}_v^{(l)})^2}{\sigma_v^3} e^{-\frac{(\operatorname{net}_v^{(l)})^2}{2\sigma_v^2}}.$$

Note in the computations carried out above that we do not have a single learning rate for all neurons as in a multi-layer perceptron. Rather we have a total of three: one learning rate for the weights of the connections to the hidden neurons (η_1), one for the radii σ of the radial basis functions (η_2), and one for the weights of the connections to the output neurons and the bias values of the output neurons (η_3). According to recommendations in Zell (1996), these learning rates should be chosen considerably smaller than the (one) learning rate for training a multi-layer perceptron. Especially the third learning rate η_3 should be small, because the weights of the connections to the output neurons and the bias values of the output neurons have a strong influence on the function that is computed by a radial basis function network. In addition, using online training is often discouraged, since it is a lot less stable than for multi-layer perceptrons. Batch training should be preferred.

6.5 Generalized Form

Up to now we always used distance functions that are either isotropic (direction independent), like the Euclidean distance, or with which deviations from isotropy are fixed by the coordinate axes, like the city block distance or the maximum distance (see Fig. 6.1 on page 84). However, if the training examples form point clouds with an "oblique" orientation in the input space, they cannot be captured well with such a distance function. In this case, one either needs a larger number of radial basis functions, which are stringed along the point cloud and increase the complexity of the network, or one has to accept that larger regions outside of the point clouds are covered as well, which can harm the performance on new data.

In such a case, one desires distance function that can describe ellipses (or generally hyper-ellipsoids) with arbitrary orientation. Such a distance function is the **Mahalanobis distance**, which is defined as

$$d(\mathbf{x}, \mathbf{y}) = \sqrt{(\mathbf{x} - \mathbf{y})^\top \Sigma^{-1} (\mathbf{x} - \mathbf{y})}.$$

Here Σ is a matrix which is called *covariance matrix* (this name is due to certain connections to statistics, which, however, we do not explain in detail here) and which describes the (direction dependence) of the distance. Note that the Mahalanobis distance is identical to the Euclidean distance if Σ is the unit matrix.

To illustrate the possibilities that result from using the Mahalanobis distance, we reconsider representing the biimplication with a radial basis function network: the Mahalanobis distance enables us to solve this task with only one neuron in the hidden layer. The corresponding network and the covariance matrix that is now necessary as an additional parameter of the network input function is shown in Fig. 6.15

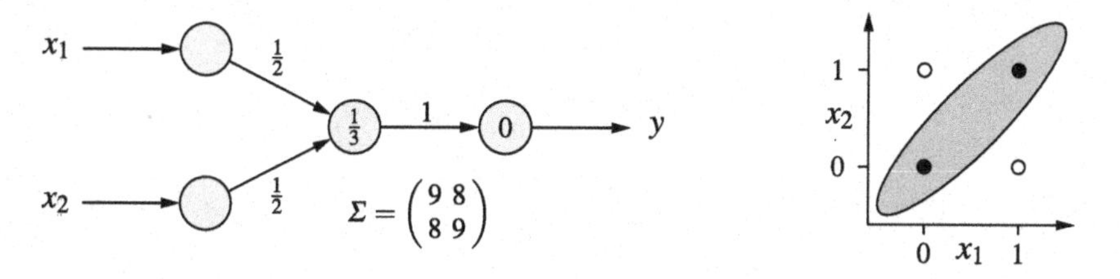

Fig. 6.15 A radial basis function network for the biimplication with Mahalanobis distance and rectangular activation function

on the left. As the activation function, we assume (like in the examples shown in Figs. 6.3 and 6.4 on page 86) a rectangular function. The computations of this network are illustrated in Fig. 6.15 on the right. Inside the gray ellipse the output is 1, while outside it is 0. This computes exactly the biimplication, as desired.

For radial basis function networks that employ the Mahalanobis distance, gradients can be derived for the shape parameter (i.e., the elements of the covariance matrix) as well. The computations follow essentially the same paths as those executed in Sect. 6.4. An explicit treatment is beyond the scope of this book.

References

A. Albert. *Regression and the Moore-Penrose Pseudoinverse*. Academic Press, New York, NY, USA, 1972

J.A. Hartigan and M.A. Wong. A k-means Clustering Algorithm. *Applied Statistics* 28:100–108. Blackwell, Oxford, United Kingdom, 1979

A. Zell. *Simulation Neuronaler Netze*. Addison-Wesley, Stuttgart, Germany, 1996

Chapter 7
Self-organizing Maps

Self-organizing maps are closely related to radial basis function networks. They can be seen as radial basis function networks without an output layer, or, rather, the hidden layer of a radial basis function network is already the output layer of a self-organizing map. This output layer also has an internal structure, since the neurons are arranged in a grid. The neighborhood relationships resulting from this grid are exploited in the training process in order to determine a topology preserving map.

7.1 Definition and Examples

Definition 7.1 A **self-organizing map** (SOM) or **Kohonen feature map** is a neural network with a graph $G = (U, C)$ that satisfies the following conditions:

1. $U_{\text{hidden}} = \emptyset$, $U_{\text{in}} \cap U_{\text{out}} = \emptyset$,
2. $C = U_{\text{in}} \times U_{\text{out}}$.

The network input function of each output neuron is a **distance function** of the input and the weight vector (cf. Definition 6.1 on page 83). The activation function of each output neuron is a **radial function** (cf. also Definition 6.1 on page 83), that is, a monotone non-increasing function

$$f : \mathbb{R}_0^+ \to [0, 1] \quad \text{with } f(0) = 1 \text{ and } \lim_{x\to\infty} f(x) = 0.$$

The output function of each output neuron is the identity. The output may be discretized according to the "winner takes all" principle: the neuron with the highest activation produces the output 1, all other neurons produce the output 0.

On the neurons of the output layer a **neighborhood relationship** is defined, which is described by a distance function

$$d_{\text{neurons}} : U_{\text{out}} \times U_{\text{out}} \to \mathbb{R}_0^+.$$

This function assigns a non-negative real number to each pair of output neurons.

R. Kruse et al., *Computational Intelligence*, Texts in Computer Science,
DOI 10.1007/978-1-4471-5013-8_7, © Springer-Verlag London 2013

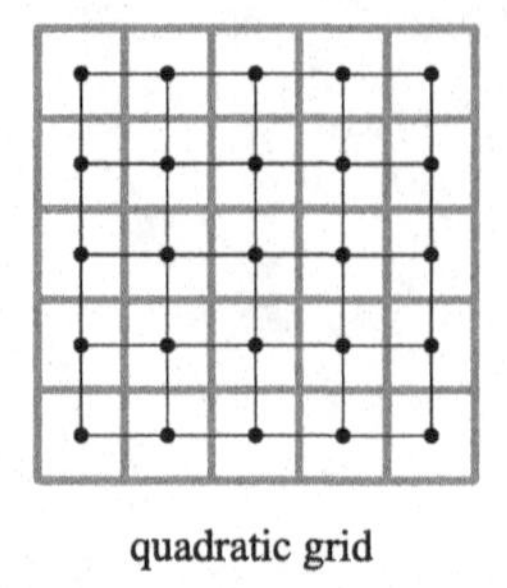

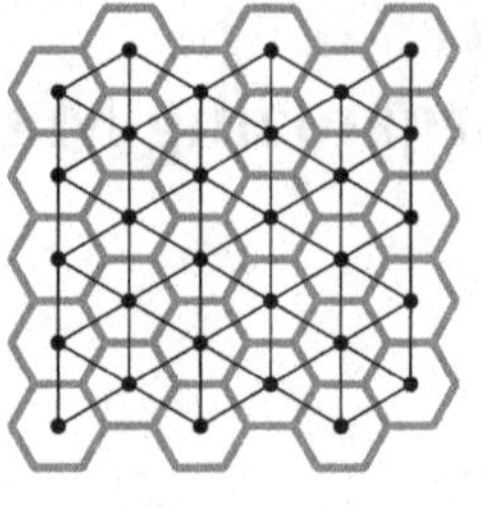

Fig. 7.1 Examples for arrangements of the output neurons of a self-organizing map. Each *dot* corresponds to an output neuron. The *lines* are meant to make the neighborhood structure more clearly visible

As a consequence, a self-organizing map is a two-layered neural network without hidden neurons. Its structure corresponds essentially to the input and the hidden layer of the radial basis function networks as they were discussed in the preceding chapter. Their alternative name *Kohonen feature map* refers to their inventor (Kohonen 1982, 1995).

In analogy to radial basis function networks, the weights of the connections from the input to the output neurons state the coordinates of a **center**, from which the distance of an input pattern is measured. In connection to self-organizing maps this center is often called a **reference vector**. The closer an input pattern is to a reference vector, the higher the activation of the corresponding neuron. Usually all output neurons have the same network input function (distance function) and the same activation function (radial function) with the same **(reference) radius** σ.

The neighborhood relationship of the output neurons is usually defined by arranging these neurons into a (usually two-dimensional) grid. Examples of such grids are shown in Fig. 7.1. Each dot stands for one output neuron. The lines connecting the dots are meant to make the neighborhood structure more easily visible by indicating the closest neighbors. The gray lines indicate a possible visualization of a self-organizing map, which we study in more detail below.

However, the neighborhood relationship may also be missing, which is formally represented by an extreme distance measure for the neurons: each neuron has the distance 0 to itself and an infinite distance to all other neurons. By choosing this distance measure, the neurons become effectively independent.

If a neighborhood relationship is missing and the output is discretized (that is, the output neuron with the highest activation produces output 1, all other neurons produce output 0) a self-organizing map describes a so-called **vector quantization** of the input space: the input space is divided into as many regions as there are output neurons. This is achieved by assigning to an output neuron all points of the input space for which the neuron yields the highest activation. If the distance and activation functions are identical for all output neurons, we may also say: to an output neuron all points of the input space are assigned that are closer to the neuron's reference vector than to the reference vector of any other output neuron. This

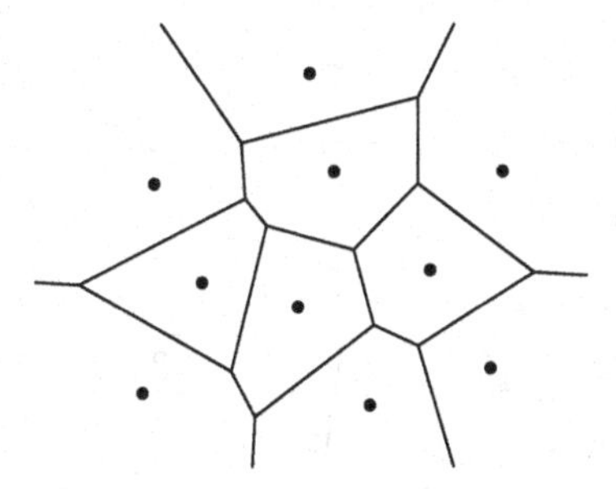

Fig. 7.2 Voronoi diagram of a vector quantization of a two-dimensional region with ten reference vectors. It depicts how the input space is divided by the reference vectors

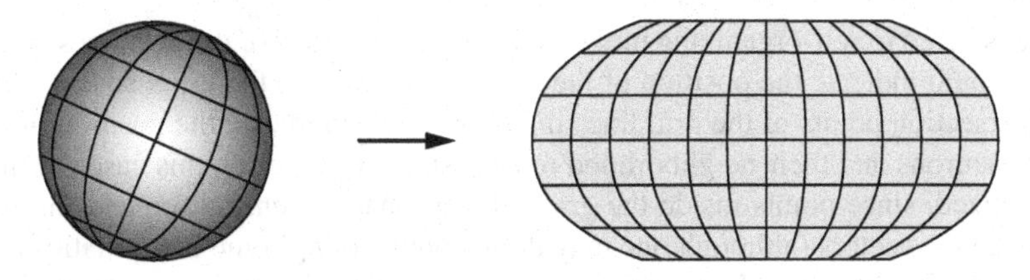

Fig. 7.3 Example of a topology preserving map: Robinson projection of the surface of a sphere onto a plane, as it is commonly used for world maps

"tessellation" into regions can be represented by a so-called **Voronoi diagram** (Aurenhammer 1991), as it is shown in Fig. 7.2 for two-dimensional inputs. The dots indicate the position of the reference vectors, the lines the division into regions.

The neighborhood relationship of the output neurons constrains the vector quantization. The objective is that reference vectors that are close to each other in the input space belong to output neurons that have a small distance from each other. That is, the neighborhood relationship of the output neurons is meant to reflect the relative position of the corresponding reference vectors in the input space (at least approximately). If this is the case, the self-organizing map describes a (quantized) **topology preserving map**, that is, a map that (approximately) preserves the position relationships between points (Greek $\tau o \pi o \varsigma$: position, location).

A well-known example of a topology preserving map is shown in Fig. 7.3, namely the so-called **Robinson projection** of the surface of a sphere onto a plane, as it is commonly used for world maps. Each point on the surface of a sphere (shown on the left) is mapped to a point of an approximately oval shape (shown on the right). With this map the position relationships between points are approximately preserved, even though the ratio of the distances between two points in the projection to the distance of its originals on the sphere is the larger, the farther these two points are from the equator of the sphere. Therefore this projection, if it is used for a world map, does not always accurately reflect the actual distances between points on the surface of the earth. Nevertheless it conveys a reasonably good impression of the relative location of cities, countries and continents.

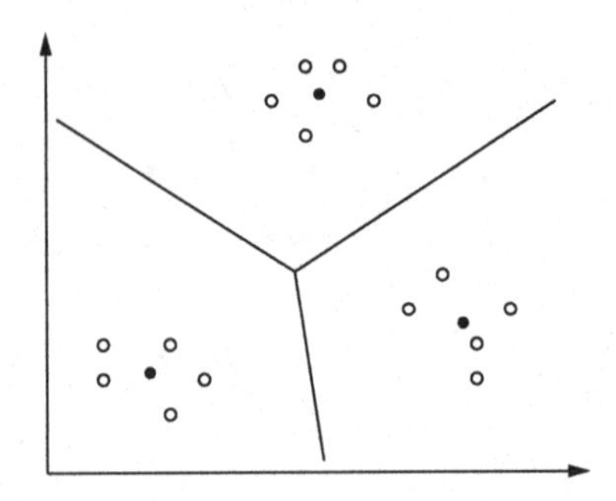

Fig. 7.4 Clustering of data by learning vector quantization: each group of data points (∘) is assigned to a reference vector (•)

Transferred to self-organizing maps the intersection points of the grid lines on the sphere could indicate the position of the reference vectors in the input space, while the intersection points of the grid lines in the projection indicate the position of the output neurons and their neighborhood relationship. However, in this case the map is quantized, since points inside the grid cells are mapped only discretely through the reference vectors (although one may think about adding some interpolation).

The advantage of topology preserving maps is that they allow us to map high-dimensional structures onto low-dimensional spaces. In particular, a map to a space with only two or three dimensions is interesting, since then we can display the image of the high-dimensional structure graphically. For this, one exploits the cell structure that is indicated in Fig. 7.1 by the gray lines. Obviously, this cell structure corresponds to a Voronoi diagram in the space of the output neurons. To each of these (two-dimensional) neuron Voronoi cells a (generally higher-dimensional) cell of the input space is mapped by the reference vector belonging to the corresponding output neuron. Hence, we can visualize the relative position of points in the input space by finding the reference vector Voronoi cells in which they lie and then, for instance, coloring the corresponding neuron Voronoi cells. An even better impression is obtained if one chooses a different color for each represented point and not only colors the single neuron Voronoi cell, in whose corresponding reference vector Voronoi cell the point lies, but colors all neuron Voronoi cells in such a way that the color saturation represents the activation of the corresponding neuron. An example of such a visualization can be found in Sect. 7.3.

7.2 Learning Vector Quantization

In order to explain the training of self-organizing maps, we first neglect the neighborhood relationship of the output neurons and thus confine ourselves to so-called **learning vector quantization** (Kohonen 1986). The objective of this method is to organize the data into clusters, similar to what we discussed for initializing radial basis function networks in Sect. 6.3 on page 97: with the help of c-means clustering we tried to find good starting points for the centers of the radial basis functions. We also mentioned that learning vector quantization is an alternative.

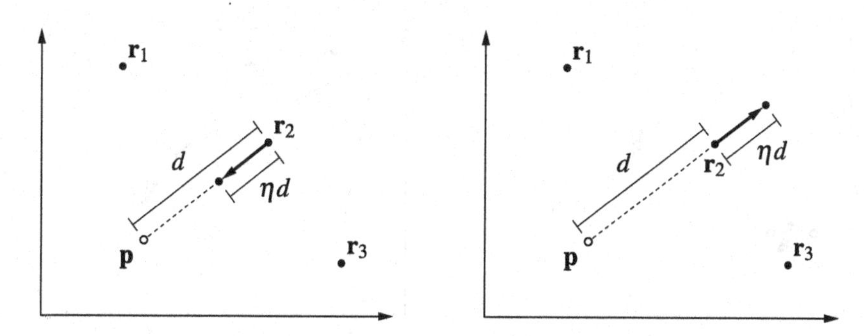

Fig. 7.5 Adaptation of reference vectors (•) with a training pattern (∘), $\eta = 0.4$. *Left*: attraction rule, *right*: repulsion rule

In both c-means clustering and learning vector quantization, the individual clusters are represented by a center (or a reference vector, but that is merely a different name for center). This center is to be positioned in such a way that it lies roughly in the middle of the data point cloud that constitutes the cluster. An example is shown in Fig. 7.4: each group of data points (depicted as ∘) is assigned to a reference vector (depicted as •). Thus the input space is divided—as indicated by the lines—in such a way that each point cloud lies in its own Voronoi cell.

The two methods mainly differ in how the cluster centers or the reference vectors are adapted. While in c-means clustering the two steps of assigning the data points to the clusters and recomputing the cluster centers as the center of gravity of the assigned data points are alternatingly executed, learning vector quantization processes the data points one by one and adapts only one reference vector per data point. The procedure is known as **competitive learning**: the training patterns (data points) are traversed one by one. For each training pattern a "competition" is carried out, which is won by the output neuron that yields the highest activation for this training pattern (if distance and activation functions are the same for all output neurons, we may say equivalently: whose reference vector is closest to the training vector). Only this "winner neuron" is adapted, namely in such a way that its reference vector is moved closer to the training pattern. Hence, the rule for adapting the reference vectors is

$$\mathbf{r}^{(\text{new})} = \mathbf{r}^{(\text{old})} + \eta\big(\mathbf{p} - \mathbf{r}^{(\text{old})}\big),$$

where $\mathbf{p}$ is the training pattern, $\mathbf{r}$ is the reference vector of the winner neuron for $\mathbf{p}$ and η is a learning rate with $0 < \eta < 1$. This rule is illustrated in Fig. 7.5 on the left: the learning rate η determines by what fraction of the distance $d = |\mathbf{p} - \mathbf{r}|$ between reference vector and training pattern the reference vector is shifted.

As for threshold logic units, multi-layer perceptrons and radial basis function networks we distinguish again between **online training** and **batch training**. In the former, the reference vector is adapted immediately and thus in the next step, that is, when processing the next training pattern, the reference vector already has its new position. In the latter (batch training), the changes are aggregated and the reference

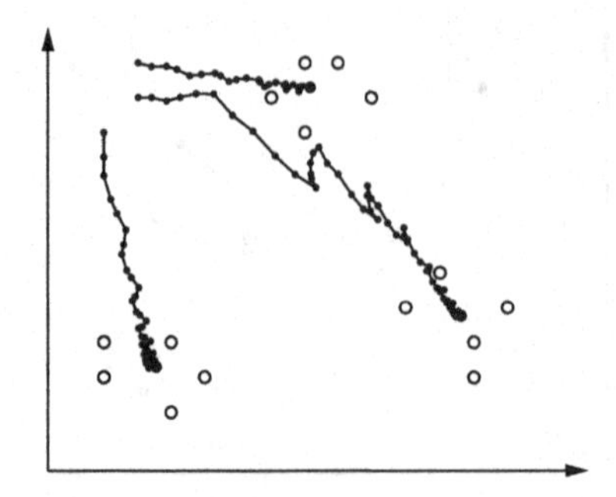
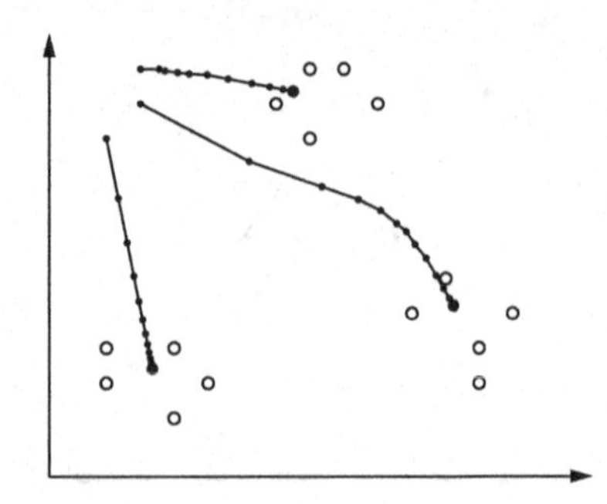

Fig. 7.6 Learning vector quantization for the data points shown in Fig. 7.4 with three reference vectors that start in the upper left corner. *Left*: online training with learning rate $\eta = 0.1$, *right*: batch training with learning rate $\eta = 0.05$

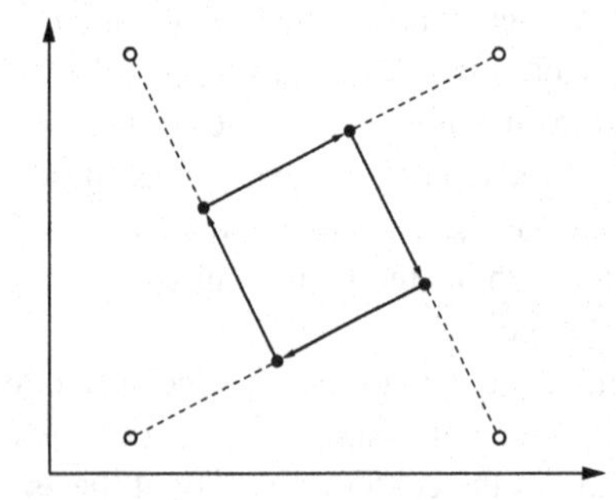
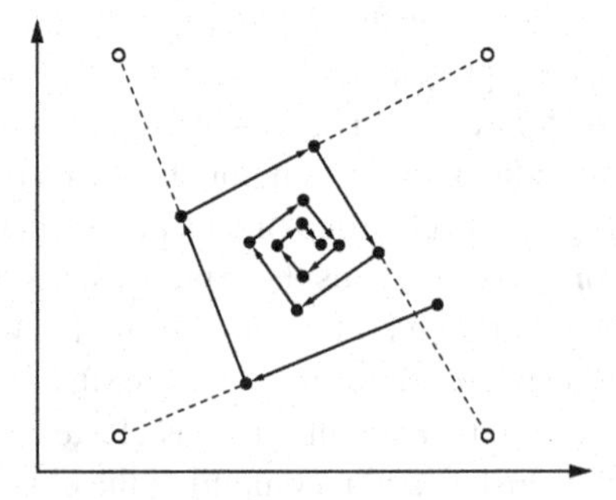

Fig. 7.7 Adaptation of a reference vector with four training patterns. *Left*: constant learning rate $\eta(t) = 0.5$, *right*: decreasing learning rate $\eta(t) = 0.6 \cdot 0.85^t$. In the first step it is $t = 0$

vectors are adapted only at the end of an epoch, that is, after all training patterns have been processed. Note that the batch mode of learning vector quantization is very similar to c-means clustering: the assignment of the data points is clearly identical, since in the batch procedure the position of the reference vectors is constant during an epoch. However, due to the learning the new position of the reference vector is not necessarily the center of gravity of the assigned data points, but generally a point between the old position and this center of gravity.

As an illustration, Fig. 7.6 shows the process of learning vector quantization for the data points shown in Fig. 7.4 with online training on the left and batch training on the right. Since only few epochs have been computed, the reference vectors have not yet reached their final positions, which are shown in Fig. 7.4. However, it is already clear that the desired clustering will finally be reached.

Up to now we used a constant learning rate η and merely required $0 < \eta < 1$. However, especially if online training is applied, a fixed learning rate can lead to problems, as shown in Fig. 7.7 on the left. In this diagram, a reference vector is repeatedly adapted with four data points, which causes the reference vector to move on a cyclic trajectory. The center of the four data points, which would be the desired result, is never reached. In order to solve this problem, a **time-dependent learning rate** is introduced, for instance,

$$\eta(t) = \eta_0 \alpha^t, \quad 0 < \alpha < 1, \quad \text{or} \quad \eta(t) = \eta_0 t^\kappa, \quad \kappa < 0.$$

This continuously decreasing learning rate turns the cyclic movement into a spiral that leads into the center (as desired), see Fig. 7.7 on the right. As a consequence, the desired destination of the reference vector is finally reached.

Although a time-dependent learning rate guarantees that the procedure converges, one should keep in mind that the learning rate must also not decrease too quickly, because otherwise the procedure may end in what is often called "starvation." That is, the adaptation steps become very small very quickly, so that the reference vectors never reach their natural destinations or may not even get close to them. On the other hand, the learning rate should not decrease too quickly, because then the learning process may converge only very slowly. As we already saw with other network types, choosing a proper learning rate is a difficult problem.

Although their main purpose is to find groups of data points, learning vector quantization can be applied not only for clustering, that is, to solve a free learning task. It can be extended in such a way that it takes classes into account that are assigned to the data points. In this way, fixed learning tasks may be solved, at least such learning tasks in which the desired output comes from a finite set of values (classes). To this end, the output neurons—and thus the reference vectors—are endowed with class labels and the adaptation rule is split into two types. If the class of the data point and the class of (the reference vector of) the winner neurons coincide, the **attraction rule** is applied, which is identical to the rule considered above:

$$\mathbf{r}^{(\text{new})} = \mathbf{r}^{(\text{old})} + \eta\big(\mathbf{p} - \mathbf{r}^{(\text{old})}\big).$$

That is, the reference vector is moved towards the training pattern (in other words: it is "attracted" by the training pattern). However, if the classes of the data point and the reference vector differ, the **repulsion rule** is applied:

$$\mathbf{r}^{(\text{new})} = \mathbf{r}^{(\text{old})} - \eta\big(\mathbf{p} - \mathbf{r}^{(\text{old})}\big).$$

That is, the reference vector is moved away from the training pattern (in other words: it is "repelled" by the training pattern), see Fig. 7.5 on the right. In this way the reference vectors move towards groups of data points that carry the same class label as they do themselves. A trained vector quantization yields for a new input vector to classify the class that is assigned to the output neuron with the highest activation (*nearest prototype classifier* if all distance and radial functions are identical).

Improved versions of learning vector quantization for fixed learning tasks do not only adapt the one reference vector that is closest to the current data point, but the two closest reference vectors (Kohonen 1990, 1995). Let $\mathbf{r}_j$ and $\mathbf{r}_k$ be these two closest reference vectors. They are adapted if the classes c_j and c_k assigned to them differ, but one of them coincides with the class z of the data point $\mathbf{p}$. Without loss of generality we assume that $c_k = z$. Then the adaptation rules read

$$\mathbf{r}_j^{(\text{new})} = \mathbf{r}_j^{(\text{old})} + \eta\big(\mathbf{p} - \mathbf{r}_j^{(\text{old})}\big) \quad \text{and}$$
$$\mathbf{r}_k^{(\text{new})} = \mathbf{r}_k^{(\text{old})} - \eta\big(\mathbf{p} - \mathbf{r}_k^{(\text{old})}\big).$$

All other reference vectors remain unchanged. However, if the classes c_j and c_k of the two closest reference vectors coincide (independent of whether they also coincide with the class z of the data point or not), no reference vector is adapted. These rules often yield good nearest prototype classifiers (Kohonen 1990).

Unfortunately, though, it was observed in practical tests that this version of learning vector quantization tends, in certain situations, to drive the reference vectors further and further apart instead of leading to a stable convergence. In order to counteract this clearly undesirable behavior (Kohonen 1990) introduced a so-called **window rule** into the adaptation: the reference vectors are adapted only if the data point $\mathbf{p}$ lies close to the classification border, that is, close to the (hyper-)surface, which separates regions in which different classes are predicted. The somewhat vague notion "close" is made formally precise by requiring

$$\min\left(\frac{d(\mathbf{p},\mathbf{r}_j)}{d(\mathbf{p},\mathbf{r}_k)}, \frac{d(\mathbf{p},\mathbf{r}_j)}{d(\mathbf{p},\mathbf{r}_k)}\right) > \theta, \quad \text{with } \theta = \frac{1-\xi}{1+\xi}.$$

Here ξ is a parameter that has to be chosen by a user. Intuitively, ξ describes the "width" of the window around the classification border in which a data point $\mathbf{p}$ has to lie in order to cause an adaptation of reference vectors. This rule prevents a divergence of the reference vectors, because the adaptations caused by a data point stop as soon as the classification border is far enough away.

One has to concede, though, that this window rule is not particularly intuitive and thus that it is desirable if one could do without it. This is indeed possible if one derives the adaptation rule for the reference vectors with a gradient descent approach for a specific objective function (Seo and Obermayer 2003). This approach starts from the assumption that the probability distribution of the data points for each class can be described sufficiently well by a mixture of (multi-dimensional) normal distributions. That is, each class consists of multiple clusters, each of which is covered by one normal distribution. Furthermore, it is assumed that all of these normal distributions have fixed and equal standard deviations σ. Intuitively this means that all clusters have the same size and a (hyper-)spherical shape. Finally, it is assumed that all clusters are equally likely, that is, that all clusters comprise (roughly) the same number of data points. With these restrictions, only the distance of a data point from a reference vector decides how it is classified.

The adaptation procedure is derived from a maximization of an objective function that describes the probability that a data point is correctly classified. That is, one tries to maximize the so-called **likelihood ratio**. As we show below, the attraction rule considered above thus results from maximizing the posterior probability of the correct class (that is, the true class that is assigned to a data point), while the repulsion rule is a consequence of minimizing the posterior probability of a wrong class (Seo and Obermayer 2003). Formally, we execute a gradient descent on the maximum likelihood ratio. Starting from the assumptions made above, we obtain for the likelihood ratio (or rather its natural logarithm, which simplifies handling the normal distributions)

$$\ln L_{\text{ratio}} = \sum_{j=1}^{n} \ln \sum_{i \in I(z_j)} \exp\left(-\frac{(\mathbf{p}_j - \mathbf{r}_i)^\top (\mathbf{p}_j - \mathbf{r}_i)}{2\sigma^2}\right)$$

$$- \sum_{j=1}^{n} \ln \sum_{i \notin I(z_j)} \exp\left(-\frac{(\mathbf{p}_j - \mathbf{r}_i)^\top (\mathbf{p}_j - \mathbf{r}_i)}{2\sigma^2}\right),$$

where $I(z)$ contains the indices of the reference vectors that are labeled as belonging to class z. Note that the normalization factors, which appear in the well-known formula for a normal distribution, cancel, since all clusters have the same standard variation (or variance, respectively). In the same way the prior probabilities of the different clusters cancel, since we assumed that they are all the same.

From this objective function, we obtain almost immediately as the online adaptation rule for a gradient descent

$$\mathbf{r}_i^{(\text{new})} = \mathbf{r}_i^{(\text{old})} + \eta \cdot \nabla_{\mathbf{r}_i} \ln L_{\text{ratio}}\big|_{\mathbf{r}_i^{(\text{alt})}}$$

$$= \mathbf{r}_i^{(\text{old})} + \eta \cdot \begin{cases} u_{ij}^{\oplus(\text{old})} \cdot (\mathbf{p}_j - \mathbf{r}_i^{(\text{old})}) & \text{if } z_j = c_i, \\ -u_{ij}^{\ominus(\text{old})} \cdot (\mathbf{p}_j - \mathbf{r}_i^{(\text{old})}) & \text{if } z_j \neq c_i, \end{cases}$$

where c_i is again the class that is assigned to the ith reference vector and z_j the class of the data point $\mathbf{p}_j$. The "membership degrees" $u_{ij}^{\oplus}$ and $u_{ij}^{\ominus}$, with which a data point $\mathbf{p}_j$ belongs to the cluster of the reference vector $\mathbf{r}_i$, are given by

$$u_{ij}^{\oplus(\text{old})} = \frac{\exp(-\frac{1}{2\sigma^2}(\mathbf{p}_j - \mathbf{r}_i^{(\text{old})})^\top (\mathbf{p}_j - \mathbf{r}_i^{(\text{old})}))}{\sum_{k \in I(z_j)} \exp(-\frac{1}{2\sigma^2}(\mathbf{p}_j - \mathbf{r}_k^{(\text{old})})^\top (\mathbf{p}_j - \mathbf{r}_k^{(\text{old})}))} \quad \text{and}$$

$$u_{ij}^{\ominus(\text{old})} = \frac{\exp(-\frac{1}{2\sigma^2}(\mathbf{p}_j - \mathbf{r}_i^{(\text{old})})^\top (\mathbf{p}_j - \mathbf{r}_i^{(\text{old})}))}{\sum_{k \notin I(z_j)} \exp(-\frac{1}{2\sigma^2}(\mathbf{p}_j - \mathbf{r}_k^{(\text{old})})^\top (\mathbf{p}_j - \mathbf{r}_k^{(\text{old})}))}.$$

The split into the two cases $z_j = c_i$ (the class of the data point coincides with the class of the reference vector) and $z_j \neq c_i$ (the reference vector and the data point belong to different classes) results from the fact that each reference vector r_i appears in only one of the two sums: either its index i is in $I(z_j)$, and then only the first sum contributes, or its index is not in $I(z_j)$, and then only the second sum contributes. The denominators of the fractions result from the derivative of the natural logarithm.

Thus we obtained a scheme for a "soft" learning vector quantization, where "soft" means that all reference vectors are adapted, but with different strength: all reference vectors with the same class as the data point are "attracted", while all reference vectors with a different class are "repelled" (Seo and Obermayer 2003).

A "hard" learning vector quantization can easily be derived from this scheme by letting the standard deviations (or variances, respectively) that are assigned to the reference vectors go to zero. In the limit we obtain a hard assignment

$$u_{ij}^{\oplus} = \delta_{i,k^{\oplus}(j)}, \quad \text{where } k^{\oplus}(j) = \underset{l \in I(z_j)}{\operatorname{argmin}}\, d(\mathbf{p}_j, \mathbf{r}_l), \quad \text{and}$$

$$u_{ij}^{\ominus} = \delta_{i,k^{\ominus}(j)}, \quad \text{where } k^{\ominus}(j) = \underset{l \notin I(z_j)}{\operatorname{argmin}}\, d(\mathbf{p}_j, \mathbf{r}_l),$$

and $\delta_{i,k}$ is the *Kronecker symbol* ($\delta_{i,k} = 1$, if $i = k$, and $\delta_{i,k} = 0$ otherwise).

Note, however, that this scheme is *not* identical to the one discussed above, which was proposed in Kohonen (1990, 1995). While in Kohonen's scheme the two closest reference vectors are determined and adapted only if they are labeled with different classes, this scheme *always* adapts two reference vectors, namely the one that is closest among those that carry the same class label (this vector is attracted) and the closest among those that carry a different class label (this vector is repelled). Note that these two need not be the two closest ones among all reference vectors: although one of them must be the closest overall, the second may be much farther away than several other reference vectors.

An advantage of this approach is that it explains why sometimes diverging behavior can be observed. (Details are beyond the scope of this book—an interested reader should consult (Seo and Obermayer 2003).) Furthermore, it suggests a method how the divergence can be avoided without having to introduce a window rule. The idea consists in a minor modification of the objective function (Seo and Obermayer 2003) (cf. page 113):

$$\begin{aligned} \ln L_{\text{ratio}} &= \sum_{j=1}^{n} \ln \sum_{i \in I(z_j)} \exp\left(-\frac{(\mathbf{x}_j - \mathbf{r}_i)^{\top}(\mathbf{x}_j - \mathbf{r}_i)}{2\sigma^2}\right) \\ &\quad - \sum_{j=1}^{n} \ln \sum_{i} \exp\left(-\frac{(\mathbf{x}_j - \mathbf{r}_i)^{\top}(\mathbf{x}_j - \mathbf{r}_i)}{2\sigma^2}\right). \end{aligned}$$

Obviously the difference consists only in the fact that the second sum now runs over all reference vectors (and not only over those with a different class label than the data point). That is, we no longer compare to the likelihood of seeing a data point with a different class, but to the likelihood of seeing a data point at all. Again we obtain an adaptation rule for a "soft" learning vector quantization:

$$\begin{aligned} \mathbf{r}_i^{(\text{new})} &= \mathbf{r}_i^{(\text{old})} + \eta_{\mathbf{r}} \cdot \nabla_{\mathbf{r}_i} \ln L_{\text{ratio}} \\ &= \mathbf{r}_i^{(\text{old})} + \eta_{\mathbf{r}} \cdot \begin{cases} (u_{ij}^{\oplus(\text{old})} - u_{ij}^{(\text{old})}) \cdot (\mathbf{p}_j - \mathbf{r}_i^{(\text{old})}) & \text{if } z_j = c_i, \\ -u_{ij}^{(\text{old})} \cdot (\mathbf{p}_j - \mathbf{r}_i^{(\text{old})}) & \text{if } z_j \neq c_i. \end{cases} \end{aligned}$$

(Note that this time the reference vectors that carry the correct class label occur in both terms of the likelihood ratio, which explains the sum $u_{ij}^{\oplus} - u_{ij}$ in the first case.) The "membership degrees" $u_{ij}^{\oplus}$ and u_{ij} are given as

$$u_{ij}^{\oplus(\text{old})} = \frac{\exp(-\frac{1}{2\sigma^2}(\mathbf{x}_j - \mathbf{r}_i^{(\text{old})})^\top(\mathbf{x}_j - \mathbf{r}_i^{(\text{old})}))}{\sum_{k \in I(z_j)} \exp(-\frac{1}{2\sigma^2}(\mathbf{x}_j - \mathbf{r}_k^{(\text{old})})^\top(\mathbf{x}_j - \mathbf{r}_k^{(\text{old})}))} \quad \text{and}$$

$$u_{ij}^{(\text{old})} = \frac{\exp(-\frac{1}{2\sigma^2}(\mathbf{x}_j - \mathbf{r}_i^{(\text{old})})^\top(\mathbf{x}_j - \mathbf{r}_i^{(\text{old})}))}{\sum_{k} \exp(-\frac{1}{2\sigma^2}(\mathbf{x}_j - \mathbf{r}_k^{(\text{old})})^\top(\mathbf{x}_j - \mathbf{r}_k^{(\text{old})}))}.$$

A "hard" variant can again be obtained by letting the standard variations σ go to zero. This leads to

$$u_{ij}^{\oplus} = \delta_{i,k^{\oplus}(j)}, \quad \text{where } k^{\oplus}(j) = \operatorname*{argmin}_{l \in I(z_j)} d(\mathbf{x}_j, \mathbf{r}_l), \quad \text{and}$$

$$u_{ij} = \delta_{i,k(j)}, \quad \text{where } k(j) = \operatorname*{argmin}_{l} d(\mathbf{x}_j, \mathbf{r}_l).$$

This adaptation rule is very similar to the one of Kohonen (1990, 1995). Intuitively, we can interpret it as follows: if the closest reference vector carries the same class label as the data point, no adaptation is carried out. However, if the class of the closest reference vector differs from the class of the data point, then this reference vector is repelled, while the closest reference vector among those with the same class label is attracted. In other words, the reference vectors are adapted only if a data point would be classified wrongly by the closest reference vector. Otherwise the current positions of the reference vectors are maintained.

For learning vector quantization with a time-dependent learning rate for classified and unclassified training patterns (though only with batch training and without any window rule or any of the improved adaptation rules), the web page

http://www.borgelt.net/lvqd.html

offers the programs `wlvq` (for Microsoft Windows™) and `xlvq` (for Linux). With them, clusters can be found for two-dimensional data (selectable from a larger number of dimensions) and the movement of the reference vectors can be followed.

7.3 Neighborhood of the Output Neurons

Up to now, we neglected the neighborhood relationship of the output neurons: the reference vectors could move independently of each other. As a consequence, learning vector quantization does not generally allow us to read anything from the (relative) position of the output neurons about the (relative) position of the reference vectors. In order to learn a topology preserving map in which the (relative) position of the output neurons reflects the (relative) position of the reference vectors, the neighborhood relationship has to be respected in the training process. Only in this case, one calls the network a **self-organizing map** (Kohonen 1982, 1995).

Self-organizing maps are trained—like vector quantization—with **competitive learning**. That is, the training patterns are visited one after the other and for each training pattern the neuron is determined that yields the highest activation. In self-organizing maps it is mandatory that all output neurons have the same distance function and the same activation function. Thus we can definitely say here equivalently: we find the output neuron, whose reference vector is closest to the training pattern. This neuron is the "winner" of the competition.

However, in contrast to learning vector quantization not only the reference vector of the winner neuron is adapted. At the end of training, the reference vectors of its neighbor neurons are supposed to be close to its reference vector. Hence, the neighboring reference vectors are adapted as well, though possibly less severely than that of the winner neuron. In this way, it is achieved that the reference vectors of neighboring neurons cannot move arbitrarily far apart, since they are adapted analogously. Thus, it can be expected that in the learning result neighboring output neurons possess reference vectors that are close to each other in the input space.

Another important difference to learning vector quantization is that self-organizing maps are almost exclusively used for free learning tasks. Since prior to training nothing is known about the relative position of different classes, it is difficult to assign meaningful classes to the reference vectors. One may assign classes to the output neurons *after* training is completed, though, namely by simply assigning the class that is most frequent among the training patterns for which the output neuron yields the highest activation. However, in this case the class information does not influence the training of the map (and thus the position of the reference vectors). Hence classifying data with the help of a self-organizing map that has been extended in this way is not necessarily recommended. Such a class assignment may, however, provide a good impression of the distribution and the relative position of different classes in the input space and thus may be a useful analysis tool.

Since only free learning tasks can reasonably be handled, there is only one adaptation rule for the reference vectors, which is analogous to the attraction rule discussed in the preceding section. This rule reads

$$\mathbf{r}_u^{(\text{new})} = \mathbf{r}_u^{(\text{old})} + \eta(t) \cdot f_{\text{nb}}\big(d_{\text{neurons}}(u, u_*), \rho(t)\big) \cdot \big(\mathbf{p} - \mathbf{r}_u^{(\text{old})}\big),$$

where $\mathbf{p}$ is the considered training pattern, $\mathbf{r}_u$ is the reference vector of neuron u, u_* is the winner neuron, $\eta(t)$ is a time-dependent learning rate and $\rho(t)$ is a time-dependent neighborhood radius. The variable d_{neurons} measures the distance of output neurons (cf. Definition 7.1 on page 105), here specifically the distance of the neuron to adapt from the winner neuron. (Since in a self-organizing map the neighbors of the winner neuron are adapted as well, we can no longer restrict the adaptation rule to the winner neuron.) How severely the reference vectors of other output neurons can be adapted depends, according to this rule, via a function f_{nb} (nb for neighbor) on the distance of the neuron from the winner neuron and the size of the radius $\rho(t)$ that determines the neighborhood.

The function f_{nb} is a radial function, that is, it is of the same type as the functions that we used to compute the activation of a neuron depending on the distance of a training patterns from a reference vector (cf. Fig. 6.2 on page 85). It assigns a number between 0 and 1 to each output neuron, which depends on its distance to the winner neuron[1] and describes the strength of the adaptation of its reference vector relative to the strength of the adaptation of the reference vector of the winner neuron. If the function f_{nb} is, for example, a rectangular function, then all output neurons in a certain radius around the winner neuron are adapted with full strength, while all other output neurons are left unchanged. Most commonly, however, a Gaussian neighborhood function is used, so that the strength of the adaptation of the reference vectors decays exponentially with the distance from the winner neuron.

A **time-dependent learning rate** is used for the same reasons as discussed for learning vector quantization, namely to avoid update cycles. Therefore, it can be defined in the same way, for example, as

$$\eta(t) = \eta_0 \alpha_\eta^t, \quad 0 < \alpha_\eta < 1, \quad \text{or} \quad \eta(t) = \eta_0 t^{\kappa_\eta}, \quad \kappa_\eta < 0.$$

Analogously a **time-dependent neighborhood radius** is defined, for instance, as

$$\varrho(t) = \varrho_0 \alpha_\varrho^t, \quad 0 < \alpha_\varrho < 1, \quad \text{or} \quad \varrho(t) = \varrho_0 t^{\kappa_\varrho}, \quad \kappa_\varrho < 0.$$

That the neighborhood radius decreases with time is reasonable, because then the self-organizing map can properly "unfold" in the first training steps (larger neighborhood), while in later training steps (smaller neighborhood) the position of the reference vectors is fitted more closely to the position of the training patterns.

As an example of a training process with the stated adaptation rule, we consider a self-organizing map with 100 output neurons, which are arranged in a square 10×10 grid. This map is trained with points that are chosen randomly (uniform distribution) from the square $[-1, 1] \times [-1, 1]$. The training procedure is depicted in Fig. 7.8. All diagrams show the input space, with the frame indicating the square $[-1, 1] \times [-1, 1]$. The grid of the output neurons is projected into this space by connecting, with a straight line, the reference vector of an output neuron to the reference vectors of its immediate neighbors. The top left diagram shows the situation directly after the reference vectors have been initialized with random weights from the interval $[-0.5, 0.5]$. Due to the randomness of the initialization, the (relative) position of the reference vectors is completely independent of the (relative) position of the output neurons, so that no grid structure can be discerned.

The following diagrams (first the top row from left to right, then the bottom row from left to right) show the state of the self-organizing map after 10, 20, 40, 80 and 160 training steps (in each training step one training pattern is processed; learning rate $\eta(t) = 0.6 \cdot t^{-0.1}$, Gaussian neighborhood function f_{nb}, neighborhood radius $\varrho(t) = 2.5 \cdot t^{-0.1}$). These diagrams show nicely how the self-organizing map slowly "unfolds" and thus adjusts itself to the input space. That the grid structure becomes visible demonstrates how the arrangement of the output neurons is transferred to the arrangement of the reference vectors in the input space.

[1] Note that the distance is computed from the grid, in which the output neurons are arranged, and *not* from the position of the reference vectors or the distance measure in the input space.

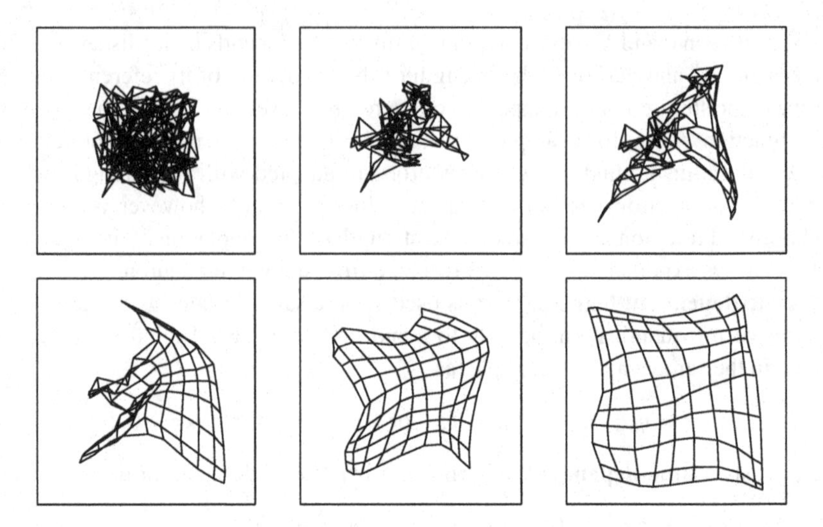

Fig. 7.8 Unfolding of a self-organizing map trained with random patterns from the square $[-1, 1] \times [-1, 1]$ (indicated by the frames). The *lines* connect the reference vectors of neighboring neurons

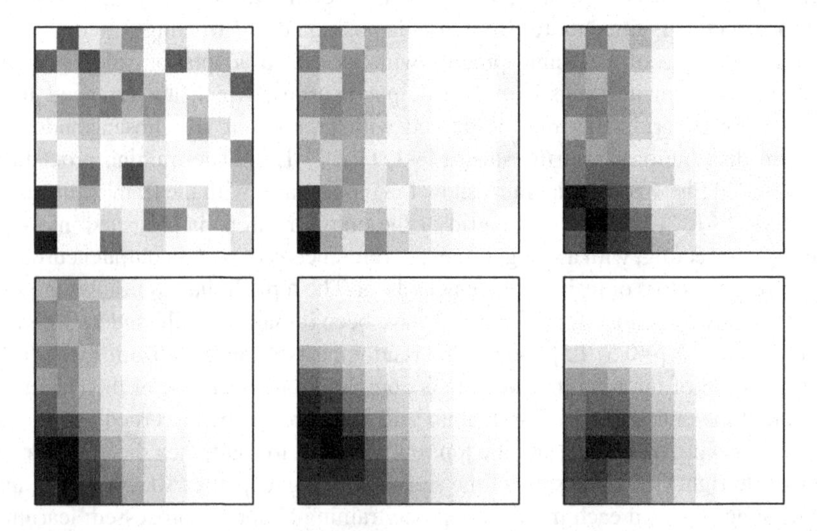

Fig. 7.9 Coloring of the training steps of the self-organizing map shown in Fig. 7.8 for the input pattern $(-0.5, -0.5)$ using a Gaussian activation function

For the same example, Fig. 7.9 shows a way to visualize a self-organizing map, which we mentioned in Sect. 7.1. All diagrams show the grid structure of the output neurons (*not* the input space as in Fig. 7.8). Each neuron is represented by a small square (cf. Fig. 7.1 on page 106). The shades of gray encode the activation of the output neurons if the pattern $(-0.5, -0.5)$ is fed into the network (Gaussian activa-

Fig. 7.10 If the initialization is unfavorable or the learning rate or the neighborhood is too small, the result can be a twisted map

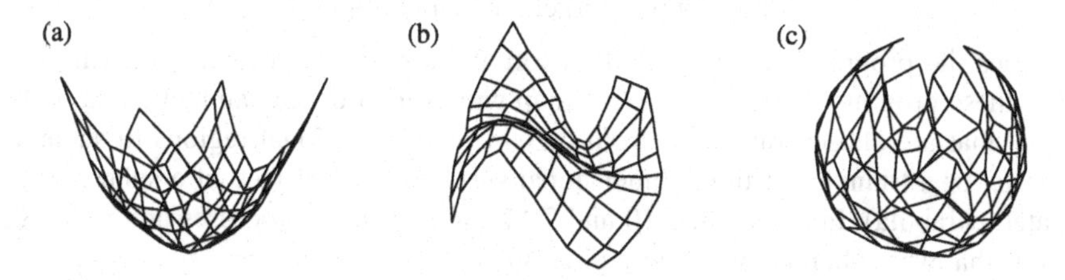

Fig. 7.11 Self-organizing maps that were trained with random points on (**a**) a rotated parabola, (**b**) a cubic function, (**c**) the surface of a sphere

tion function): the darker a square, the higher the activation. With this representation the training can also be followed well. After the initialization, the heavily activated neurons are still randomly distributed on the map. However, the further training proceeds, the better they group together. Note the activation structures after 20 training patterns (3. diagram) and after 40 training patterns (4. diagram) and compare them to the corresponding diagrams of the self-organizing map in the input space shown in Fig. 7.8: since in these phases the map is not well unfolded on the left side, many neurons on the left are strongly activated.

The example we just studied shows an almost ideal training process of a self-organizing map. After only few training steps the map is already unfolded and very well adapted to the training patterns. Further training stretches the map somewhat more, until it covers the region of the training patterns almost uniformly (even though the projection of the neuron grids, as can easily be worked out, can never reach the edges of the square $[-1, 1] \times [-1, 1]$).

However, training is not always so successful. If, by accident, the map is initialized in an unfavorable way, and particularly if the learning rate or the neighborhood radius are chosen too small or decrease too quickly, "twisted" maps can result. An example of the result of a training process that failed in this way is shown for the square example in Fig. 7.10. The map did not properly unfold. The corners of the unit square have been assigned "wrongly" to the corners of the grid, so that the middle of the map exhibits a kind of "knot." In general, such a twist cannot be amended, regardless of how long training is continued. However, most of the time it can be avoided by starting with a large learning rate and particularly a large neighborhood radius (in the order of the grid side length of the self-organizing map).

In order to illustrate the dimension reduction that can be achieved with a (quantized) topology preserving map, as it is implemented by a self-organizing map, Fig. 7.11 shows the projection of the neuron grid of a self-organizing map with 10×10 output neurons into a three-dimensional input space. The map shown on the left was trained with random points from a rotated parabola, the map in the middle with random points from a binary cubic function, and the map on the right with random points from the surface of a sphere. Since in these cases the input space is actually two-dimensional (all training patterns lie on, though curved, surfaces), a self-organizing map can adjust very well to the training patterns.

As a further illustration of the training of self-organizing maps the web page

http://www.borgelt.net/somd.html

provides the programs `wsom` (for Microsoft Windows™) and `xsom` (for Linux). With these programs, a self-organizing map with a square grid can be trained with points that are chosen randomly from certain two-dimensional regions (a square, a circle, or a triangle) or three-dimensional surfaces (e.g. surface of a sphere or a rotated parabola). Figures 7.8, 7.10 and 7.11 show training processes and training results that were obtained with these programs.

With these programs, it can also be studied what happens if the training patterns truly have a higher-dimensional structure, so that it is not possible to map them onto a two-dimensional grid with only little loss. As an example, one may train with one of these programs a self-organizing map with at least 30×30 output neurons for training patterns that are randomly selected from a cube (volume, not surface). The self-organizing map will be folded in several places in order to fill the space as uniformly as possible. Although self-organizing maps are still useful in such cases, one has to pay attention to the fact that due to the folds it may happen that an input pattern activates output neurons, which are fairly far away from each other in the grid structure of the map, simply because they lie on the two sides of a fold.

References

F. Aurenhammer. Voronoi Diagrams—A Survey of a Fundamental Geometric Data Structure. *ACM Computing Surveys* 23(3):345–405. ACM Press, New York, NY, USA, 1991

T. Kohonen. Self-organized Formation of Topologically Correct Feature Maps. *Biological Cybernetics*. Springer-Verlag, Heidelberg, Germany, 1982

T. Kohonen. *Learning Vector Quantization for Pattern Recognition*. Technical Report TKK-F-A601. Helsinki University of Technology, Finland, 1986

T. Kohonen. Improved Versions of Learning Vector Quantization. *Proc. Int. Joint Conference on Neural Networks (IJCNN 1990, San Diego, CA)*, 1:545–550. IEEE Press, Piscataway, NJ, USA, 1990

T. Kohonen. *Self-Organizing Maps*. Springer-Verlag, Heidelberg, Germany, 1995 (3rd ext. edition 2001)

S. Seo and K. Obermayer. Soft Learning Vector Quantization. *Neural Computation* 15(7):1589–1604. MIT Press, Cambridge, MA, USA, 2003

Chapter 8
Hopfield Networks

In the preceding Chaps. 5 to 7, we studied so-called *feed forward networks*, that is, networks with an acyclic graph (no directed cycles). In this and the next chapter, however, we turn to so-called *recurrent networks*, that is, networks, the graph of which may contain (directed) cycles. We start with one of the simplest forms, the so-called **Hopfield networks** (Hopfield 1982, 1984), which originated as physical models to describe magnetism. Hopfield networks are indeed closely related to the Ising model of magnetism Ising (1925) (see below).

8.1 Definition and Examples

Definition 8.1 A **Hopfield network** is a neural network with a graph $G = (U, C)$ that satisfies the following conditions:

1. $U_{\text{hidden}} = \emptyset$, $U_{\text{in}} = U_{\text{out}} = U$,
2. $C = U \times U - \{(u, u) \mid u \in U\}$.

The connection weights are symmetric, that is, we have $\forall u, v \in U, u \neq v\colon w_{uv} = w_{vu}$. The network input function of each neuron u is the weighted sum of the outputs of all other neurons, that is,

$$\forall u \in U: \quad f_{\text{net}}^{(u)}(\mathbf{w}_u, \mathbf{in}_u) = \mathbf{w}_u \mathbf{in}_u = \sum_{v \in U - \{u\}} w_{uv} \, \text{out}_v \, .$$

The activation function of each neuron u is a threshold function

$$\forall u \in U: \quad f_{\text{act}}^{(u)}(\text{net}_u, \theta_u) = \begin{cases} 1 & \text{if } \text{net}_u \geq \theta_u, \\ -1 & \text{otherwise.} \end{cases}$$

The output function of each neurons is the identity, that is,

$$\forall u \in U: \quad f_{\text{out}}^{(u)}(\text{act}_u) = \text{act}_u \, .$$

R. Kruse et al., *Computational Intelligence*, Texts in Computer Science,
DOI 10.1007/978-1-4471-5013-8_8, © Springer-Verlag London 2013

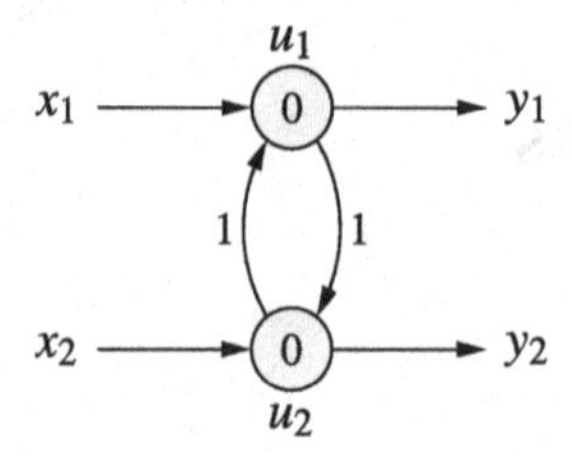

Fig. 8.1 A simple Hopfield network that can oscillate if the activations of the two neurons are updated in parallel, but reaches a stable state if the neurons are updated alternatingly

Note that there are no loops in a Hopfield network, that is, no neuron receives its own output as input. All feedback loops run through other neurons: a neuron u receives the outputs of all other neurons as its input and all other neurons receive the output of the neuron u as input.

The neurons of a Hopfield network work exactly like the threshold logic units that we studied in Chap. 3: depending on whether the weighted sum of the inputs exceeds a threshold θ_u or not, their activation is set to 1 or -1. Although in Chap. 3 the inputs and activations usually had the values 0 and 1, we mentioned in Sect. 3.6 the ADALINE variant that uses the values -1 and 1 instead. Section 10.3 shows how the two versions can be transformed into each other.

Sometimes the activation function of the neurons of a Hopfield network is defined, drawing on the old activation act_u, as follows:

$$\forall u \in U: \quad f_{\text{act}}^{(u)}(\text{net}_u, \theta_u, \text{act}_u) = \begin{cases} 1 & \text{if } \text{net}_u > \theta_u, \\ -1 & \text{if } \text{net}_u < \theta_u, \\ \text{act}_u & \text{if } \text{net}_u = \theta_u. \end{cases}$$

This is certainly advantageous for the physical interpretation of a Hopfield network (see below) and also simplifies a proof which we present in the next section. Nevertheless, we stick to the definition given above, because it has other advantages.

To carry out the computations in this section, it is beneficial to represent the connection weights in a weight matrix (cf. also Chaps. 4 and 5). In order to do so, we set the missing weights $w_{uu} = 0$ (feedback loops of a neuron to itself), as this is, due to the special network input function of Hopfield neurons, equivalent to a missing connection. Because of the symmetric weights the weight matrix is obviously symmetric (that is, it is equal to its transpose) and because of the missing feedback loops of neurons to themselves its diagonal is 0. That is, a Hopfield network with n neurons $u_1, \ldots, u_n$ can be described by the $n \times n$ matrix

$$\mathbf{W} = \begin{pmatrix} 0 & w_{u_1u_2} & \ldots & w_{u_1u_n} \\ w_{u_1u_2} & 0 & \ldots & w_{u_2u_n} \\ \vdots & \vdots & & \vdots \\ w_{u_1u_n} & w_{u_2u_n} & \ldots & 0 \end{pmatrix}.$$

As a first example for a Hopfield network, we consider the network with two neurons that is shown in Fig. 8.1. The weight matrix of this network is

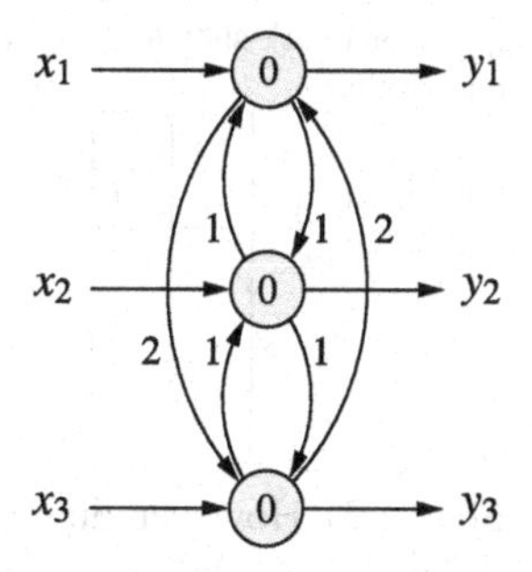

Fig. 8.2 A simple Hopfield network with three neurons u_1, u_2 and u_3 (*from top to bottom*)

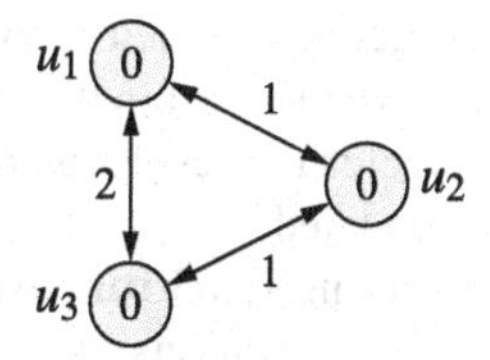

Fig. 8.3 Simplified representation of the Hopfield network shown in Fig. 8.2 that exploits the symmetry of the weights

$$\mathbf{W} = \begin{pmatrix} 0 & 1 \\ 1 & 0 \end{pmatrix}.$$

Both neurons have the threshold 0. Like we did for the threshold logic units studied in Chap. 3, we write this threshold into the circle that represents the corresponding neuron. A second example of a simple Hopfield network is shown in Fig. 8.2. The weight matrix of this network is

$$\mathbf{W} = \begin{pmatrix} 0 & 1 & 2 \\ 1 & 0 & 1 \\ 2 & 1 & 0 \end{pmatrix}.$$

Again all neurons have threshold 0. This example shows that the large number of connections can make a graphical representation difficult to read. In order to obtain a simpler representation, we exploit that every neuron is both input as well as output neuron. Therefore we need not draw input and output arrows explicitly, since these only serve the purpose to point out the input and output neurons. Furthermore, we know that the weights must be symmetric. Hence, we can combine the connections of two neurons into a double arrow, which is labeled only once with the corresponding weight. Thus we reach the representation shown in Fig. 8.3.

Let us now turn to the computations of a Hopfield network. We start with the Hopfield network with two neurons shown in Fig. 8.1. We assume that this network is fed with the values $x_1 = -1$ and $x_2 = 1$. This means that in the input phase the activation of the neuron u_1 is set to -1 (i.e., $\text{act}_{u_1} = -1$) and the activation of the neuron u_2 is set to 1 (i.e., $\text{act}_{u_2} = 1$). This is in no way different from how we always proceeded up to now (according to the general description of the operation

Table 8.1 Computations of the Hopfield network shown in Fig. 8.1 for the inputs $x_1 = -1$ and $x_2 = 1$ if the activations are updated in parallel

	u_1	u_2
input phase	**−1**	**1**
work phase	**1**	**−1**
	−1	**1**
	1	**−1**
	−1	**1**

of a neuron, as it was given in Sect. 4.2). However, due to the cycle in the graph that underlies this network, we face the question how the activations of the neurons are to be recomputed in the work phase. Up to now, there was no need to consider this question, because in a feed forward network the update order is irrelevant: regardless of the order in which the neurons recompute their activation and output, we always obtain the same result. However, as we know from the example discussed in Sect. 4.2 (see page 42), in a network with cycles the result can depend on the order in which the activations of the neurons are updated.

We try first to update the activations **synchronously** (at the same time, in parallel). That is, we compute the new activations and new outputs of both neurons from their old outputs. This leads to the computations shown in Table 8.1. Obviously, no stable activation state is reached, but the network oscillates between the states $(-1, 1)$ and $(1, -1)$. However, if we update the activations **asynchronously**, that is, if we update the activation and output only for one neuron at a time and already use the new output in subsequent updates (of other neurons), the network always reaches a stable state. As an illustration, Table 8.2 shows the two possible update sequences, in which both neurons are updated alternatingly. In both cases a stable state is reached. However, which state is reached depends on which neuron is updated first. Arguments of symmetry easily show that for any other inputs one of these two stable states is reached as well.

A similar observation can be made for the Hopfield network with three neurons that is shown in Fig. 8.2. If the input pattern $(-1, 1, 1)$ is fed into the network, a synchronous update of the neurons lets the network oscillate between the states $(-1, 1, 1)$ and $(1, 1, -1)$, while an asynchronous update leads either to the stable state $(1, 1, 1)$ or to the stable state $(-1, -1, -1)$.

For any other inputs, one of the same two stable states is finally reached as well, regardless of the order we choose to update the neurons. This is most easily seen with the help of a **state graph** as it is shown in Fig. 8.4. Each state (that is, each

Table 8.2 If the activations of the neurons of the Hopfield network shown in Fig. 8.1 are updated alternatingly, a stable state is reached. The result depends, though, on which neuron is updated first

	u_1	u_2
input phase	**−1**	**1**
work phase	**1**	1
	1	**1**
	1	1
	1	**1**

	u_1	u_2
input phase	**−1**	**1**
work phase	−1	**−1**
	−1	−1
	−1	**−1**
	−1	−1

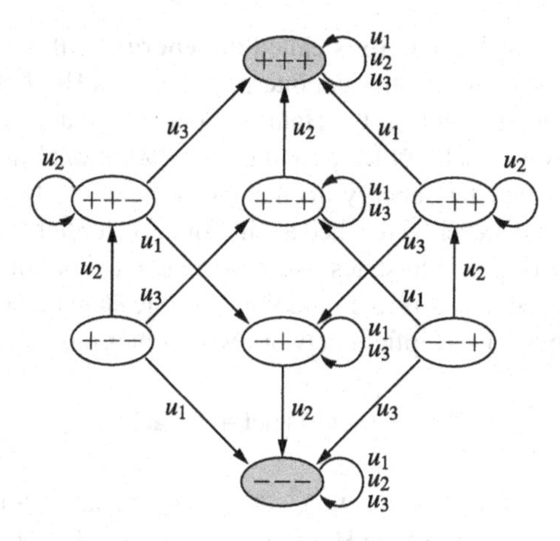

Fig. 8.4 State graph of the Hopfield network shown in Fig. 8.2. The *arrows* are labeled with the neurons, an update of which causes the corresponding state transition. The two stable states are drawn in *gray*

combination of activations of the neurons) is represented by an ellipse, into which the signs of the activations of the three neurons u_1, u_2 and u_3 are written (from left to right). The arrows between the states are labeled with the neuron, an update of which causes the corresponding state transition. Since for each state the caused transitions are stated for an update of each of the three neurons, we can easily read from this graph which state sequences are traversed for any order in which the activations of the neurons are updated. It is easy to check that finally one of the two stable states $(+1, +1, +1)$ or $(-1, -1, -1)$ will be reached.

8.2 Convergence of the Computations

As we saw from the examples of the preceding sections, the network may oscillate between different activation states if the activations of the different neurons are updated synchronously. However, if we update them asynchronously, a stable state is always reached in these examples. Indeed, one can show that an asynchronous update of the activations cannot lead to oscillations and must lead to a stable state.

Theorem 8.1 (Convergence Theorem for Hopfield Networks) *If the activations of the neurons of a Hopfield network are updated asynchronously, a stable state is reached in a finite number of steps. If the neurons are traversed in an arbitrary, but fixed cyclic fashion, at most $n \cdot 2^n$ steps (updates of individual neurons) are needed, where n is the number of neurons of the network.*

Proof This theorem is proven with a method that one may call (in analogy to Fermat's method of infinite descent) the *method of finite descent*. We define a function that maps every state of a Hopfield network to a real-valued number, which is reduced with every state transition or at least stays the same. This function is commonly called the **energy function** of the Hopfield network. The number this function maps an activation state to is called the **energy** of this state. (The reason for these names stems from the physical interpretation of a Hopfield network, since the energy function corresponds to the Hamilton operator that describes the energy of the magnetic field; see below.) By drawing on an additional insight in case of a transition that leaves the state energy unchanged, we can easily show that a state, once it has been left, can never be reached again. Since a Hopfield network possesses only a finite number of possible states, we must reach a situation in which further transitions cannot descend any further and thus a stable state must be reached.

The energy function of a Hopfield network with n neurons $u_1, \dots, u_n$ is

$$E = -\frac{1}{2}\mathbf{act}^\top \mathbf{W}\mathbf{act} + \boldsymbol{\theta}^T \mathbf{act},$$

where the vector $\mathbf{act} = (\mathrm{act}_{u_1}, \dots, \mathrm{act}_{u_n})^\top$ describes the activation state of the network, $\mathbf{W}$ is the weight matrix of the Hopfield network and $\boldsymbol{\theta} = (\theta_{u_1}, \dots, \theta_{u_n})^\top$ collects the thresholds of the neurons in a vector. If it is spelled out with individual connection weights, this energy function reads

$$E = -\frac{1}{2} \sum_{u,v \in U, u \neq v} w_{uv}\, \mathrm{act}_u\, \mathrm{act}_v + \sum_{u \in U} \theta_u\, \mathrm{act}_u\,.$$

In this form we can easily understand the reason for the factor $\frac{1}{2}$ in front of the first sum. Due to the symmetry of the weights, every term in the first sum occurs twice, which is compensated by the factor $\frac{1}{2}$.

We now show that the energy cannot increase in a state transition caused by updating a neuron. Since the neurons are updated asynchronously, a state transition means updating the activation of only one neuron u. We assume that due to this update it changes its activation from $\mathrm{act}_u^{(\mathrm{old})}$ to $\mathrm{act}_u^{(\mathrm{new})}$. The difference of the energy of the old and the new activation state consists of all terms that contain the activation act_u. All other terms cancel, because they are contained in both the old and the new energy. Hence, we have

$$\begin{aligned}\Delta E = E^{(\mathrm{new})} - E^{(\mathrm{old})} &= \Bigg(- \sum_{v \in U - \{u\}} w_{uv}\, \mathrm{act}_u^{(\mathrm{new})}\, \mathrm{act}_v + \theta_u\, \mathrm{act}_u^{(\mathrm{new})}\Bigg) \\ &\quad - \Bigg(- \sum_{v \in U - \{u\}} w_{uv}\, \mathrm{act}_u^{(\mathrm{old})}\, \mathrm{act}_v + \theta_u\, \mathrm{act}_u^{(\mathrm{old})}\Bigg).\end{aligned}$$

The factor $\frac{1}{2}$ vanishes because of the symmetry of the weights, due to which every term of the sum occurs twice. From the above sums, we can extract the new and the old activation of the neuron u and thus reach

$$\Delta E = \big(\mathrm{act}_u^{(\mathrm{old})} - \mathrm{act}_u^{(\mathrm{new})}\big)\underbrace{\Bigg(\sum_{v \in U - \{u\}} w_{uv}\, \mathrm{act}_v - \theta_u\Bigg)}_{= \mathrm{net}_u}.$$

We now have to distinguish two cases. If $\text{net}_u < \theta_u$, then the second factor is less than 0. In addition, it is $\text{act}_u^{(\text{new})} = -1$, and since we assumed that the activation changed due to the update, we know $\text{act}_u^{(\text{old})} = +1$. Therefore, the first factor is greater than 0 and hence $\Delta E < 0$. However, if $\text{net}_u \geq \theta_u$, then the second factor is no less than 0. In addition, we have $\text{act}_u^{(\text{new})} = +1$, which implies $\text{act}_u^{(\text{old})} = -1$. It follows that the first factor is less than 0 and hence $\Delta E \leq 0$.

If a state transition reduced the energy, the original state cannot be reached anymore, because this would require an (impossible) energy increase. However, the second case allows for transitions that keep the energy constant. Hence we have to exclude the possibility of cycles of states having the same energy. Fortunately, such a transition increases the number of $+1$ activations (because it can occur only if the activation of the updated neuron changed from -1 to $+1$, see above). Therefore the original state cannot be reached again in this situation either. As a consequence, every state transition reduces the number of reachable states, and since there are only finitely many states, we must finally reach a stable state.

It should be noted that the additional criterion (number of $+1$ activations) is not needed if the activation function is defined as stated on page 122 as an alternative, that is, if the old activation is maintained if the network input coincides with the threshold. In this case the activation only changes from -1 to $+1$ if $\text{net}_u > \theta_u$. As a consequence we obtain for the second case $\Delta E < 0$ as well and therefore it is sufficient to study the behavior of the energy of the Hopfield network.

We remark here also that it is only guaranteed that the computations converge to a state of (locally) minimal energy if no neurons are excluded from updates of their activation after a certain point in time. Otherwise we could update, for instance, the same neuron over and over again, which does not cause the current activation state to be left. That no neuron is excluded from an activation update is certain if the neurons are updated in an arbitrary, but fixed cyclic order. In this case we can derive the following corollary: either a traversal of all neurons did not change any activation. Then we obviously reached a stable state. Or at least one activation changed. Then (at least) one of the 2^n possible activation states (n neurons, each with two possible activations) has been excluded, since the old state cannot be reached again as we proved above. Therefore a stable state must be reached after at most 2^n traversals of the neurons, that is, after at most $n \cdot 2^n$ updates of neuron activations. □

The energy function we introduced in the proof of the above theorem plays an important role in the following. Therefore, but also to illustrate the above theorem, we consider it for the example of the simple Hopfield network shown in Fig. 8.2. The energy function of this network is

$$E = -\text{act}_{u_1}\,\text{act}_{u_2} - 2\,\text{act}_{u_1}\,\text{act}_{u_3} - \text{act}_{u_2}\,\text{act}_{u_3}\,.$$

If we arrange the activation states of this network (cf. Fig. 8.4) according to their energy, discarding the (self-)loops and the edge labels for reasons of clarity, we obtain Fig. 8.5, in which the two stable states are clearly visible as the two states with the lowest energy. Note that there are no transitions from a state lower in the diagram

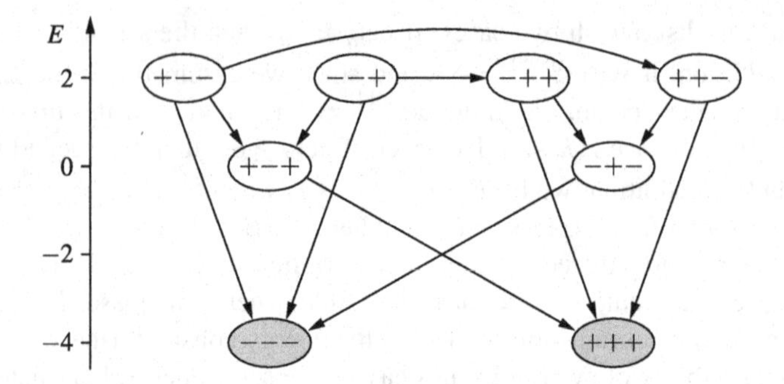

Fig. 8.5 (Simplified) state graph of the Hopfield network shown in Fig. 8.2, in which the states are arranged according to their energy. The two stable states are shown in *gray*

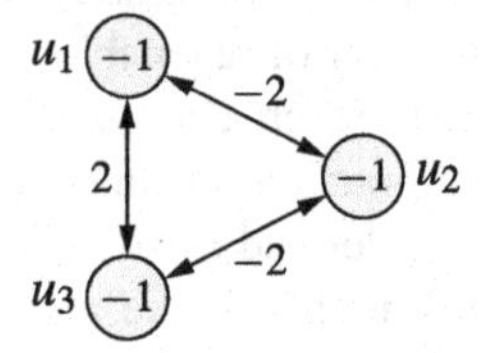

Fig. 8.6 A Hopfield network with three neurons and non-vanishing thresholds

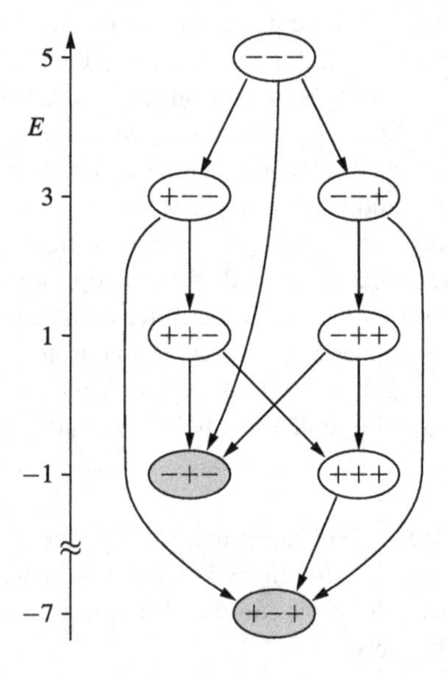

Fig. 8.7 (Simplified) state graph of the Hopfield network shown in Fig. 8.6, in which the states are arranged according to their energy. The two stable state are drawn in *gray*. Note that the energy scale is broken between -1 and -7 and thus the bottom state lies actually much further down

to a state higher up in the diagram, which would correspond to an increase of energy. Note also that all transitions between states having the same energy increase the number of $+1$ activations. This illustrates the arguments of the above proof.

Table 8.3 Physical interpretation of a Hopfield network as a (microscopic) model of magnetism (Ising model, Ising 1925)

physical	neural
atom	neuron
magnetic moment (spin)	activation state
strength of the external magnetic field	threshold
magnetic coupling of the atoms	connection weights
Hamilton operator of the magnetic field	energy function

It is, of course, not necessary that the state graph is as highly symmetric as this, even if the network exhibits a lot of strong symmetries. As an example, we consider the Hopfield network shown in Fig. 8.6. Even though this network has essentially the same symmetry structure as the one shown in Fig. 8.3, it possesses, due to the non-vanishing thresholds, a very different state graph. Here we only study the form in which the states are arranged according to the energy function

$$E = 2\,\mathrm{act}_{u_1}\,\mathrm{act}_{u_2} - 2\,\mathrm{act}_{u_1}\,\mathrm{act}_{u_3} + 2\,\mathrm{act}_{u_2}\,\mathrm{act}_{u_3} - \mathrm{act}_{u_1} - \mathrm{act}_{u_2} - \mathrm{act}_{u_3}$$

of this network. This state graph is shown in Fig. 8.7. Note that the asymmetries of this graph are basically the effect of the non-vanishing thresholds.

To conclude this section, we remark that the energy function of a Hopfield network establishes the connection to physics, which we already mentioned at the beginning of this chapter. In physics, Hopfield networks are used as (microscopic) models of magnetism, based on the relations between physical and neural notions that are shown in Table 8.3. More specifically, a Hopfield network corresponds to the so-called Ising model of Magnetism (Ising 1925). This physical analogy also provides (another) reason why the activation function of the neurons of a Hopfield network is sometimes defined in such a way that the neuron does not change its activation if its network input equals its threshold (see page 122): if the effects of the external magnetic field and the magnetic coupling of the atoms cancel, the atom should maintain its magnetic moment.

8.3 Associative Memory

Hopfield networks are very well suited to implement so-called **associative memory**, that is, a kind of memory that is addressed by its contents. If a pattern is presented to an associative memory, the memory returns whether this pattern coincides with one of the stored patterns. This coincidence need not be exact. An associative memory may also return a stored pattern that is as similar as possible to the presented pattern. In this way, "noisy" input patterns may be recognized as well.

Hopfield networks are employed as associative memory by exploiting their stable states, of which one is eventually reached in the work phase. If we determine the weights and the thresholds of a Hopfield network in such a way that the patterns to store are exactly the stable states, the normal update procedure of a Hopfield network finds for any input pattern a similar stored pattern. In this way "noisy" patterns may be corrected or disturbed patterns may still be recognized.

In order to simplify the following computations, we start by considering how a single pattern $\mathbf{p} = (\mathrm{act}_{u_1}, \ldots, \mathrm{act}_{u_n})^\top \in \{-1, 1\}^n$, $n \geq 2$, can be stored in a Hopfield network. To this end we have to determine the weights and the thresholds in such a way that the pattern becomes a stable state (also: an attractor) of the Hopfield network. Therefore we need to ensure

$$S(\mathbf{W}\mathbf{p} - \boldsymbol{\theta}) = \mathbf{p},$$

where $\mathbf{W}$ is the weight matrix of the Hopfield network, $\boldsymbol{\theta} = (\theta_{u_1}, \ldots, \theta_{u_n})^\top$ is the threshold vector and S is a function

$$S : \mathbb{R}^n \to \{-1, 1\}^n,$$
$$\mathbf{x} \mapsto \mathbf{y},$$

where the vector $\mathbf{y}$ is determined by

$$\forall i \in \{1, \ldots, n\} : \quad y_i = \begin{cases} 1 & \text{if } x_i \geq 0, \\ -1 & \text{otherwise.} \end{cases}$$

That is, the function S is a kind of element-wise threshold function.

If we set $\boldsymbol{\theta} = \mathbf{0}$, that is, if we choose all thresholds to be 0, a suitable matrix $\mathbf{W}$ can easily be found, since it obviously suffices if

$$\mathbf{W}\mathbf{p} = c\mathbf{p} \quad \text{with } c \in \mathbb{R}^+$$

holds. Algebraically: we have to find a matrix $\mathbf{W}$ that possesses w.r.t. $\mathbf{p}$ a positive eigenvalue c.[1] We choose now

$$\mathbf{W} = \mathbf{p}\mathbf{p}^T - \mathbf{E}$$

with the $n \times n$ unit matrix $\mathbf{E}$. The term $\mathbf{p}\mathbf{p}^T$ is the so-called **outer product** (or **matrix product**) of the vector $\mathbf{p}$ with itself. It yields a symmetric $n \times n$ matrix. The unit matrix $\mathbf{E}$ has to be subtracted from this matrix in order to ensure that the diagonal of the weight matrix becomes 0, since there are no (self-)loops in a Hopfield network. With this matrix $\mathbf{W}$ we have for the pattern $\mathbf{p}$:

$$\mathbf{W}\mathbf{p} = (\mathbf{p}\mathbf{p}^T)\mathbf{p} - \underbrace{\mathbf{E}\mathbf{p}}_{=\mathbf{p}} \overset{(*)}{=} \mathbf{p}\underbrace{(\mathbf{p}^T\mathbf{p})}_{=|\mathbf{p}|^2=n} - \mathbf{p}$$
$$= n\mathbf{p} - \mathbf{p} = (n-1)\mathbf{p}.$$

The equality $(*)$ holds, because matrix and vector multiplications are associative and thus we may change the order of the operations (in other words: may place the parentheses in the expressions differently). In this form we first have to compute the

[1]In linear algebra one usually studies the inverse problem, that is, given a matrix, one tries to find the eigenvalues and eigenvectors.

inner product (or **scalar product**) of the vector $\mathbf{p}$ with itself. This yields its squared length. Since we know that $\mathbf{p} \in \{-1, 1\}^n$, we derive that $\mathbf{p}^T\mathbf{p} = |\mathbf{p}|^2 = n$. Because we assumed $n \geq 2$, it is $c = (n-1) > 0$, as required. Therefore, the pattern $\mathbf{p}$ is a stable state of the Hopfield network.

If we write the computations with individual weights, we obtain:

$$w_{uv} = \begin{cases} 0 & \text{if } u = v, \\ 1 & \text{if } u \neq v,\ \mathrm{act}_u^{(p)} = \mathrm{act}_v^{(p)}, \\ -1 & \text{otherwise.} \end{cases}$$

This rule is also known as the **Hebbian learning rule** (Hebb 1949). Originally, this rule was derived from a biological analogy: the connection between two neurons that are synchronously active is strengthened ("cells that fire together, wire together").

Note, however, that with this method the pattern $-\mathbf{p}$, which is complementary to the pattern $\mathbf{p}$, becomes a stable state as well. The reason is that with

$$\mathbf{W}\mathbf{p} = (n-1)\mathbf{p} \quad \text{we also have} \quad \mathbf{W}(-\mathbf{p}) = (n-1)(-\mathbf{p}).$$

Unfortunately, it is impossible to avoid that the complementary pattern is also stored.

If multiple patterns $\mathbf{p}_1, \ldots, \mathbf{p}_m$, $m < n$, are to be stored, we determine a matrix $\mathbf{W}_i$ for each pattern $\mathbf{p}_i$ in the way described above. The weight matrix $\mathbf{W}$ is then computed as the sum of these matrices, that is,

$$\mathbf{W} = \sum_{i=1}^{m} \mathbf{W}_i = \left(\sum_{i=1}^{m} \mathbf{p}_i \mathbf{p}_i^T \right) - m\mathbf{E}.$$

If the patterns to be stored are pairwise orthogonal (that is, if the corresponding vectors are perpendicular to each other), we have with this matrix $\mathbf{W}$ for an arbitrary pattern $\mathbf{p}_j$, $j \in \{1, \ldots, m\}$:

$$\begin{aligned} \mathbf{W}\mathbf{p}_j = \sum_{i=1}^{m} \mathbf{W}_i \mathbf{p}_j &= \left(\sum_{i=1}^{m} (\mathbf{p}_i \mathbf{p}_i^T)\mathbf{p}_j \right) - m \underbrace{\mathbf{E}\mathbf{p}_j}_{=\mathbf{p}_j} \\ &= \left(\sum_{i=1}^{m} \mathbf{p}_i (\mathbf{p}_i^T \mathbf{p}_j) \right) - m\mathbf{p}_j. \end{aligned}$$

Since we assumed that the patterns are pairwise orthogonal, it is

$$\mathbf{p}_i^T \mathbf{p}_j = \begin{cases} 0 & \text{if } i \neq j, \\ n & \text{if } i = j, \end{cases}$$

because the scalar product of orthogonal vectors vanishes, while the scalar product of a vector with itself yields the squared length of the vector. This length equals n, because $\mathbf{p}_j \in \{-1, 1\}^n$ (see above). Therefore, we obtain

$$\mathbf{W}\mathbf{p}_j = (n-m)\mathbf{p}_j,$$

and hence $\mathbf{p}_j$ is a stable state of the Hopfield network if $m < n$. Note that in this case the pattern $-\mathbf{p}_j$, which is complementary to the pattern $\mathbf{p}_j$ is again a stable state as well, since due to

$$\mathbf{W}\mathbf{p}_j = (n-m)\mathbf{p}_j \quad \text{we also have} \quad \mathbf{W}(-\mathbf{p}_j) = (n-m)(-\mathbf{p}_j).$$

Although we can choose n pairwise orthogonal vectors in an n-dimensional space, this method allows to store only $n-1$ patterns (and their complements), because we have to satisfy $n-m > 0$ (see above). Compared to the number of possible states (2^n, since we have n neurons with two possible states each) the storage capacity of a Hopfield network is fairly small.

If the patterns are not pairwise orthogonal, as it is often the case in practice, we have for an arbitrary pattern $\mathbf{p}_j$, $j \in \{1, \dots, m\}$:

$$\mathbf{W}\mathbf{p}_j = (n-m)\mathbf{p}_j + \underbrace{\sum_{\substack{i=1 \\ i \neq j}}^{m} \mathbf{p}_i \left(\mathbf{p}_i^T \mathbf{p}_j\right)}_{\text{"disturbance term"}}.$$

Now the state $\mathbf{p}_j$ may be stable nevertheless, namely if $n-m > 0$ and the "disturbance term" is sufficiently small. This is the case if the patterns $\mathbf{p}_i$ are "approximately" orthogonal, because then the scalar products $\mathbf{p}_i^T \mathbf{p}_j$ are small. Clearly, the larger the number of patterns to store, the smaller this disturbance term must be, because a growing m means a reducing $n-m$, which makes the state more "susceptible" to disturbances. Hence, in practice the theoretical limit for the storage capacity of a Hopfield network is never reached.

To illustrate the discussed method, we determine the weight matrix of a Hopfield network with four neurons that stores the two patterns $\mathbf{p}_1 = (+1, +1, -1, -1)^\top$ and $\mathbf{p}_2 = (-1, +1, -1, +1)^\top$. It is

$$\mathbf{W} = \mathbf{W}_1 + \mathbf{W}_2 = \mathbf{p}_1\mathbf{p}_1^T + \mathbf{p}_2\mathbf{p}_2^T - 2\mathbf{E}$$

with the individual matrices

$$\mathbf{W}_1 = \begin{pmatrix} 0 & 1 & -1 & -1 \\ 1 & 0 & -1 & -1 \\ -1 & -1 & 0 & 1 \\ -1 & -1 & 1 & 0 \end{pmatrix}, \qquad \mathbf{W}_2 = \begin{pmatrix} 0 & -1 & 1 & -1 \\ -1 & 0 & -1 & 1 \\ 1 & -1 & 0 & -1 \\ -1 & 1 & -1 & 0 \end{pmatrix}.$$

The weight matrix of the Hopfield network thus reads

$$\mathbf{W} = \begin{pmatrix} 0 & 0 & 0 & -2 \\ 0 & 0 & -2 & 0 \\ 0 & -2 & 0 & 0 \\ -2 & 0 & 0 & 0 \end{pmatrix}.$$

It is easy to check that with this matrix we have

$$\mathbf{W}\mathbf{p}_1 = (+2, +2, -2, -2)^\top \quad \text{and} \quad \mathbf{W}\mathbf{p}_1 = (-2, +2, -2, +2)^\top.$$

Therefore both patterns are indeed stable states. However, their complements—that is, the patterns $-\mathbf{p}_1 = (-1, -1, +1, +1)$ and $-\mathbf{p}_2 = (+1, -1, +1, -1)$—are stable states as well, as an analogous computation shows.

Another possibility to determine the parameters of a Hopfield network consists in mapping the network to a threshold logic unit, which is then trained with the delta rule (Rojas 1996). This approach works as follows: if a pattern $\mathbf{p} = (\mathrm{act}_{u_1}^{(p)}, \ldots, \mathrm{act}_{u_n}^{(p)}) \in \{-1, 1\}^n$ is to be a stable state of a Hopfield network, the following n equations must hold

$$\begin{array}{lllll}
s\big(0 & + w_{u_1u_2}\,\mathrm{act}_{u_2}^{(p)} & + \cdots + w_{u_1u_n}\,\mathrm{act}_{u_n}^{(p)} & - \theta_{u_1}\big) & = \mathrm{act}_{u_1}^{(p)}, \\
s\big(w_{u_2u_1}\,\mathrm{act}_{u_1}^{(p)} & + 0 & + \cdots + w_{u_2u_n}\,\mathrm{act}_{u_n}^{(p)} & - \theta_{u_2}\big) & = \mathrm{act}_{u_2}^{(p)}, \\
\vdots & \vdots & \vdots & \vdots & \vdots \\
s\big(w_{u_nu_1}\,\mathrm{act}_{u_1}^{(p)} & + w_{u_nu_2}\,\mathrm{act}_{u_2}^{(p)} & + \cdots + 0 & - \theta_{u_n}\big) & = \mathrm{act}_{u_n}^{(p)}.
\end{array}$$

Here s is the usual threshold function

$$s(x) = \begin{cases} 1 & \text{if } x \geq 0, \\ -1 & \text{otherwise.} \end{cases}$$

For training, we transform the weight matrix into a weight vector, namely by traversing the rows of the upper triangle of the matrix (excluding the diagonal; the lower triangle is neglected, because the weights are symmetric). This weight vector is then extended by appending the thresholds:

$$\begin{array}{rlllll}
\mathbf{w} = (& w_{u_1u_2}, & w_{u_1u_3}, & \ldots, & w_{u_1u_n}, & \\
& & w_{u_2u_3}, & \ldots, & w_{u_2u_n}, & \\
& & & \ddots & \vdots & \\
& & & & w_{u_{n-1}u_n}, & \\
& -\theta_{u_1}, & -\theta_{u_2}, & \ldots, & -\theta_{u_n} &).
\end{array}$$

For this weight vector, we can find input vectors $\mathbf{z}_1, \ldots, \mathbf{z}_n$ such that the arguments of the threshold function appearing in the above equations can be written as scalar products $\mathbf{w}\mathbf{z}_i$. For example, we may choose

$$\mathbf{z}_2 = \big(\mathrm{act}_{u_1}^{(p)}, \underbrace{0, \ldots, 0}_{n-2 \text{ zeros}}, \mathrm{act}_{u_3}^{(p)}, \ldots, \mathrm{act}_{u_n}^{(p)}, \ldots 0, 1, \underbrace{0, \ldots, 0}_{n-2 \text{ zeros}}\big).$$

In this way, we can map training a Hopfield network to training a threshold logic unit with the threshold 0 and the weight vector $\mathbf{w}$ for the training patterns $l_i = (\mathbf{z}_i, \mathrm{act}_{u_i}^{(p)})$. This threshold logic unit may now be trained, for instance, with the delta rule (cf. Sect. 3.5). If multiple patterns are to be stored, we obtain more input

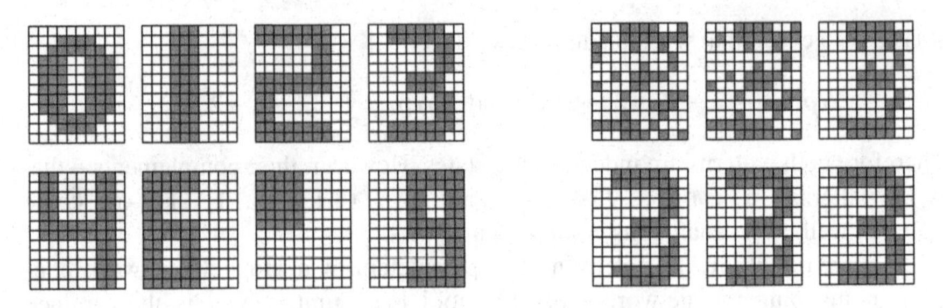

Fig. 8.8 Example patterns that are stored in a Hopfield network (*left*) and the reconstruction of a pattern from disturbed input (*right*)

patterns $\mathbf{z}_i$. Note, though, that this option is more of theoretical interest than of practical value.

To illustrate how a Hopfield network can be used to recognize patterns, we consider a simple number recognition example (derived from Haykin 2008). The patterns (two-dimensional black and white pictures) shown in Fig. 8.8 are stored in a Hopfield network with $10 \times 12 = 120$ neurons, encoding a dark field as $+1$ and a white one as -1. The resulting patterns are not exactly, but sufficiently orthogonal, so that they can be turned into stable states of a Hopfield network with the method described above. If a pattern is presented to the Hopfield network determined in this way, the computations of the network reconstruct one of these stored patterns, as shown in Fig. 8.8 on the right. Note, however, that there are multiple update steps between two consecutive diagrams in this figure.

In order to better understand this example, the web page

http://www.borgelt.net/hfnd.html

provides the programs `whopf` (for Microsoft Windows™) and `xhopf` (for Linux). With them one can store two-dimensional patterns in a Hopfield network and then retrieve them again. The patterns shown in Fig. 8.8 are available in a file.

However, with these programs some of the problems of the discussed method become obvious as well. We already know that this method also stores the complementary patterns. As a consequence, they may be produced as the result of the computations. Besides these patterns there are additional stable states, which differ only marginally from those stored. Among other things, these problems result from the fact that the patterns are not exactly orthogonal.

8.4 Solving Optimization Problems

By exploiting their energy function, Hopfield networks can be used to solve optimization problems. The core idea is as follows: by updating a Hopfield network a (local) minimum of its energy function is reached. If we can rewrite a given function to optimize in such a way that it can be interpreted as an energy function (to minimize) of a Hopfield network, we can construct a Hopfield network by reading the weights and thresholds from its terms. Then this Hopfield network is initialized randomly—that is, it is placed into a random activation state—and the update computations are carried out as usual. The network eventually reaches a stable state, which corresponds to a minimum of the energy function and thus an optimum of the function to optimize. Note, however, that this optimum may only be a local one.

This principle is obviously very simple. The only difficulty we face is that, if we try to solve an optimization problem, we often have to respect certain additional constraints. For example, it may be that the arguments of the function to optimize must not leave certain ranges of values. In such a case, it is not sufficient to simply turn the function to optimize into an energy function of a Hopfield network, but we also have to take precautions that the additional constraints are respected, so that the solution found with the help of a Hopfield network is actually valid.

In order to incorporate the additional constraints, we proceed in essentially the same way as for optimizing the objective function. For each additional constraint we construct a function, which is minimized if the constraint is respected and then transform this function into an energy function of a Hopfield network. Finally, we combine the energy function that describes the objective function and the energy functions that result from the additional constraints, exploiting the following lemma.

Lemma 8.1 *Suppose we have two Hopfield networks defined on the same set U of neurons with weights $w_{uv}^{(i)}$, thresholds $\theta_u^{(i)}$ and energy functions*

$$E_i = -\frac{1}{2}\sum_{u\in U}\sum_{v\in U-\{u\}} w_{uv}^{(i)} \operatorname{act}_u \operatorname{act}_v + \sum_{u\in U} \theta_u^{(i)} \operatorname{act}_u$$

for $i = 1, 2$. (The index i states which of the two networks the quantities refer to.) Furthermore, let $a, b \in \mathbb{R}$. Then $E = aE_1 + bE_2$ is the energy function of the Hopfield network with neurons in U that has the weights $w_{uv} = aw_{uv}^{(1)} + bw_{uv}^{(2)}$ and the thresholds $\theta_u = a\theta_u^{(1)} + b\theta_u^{(2)}$.

This lemma allows us to construct the energy function that is to be optimized by a Hopfield network as a linear combination of several energy function. Proving this lemma is trivial (simply calculate the expression $E = aE_1 + bE_2$) and thus the proof is not demonstrated in detail here.

As an example for the described procedure we consider how the well-known traveling salesman problem (TSP) can be solved (approximately) with the help of a Hopfield network. This problem consists in the task to find for a traveling salesman the best tour through a given set of n cities, so that each city is visited exactly once. In order to solve this problem with the help of a Hopfield network, we use the activations 0 and 1 for the neurons, because this simplifies setting up the energy functions. That we may deviate from the original definition in this way, because we can always transform the weights and thresholds into those needed for a Hopfield network with activations 1 and -1, is shown in Sect. 10.3, in which the needed transformation formulas are derived.

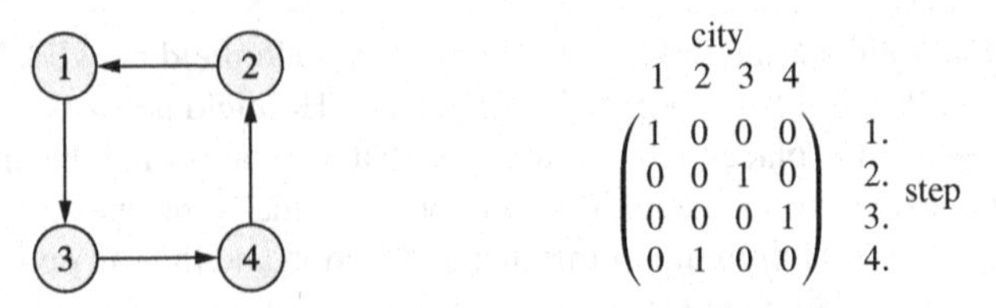

Fig. 8.9 A tour through four cities and a 4 × 4 matrix representing it

A tour through n given cities is encoded as follows: we set up a binary $n \times n$ matrix $\mathbf{M} = (m_{ij})$, the columns of which correspond to the cities and the rows of which correspond to the steps of the tour. We write a 1 into row i and column j of this matrix ($m_{ij} = 1$) if the city j is visited in the ith step of the tour. Otherwise we write a 0 into this element ($m_{ij} = 0$). For example, the matrix shown in Fig. 8.9 on the right describes the tour through the cities 1 to 4 shown in the same figure on the left. Note that cyclic permutation of the steps (rows) describes the same tour, since we did not define any city as the one to start from.

The Hopfield network we have to construct possesses one neuron for each element of this $n \times n$ matrix, which we denote by the coordinates (i, j) of the corresponding matrix element and whose activation corresponds to the value of this matrix element. After the update process has reached a stable state we can thus read the found tour from the activations of the neurons. Note that in the following we always use index i to refer to steps and index j to refer to cities.

With the help of the matrix $\mathbf{M}$, we can describe the objective function that has to be minimized in order to solve the traveling salesman problem as

$$E_1 = \sum_{j_1=1}^{n} \sum_{j_2=1}^{n} \sum_{i=1}^{n} d_{j_1 j_2} \cdot m_{ij_1} \cdot m_{(i \bmod n)+1, j_2}.$$

Here $d_{j_1 j_2}$ is the distance between city j_1 and city j_2. With the two factors that refer to the matrix $\mathbf{M}$, we ensure that only distances between cities are summed that are visited consecutively on the tour, that is, if the city j_1 is visited in the ith step and the city j_2 is visited in the $((i \bmod n) + 1)$th step of the tour. Only in this case both matrix elements are 1. On the other hand, if the cities are not visited consecutively in the tour, at least one of the matrix elements and thus the term is 0.

Following the plan laid down above, we now have to transform the function E_1 in such a way that it takes the form of the energy function of a Hopfield network with the neurons (i, j), where the matrix elements m_{ij} play the role of the activations of the neurons. In order to do so, we have to introduce a second sum over the steps of the tour (index i). We achieve this by using two indices i_1 and i_2 for the steps in which the cities j_1 and j_2 are visited. In addition, we ensure by an additional factor that only such terms are formed in which these two indices are related in the desired way (that is, i_2 follows i_1). Thus we obtain

$$E_1 = \sum_{(i_1,j_1)\in\{1,\dots,n\}^2} \sum_{(i_2,j_2)\in\{1,\dots,n\}^2} d_{j_1 j_2} \cdot \delta_{(i_1 \bmod n)+1, i_2} \cdot m_{i_1 j_1} \cdot m_{i_2 j_2},$$

where δ_{ab} is the so-called *Kronecker symbol* ($\delta_{ab} = 1$ if $a = b$ and $\delta_{ab} = 0$ otherwise).

All that is still missing now to reach the form of an energy function is the factor $-\frac{1}{2}$ in front of the sums. This factor can easily be introduced, for example, by moving a factor -2 into the sums. However, it seems to be more appropriate to move only a factor of -1 into the sums and to obtain the factor 2 by making the factor with the Kronecker symbol symmetric. Clearly, it is irrelevant whether i_2 follows i_1 or vice versa: both cases describe the same relation between the cities (only the tour is reversed). If we allow for both orders, every distance between two cities is automatically considered twice. Thus, we finally arrive at

$$E_1 = -\frac{1}{2} \sum_{\substack{(i_1,j_1)\in\{1,\ldots,n\}^2 \\ (i_2,j_2)\in\{1,\ldots,n\}^2}} -d_{j_1 j_2} \cdot (\delta_{(i_1 \bmod n)+1, i_2} + \delta_{i_1,(i_2 \bmod n)+1}) \cdot m_{i_1 j_1} \cdot m_{i_2 j_2}.$$

This function has the form of the energy function of a Hopfield network. However, we cannot use it directly, because it is obviously minimized if and only if all $m_{ij} = 0$, regardless of the distances between the cities. Indeed, we have to respect two constraints when minimizing the above function, namely:

- Each city is visited in exactly one step of the tour:

$$\forall j \in \{1,\ldots,n\}: \quad \sum_{i=1}^{n} m_{ij} = 1,$$

  that is, every column of the matrix contains exactly one 1.
- In each step of the tour exactly one city is visited:

$$\forall i \in \{1,\ldots,n\}: \quad \sum_{j=1}^{n} m_{ij} = 1,$$

  that is, every row of the matrix contains exactly one 1.

These two constraints exclude the trivial solution (all $m_{ij} = 0$). Since these constraints have the same structure, we demonstrate in detail only how the first is turned into an energy function. Clearly, the first constraint is satisfied if and only if

$$E_2^* = \sum_{j=1}^{n} \left(\sum_{i=1}^{n} m_{ij} - 1 \right)^2 = 0.$$

Since E_2^* cannot be negative (due to the squared terms), the first constraint is satisfied if and only if E_2^* is minimized. Simply computing the square yields

$$E_2^* = \sum_{j=1}^{n} \left(\left(\sum_{i=1}^{n} m_{ij} \right)^2 - 2 \sum_{i=1}^{n} m_{ij} + 1 \right)$$

$$= \sum_{j=1}^{n}\left(\left(\sum_{i_1=1}^{n} m_{i_1 j}\right)\left(\sum_{i_2=1}^{n} m_{i_2 j}\right) - 2\sum_{i=1}^{n} m_{ij} + 1\right)$$

$$= \sum_{j=1}^{n}\sum_{i_1=1}^{n}\sum_{i_2=1}^{n} m_{i_1 j} m_{i_2 j} - 2\sum_{j=1}^{n}\sum_{i=1}^{n} m_{ij} + n.$$

The constant term n can be neglected, because it does not change the location of the minimum. In order to obtain the form of an energy function, we merely have to duplicate the sum over the cities (index j), using the same principle that we already applied to derive the objective function E_1. This leads to

$$E_2 = \sum_{(i_1,j_1)\in\{1,\dots,n\}^2} \sum_{(i_2,j_2)\in\{1,\dots,n\}^2} \delta_{j_1 j_2} \cdot m_{i_1 j_1} \cdot m_{i_2 j_2} - 2 \sum_{(i,j)\in\{1,\dots,n\}^2} m_{ij}.$$

By moving the factors -2 into both sums, we finally arrive at

$$E_2 = -\frac{1}{2} \sum_{\substack{(i_1,j_1)\in\{1,\dots,n\}^2 \\ (i_2,j_2)\in\{1,\dots,n\}^2}} -2\delta_{j_1 j_2} \cdot m_{i_1 j_1} \cdot m_{i_2 j_2} + \sum_{(i,j)\in\{1,\dots,n\}^2} -2m_{ij}$$

and thus the form of the energy function of a Hopfield network. In a completely analogous fashion, we obtain from the second constraint

$$E_3 = -\frac{1}{2} \sum_{\substack{(i_1,j_1)\in\{1,\dots,n\}^2 \\ (i_2,j_2)\in\{1,\dots,n\}^2}} -2\delta_{i_1 i_2} \cdot m_{i_1 j_1} \cdot m_{i_2 j_2} + \sum_{(i,j)\in\{1,\dots,n\}^2} -2m_{ij}.$$

Finally, we combine the three energy functions E_1 (objective function), E_2 (first constraint) and E_3 (second constraint) to obtain the total energy function

$$E = aE_1 + bE_2 + cE_3.$$

The factors $a, b, c \in \mathbb{R}^+$ have to be chosen in such a way that it is not possible to reduce the value of the energy function by violating the constraints. This is certainly the case if

$$\frac{b}{a} = \frac{c}{a} > 2 \max_{(j_1,j_2)\in\{1,\dots,n\}^2} d_{j_1 j_2},$$

that is, if the maximum improvement achievable by a (local) modification of the tour is less than the minimum degradation that results from a violation of a constraint.

Since the matrix elements m_{ij} correspond to the activations $\mathrm{act}_{(i,j)}$ of the neurons (i, j) of the Hopfield network, we read from the total energy function E the following weights and thresholds:

$$w_{(i_1,j_1)(i_2,j_2)} = \underbrace{-ad_{j_1 j_2} \cdot (\delta_{(i_1 \bmod n)+1,i_2} + \delta_{i_1,(i_2 \bmod n)+1})}_{\text{from } E_1} \underbrace{-2b\delta_{j_1 j_2}}_{\text{from } E_2} \underbrace{-2c\delta_{i_1 i_2}}_{\text{from } E_3},$$

$$\theta_{(i,j)} = \underbrace{0a}_{\text{from } E_1} \quad \underbrace{-2b}_{\text{from } E_2} \quad \underbrace{-2c}_{\text{from } E_3} = -2(b+c).$$

The resulting Hopfield network is now initialized randomly (that is, the activations of the neurons are set to randomly chosen values from $\{0, 1\}$) and then these activations are repeatedly updated until a stable state is reached. The solution—that is, the found tour—can then be read from this state.

Note, however, that the presented approach to solve the traveling salesman problem is, despite its plausibility, of very limited use for practical purposes. One of the main problems is that the Hopfield network is unable to switch from a found tour to another with a lower total length. The reason is that transforming a matrix that represents a tour into another matrix that represents a different tour requires that at least four neurons (matrix elements) change their activations. (For example, if the steps, in which two cities are visited, are exchanged, two neurons must change their activation from 1 to 0 and two others must change their activations from 0 to 1.) However, each of these changes, executed individually, violates at least one of the two constraints and thus leads to an increase of energy. Only all four changes together can result in a smaller energy, but cannot be executed together due to the asynchronous update. Therefore, the normal activation updates can never change an already found tour into another, even if this requires only a marginal change of the tour. As a consequence, it is highly likely that the Hopfield network gets stuck in a local minimum of the energy function. Although it can never be made absolutely sure that the process does not get stuck in a local minimum, this problem is particularly annoying here, since it occurs even in situations in which the modifications, which are necessary to improve the tour, are, so to say, "obvious" (for example, exchanging one city with another or reversing a sub-tour).

Unfortunately, the situation is actually even worse. Although we introduced the energy functions E_1 and E_2 to represent the constraints that describe a valid tour, it is not guaranteed that the resulting stable state actually represents a valid tour. The reason is that there are also situations, in which a matrix that does *not* describe a valid tour can be changed into a matrix that *does* describe a valid tour only through intermediate states that possess a higher energy. For example, if a column of the matrix contains two ones (and thus the first constraint is violated), but these two ones are the only ones in their respective columns, the violation of the first constraint can only be resolved by, at least temporarily, violating the second constraint. Since both constraints are equivalent, the activations remain unchanged.

The above considerations can be checked with the program `tsp.c`, which is available on the web page for this book. This program tries to solve the very simple traveling salesman problem with 5 cities that is shown in Fig. 8.10 with the help of a Hopfield network. The found solution is not always a valid tour and even if it is valid, it appears to be chosen entirely at random.

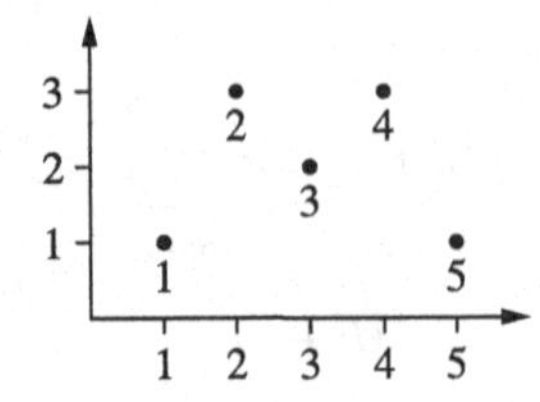

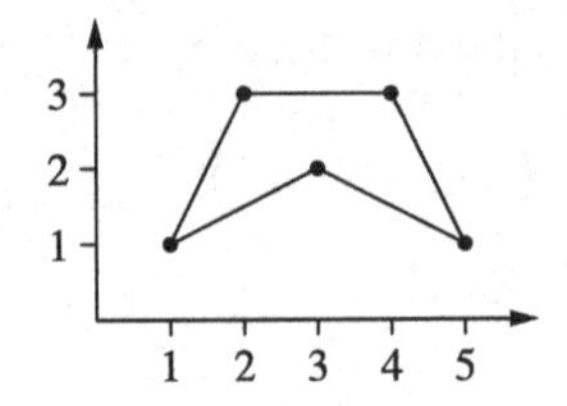

Fig. 8.10 A very simple traveling sales man problem with 5 cities and its solution

Therefore it cannot really be recommended to use a Hopfield network to solve the traveling salesman problem. We considered this problem nevertheless, because it allowed us to explain the procedure of setting up the energy functions in a very clear and straightforward way. Note, however, that the problems encountered here have to be taken into account also if Hopfield networks are applied to other optimization problems, which may render Hopfield networks a sub-optimal choice.

A certain improvement is possible if one does not use **discrete Hopfield networks** with only two possible activations per neuron as we studied them up to now, but rather **continuous Hopfield networks**, in which the activation of a neuron can be any number from $[-1, 1]$ (or $[0, 1]$, respectively). This transition corresponds roughly to the generalization of the activation function that led us from threshold logic units to multi-layer perceptrons (see Fig. 5.2 on page 49). With continuous Hopfield networks, which also have the advantage they are better suited for being implemented in hardware with the help of an (analog) electrical circuit, solving the traveling salesman problem is more successful (Hopfield and Tank 1985).

8.5 Simulated Annealing

The problems discussed in the preceding section, which are encountered if we try to use Hopfield networks to solve optimization problems, are mainly caused by the possibility that the process gets stuck in a local minimum of the energy function. Since the same problem is—not surprisingly—also encountered with other optimization methods, the idea suggests itself to transfer approaches to handle this problem, which have been developed for other optimization methods, to Hopfield networks. One such approach is so-called **simulated annealing**.

The idea of simulated annealing (Metropolis et al. 1953; Kirkpatrick et al. 1983) is to start with a randomly generated candidate solution of the optimization problem and to evaluate it. In every later step, the current candidate solution is modified and re-evaluated. If the new solution is better than the old, it is accepted and replaces the old solution. However, if it is worse, it is accepted only with a certain probability that depends on how much worse the new solution is. In addition, this probability is reduced over time, so that in the limit new candidate solutions are only accepted if they are actually better than the old solution. Furthermore, the best solution found so far is usually recorded in parallel.

The reason why a worse candidate solution is sometimes accepted is that the method would otherwise be very similar to a gradient descent. The only difference would be that the descent direction is not determined by computing a gradient, but is found by trial and error. However, we saw in Chap. 5 that a gradient descent can easily get stuck in a local minimum (see Fig. 5.23 on page 71). However, if solutions that are actually worse than the current candidate are accepted at least sometimes, this undesired behavior can be counteracted to some degree. Intuitively, it allows to overcome "barriers" (regions of the search space in which candidate solutions have low quality) that separate local minima from the global minimum. Only later, when the probability for accepting candidate solutions that are worse has been decreased, the objective function is locally optimized.

The name "simulated annealing" for this approach originates from the fact that it is very similar to the physical minimization of the lattice energy of the atoms if a heated piece of metal is cooled slowly. This process is usually called "annealing" and serves the purpose to make a metal easier to work or to machine by relieving tensions and correcting lattice malformations. Seen from a physical perspective, the thermal energy of the atoms prevents them from settling in a configuration that is only a local minimum of the lattice energy. They rather "jump out" of this configuration. However, the "deeper" the (local) energy minimum is, the harder it is for the atoms to abandon the configuration. Therefore, it is likely that they finally settle in a configuration of very low lattice energy, the optimum of which, in the case of a metal, is a mono-crystalline structure.

It should be clear, though, that it cannot be guaranteed that the global minimum of the lattice energy is reached. Especially, if the metal is not heated long enough or is cooled down too quickly, it is likely that the atoms settle in a configuration that is only a local minimum of the energy (which in the case of the metal is a polycrystalline structure). Hence, it is important that the temperature is reduced slowly, so that the probability that local minima are abandoned again, is sufficiently large.

This energy minimization can also be illustrated by a ball that rolls around on a curved surface. In this case, the function to minimize is the potential energy of the ball. In the beginning, the ball is endowed with a certain kinetic energy, which enables it to roll up some slopes of the surface. In the course of time, however, friction reduces the kinetic energy of the ball more and more, so that it finally comes to a rest in a valley of the surface (a minimum of the potential energy). Since a larger kinetic energy is needed to escape from a deep valley than to escape from a shallow one, it is likely that the point at which the ball finally stops is located in a fairly deep valley, possibly even in the deepest one available (the global minimum).

The reduction of the thermal energy of the atoms in the simulated annealing process or the reduction of the kinetic energy of the ball in the illustration is modeled by the reducing probability for accepting a worse solution. Often an explicit temperature parameter is introduced, with the help of which the probability is computed. Since the probability distribution over the velocities of atoms is often an exponential distribution (for example, the Maxwell distribution, which describes the velocity distribution for an ideal gas (Greiner et al. 1987)), a function like

$$P(\text{accept the solution}) = ce^{-\frac{\Delta Q}{T}}$$

is a common choice to compute the probability of accepting a solution that is worse than the current one. ΔQ is the difference in quality between the current and the new candidate solution, T the temperature parameter, which is reduced in the course of time, and c is a normalization constant.

Applying simulated annealing to Hopfield networks is very simple: after the activations have been initialized randomly, the neurons of the Hopfield network are traversed (for example, in some random order) and it is determined whether an activation change leads to a reduction of the network energy or not. An activation change that reduces the network energy is always accepted (in the normal update process, only such changes occur, see above). However, if an activation change increases the network energy, it is accepted with a probability that is computed with the formula stated above. Note that in this case we have simply

$$\Delta Q = \Delta E = |\,\mathrm{net}_u - \theta_u\,|$$

(cf. the proof of Theorem 8.1 on page 125).

References

J.A. Anderson and E. Rosenfeld. *Neurocomputing: Foundations of Research*. MIT Press, Cambridge, MA, USA, 1988

W. Greiner, L. Neise and H. Stöcker. *Thermodynamik und Statistische Mechanik (Series: Theoretische Physik)*. Verlag Harri Deutsch, Thun/Frankfurt am Main, Germany, 1987. English edition: *Thermodynamics and Statistical Physics*. Springer-Verlag, Berlin, Germany, 2000

S. Haykin. *Neural Networks and Learning Machines*. Prentice Hall, Englewood Cliffs, NJ, USA, 2008

D.O. Hebb. *The Organization of Behaviour*. J. Wiley & Sons, New York, NY, USA, 1949. Chapter 4: "The First Stage of Perception: Growth of an Assembly" reprinted in Anderson and Rosenfeld (1988), 45–56

J.J. Hopfield. Neural Networks and Physical Systems with Emergent Collective Computational Abilities. *Proceedings of the National Academy of Sciences* 79:2554–2558, USA, 1982

J.J. Hopfield. Neurons with Graded Response Have Collective Computational Properties Like Those of Two-State Neurons. *Proceedings of the National Academy of Sciences* 81:3088–3092, USA, 1984

J. Hopfield and D. Tank. "Neural" Computation of Decisions in Optimization Problems. *Biological Cybernetics* 52:141–152. Springer-Verlag, Heidelberg, Germany, 1985

E. Ising. Beitrag zur Theorie des Ferromagnetismus. *Zeitschrift für Physik* 31(253), 1925

S. Kirkpatrick, C.D. Gelatt and M.P. Vercchi. Optimization by Simulated Annealing. *Science* 220:671–680. High Wire Press, Stanford, CA, USA, 1983

N. Metropolis, N. Rosenblut, A. Teller and E. Teller. Equation of State Calculations for Fast Computing Machines. *Journal of Chemical Physics* 21:1087–1092. American Institute of Physics, Melville, NY, USA, 1953

R. Rojas. *Theorie der neuronalen Netze—Eine systematische Einführung*. Springer-Verlag, Berlin, Germany, 1996

Chapter 9
Recurrent Networks

The Hopfield networks that we discussed in the preceding chapter are special recurrent networks, which have a very constrained structure. In this chapter, however, we lift all restrictions and consider recurrent networks without any constraints. Such general recurrent networks are well suited to represent **differential equations** and to solve them (approximately) in a numerical fashion. If the type of differential equation is known that describes a given system, but the values of the parameters appearing in it are unknown, one may also try to train a suitable recurrent network with the help of example patterns in order to determine the system parameters.

9.1 Simple Examples

In contrast to all preceding chapters, we do not start with a definition. The reason is that all special types of neural networks we studied were obtained from the general definition in Chap. 4 by introducing specific restrictions. In this chapter, however, all constraints are lifted. Hence, we turn immediately to examples.

As a first example, we consider the cooling (or warming) of a body with the temperature ϑ_0 that is placed into a medium with the temperature ϑ_a (ambient temperature), which is held constant. Depending on whether the initial temperature of the body is higher or lower than the ambient temperature, the body will dissipate heat to the medium or absorb heat from it until its temperature reaches the ambient temperature ϑ_a. It is plausible that the amount of heat dissipated or absorbed per unit of time—and thus the temperature change—is proportional to the difference between the current temperature $\vartheta(t)$ of the body and the ambient temperature ϑ_a. That is, we assume that

$$\frac{\mathrm{d}\vartheta}{\mathrm{d}t} = \dot{\vartheta} = -k(\vartheta - \vartheta_a).$$

This equation is called **Newton's Cooling Law** (Heuser 1989). The minus sign in front of the (positive) **cooling constant** k, which depends on the considered body, indicates that the temperature change reduces the temperature difference.

R. Kruse et al., *Computational Intelligence*, Texts in Computer Science,
DOI 10.1007/978-1-4471-5013-8_9, © Springer-Verlag London 2013

Of course, a differential equation as simple as this can be solved analytically. With standard methods we obtain (see, for example, Heuser 1989)

$$\vartheta(t) = \vartheta_a + (\vartheta_0 - \vartheta_a)e^{-k(t-t_0)}$$

with the initial temperature $\vartheta_0 = \vartheta(t_0)$ of the body. Here, however, we consider a numerical approximation of the solution, namely with the help of the so-called **Euler–Cauchy polygonal course** (Heuser 1989). The idea of this method consists in the insight that with the differential equation we can compute the derivative $\dot{\vartheta}(t)$ of the function $\vartheta(t)$ for arbitrary points in time t, that is, we know the course of the function $\vartheta(t)$ locally. With a given initial value $\vartheta_0 = \vartheta(t_0)$, we can thus compute any value $\vartheta(t)$ *approximately* as follows: we divide the interval $[t_0, t]$ into n parts of equal length $\Delta t = \frac{t-t_0}{n}$. The split points are given as

$$\forall i \in \{0, 1, \dots, n\}: \quad t_i = t_0 + i\Delta t.$$

We proceed from the starting point $P_0 = (t_0, \vartheta_0)$ along a *straight line* with the slope $\dot{\vartheta}(t_0)$ given by the differential equation until we reach the time t_1 and thus the point $P_1 = (t_1, \vartheta_1)$. It is

$$\vartheta_1 = \vartheta(t_1) = \vartheta(t_0) + \dot{\vartheta}(t_0)\Delta t = \vartheta_0 - k(\vartheta_0 - \vartheta_a)\Delta t.$$

In this point P_1 the slope of the function is described by the differential equation as $\dot{\vartheta}(t_1)$. Again we proceed on a straight line with this slope until we reach the point $P_2 = (t_2, \vartheta_2)$ for the time t_2. Then we have

$$\vartheta_2 = \vartheta(t_2) = \vartheta(t_1) + \dot{\vartheta}(t_1)\Delta t = \vartheta_1 - k(\vartheta_1 - \vartheta_a)\Delta t.$$

By repeating this procedure, we compute step by step the points $P_k = (t_k, \vartheta_k)$, $k = 1, \dots, n$, the second coordinate ϑ_k of which can always be found with the recursion

$$\vartheta_i = \vartheta(t_i) = \vartheta(t_{i-1}) + \dot{\vartheta}(t_{i-1})\Delta t = \vartheta_{i-1} - k(\vartheta_{i-1} - \vartheta_a)\Delta t.$$

Finally, we reach the point $P_n = (t_n, \vartheta_n)$ and the desired approximation $\vartheta_n = \vartheta(t_n)$.

Intuitively, the described method approximates the function $\vartheta(t)$ by a *polygonal course*, since we always proceed on a *straight line* from one point to the next (hence the name *Euler–Cauchy polygonal course*). More formally, we can derive the above recursion formula for the values ϑ_i by approximating the differential quotient by a difference quotient, that is, by using

$$\frac{d\vartheta(t)}{dt} \approx \frac{\Delta\vartheta(t)}{\Delta t} = \frac{\vartheta(t+\Delta t) - \vartheta(t)}{\Delta t}$$

with a sufficiently small Δt. In this case, we clearly have

$$\vartheta(t+\Delta t) - \vartheta(t) = \Delta\vartheta(t) \approx -k\big(\vartheta(t) - \vartheta_a\big)\Delta t,$$

from which the recursion formula follows directly.

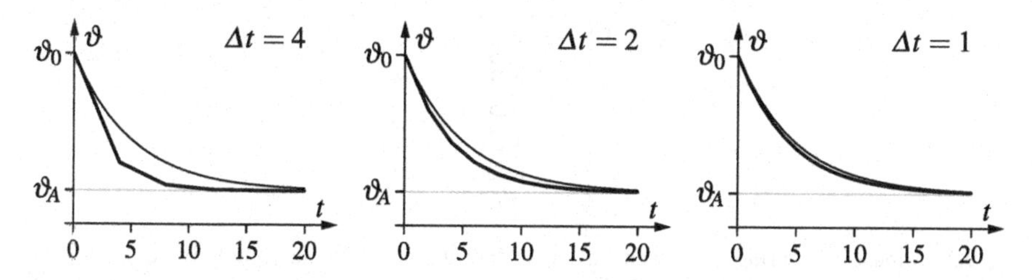

Fig. 9.1 Euler–Cauchy polygonal courses as approximate solutions of Newton's cooling law for different step widths Δt. The *thin curve* is the exact solution

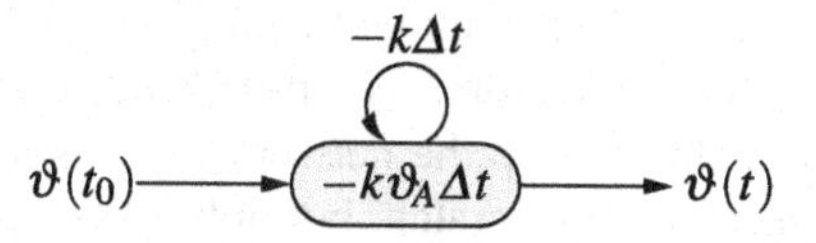

Fig. 9.2 A neural network for Newton's cooling law

It should be clear that the accuracy of the computed approximation is the better, the smaller the **step width** Δt is chosen, because then the computed polygonal course will differ less from the actual course of the function $\vartheta(t)$ under consideration. To illustrate this, Fig. 9.1 shows, for an ambient temperature $\vartheta_a = 20$, a cooling constant $k = 0.2$ and the initial values $t_0 = 0$ and $\vartheta_0 = 100$ the exact solution $\vartheta(t) = \vartheta_a + (\vartheta_0 - \vartheta_a)e^{-k(t-t_0)}$ (thin line) as well as its approximation by Euler–Cauchy polygonal courses with step widths $\Delta t = 4, 2, 1$ in the interval $[0, 20]$. Compare the deviation from the exact solution, for example, for $t = 8$ or $t = 12$.

To represent the recursion formula derived above by a recurrent neural network, we merely have to expand the right hand side. Thus, we obtain

$$\vartheta(t + \Delta t) - \vartheta(t) = \Delta\vartheta(t) \approx -k\Delta t\vartheta(t) + k\vartheta_a\Delta t$$

and therefore

$$\vartheta_i \approx \vartheta_{i-1} - k\Delta t\vartheta_{i-1} + k\vartheta_a\Delta t.$$

The form of this equation corresponds exactly to the computations of a single neuron with a feedback loop. As a consequence, we can approximate the function $\vartheta(t)$ with the help of a neural network with a single neuron u with the network input function

$$f_{\text{net}}^{(u)}(w, x) = -k\Delta tx$$

and the activation function

$$f_{\text{act}}^{(u)}(\text{net}_u, \text{act}_u, \theta_u) = \text{act}_u + \text{net}_u - \theta_u$$

with $\theta_u = -k\vartheta_a\Delta t$. This network is shown in Fig. 9.2, where—as usual—the bias value θ_u is written into the neuron.

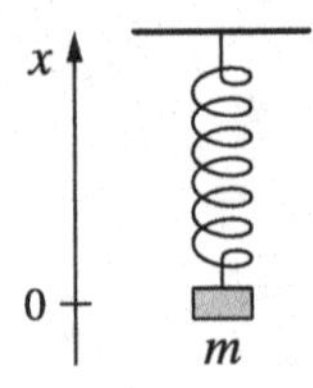

Fig. 9.3 A mass on a spring. Its movement can be described by a simple differential equation

Note that this network actually contains two feedback loops, namely (1) the feedback that is shown explicitly, which describes the temperature change as a function of the current temperature, and (2) the implicit feedback which results from the fact that the current activation of the neuron u is a parameter of its activation function. Because of this second feedback loop the network input is not used to recompute the activation of the neuron u from scratch, but only to compute the change of its activation (cf. Fig. 4.2 on page 40 and the accompanying explanations about the general structure of a generalized neuron).

Alternatively, we could have used the network input function

$$f_{\text{net}}^{(u)}(x, w) = (1 - k\Delta t)x$$

(that is, the connection weight $w = 1 - k\Delta t$) and the activation function

$$f_{\text{act}}^{(u)}(\text{net}_u, \theta_u) = \text{net}_u - \theta_u$$

(again with $\theta_u = -k\vartheta_a \Delta t$). In this way, we could have avoided the implicit feedback. However, the first form corresponds better to the structure of the differential equation and thus we prefer it here.

As a second example, we consider a mass on a spring as it is shown in Fig. 9.3. The height $x = 0$ denotes the equilibrium position of the mass m. We assume now that the mass m is lifted by a certain distance $x(t_0) = x_0$ and then dropped (that is, it has the initial velocity $v(t_0) = 0$). Since the gravitational force acting on the mass m is the same on all heights x, we can ignore its influence. The spring force is governed by **Hooke's Law** (Feynman et al. 1963; Heuser 1989), according to which the exerted force F is proportional to the length change Δl of the spring and directed opposite to the direction of this change. That is, we have

$$F = c\Delta l = -cx,$$

where c is a constant that depends on the spring. According to **Newton's Second Law** $F = ma = m\ddot{x}$ this force causes an acceleration $a = \ddot{x}$ of the mass m. Therefore we arrive at the differential equation

$$m\ddot{x} = -cx, \quad \text{or} \quad \ddot{x} = -\frac{c}{m}x.$$

Of course, this differential equation can easily be solved analytically. The general solution, which can be obtained with standard methods, is

$$x(t) = a\sin(\omega t) + b\cos(\omega t)$$

with the parameters

$$\omega = \sqrt{\frac{c}{m}}, \qquad \begin{aligned} a &= x(t_0)\sin(\omega t_0) + v(t_0)\cos(\omega t_0), \\ b &= x(t_0)\cos(\omega t_0) - v(t_0)\sin(\omega t_0). \end{aligned}$$

With the given initial values $x(t_0) = x_0$ and $v(t_0) = 0$ and the additional definition $t_0 = 0$ we obtain the simple expression

$$x(t) = x_0 \cos\left(\sqrt{\frac{c}{m}}t\right).$$

In order to construct a recurrent neural network that approximates this solution numerically, we rewrite the differential equation, which is of second order, as a system of two coupled differential equations of first order. We achieve this by introducing the velocity v of the mass as an intermediary quantity and obtain

$$\dot{x} = v \quad \text{and} \quad \dot{v} = -\frac{c}{m}x.$$

Next, we approximate, as for Newton's cooling law, the differential quotient by a difference quotient, which leads to

$$\frac{\Delta x}{\Delta t} = \frac{x(t+\Delta t) - x(t)}{\Delta t} = v \quad \text{and} \quad \frac{\Delta v}{\Delta t} = \frac{v(t+\Delta t) - v(t)}{\Delta t} = -\frac{c}{m}x.$$

From these equations, we obtain the recursion formulas

$$\begin{aligned} x(t_i) &= x(t_{i-1}) + \Delta x(t_{i-1}) = x(t_{i-1}) + \Delta t \cdot v(t_{i-1}) \quad \text{and} \\ v(t_i) &= v(t_{i-1}) + \Delta v(t_{i-1}) = v(t_{i-1}) - \frac{c}{m}\Delta t \cdot x(t_{i-1}). \end{aligned}$$

Now we only have to create a neuron for each of these formulas and to read the connection weights from the formulas. This yields the neural network that is shown in Fig. 9.4. The network input function and the activation function of the top neuron u_1 are

$$\begin{aligned} f_{\text{net}}^{(u_1)}(v, w_{u_1u_2}) &= w_{u_1u_2}v = \Delta t v \quad \text{and} \\ f_{\text{act}}^{(u_1)}(\text{act}_{u_1}, \text{net}_{u_1}, \theta_{u_1}) &= \text{act}_{u_1} + \text{net}_{u_1} - \theta_{u_1}. \end{aligned}$$

The corresponding functions of the bottom neuron u_2 are

$$\begin{aligned} f_{\text{net}}^{(u_2)}(x, w_{u_2u_1}) &= w_{u_2u_1}x = -\frac{c}{m}\Delta t x \quad \text{and} \\ f_{\text{act}}^{(u_2)}(\text{act}_{u_2}, \text{net}_{u_2}, \theta_{u_2}) &= \text{act}_{u_2} + \text{net}_{u_2} - \theta_{u_2}. \end{aligned}$$

The output function of both neurons is the identity. Obviously, these choices implement exactly the recursion formulas stated above.

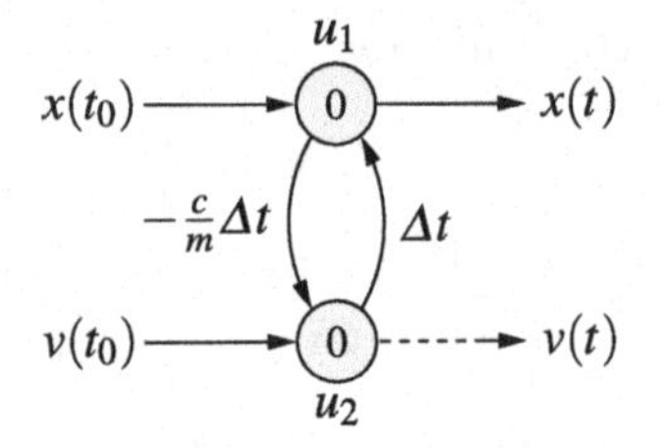

Fig. 9.4 Recurrent neural network that computes the movement of a mass on a spring (governed by Hooke's law)

t	v	x
0.0	0.0000	1.0000
0.1	−0.5000	0.9500
0.2	−0.9750	0.8525
0.3	−1.4012	0.7124
0.4	−1.7574	0.5366
0.5	−2.0258	0.3341
0.6	−2.1928	0.1148

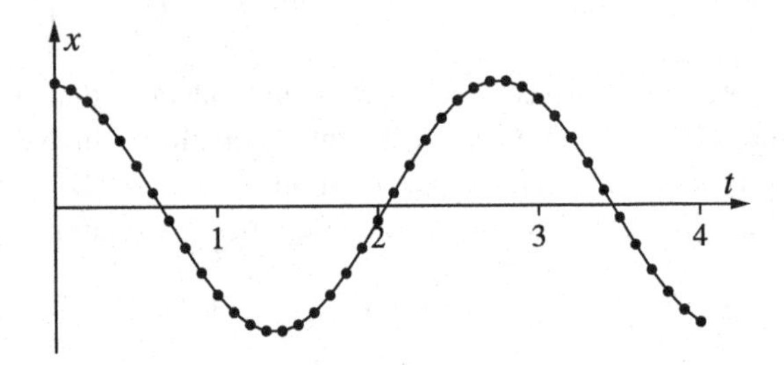

Fig. 9.5 The first computation steps of the neural network shown in Fig. 9.4 and the computed movement of a mass on a spring

Note that this network not only produces approximations for the location $x(t)$ (output of the neuron u_1), but also for the velocity $v(t)$ (output of the neuron u_2). Note also that we may draw on the following consideration (Feynman et al. 1963): the approximation becomes more accurate if we do not compute the velocity $v(t)$ for the times $t_i = t_0 + i\,\Delta t$, but for the midpoints of the intervals, that is for the times $t'_i = t_0 + i\,\Delta_t + \frac{\Delta t}{2}$. In this case the bottom neuron does not receive $v(t_0)$, but $v(t_0 + \frac{\Delta t}{2}) \approx v_0 - \frac{c}{m}\frac{\Delta t}{2}x_0$ as input.

Example computations of the neural network shown in Fig. 9.4 for the parameters $\frac{c}{m} = 5$ and $\Delta t = 0.1$ are shown in the table and the diagram in Fig. 9.5. The table contains in the columns labeled t and x the coordinates of the first seven points of the diagram. The update starts with the neuron u_2, which receives $v(t_0)$ as input.

9.2 Representing Differential Equations

From the examples, we discussed in the preceding section we can derive a simple principle how arbitrary explicit differential equations[1] can be represented by recurrent neural networks: A given explicit differential equation of nth order

[1] Due to the special operation scheme of neural networks it is not possible to solve arbitrary differential equations numerically with the help of recurrent networks. It suffices, however, if the differential equation can be solved for the independent variable or for one of the occurring derivatives of the dependent variable. Here we consider, as an example, the special case in which the differential equation can be written with the highest occurring derivative isolated on one side.

$$x^{(n)} = f\left(t, x, \dot{x}, \ddot{x}, \ldots, x^{(n-1)}\right)$$

($\dot{x}$ denotes the first, $\ddot{x}$ the second and $x^{(i)}$ the ith derivative of x w.r.t. t) is transformed into a system of n coupled differential equation of first order by introducing $n-1$ intermediary quantities

$$y_1 = \dot{x}, \quad y_2 = \ddot{x}, \quad \ldots \quad y_{n-1} = x^{(n-1)}.$$

This yields the system of differential equations

$$\begin{aligned}
\dot{x} &= y_1, \\
\dot{y}_1 &= y_2, \\
&\vdots \\
\dot{y}_{n-2} &= y_{n-1}, \\
\dot{y}_{n-1} &= f(t, x, y_1, y_2, \ldots, y_{n-1}).
\end{aligned}$$

In analogy to the examples studied in the preceding section, each differential quotient that occurs in these equations is replaced by a difference quotient. This yields the n recursion formulas

$$\begin{aligned}
x(t_i) &= x(t_{i-1}) + \Delta t \cdot y_1(t_{i-1}), \\
y_1(t_i) &= y_1(t_{i-1}) + \Delta t \cdot y_2(t_{i-1}), \\
&\vdots \\
y_{n-2}(t_i) &= y_{n-2}(t_{i-1}) + \Delta t \cdot y_{n-3}(t_{i-1}), \\
y_{n-1}(t_i) &= y_{n-1}(t_{i-1}) + f\left(t_{i-1}, x(t_{i-1}), y_1(t_{i-1}), \ldots, y_{n-1}(t_{i-1})\right).
\end{aligned}$$

For each of these equations we create a neuron, which extrapolates the quantity on the left hand side with the help of the right-hand side. If the differential equation depends explicitly on t (and not only indirectly through the quantities x, $\dot{x}$ etc. which depend on t), an additional neuron is necessary, which updates the value of t with the help of the simple formula

$$t_i = t_{i-1} + \Delta t.$$

This produces the recurrent neural network that is shown in Fig. 9.6. The bottom neuron advances only the time t by subtracting in each step the bias value $-\Delta t$ from the current activation. The top $n-1$ neurons have the network input function

$$f_{\text{net}}^{(u)}(z, w) = wz = \Delta t\, z,$$

the activation function

$$f_{\text{act}}^{(u)}(\text{act}_u, \text{net}_u, \theta_u) = \text{act}_u + \text{net}_u - \theta_u$$

and the identity as their output function. The weights of the connections to the second neuron from the bottom, its bias value as well as its network input, activation and output function depend on the form of the differential equation. For example, if the differential equation is linear and has constant coefficients, then the network

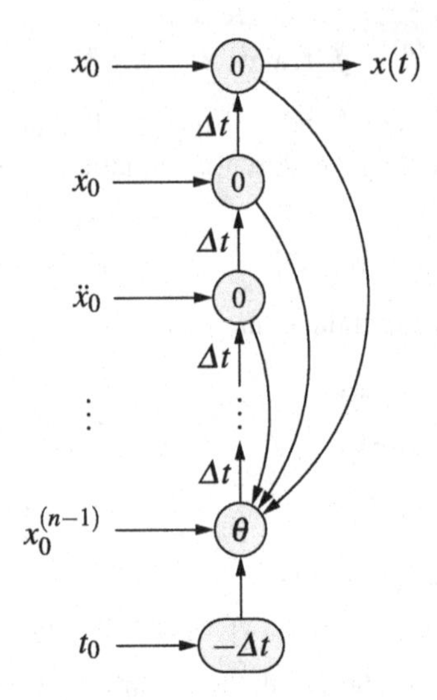

Fig. 9.6 General structure of a recurrent neural network representing an explicit differential equation of nth order. The weights of the feedback loops and the input function of the second neuron from the bottom depend on the form of the differential equation. Of course, this network not only produces approximations for $x(t)$, but also for $\dot{x}(t)$, $\ddot{x}(t)$ etc.

input function is the weighted sum (as for the neurons of a multi-layer perceptron), the activation function is a linear function and the output function is the identity.

9.3 Vectorial Neural Networks

Up to now we considered only differential equations of a function $x(t)$. However, in practice one often meets systems of differential equations with more than one function. A simple example are the differential equations of a two-dimensional movement, for example, of an oblique throw: a (punctiform) body is thrown at time t_0 from the point (x_0, y_0) of a coordinate system with horizontal x-axis and vertical y-axis, namely with the initial velocity $v_0 = v(t_0)$ and with the (upward) angle φ, $0 \leq \varphi \leq \frac{\pi}{2}$, w.r.t. the x-axis (see Fig. 9.7). In this case, we have to compute the functions $x(t)$ and $y(t)$, which describe the position of the body at time t. If we ignore air friction, we have the two equations

$$\ddot{x} = 0 \quad \text{and} \quad \ddot{y} = -g,$$

where $g = 9.81\,\text{ms}^{-2}$ is the gravitational acceleration on the surface of the earth. That is, the body moves uniformly in horizontal direction (no acceleration) and in vertical direction accelerated by the gravitational attraction of the earth. In addition, we have the initial conditions $x(t_0) = x_0$, $y(t_0) = y_0$, $\dot{x}(t_0) = v_0 \cos\varphi$ and $\dot{y}(t_0) = v_0 \sin\varphi$. By introducing—according to the general principle described in

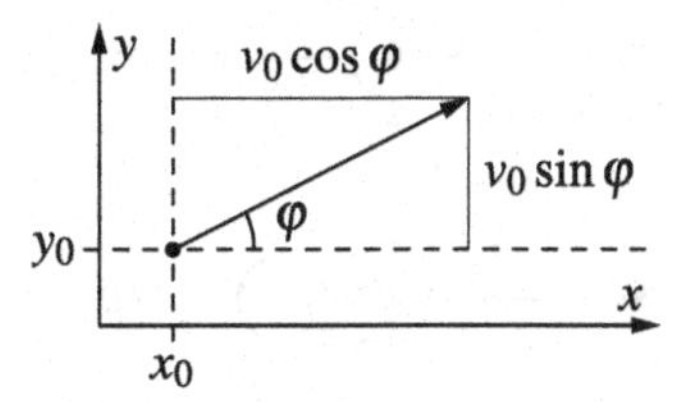

Fig. 9.7 Oblique throw of a body

the preceding section—the intermediary quantities $v_x = \dot{x}$ and $v_y = \dot{y}$, we arrive at the system of differential equations

$$\dot{x} = v_x, \qquad \dot{v}_x = 0,$$
$$\dot{y} = v_y, \qquad \dot{v}_y = -g,$$

from which we can derive the recursion formulas

$$x(t_i) = x(t_{i-1}) + \Delta t\, v_x(t_{i-1}), \qquad v_x(t_i) = v_x(t_{i-1}),$$
$$y(t_i) = y(t_{i-1}) + \Delta t\, v_y(t_{i-1}), \qquad v_y(t_i) = v_y(t_{i-1}) - \Delta t\, g.$$

The result is a recurrent neural network with two independent sub-networks consisting of two neurons each, one of which updates the position coordinate, while the other updates the corresponding velocity.

However, it appears to be more natural to combine the two coordinates x and y into a position vector $\mathbf{r}$ of the body. Since the differentiation rules transfer directly from scalar functions to vector functions (see, for instance, Greiner 1989), we may treat the derivatives of this position vector in the same way as those of a scalar quantity. The differential equation, from which we start in this case, is

$$\ddot{\mathbf{r}} = -g\mathbf{e}_y.$$

Here $\mathbf{e}_y = (0, 1)$ is the unit vector in y-direction, with which we describe the direction of the gravitational force. The initial conditions are $\mathbf{r}(t_0) = \mathbf{r}_0 = (x_0, y_0)$ and $\dot{\mathbf{r}}(t_0) = \mathbf{v}_0 = (v_0 \cos\varphi, v_0 \sin\varphi)$. As before, we introduce a (now vectorial) intermediary quantity $\mathbf{v} = \dot{\mathbf{r}}$ to obtain the system

$$\dot{\mathbf{r}} = \mathbf{v}, \qquad \dot{\mathbf{v}} = -g\mathbf{e}_y$$

of coupled differential equations. From this system, we read the recursion formulas

$$\mathbf{r}(t_i) = \mathbf{r}(t_{i-1}) + \Delta t\, \mathbf{v}(t_{i-1}),$$
$$\mathbf{v}(t_i) = \mathbf{v}(t_{i-1}) - \Delta t\, g\mathbf{e}_y,$$

which can be represented by two vectorial neurons.

The advantages of such a vectorial representation, which may appear fairly small at this point, become obvious if we refine our model by taking air friction into account. If a body moves in a medium (for example, in air), one distinguishes two types of friction: Stokesian friction, which is proportional to the velocity of the body, and Newtonian friction, which is proportional to the squared velocity of the

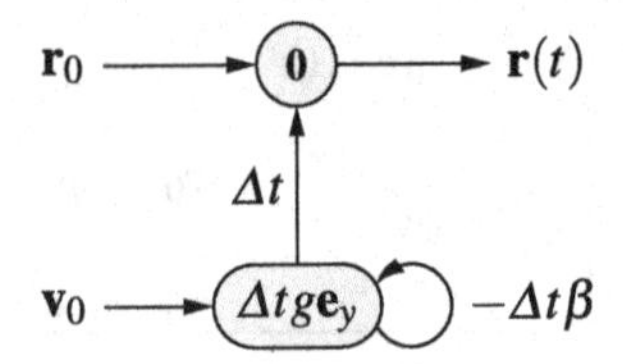

Fig. 9.8 A vectorial recurrent neural network to compute an oblique throw, taking Stokesian friction into account

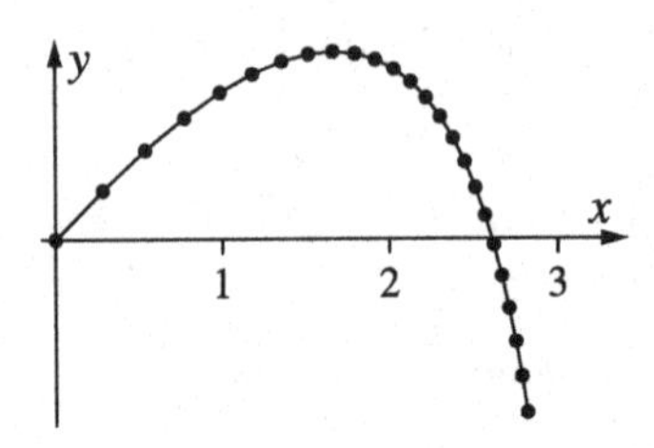

Fig. 9.9 Trajectory of an oblique throw of a body that is computed by the recurrent neural network shown in Fig. 9.8

body (Greiner 1989). If the occurring velocities are low, one usually neglects the Newtonian, if they are high, one neglects the Stokesian friction. As an example, we consider here only Stokesian friction. In this case the equation

$$\mathbf{a} = -\beta \mathbf{v} = -\beta \dot{\mathbf{r}}$$

describes the deceleration of the body that is caused by air friction, where β is a constant that depends on the shape and the volume of the body. In total we thus have the differential equation

$$\ddot{\mathbf{r}} = -\beta \dot{\mathbf{r}} - g\mathbf{e}_y.$$

With the help of the intermediary quantity $\mathbf{v} = \dot{\mathbf{r}}$ we obtain

$$\dot{\mathbf{r}} = \mathbf{v}, \qquad \dot{\mathbf{v}} = -\beta \mathbf{v} - g\mathbf{e}_y,$$

from which we derive the recursion formulas

$$\mathbf{r}(t_i) = \mathbf{r}(t_{i-1}) + \Delta t\, \mathbf{v}(t_{i-1}) \quad \text{and}$$
$$\mathbf{v}(t_i) = \mathbf{v}(t_{i-1}) - \Delta t\, \beta \mathbf{v}(t_{i-1}) - \Delta t\, g\mathbf{e}_y.$$

The corresponding network is shown in Fig. 9.8. Stokesian friction is taken into account by the feedback loop of the bottom neuron.

An example computation with $v_0 = 8$, $\varphi = 45°$, $\beta = 1.8$ and $\Delta t = 0.05$ is shown in Fig. 9.9. Note the steeper right branch of the trajectory, which demonstrates the decelerating effect of Stokesian friction. Without this friction, the trajectory would be a parabola (equal slopes on the left and right branch at the same height).

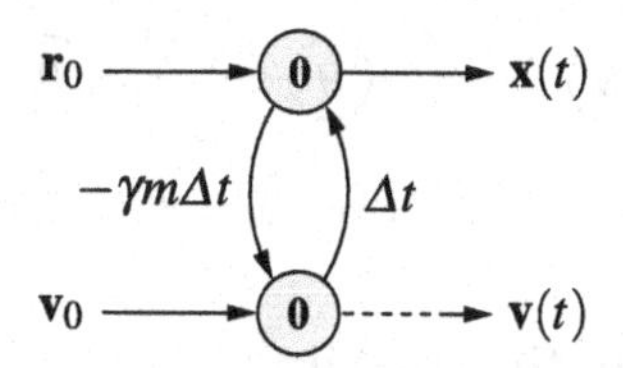

Fig. 9.10 A vectorial recurrent neural network to compute the orbit of a planet

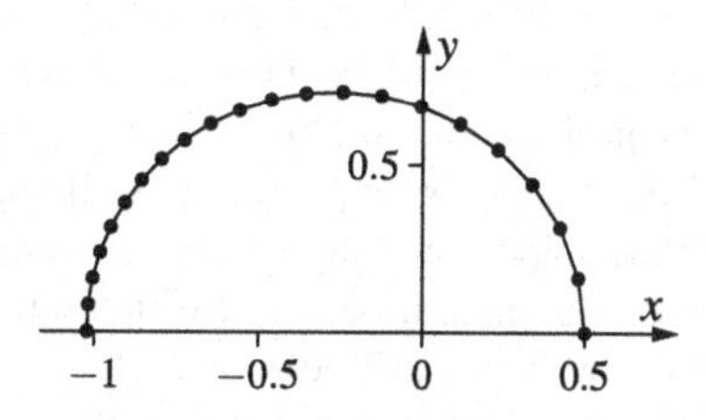

Fig. 9.11 Orbit of a planet that is computed by the recurrent neural network that is shown in Fig. 9.10. The sun is located at the origin of the coordinate system

As a second example, we consider the orbit of a planet (Feynman et al. 1963). The movement of a planet around a central body (sun) of mass m at the origin of the coordinate system can be described by the vectorial differential equation

$$\ddot{\mathbf{r}} = -\gamma m \mathbf{r} |\mathbf{r}|^{-3},$$

where $\gamma = 6.672 \cdot 10^{-11}\ \mathrm{m}^3\,\mathrm{kg}^{-1}\,\mathrm{s}^{-2}$ is the gravitational constant. This equation describes the acceleration of the planet that is caused by the mass attraction between the sun and the planet. As before, we introduce the vectorial velocity $\mathbf{v} = \dot{\mathbf{r}}$ as an intermediary quantity and thus arrive at the system of differential equations

$$\dot{\mathbf{r}} = \mathbf{v}, \qquad \dot{\mathbf{v}} = -\gamma m \mathbf{r} |\mathbf{r}|^{-3}.$$

From this system, we derive the vectorial recursion formulas

$$\mathbf{r}(t_i) = \mathbf{r}(t_{i-1}) + \Delta t\, \mathbf{v}(t_{i-1}) \qquad \text{and}$$
$$\mathbf{v}(t_i) = \mathbf{v}(t_{i-1}) - \Delta t\, \gamma m \mathbf{r}(t_{i-1}) \big|\mathbf{r}(t_{i-1})\big|^{-3},$$

which can be represented by two vectorial neurons as shown in Fig. 9.10. Note, however, that in this case the bottom neurons needs a somewhat unusual network input function (compared to the previous examples): simply multiplying the output of the top neuron with the connection weight is no longer enough.

An example computation with $\gamma m = 1$, $\mathbf{r}_0 = (0.5, 0)$ and $\mathbf{v}_0 = (0, 1.63)$ (following an example of Feynman et al. (1963)) is shown in Fig. 9.11. The diagram nicely shows the elliptical course on which the planet moves faster if it close to the sun (perihelion) than farther away from the sun (aphelion). This illustrates the content of the first two of **Kepler's Laws**, according to which the orbit of a planet is an ellipse (first law) and a radius from the sun to the planet sweeps out equal areas in equal intervals of time (second law).

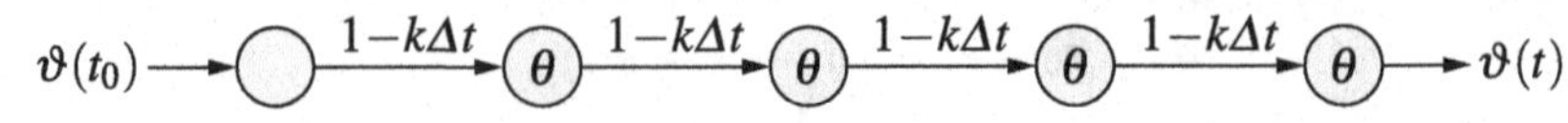

$\vartheta(t_0) \rightarrow \bigcirc \xrightarrow{1-k\Delta t} (\theta) \xrightarrow{1-k\Delta t} (\theta) \xrightarrow{1-k\Delta t} (\theta) \xrightarrow{1-k\Delta t} (\theta) \rightarrow \vartheta(t)$

Fig. 9.12 Unfolding in time of the recurrent neural network shown in Fig. 9.2 (four steps)

9.4 Error Backpropagation in Time

Computations like those executed in the preceding sections are, of course, only possible if one knows both the differential equation that describes the considered (physical) system as well as the values of the parameters appearing in it. However, in practice we often face the problem that we know the form of the differential equation, but not the values of the parameters appearing in it. If measurement data about the considered system are available, one may try in such a case to find the system parameters by training a recurrent neural network, which represents the differential equation. Since the weights and bias values of the neural network are functions of the system parameters, the actual parameter values can be read from them.

In principle, recurrent neural networks are trained in the same way as multi-layer perceptrons, namely by error backpropagation (see Figs. 5.4 and 5.5). However, due to the feedback loops, this method cannot be applied directly, since these loops would propagate the error signals in a cyclic fashion. This problem is solved by **unfolding** the network **in time** between two training patterns. This special form of error backpropagation is called **error backpropagation through time**.

We illustrate here only the basic principle with the help of Newton's cooling law, which we studied in Sect. 9.1 (see page 143). We assume that we have measurement values from the cooling (or warming) of a body at our disposal, which state the temperature of the body at different points in time. In addition, we assume that the temperature ϑ_a of the medium is known, into which the body is placed (ambient temperature). From these measurement values, we desire to determine the value of the (unknown) cooling constant k of the body.

Like for the training of multi-layer perceptrons, the weight and the bias value are initialized randomly. The time between two consecutive measurement values is divided—in analogy to Sect. 9.1—into a certain number of intervals. According to the chosen number of intervals, the feedback loop of the network is then "unfolded." For example, if there are four intervals between one measurement value and the next, that is, if $t_{j+1} = t_j + 4\Delta t$, we obtain the network shown in Fig. 9.12, which possesses five neurons. Note that the neurons of this network do not possess feedback loops, neither explicit nor implicit ones. As a consequence, the connection weights are $1 - k\Delta t$: the 1 represents the implicit feedback loop of the network in Fig. 9.2 (cf. the explanations given on page 146).

If a measurement value ϑ_j (temperature of the body at time t_j) is fed into this network, it computes—with the current values of the weight and the bias values—an approximation for the next measurement value ϑ_{j+1} (temperature of the body at time $t_{j+1} = t_j + 4\Delta t$). By comparing this value with the actual value ϑ_{j+1} we obtain an error signal, which is propagated with the known formulas of error backpropagation and thus causes adaptations of the weights and the bias values.

However, we have to pay attention to the fact that the network shown in Fig. 9.12 actually possesses only one weight and one bias value, since all weights refer to the same feedback loop and all bias values refer to the same neuron. Hence, the derived adaptations have to aggregated and must only be applied at the end of the procedure to change the one connection weight and the one bias value that actually exit. Furthermore, note that both the weight as well as the bias value contain the unknown parameter k, but only known constants apart from it. Therefore, it is advisable to transform the weight and bias changes computed by error backpropagation into changes of this single free parameter k. Thus, only one quantity needs to be adapted, from which both weight and bias value are then computed.

It should be clear that in practice one does not proceed in the described manner for such a simple differential equation as Newton's cooling law. Since this equation can be solved analytically, better and more direct methods are available to determine the value of the unknown parameter k. Starting from the analytical solution of the differential equation, the problem may be solved, for instance, with the regression methods studied in Sect. 5.3: with a suitable transformation of the measurement data the problem is reduced to the task of finding a best fit line (regression line) through the origin, the only parameter of which is the cooling constant k.

However, there are many practical problems for which it makes sense to find unknown system parameters by training a recurrent neural network. In general, this is always the case if the differential equations describing the system under consideration cannot be solved analytically. As an example, we mention here the problem of finding tissue parameters for virtual surgery, especially virtual laparoscopy[2] (Radetzky and Nürnberger 2002). The systems of coupled differential equations occurring in this application are too complex (mainly due to the high number of equations), so that they cannot be solved analytically. By training recurrent neural networks, however, remarkable results could be achieved.

References

R.P. Feynman, R.B. Leighton and M. Sands. *The Feynman Lectures on Physics, Vol. 1: Mechanics, Radiation, and Heat*. Addison-Wesley, Reading, MA, USA, 1963

W. Greiner. *Mechanik, Teil 1 (Series: Theoretische Physik)*. Verlag Harri Deutsch, Thun/Frankfurt am Main, Germany, 1989. English edition: *Classical Mechanics*. Springer-Verlag, Berlin, Germany, 2002

H. Heuser. *Gewöhnliche Differentialgleichungen*. Teubner, Stuttgart, Germany, 1989

A. Radetzky and A. Nürnberger. Visualization and Simulation Techniques for Surgical Simulators Using Actual Patient's Data. *Artificial Intelligence in Medicine* 26:3, 255–279. Elsevier Science, Amsterdam, Netherlands, 2002

[2] A laparoscope is a medical instrument with which a physician can examine the abdominal cavity through small incisions. In virtual laparoscopy an examination of the abdominal cavity is simulated with a computer and a force-feedback device in the shape of a laparoscope in order to teach medical students how to use this instrument properly.

Chapter 10
Mathematical Remarks

The following sections treat mathematical topics that were presupposed in the text (Sect. 10.1 on straight line equations and Sect. 10.2 on regression), or side remarks, which would have disturbed the flow of the exposition (Sect. 10.3 on activation transformation in a Hopfield network).

10.1 Equations for Straight Lines

In this section, a few important facts about straight lines and their equations have been collected, which are used in Chap. 3 on threshold logic units. More extensive explanations can be found in any textbook on linear algebra.

Straight lines are commonly described in one of the following forms:

explicit form:	$g \equiv x_2 = bx_1 + c$
implicit form:	$g \equiv a_1x_1 + a_2x_2 + d = 0$
point-direction form:	$g \equiv \mathbf{x} = \mathbf{p} + k\mathbf{r}$
normal form:	$g \equiv (\mathbf{x} - \mathbf{p})\mathbf{n} = 0$

with the parameters

b : slope of the line
c : intercept
$\mathbf{p}$: position vector of a point of the line (support vector)
$\mathbf{r}$: direction vector of the line
$\mathbf{n}$: normal vector of the line.

It is a disadvantage of the explicit form that straight lines that are parallel to the x_2-axis cannot be represented. All other forms can represent arbitrary lines.

The implicit form and the normal form are closely related to each other, because the coefficients a_1 and a_2 of the variables x_1 and x_2, respectively, are the coordinates of a normal vector of the line. That is, we may use $\mathbf{n} = (a_1, a_2)$ in the normal form. Expanding the normal form also shows that $d = -\mathbf{pn}$.

R. Kruse et al., *Computational Intelligence*, Texts in Computer Science,
DOI 10.1007/978-1-4471-5013-8_10, © Springer-Verlag London 2013

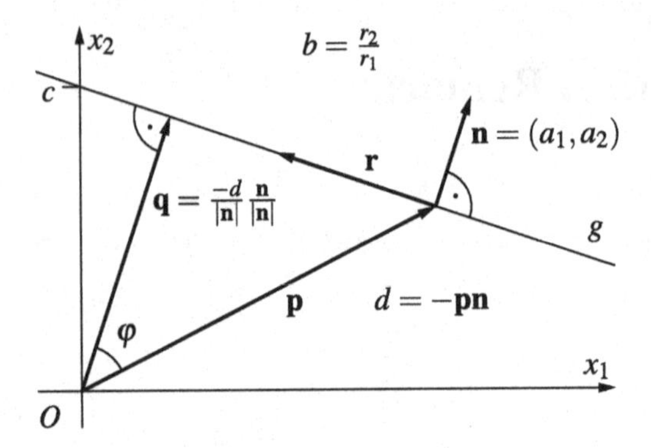

Fig. 10.1 A straight line and the parameters describing it

The relations between the parameters of the different forms of stating a straight line are shown in Fig. 10.1. Particularly important is the vector $\mathbf{q}$, which provides an interpretation for the parameter d of the implicit form. The vector $\mathbf{q}$ is obtained by projecting the support vector $\mathbf{p}$ onto the direction normal to the straight line. This is achieved with the scalar product. It is

$$\mathbf{pn} = |\mathbf{p}||\mathbf{n}|\cos\varphi.$$

From the diagram, we see that $|\mathbf{q}\,| = |\mathbf{p}\,|\cos\varphi$. There fore we have

$$|\mathbf{q}\,| = \frac{|\mathbf{pn}|}{|\mathbf{n}|} = \frac{|d|}{|\mathbf{n}|}.$$

Hence, $|d|$ measures the distance of the straight line from the origin of the coordinate system relative to the length of the normal vector. If $\sqrt{a_1^2 + a_2^2} = 1$, that is, if the normal vector has unit length, then $|d|$ measures this distance directly. In this case the normal form is called the **Hessian normal form** of the line equation.

If one takes into account that $\mathbf{pn}$ becomes negative if $\mathbf{n}$ does not point away from the origin (as in the diagram), but towards it, one finally obtains:

$$\mathbf{q} = \frac{\mathbf{pn}}{|\mathbf{n}|}\frac{\mathbf{n}}{|\mathbf{n}|} = \frac{-d}{|\mathbf{n}|}\frac{\mathbf{n}}{|\mathbf{n}|}.$$

Note that $\mathbf{q}$ always points from the origin to the straight line, regardless of whether $\mathbf{n}$ points towards the origin or away from it. Therefore, we can read the location of the origin from the sign of d:

$d = 0$: The straight line contains the origin,
$d < 0$: $\mathbf{n} = (a_1, a_2)$ points away from the origin,
$d > 0$: $\mathbf{n} = (a_1, a_2)$ points towards the origin.

Of course, we can carry out these computations not only for a support vector $\mathbf{p}$ of the straight line, but for an arbitrary vector $\mathbf{x}$ (see Fig. 10.2). Thus, we obtain a

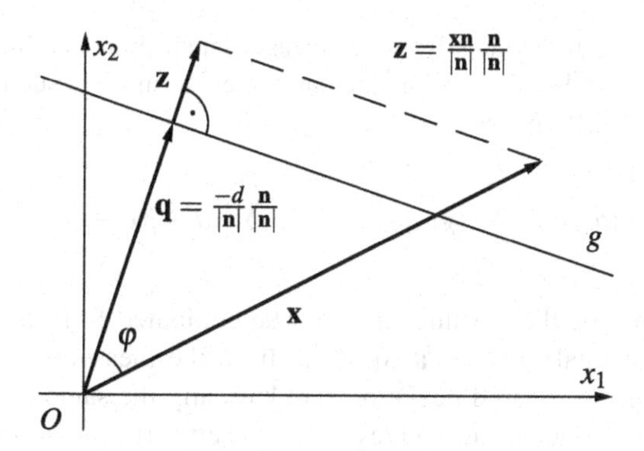

Fig. 10.2 Determining the side of straight line on which a point **x** lies

vector **z** that is the projection of the vector **x** onto the direction normal to the line. By comparing this vector to the vector **q** considered above, we can determine on which side of the straight line the point lies that has the position vector **x**:

A point with position vector **x** *lies on the side of the straight line to which the normal vector* **n** *points, if* $\mathbf{xn} > -d$*, and on the other side, if* $\mathbf{xn} < -d$*. If* $\mathbf{xn} = -d$*, the point lies on the straight line.*

It should be clear that these considerations are not restricted to straight lines, but can be transferred immediately to planes and hyperplanes. Thus, we can easily determine for them as well on which side a point with given position vector lies.

10.2 Regression

This section recalls the **method of least squares**, also known as **regression**, which is well known in calculus and statistics. It is used to determine best fit straight lines (regression lines) and generally best fit polynomials (regression polynomials). The following exposition follows mainly (Heuser 1988).

(Physical) measurement data rarely show the exact relationship of the measured quantities as it is described by physical laws, since they are inevitably afflicted by errors. If one wants to determine the relationship of the quantities nevertheless (at least approximately), one faces the task to find a function that fits the measurement points as well as possible, so that the measurement errors are somehow "balanced." Naturally, in order to achieve this, we should have a hypothesis about the type of relationship, so that we can choose a function class and thus reduce the problem to the selection of the parameters of a function of a specific type.

For example, if we expect two quantities x and y to exhibit a linear dependence (for instance, because a scatter plot of the measurement points suggests such a relationship), we have to determine the parameters a and b of the straight line $y = g(x) = a + bx$. However, due to the inevitable measurement errors it will generally not be possible to find a straight line in such a way that all n given measurement points (x_i, y_i), $1 \le i \le n$, lie exactly on this straight line. Rather we have to try to find a straight line that deviates from the measurement points as little as possible. Therefore, it is plausible to determine the parameters a and b in such a way that the sum of the squared differences

$$F(a,b) = \sum_{i=1}^{n} \left(g(x_i) - y_i\right)^2 = \sum_{i=1}^{n} (a + bx_i - y_i)^2$$

is minimized. That is, the y-values that can be computed from the line equation should deviate (in total) as little as possible from the measured values. The reasons for choosing the squared deviations are basically the same as those given in Sect. 4.3: in the first place using squares makes the error functions continuously differentiable everywhere. In contrast to this, the derivative of the absolute value, which would be an obvious alternative, does not exist/is not continuous at 0. Secondly, squaring the deviations weights large deviations more heavily than small ones, so that there is a tendency to avoid individual large deviations from the measured data.[1]

A necessary condition for a minimum of the error function $F(a,b)$ defined above is that the partial derivatives of this function w.r.t. the parameters a and b vanish:

$$\frac{\partial F}{\partial a} = \sum_{i=1}^{n} 2(a + bx_i - y_i) = 0 \quad \text{and}$$

$$\frac{\partial F}{\partial b} = \sum_{i=1}^{n} 2(a + bx_i - y_i)x_i = 0.$$

From these equations we obtain, after a few simple transformations, the so-called **normal equations**

$$na + \left(\sum_{i=1}^{n} x_i\right) b = \sum_{i=1}^{n} y_i,$$

$$\left(\sum_{i=1}^{n} x_i\right) a + \left(\sum_{i=1}^{n} x_i^2\right) b = \sum_{i=1}^{n} x_i y_i,$$

that is, a linear equation system with two equations and two unknowns a and b. It can be shown that this equation system has a unique solution unless the x-values of all measurement points are identical (that is, unless $x_1 = x_2 = \cdots = x_n$), and that this solution indeed describes a minimum of the function F (Heuser 1988). The straight line $y = g(x) = a + bx$ determined in this way is called the **best fit (straight) line** or the **regression line** for the data set $(x_1, y_1), \ldots, (x_n, y_n)$.

[1] Note, however, that this can also be a disadvantage. If the data set contains "outliers" (that is, measurement values that due to random, disproportionally large measurement errors deviate strongly from the true value), the position of the regression line may be influenced heavily by very few measurement points (precisely the outliers), which can lead to an unusable result.

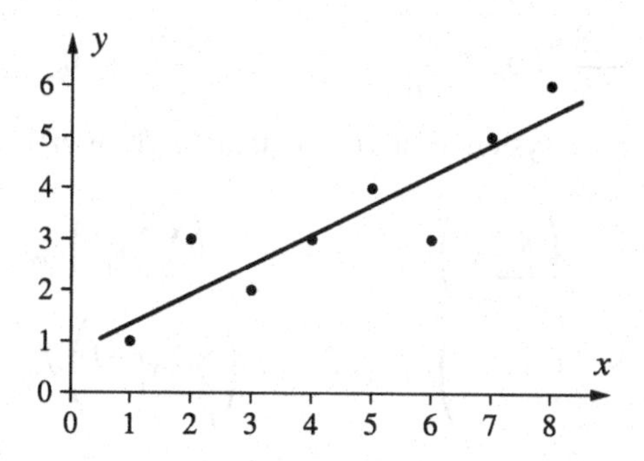

Fig. 10.3 An example data set and a regression line that was computed with the method of least squares

To illustrate the procedure, we consider a simple example. Let the data set consisting of eight data points $(x_1, y_1), \dots, (x_8, y_8)$ be given that is shown in the following table (Heuser 1988) (see also Fig. 10.3):

x	1	2	3	4	5	6	7	8
y	1	3	2	3	4	3	5	6

In order to set up the system of normal equations, we compute

$$\sum_{i=1}^{8} x_i = 36, \qquad \sum_{i=1}^{8} x_i^2 = 204, \qquad \sum_{i=1}^{8} y_i = 27, \qquad \sum_{i=1}^{8} x_i y_i = 146.$$

Thus we obtain the equation system (normal equations)

$$\begin{aligned} 8a + \ \ 36b &= \ \ 27, \\ 36a + 204b &= 146, \end{aligned}$$

which possesses the solution $a = \frac{3}{4}$ and $b = \frac{7}{12}$. Therefore the regression line is

$$y = \frac{3}{4} + \frac{7}{12}x.$$

This line is shown, together with the data points we started from, in Fig. 10.3.

The method we just considered is, of course, not limited to straight lines, but can be extended at least to polynomials. In this case one tries to find a polynomial

$$y = p(x) = a_0 + a_1 x + \cdots + a_m x^m$$

with a given, fixed degree m that approximates the n data points $(x_1, y_1), \dots, (x_n, y_n)$ as well as possible. In this case we have to minimize

$$F(a_0, a_1, \dots, a_m) = \sum_{i=1}^{n} \bigl(p(x_i) - y_i\bigr)^2 = \sum_{i=1}^{n} \bigl(a_0 + a_1 x_i + \cdots + a_m x_i^m - y_i\bigr)^2.$$

Necessary conditions for a minimum are again that the partial derivatives w.r.t. the parameters a_0 to a_m vanish, that is,

$$\frac{\partial F}{\partial a_0} = 0, \quad \frac{\partial F}{\partial a_1} = 0, \quad \ldots, \quad \frac{\partial F}{\partial a_m} = 0.$$

In this way, we obtain the system of normal equations (Heuser 1988)

$$\begin{array}{ccccccc}
n a_0 & + & \left(\sum_{i=1}^{n} x_i\right) a_1 & + \cdots + & \left(\sum_{i=1}^{n} x_i^m\right) a_m & = & \sum_{i=1}^{n} y_i \\
\left(\sum_{i=1}^{n} x_i\right) a_0 & + & \left(\sum_{i=1}^{n} x_i^2\right) a_1 & + \cdots + & \left(\sum_{i=1}^{n} x_i^{m+1}\right) a_m & = & \sum_{i=1}^{n} x_i y_i \\
\vdots & & \vdots & & \vdots & & \vdots \\
\left(\sum_{i=1}^{n} x_i^m\right) a_0 & + & \left(\sum_{i=1}^{n} x_i^{m+1}\right) a_1 & + \cdots + & \left(\sum_{i=1}^{n} x_i^{2m}\right) a_m & = & \sum_{i=1}^{n} x_i^m y_i,
\end{array}$$

from which the parameters a_0 to a_m can be derived with the usual methods of linear algebra (Gaussian elimination, Cramer's rule, inverting the coefficient matrix etc.). The resulting polynomial $p(x) = a_0 + a_1 x + a_2 x^2 + \cdots + a_m x^m$ is called **best fit polynomial** or **regression polynomial** of degree m for the data set $(x_1, y_1), \ldots, (x_n, y_n)$.

Furthermore, the method of least squares cannot only be used, as considered up to now, to compute regression polynomials, but may as well be employed to fit functions with more than one argument. This case is called **multiple** or **multivariate regression**. We consider, as an example, only the special case of **multilinear regression** and confine ourselves to a function with two arguments. That is, we consider, how one can find a best fitting function of the form

$$z = f(x, y) = a + bx + cy$$

for a given data set $(x_1, y_1, z_1), \ldots, (x_n, y_n, z_n)$ in such a way that the sum of squared errors is minimized. In this case, the normal equations are derived in a perfectly analogous way. We have to minimize

$$F(a, b, c) = \sum_{i=1}^{n} \big(f(x_i, y_i) - z_i\big)^2 = \sum_{i=1}^{n} (a + bx_i + cy_i - z_i)^2.$$

Necessary conditions for a minimum are

$$\begin{aligned}
\frac{\partial F}{\partial a} &= \sum_{i=1}^{n} 2(a + bx_i + cy_i - z_i) &= 0, \\
\frac{\partial F}{\partial b} &= \sum_{i=1}^{n} 2(a + bx_i + cy_i - z_i)x_i &= 0, \\
\frac{\partial F}{\partial c} &= \sum_{i=1}^{n} 2(a + bx_i + cy_i - z_i)y_i &= 0.
\end{aligned}$$

Therefore we obtain the system of normal equations

$$\begin{aligned}
na + \left(\sum_{i=1}^{n} x_i\right)b + \left(\sum_{i=1}^{n} y_i\right)c &= \sum_{i=1}^{n} z_i,\\
\left(\sum_{i=1}^{n} x_i\right)a + \left(\sum_{i=1}^{n} x_i^2\right)b + \left(\sum_{i=1}^{n} x_i y_i\right)c &= \sum_{i=1}^{n} z_i x_i,\\
\left(\sum_{i=1}^{n} y_i\right)a + \left(\sum_{i=1}^{n} x_i y_i\right)b + \left(\sum_{i=1}^{n} y_i^2\right)c &= \sum_{i=1}^{n} z_i y_i
\end{aligned}$$

from which a, b and c can easily be computed.

It should be immediately clear that the method of least squares can also be extended to polynomials in multiple variables. How it may also be extended, under certain conditions, to other function classes is demonstrated in Sect. 5.3 with the help of the example of **logistic regression**.

A program for multi-variate polynomial regression that uses ideas from dynamic programming to quickly compute the needed power products can be found at

http://www.borgelt.net/regress.html

10.3 Activation Transformation

In this section we demonstrate how the weights and thresholds of a Hopfield network that works with activations 0 and 1 can be transformed into the corresponding parameters of a Hopfield network that works with the activations -1 and $+1$ (and vice versa). This shows that the two network types are essentially equivalent, and thus that it was justified to choose in Chap. 8 whichever form was more suitable for the specific task under consideration.

In the following we indicate by an upper index of the considered quantities what the range of activation values of the neural network is, to which they refer:

$$\begin{aligned}
{}^{0} &: \text{quantity of a network with } \mathrm{act}_u \in \{\ \ 0, 1\},\\
{}^{-} &: \text{quantity of a network with } \mathrm{act}_u \in \{-1, 1\}.
\end{aligned}$$

Clearly, we must always have

$$\mathrm{act}_u^0 = \frac{1}{2}(\mathrm{act}_u^- + 1) \quad \text{and} \quad \mathrm{act}_u^- = 2\,\mathrm{act}_u^0 - 1.$$

That is, the neuron u either has activation 1 in both networks or it has activation 0 in one network and activation -1 in the other. In order to achieve that both network types exhibit the same behavior, it must also hold that:

$$s(\mathrm{net}_u^- - \theta_u^-) = s\big(\mathrm{net}_u^0 - \theta_u^0\big),$$

where

$$s(x) = \begin{cases} 1, & \text{if } x \geq 0, \\ -1, & \text{otherwise.} \end{cases}$$

Only if this is the case the activation changes are the same in both networks. The above equation clearly holds if

$$\text{net}_u^- - \theta_u^- = \text{net}_u^0 - \theta_u^0.$$

(Note that this is a sufficient, but not a necessary condition.) Using the relations between the activations stated above, we obtain from this equation

$$\begin{aligned}
\text{net}_u^- - \theta_u^- &= \sum_{v \in U - \{u\}} w_{uv}^- \, \text{act}_u^- - \theta_u^- \\
&= \sum_{v \in U - \{u\}} w_{uv}^- (2 \, \text{act}_u^0 - 1) - \theta_u^- \\
&= \sum_{v \in U - \{u\}} 2 w_{uv}^- \, \text{act}_u^0 - \sum_{v \in U - \{u\}} w_{uv}^- - \theta_u^- \\
&\stackrel{!}{=} \text{net}_u^0 - \theta_u^0 \\
&= \sum_{v \in U - \{u\}} w_{uv}^0 \, \text{act}_u^0 - \theta_u^0.
\end{aligned}$$

This equation holds if we choose

$$\begin{aligned}
w_{uv}^0 &= 2 w_{uv}^- \quad \text{and} \\
\theta_u^0 &= \theta_u^- + \sum_{v \in U - \{u\}} w_{uv}^-.
\end{aligned}$$

For the opposite direction, we obtain

$$\begin{aligned}
w_{uv}^- &= \frac{1}{2} w_{uv}^0 \quad \text{and} \\
\theta_u^- &= \theta_u^0 - \sum_{v \in U - \{u\}} w_{uv}^- = \theta_u^0 - \frac{1}{2} \sum_{v \in U - \{u\}} w_{uv}^0.
\end{aligned}$$

References

H. Heuser. *Lehrbuch der Analysis, Teil 1+2*. Teubner, Stuttgart, Germany, 1988

Part II
Evolutionary Algorithms

Chapter 11
Introduction to Evolutionary Algorithms

Evolutionary algorithms comprise a class of optimization techniques that imitate principles of biological evolution. They belong to the family of **metaheuristics**, which also includes, for example, particle swarm (Kennedy and Eberhart 1995) and ant colony optimization (Dorigo and Stützle 2004), which are inspired by other biological structures and processes, as well as classical methods like simulated annealing (Metropolis et al. 1953; Kirkpatrick et al. 1983), which is inspired by a thermodynamical process. The core principle of evolutionary algorithms is to apply evolution principles like mutation and selection to populations of candidate solutions in order to find a (sufficiently good) solution for a given optimization problem.

11.1 Metaheuristics

Metaheuristics are fairly general computational techniques that are typically used to solve numerical and combinatorial optimization problems approximately in several iterations (as opposed to analytically and exactly in a single step). Metaheuristics are generally defined as an abstract sequence of operations on certain objects and are applicable to essentially arbitrary problems. However, the objects operated on and the steps to be carried out must be adapted to the specific problem at hand. Thus the core task is usually to find a proper mapping of a given problem to the abstract structures and operations that constitute the metaheuristic.

Metaheuristics are usually applied to problems for which no efficient solution algorithm is known, that is, problems, for which all known algorithms have an (asymptotic) time complexity that is exponential in the problem size. In practice, such problems can rarely be solved exactly, due to the prohibitively high demands on computing time and/or computing power. As a consequence, approximate solutions have to be accepted, and this is what metaheuristics can provide. Although there is no guarantee that they will find the optimal solution or even a solution of a given minimum quality (although this is not impossible either), they usually offer good chances of finding a "sufficiently good" solution.

R. Kruse et al., *Computational Intelligence*, Texts in Computer Science,
DOI 10.1007/978-1-4471-5013-8_11,

The success and the execution time of metaheuristics depend critically on a proper mapping of the problem to the steps of the metaheuristic and the efficient implementation of every single step. Many metaheuristics work by iteratively improving a set of so-called candidate solutions. They differ in the methods they employ to vary solutions in order to possibly improve them, in the principles by which partial solutions are combined or elements of found solutions are exploited to find new solutions, as well as in the principles by which a new set of candidate solutions is selected from the previously created ones. However, they share that they usually carry out a **guided random search** in the space of solution candidates. That is, these algorithms carry out a search that contains certain random elements to explore the search space, but they are also guided by some measure of the solution quality, which governs which (parts of) solution candidates are focused on or at least kept for further exploration and which are discarded, because they are not promising.

An important advantage of metaheuristics is the fact that they can usually be terminated after any iteration step (so-called **anytime algorithms**), because they have, at any point in time, at least some solution candidates available. From these the best solution candidate found so far is then retrieved and returned, regardless of whether some other termination criterion is met or not. However, it should be clear that the solution quality is usually the better, the longer the search can run.

A wide range of metaheuristics based on various principles has been proposed, many of which are nature-inspired. While evolutionary algorithms rely on principles of biological evolution in various forms, (particle) swarm optimization (Kennedy and Eberhart 1995) mimics the behavior of swarms of animals (like fish or birds) that search for food in schools or flocks. Ant colony optimization (Dorigo 1992; Dorigo and Stützle 2004) mimics the path finding behavior of ants and termites. Other biological entities that inspired metaheuristics include, among others, honey bees (Nakrani and Tovey 2004) or the immune system of vertebrates (Farmer et al. 1986). Alternatives are algorithms that draw on physical rather than biological analogies, like simulated annealing (Metropolis et al. 1953; Kirkpatrick et al. 1983), which mimics annealing processes, threshold accepting (Dueck and Scheuer 1990) or the deluge algorithm (Dueck 1993).

11.2 Biological Evolution

Evolutionary algorithms are among the oldest and most popular metaheuristics. They are essentially based on Charles Darwin's theory of biological evolution, which he proposed in his seminal book "The Origin of Species" (Darwin 1859).[1] This theory explains the diversity and complexity of all forms of living organisms and allows us to unify all biological disciplines. Non-technical modern introductions to the theory of biological evolution can be found in the popular and very readable books (Dawkins 1976, 1986, 2009).

[1] The full title "The Origin of Species by Means of Natural Selection, or the Preservation of Favoured Races in the Struggle for Life" is usually shortened to merely "The Origin of Species."

The core principle of biological evolution can be formulated as:

Beneficial traits resulting from random variation are favored by natural selection.

That is, individuals with beneficial traits have better chances to procreate and multiply, which may also be captured by the expression **differential reproduction**.

New or at least modified traits may be created by various processes. There is, in the first place, the both blind and purely random modification of genes, that is, **mutation**, which affects both sexually and asexually reproducing life forms. Mutations may occur due to exposure to radioactivity (e.g., caused by earth or cosmic radiation or nuclear reactor disasters) or to so-called *mutagens* (i.e., chemical compounds that disturb the copying process of the genetic information), but may also happen simply naturally, due to an unavoidable susceptibility of the complex genetic copying process to errors. In sexual reproduction, equally blindly and purely randomly selected halves of the (diploid) chromosome sets of the parents are (re-)combined, thus creating new combinations of traits and physical characteristics. In addition, during *meiosis* (i.e., the cell division process that produces the germ cells or *gametes*), parts of (homologous) chromosomes, in a process called **crossover** (or *crossing over*), may cross each other, break, and rejoin in a modified fashion, thus exchanging genetic material between (homologous) chromosomes. As a result offspring with new or at least modified genetic plans and thus physical traits is created.

The vast majority of these (genetic) modifications are unfavorable or even harmful, in the worst case rendering the resulting individual unable to live. However, there is a (small) chance that some of these modifications result in (small) improvements, endowing the individual with traits that help it to survive. For example, they may make it better able to find food, to defend itself against predators or at least to hide or run from them, to attract mates for reproduction etc. Generally speaking, each individual is put to the test in its natural environment where it either proves to have a high fitness, making it more likely to procreate and multiply, or where it fails to survive or mate, thus causing it or its traits to disappear.

Note that the natural selection process is driven by both the natural environment and the individual traits, leading to different reproduction rates or probabilities. Life forms with traits that are better fitted to their environment usually have more offspring on average. Consequently their traits become more frequent with each generation of individuals. On the other hand, life forms with traits less favorable in their environment usually have less offspring on average and thus might even become extinct after some generations (at least in this environment).

It is important to understand that a trait is not beneficial or harmful in itself, but only w.r.t. the environment. For example, while the dark skin color of many Africans protects their skin against the intense sun in regions close to the equator, their skin pigmentation can turn out to be a disadvantage in regions where sunlight is scarce, because it increases the risk of vitamin D deficiency (Harris 2006), as vitamin D is produced in the skin under the influence of ultraviolet light. On the other hand, humans with little skin pigmentation may be less likely to suffer from vitamin D deficiency, but may have to take more care to protect their skin against sunlight to reduce the risk of premature skin aging and skin cancer (Brenner and Hearing 2008). Another example is sickle cell anemia, which is a usually harmful misshaping of hemoglobin (red blood cells), because it reduces the capability of hemoglobin to transport oxygen. However, it has certain protective effects against malaria and thus can be an advantage in regions in which malaria is common (Aidoo et al. 2002).

Note also that it is the interaction of random variation and natural selection that explains why there are so many complex life forms, even though these life forms are extremely unlikely to come into existence as the result of pure chance. However, evolution is *not* just pure chance. Although the variations are blind and random, their selection is not, but strictly driven by their benefit for the survival and procreation of the individuals endowed with them. As a consequence, small improvements can accumulate over many generations and finally lead to surprising complexity and strikingly fitting adaptations to an environment.

One of the reasons why we tend to make wrong estimates of the probability of complex and adapted life forms, seeing them as (much) less likely than they actually are, is that we tend to use the following model: consider a box full of scrap metal parts collected from a junkyard, all of which are car parts. If we shake this box long enough there is a certain (though extremely small) probability that after some time a drivable car is assembled in the box. This model represents creating something complex by mere chance. It clearly makes it effectively impossible that anything even remotely complex, let alone a living organisms, comes into existence.

In evolution, however, each small variation is immediately put to the test w.r.t. an environment and only the beneficial variations are kept and extended. Therefore, a better model is the following thought experiment, suggested by B.F. Skinner (Dawkins 1986): Suppose we want to convince people that we can predict the outcome of horse races. We send a letter (or nowadays maybe rather an email) to 10,000 people, predicting the winning horse in the next race. There are ten horses and since we actually have no clue which one will win, we predict to the first 1000 people that horse number 1 will win, to the next 1000 people that horse number 2 will win etc. Thus, after the race has taken place, we are sure to have 1000 people to whom we predicted the correct horse, while we forget about the 9000 others. We repeat the process with another horse race, predicting horse number 1 to the first 100 people (of the 1000 that remained after the first race), horse number 2 to the second 100 etc. After the race, we have 100 people to whom we predicted the correct horse twice, while we forget about the 900 others, to whom we predicted the wrong horse. Doing another round in the same manner, we end up with 10 people to whom we predicted the winning horse in three consecutive races. In another letter (or email) to these 10 people, we propose to them to predict the winning horse in yet another race, only that this time we charge a fee. Consider the situation from the point of view of these 10 people: they know that we correctly predicted the winning horse in three consecutive races. The chance of achieving such accuracy by mere guessing is 1 in 1000 and thus highly unlikely. Therefore, these 10 people may be inclined to think that we have some insider knowledge that allows us to predict the winning horse much better than by pure chance and thus may be willing to pay the fee.

However, from the process in which these 10 people were selected, we know that it was certain that we would end up with 10 people to which we predicted the winning horse three times in a row. We did not (and need not) know anything about the horses or the races. All we did was to select the successes, that is, the people to whom we made the correct prediction, and ignore the failures, that is, the people to whom we sent a wrong prediction. Evolution works in essentially the same manner: it focuses on the successes (the life forms that survive and procreate) and forgets about the failures (the life forms that go extinct). Since we tend to ignore the failures—simply because we do not see them as they do not exist anymore—we underestimate the probability of seeing an evolved complex individual.

In addition, we have trouble imagining the time—actually billions of years—that has passed since the first, extremely simple cells assembled. There was so much time for variation and selection, for tiny improvements to accumulate, that complex life forms may not even be unlikely, but may actually be almost unavoidable once the process started (at least according to some authors like Dawkins 1986).

Up to now we focused on variation (mutation and recombination) and selection as the fundamental principles of (biological) evolution. These may indeed be the most important constituents, and often a description of evolutionary processes is reduced to these core elements. However, a more detailed analysis reveals many more principles, of which we list some of the more important in the following. A more detailed discussion can be found in Vollmer (1995) or Hartl and Clark (2007).

- **Diversity**
  All forms of life—even organisms of the same species—*differ* from each other, and not just physically, but already in their genetic material (*biological diversity* or *diversity of species*). Nevertheless, the currently actually existing life forms are only a tiny fraction of the theoretically possible ones.
- **Variation**
  Mutation and genetic recombination (in sexual reproduction) continuously create *new variants*. These new variants may exhibit a new combination of already existing traits or may introduce a modified and thus new trait.
- **Inheritance**
  As long as variations enter the germ line, they are *heritable*, that is, they are genetically passed on to the next generation. However, there is generally no inheritance of acquired traits (so-called *Lamarckism* (Lamarck 1809)).
- **Speciation**
  Individuals and populations *diverge genetically*. Thus new *species* are created once their members cannot crossbreed any longer. Speciation gives the phylogenetic "pedigree" its characteristic branching structure.
- **Birth surplus / Overproduction**
  Nearly all life forms produce *more offspring* than can ever become mature enough to procreate themselves.
- **Adaptation / Natural Selection / Differential Reproduction**
  On average, the survivors of a population exhibit such hereditary variations which *increase* their *adaptation* to the local environment. Herbert Spencer's expression

"survival of the fittest" is rather misleading, though. We prefer to speak of *differential reproduction due to different fitness*, because mere survival without offspring is obviously without effect in the long run, especially since the life span of most organisms is limited.

- **Randomness / Blind Variation**
  Variations are *random*. Although *triggered*, *initiated*, or *caused by* something, they do not favor certain traits or beneficial adaptions. In this sense they are *non-teleological* (from the Greek $\tau\varepsilon\lambda o\varsigma$: goal, purpose, objective).
- **Gradualism**
  Variations happen in comparatively *small steps* as measured by the complete information content (entropy) or the complexity of an organism. Thus phylogenetic changes are *gradual* and relatively slow. (In contrast to this *saltationism*—from Latin *saltare*: to jump—means fairly large and sudden changes in development.)
- **Evolution / Transmutation / Inheritance with Modification**
  Due to the adaptation to the environment, species are not immutable. They rather *evolve* in the course of time. Hence the theory of evolution opposes *creationism*, which claims that species are immutable.
- **Discrete Genetic Units**
  The genetic information is stored, transferred and changed in discrete ("atomic", from the Greek $\breve{\alpha}\tau o\mu o\varsigma$: indivisible) units. There is no continuous blend of hereditary traits. Otherwise we might see the so-called *Jenkins Nightmare*, that is, a complete disappearance of any differences in a population due to averaging.
- **Opportunism**
  The processes of evolution are extremely opportunistic. They work exclusively with what is present and not with what once was or could be. Better or optimal solutions are not found if the intermediary stages (that are evolutionarily necessary to build these solutions) exhibit certain fitness handicaps.
- **Evolution-strategic Principles**
  Not only organisms are optimized, but also the mechanisms of evolution. These include parameters like reproduction and mortality rates, life spans, vulnerability to mutations, mutation step sizes, evolutionary speed etc.
- **Ecological Niches**
  Competitive species can tolerate each other if they occupy different ecological niches ("biospheres" in a wider sense) or even create them themselves. This is the only way the observable biological diversity of species is possible in spite of competition and natural selection.
- **Irreversibility**
  The course of evolution is irreversible and unrepeatable.
- **Unpredictability**
  The course of evolution is neither determined, nor programmed, nor purposeful and thus not predictable.
- **Increasing Complexity**
  Biological evolution has led to increasingly more complex systems. However, an open problem in evolutionary biology is the question how we can actually measure the complexity of life forms.

11.3 Simulated Evolution

Given that biological evolution has created complex life forms and solved difficult adaptation problems, it is reasonable to assume that the same optimization principles can be used to find good solutions for (complex) optimization problems. Hence, we start by formally defining optimization problems in Sect. 11.3.1 and consider some of their main properties. In Sect. 11.3.2, we then transfer some basic notions of biological evolution to simulated evolution, pointing out what requirements have to be fulfilled in order to make an evolutionary algorithm approach worthwhile. In Sect. 11.3.3, we then present the building blocks of an evolutionary algorithm in an abstract fashion, before we turn in Sect. 11.4 to a concrete illustrative example.

11.3.1 Optimization Problems

Definition 11.1 (Optimization problem) An *optimization problem* is a pair (Ω, f) consisting of a (search) space Ω of potential solutions and an evaluation function $f : \Omega \to \mathbb{R}$ that assigns a quality assessment $f(\omega)$ to each candidate solution $\omega \in \Omega$. An element $\omega \in \Omega$ is called an (exact) **solution** of the optimization problem (Ω, f) if and only if it is a **global maximum** of f, that is, if $\forall \omega' \in \Omega : f(\omega') \leq f(\omega)$.

Note that, even though the above definition requires a solution to maximize the value of the evaluation function f, this is no actual restriction. If we have a function that yields smaller values for better solutions, we may simply use $-f$ in the definition. In order to capture this, we will speak of a solution as a *global optimum*, thus covering maxima and minima. Note also that a solution need not be unique. There may be multiple elements of Ω for which f yields the same (optimal) value.

As a simple example of an optimization problem consider the task of finding the side lengths of a box with fixed surface area S such that its volume is maximized. Here the search space Ω is the set of all triples (x, y, z), that is, the three side lengths, with $x, y, z \in \mathbb{R}^+$ (i.e., the set of all positive real numbers) and $2xy + 2xz + 2yz = S$, while the evaluation function f is simply $f(x, y, z) = xyz$. In this case, the solution is unique, namely $x = y = z = \sqrt{S/6}$, that is, the box is a cube.

Note that this example already exhibits an important feature that will be considered in more detail later, namely that the search space is **constrained**: we do not consider all triples of real numbers, but only those with positive elements that satisfy $2xy + 2xz + 2yz = S$. In this example, this is even necessary to have a well-defined solution: if x, y and z could be arbitrary (positive) real numbers, there would be no (finite) optimum. The problem thus consists in making sure that in the search for a solution we never leave the search space, that is, that we never consider as solution candidates objects that do not satisfy the constraints.

Optimization problems occur in many applications areas of which the following (certainly incomplete) list gives only a limited impression:

- **Parameter Optimization**
  In many situations a set of (suitable) parameters has to be found such that a given real-valued function is optimized. Such parameters can be, for example, the angle and curvature of the air intake and exhaust pipes for automobile motors to maximize power, the relative quantities of the ingredients of a rubber mixture for tires to maximize grip under different conditions, and the temperatures in different components of a thermal power plant to maximize energy efficiency.
- **Routing Problems**
  The most famous routing problem is clearly the *traveling salesman problem*, which occurs in practice, for instance, if holes have to be drilled in printed circuit board and the distance the drill is to be moved (and thus the movement time) is to be minimized. Other examples include the optimization of delivery routes from a central storage to individual shops or the arrangement of printed circuit board tracks with the objective to minimize length and number of layers.
- **Packing and Cutting Problems**
  Classical examples of packing problems include the *knapsack* (or *backpack* or *rucksack*) *problem*, in which a knapsack of a given (maximum) capacity is to be filled with given goods of known value and size (or weight) such that the total value is maximized, the *bin packing problem*, in which given objects of known size and shape are to be packed into boxes of given size and shape such that the number of boxes is minimized, and the *cutting stock problem* in its various forms, in which geometrical shapes are to be arranged in such a way as to minimize waste (e.g. the wasted cloth after the parts of a garment have been cut out).
- **Arrangement and Location Problems**
  An well-known example of this problem type is the so-called facility location problem, which consists in finding the best placement of multiple facilities (e.g., the distribution nodes in a telephone network), usually under certain constraints. It is also known as Steiner's problem, because certain specific cases are equivalent to the introduction of so-called Steiner points to minimize the length of a spanning tree in a geometric planar graph.
- **Scheduling Problems**
  Job shop scheduling in its various forms, in which jobs have to be assigned to resources at certain times in order to minimize the time to complete all jobs, is a common optimization problem. A special case is the reordering of instructions by a compiler in order to maximize the execution speed of a program. This problem type also includes setting up timetables for schools (where constraints like the number of classrooms and the need to avoid skip hours at least for the lower grades complicate the problem) or trains (in which the number of tracks available on certain lines and the different speeds of the trains render the problem difficult).
- **Strategy Problems**
  Finding optimal strategies of how to behave in the (iterated) prisoner's dilemma and other models of game theory is a common problem in economics. A related goal is to simulate the behavior of actors in economic life, where not only strategies are optimized, but also their prevalence in a population. We discuss a specific (and fairly simple) behavioral simulation in more detail in Sect. 14.1.

As a side remark, we mention that (not surprisingly) evolutionary algorithms can also be used for **biological modeling**. An example is the "Netspinner" program (Krink and Vollrath 1997), which describes the web building behavior of certain spiders by parametrized rules (e.g., number of spokes, angle of the spiral etc.). With the help of an evolutionary algorithm the program optimizes the rule parameters based on an evaluation function that takes both the metabolic cost of building the web as well as the chances of catching prey with it into account. The obtained results mimic the behavior observed in real spiders very well and thus help to understand the forces that cause spiders to build their webs the way they do.

Optimization problems can be tackled in many different ways, but all possible approaches can essentially be categorized into four classes:

- **Analytical Solution**
  Some optimization problems can be solved analytically. For example, the problem of finding the side lengths of a box with given surface area S such that its volume is maximized (see above) can easily be solved with the method of Lagrange multipliers, setting the partial derivatives of the constructed Lagrange functional w.r.t. the parameters equal to zero and solving the resulting equation system. If an analytical approach exists, it is often the method of choice, because it usually guarantees that the solution is actually optimal and that it can be found in a fixed number of steps. However, for many practical problems no (efficient) analytical methods exists, either because the problem is not yet understood well enough or because it is too difficult in a fundamental way (e.g., because it is NP-hard).
- **Complete/Exhaustive Exploration**
  Since the definition of an optimization problem already contains all candidate solutions in the form of the (search) space Ω, one may consider simply enumerating and evaluating all of its elements. Although this approach certainly guarantees that the optimal solution will be found, it can be extremely inefficient and thus is usually applicable only to (very) small search spaces Ω. It is clearly infeasible for parameter optimization problems over real domains, since then Ω is infinite and thus can not possibly be explored exhaustively.
- **(Blind) Random Search**
  Instead of enumerating all elements of the search space Ω (which may not be efficiently possible anyway), we may consider picking and evaluating random elements, always keeping track of the best solution (candidate) found so far. This approach is efficient and has the advantage that it can be stopped at any time, but suffers from the severe drawback that it depends on pure luck whether we obtain a reasonably good solution. This approach corresponds to the model of a box with car parts that is shook to obtain a drivable car, as we discussed it above. It usually offers only very low chances of obtaining a satisfactory solution.
- **Guided (Random) Search**
  Instead of blindly picking random elements from the search space Ω, we may try to exploit the structure of the search space and how the evaluation function f assesses similar elements to control the search. The fundamental idea is to exploit information that has been gained from evaluating certain solution candidates to guide the choice of the next solution candidates to examine. Of course, for this to

be possible, the evaluation of similar elements of the search space must be similar. Otherwise there is no basis on which we may transfer gathered information. Note that the choice of the next solution candidates to examine may still contain a random element (non-deterministic choice), though, but that the choice is constrained by the evaluation of formerly examined solution candidates.

All metaheuristics, including evolutionary algorithms, fall into the last category. They differ, as already pointed out above, mainly in how the gathered information is represented and exploited for picking the next solution candidates to evaluate. Although metaheuristics thus provide fairly good chances of obtaining a satisfactory solution, it should always be kept in mind that they *cannot guarantee that the optimal solution is found*. That is, the solution candidate they return may have a high quality, and this quality may be high enough for many practical purposes, but there might still be room for improvement. If the problem at hand requires to find a truly (guaranteed) optimal solution, evolutionary algorithms are not suited for the task. In such a case one has to opt for an analytical solution or an exhaustive exploration.

It is also important to keep in mind that metaheuristics require that the evaluation function allows for **gradual improvement** (similar solution candidates have similar quality). Although in evolutionary algorithms the evaluation function is motivated by the biological fitness or aptitude in an environment and thus must differentiate better and worse candidate solutions, it must not possess large jumps at random points in the search space. Consider, for example, an evaluation function that assigns a value of 1 to exactly one solution candidate and 0 to all others. In this case, any evolutionary algorithm (actually any metaheuristic) cannot perform better than (blind) random search, because the quality assessment of non-optimal solution candidates does not provide any information about the location of the actual optimum.

11.3.2 Basic Notions and Concepts

In this section we introduce the basic notions and concepts of evolutionary algorithms by transferring them from their biological counterparts, see Table 11.1.

An **individual**, which is a living organism in biology, corresponds to a candidate solution in computer science. Individuals are the entities to which a fitness is assigned and which are subject to the (natural) selection process. In both domains an individual is described by a **chromosome** (from the Greek $\chi\rho\omega\mu\alpha$: color and $\sigma\omega\mu\alpha$: body, thus "colored body", because they are the colorable substance in a cell nucleus), which is the carrier of the genetic information. In biology, a chromosome consists of deoxyribonucleic acid (DNA) and many histone proteins, while in computer science the genetic information is encoded as a sequence of computational objects like bits, characters, numbers etc. A chromosome represents the "genetic blueprint" and encodes (parts of) traits of an individual. Most living organisms have several chromosomes, for example, humans have 46 chromosomes, which come in 23 so-called *homologous* pairs. In computer science, however, this complication is ignored and all genetic information is combined in a single chromosome.

A **gene** is the fundamental unit of inheritance as it determines (a part of) a trait or characteristic of an individual. An **allele** (from the Greek $\alpha\lambda\lambda\eta\lambda\omega\nu$: "each other",

Table 11.1 Basic evolutionary notions in biology and computer science

notion	biology	computer science
individual	living organism	solution candidate
chromosome	DNA histone protein strand	sequence of computational objects
	describes the "construction plan" and thus (some of the) traits of an individual in encoded form	
	usually multiple chromosomes per individual	usually only one chromosome per individual
gene	part of a chromosome	computational object (e.g., a bit, character, number etc.)
	is the fundamental unit of inheritance, which determines a (partial) characteristic of an individual	
allele (allelomorph)	form or "value" of gene	value of a computational object
	in each chromosome there is at most one form/value of a gene	
locus	position of a gene	position of a computational object
	at each position in a chromosome there is exactly one gene	
phenotype	physical appearance of a living organism	implementation / application of a solution candidate
genotype	genetic constitution of a living organism	encoding of a solution candidate
population	set of living organisms	bag / multiset of chromosomes
generation	population at a point in time	population at a point in time
reproduction	creating offspring of one or multiple (usually two) (parent) organisms	creating (child) chromosomes from one or multiple (parent) chromosomes
fitness	aptitude / conformity of a living organism	aptitude / quality of a solution candidate
	determines chances of survival and reproduction	

"mutual", because initially mainly two-valued genes were considered) refers to a possible form of a gene in biology. For example, a gene may represent the color of the iris in the eye of a human. This gene has alleles that code for blue, brown, green, gray etc. irises. In computer science, an allele is simply the value of a computational object, which selects one of several possible properties of a solution candidate that the gene stands for. Note that in a given chromosome there is exactly one allele per gene. That is, the iris color gene may code for blue or brown or green or gray eyes, but only one of these possibilities, as specified by the corresponding allele, is present in a concrete chromosome. The **locus** is the position of a gene in its chromosome. At any locus in a chromosome there is exactly one gene. Usually a gene can be identified by its locus. That is, a specific position (or a specific section) of a chromosome codes for a specific trait of the individual.

In biology, **phenotype** refers to the physical appearance of an organism, that is, the shape, structure and organization of its body. Note that the phenotype is what interacts with the environment and hence that it is the phenotype that actually determines the fitness of the individual. Likewise, in computer science, the phenotype is the implementation or application of a candidate solution, from which the fitness of the corresponding individual can be read. In contrast to this, the **genotype** is the genetic configuration of an organism or the encoding of a candidate solution, respectively. Note that the genotype determines the fitness of an individual only indirectly through the phenotype it encodes and that, at least in biology, the phenotype also comprises acquired traits that are not represented in the genotype (for example, learned behavior and bodily changes like a limb lost due to an accident).

A **population** is a simple set of organisms, usually of the same species. Due to the complexity of biological genomes it is usually safe to assume that no two individuals from a population share exactly the same genetic configuration—homozygous twins being the only exception. In addition, even genetically identical individuals differ due to acquired traits, which are never perfectly identical (even for homozygous twins) and thus lead to different phenotypes. In computer science, however, due to the usually much more limited variability of a chromosome as they are used in evolutionary algorithms and the lack of acquired traits, we must allow for the possibility of identical individuals. As a consequence, a population of an evolutionary algorithm is a *bag* or a *multiset* of individuals. In both biological and simulated evolution, **generation** refers to the population at a certain point in time.

A new generation is created by **reproduction**, that is, by the generation of offspring from one or more (in biology: if more than one, then usually two) organisms, in which genetic material of the parent individuals may be recombined. The same holds for computer science, only that the child creation process works directly on the chromosomes and that the number of parents may exceed two.

Finally, the **fitness** of an individual measures how high its chances of survival and reproduction are due to its adaptation to its environment. Since the quality of a biological organism w.r.t. its environment is difficult to assess objectively and simply defining fitness as the ability to survive can lead to a tautological "survival of the survivor," a formally more precise notion defines the fitness of an organism as the number of its offspring organisms that procreate themselves, thus linking (biological) fitness directly to the concept of differential reproduction. In computer science, the situation is simpler, because we are given an optimization problem that directly provides a fitness function with which solution candidates are to be evaluated.

It should be noted that, even though there are many parallels, simulated evolution is (usually) much simpler than biological evolution. For example, there are principles of biological evolution, e.g. speciation, that are usually not implemented in an evolutionary algorithm. On the genetic level, we already pointed out that in most life forms the genetic information is spread over multiple chromosomes, which often even come in so-called *homologous* pairs. These are pairs of chromosomes comprising the same genes, but possibly with different alleles, of which both or only one determine the corresponding phenotypical trait. Although such complications have their purpose in biological evolution, they are usually not simulated in a computer.

11.3.3 Building Blocks of an Evolutionary Algorithm

The general idea of an evolutionary algorithm is to employ evolution principles to generate increasingly better solution candidates for the optimization problem to solve. Essentially, this is achieved by evolving a population of solution candidates with the help of random variation and fitness-based selection of the next generation.

An evolutionary algorithm requires the following ingredients:

- an **encoding** for the solution candidates,
- a method to create an **initial population**,
- an **fitness function** to evaluate the individuals,
- a **selection method** on the basis of the fitness function,
- a set of **genetic operators** to modify chromosomes,
- a **termination criterion** for the search, and
- values for various **parameters**.

Since we want to evolve a population of solution candidates, we need a way of representing them as chromosomes, that is, we have to encode them, essentially as sequences of computational objects (like bits, characters, numbers etc.). Such an encoding may be so direct that the distinction between the genotype, as it is represented by the chromosome, and the phenotype, which is the actual solution candidate, becomes blurred. For example, for the problem of finding the side lengths of a box with given surface area that has maximum volume (which we considered above), we may use the triples (x, y, z) of the side lengths, which are the solution candidates, directly as the chromosomes. In other cases there is a clear distinction between the solution candidate and its encoding, for example, if we have to turn the chromosome into some other structure (the phenotype) before we can evaluate its fitness. We will see several examples of such cases in later chapters.

Generally, the encoding of the solution candidates is highly problem-specific and there are no general rules. However, in Sect. 12.1 we discuss several aspects that attention should be paid to when choosing an encoding for a given problem. An inappropriate choice can severely reduce the effectiveness of the evolutionary algorithm or may even make it impossible to find a sufficiently good solution. Depending on the problem to solve, it is therefore highly recommended to spent considerable effort on finding a good encoding of the solution candidates.

Once we have decided on an encoding, we can create an initial population of solution candidates in the form of chromosomes representing them. Since chromosomes are simple sequences of computational objects, an initial population is commonly created by simply generating random sequences. However, depending on the problem to solve and the chosen encoding, more complex methods may be needed, especially, if the solution candidates have to satisfy certain constraints.

In order to mimic the influence of the environment in biological evolution, we need a fitness function with which we can evaluate the individuals of the created population. In many cases this fitness function is simply identical to the function to optimize, which is given by the optimization problem to solve. However, the fitness function may also contain additional elements that represent constraints that have to be satisfied in order for a solution candidate to be acceptable or that introduce a tendency towards certain additionally desired properties of a solution.

The (natural) selection process of biological evolution is simulated by a method to select candidate solutions according to their fitness. This method is used to choose the parents of offspring we want to create or to select those individuals that are transferred to the next generation without change. Such a selection method may simply transform the fitness values into a selection probability, such that better individuals have higher chances of getting chosen for the next generation.

The random variation of chromosomes is simulated by so-called genetic operators that modify and recombine chromosomes, for example, **mutation**, which randomly changes individual genes, and **crossover**, which exchanges parts of the chromosomes of parent individuals to produce offspring. Depending on the problem and the chosen encoding, the genetic operators can be very generic or highly problem-specific. The choice of the genetic operators is another element that effort should be spent on, especially in connection with the chosen encoding.

The ingredients described up to now allow us to generate a sequence of populations of (hopefully) increasingly better quality. However, while biological evolution is unbounded, we need a criterion when to stop the process in order to retrieve a final solution. Such a criterion may be, for example, that the algorithm is terminated (1) after a user-specified number of generations have been created, (2) there has been no improvement (of the best solution candidate) for a user-specified number of generations, or (3) a user-specified minimum solution quality has been obtained.

To complete the specification of an evolutionary algorithm, we have to choose the values of several parameters, which include, for example, the size of the population to evolve, the fraction of individuals that is chosen from each population to produce offspring, the probability of a mutation occurring in an individual etc.

More formally, the procedure of an evolutionary algorithm looks like this:

Algorithm 11.1 (General Scheme of an Evolutionary Algorithm)

```
procedure evoalg;
begin
  t ← 0;                                (* initialize the generation counter *)
  initialize pop(t);                    (* create the initial population *)
  evaluate pop(t);                      (* and evaluate it (compute fitness) *)
  while not termination criterion do    (* loop until termination *)
    t ← t + 1;                          (* count the created generation *)
    select pop(t) from pop(t − 1);      (* select individuals based on fitness *)
    alter pop(t);                       (* apply genetic operators *)
    evaluate pop(t);                    (* evaluate the new population *)
  end                                   (* (compute new fitness) *)
end
```

That is, after having created and evaluated an initial population of solution candidates (in the form of chromosomes), a sequence of generations of solution candidates is computed. Each new generation is created by selecting individuals based

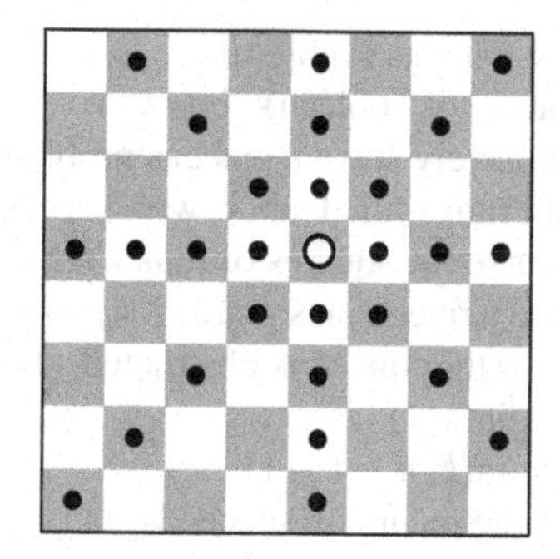
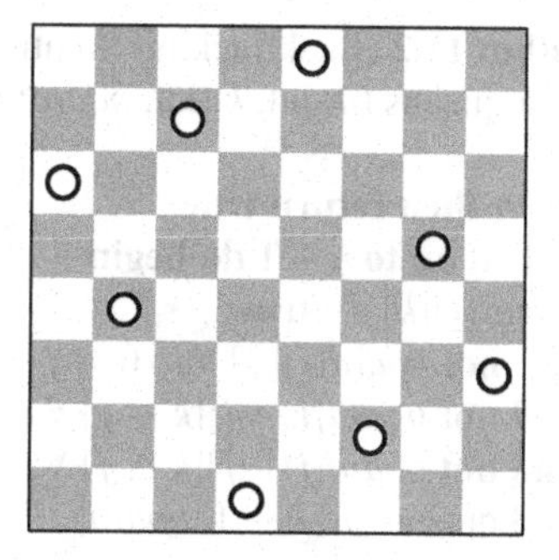

Fig. 11.1 Possible moves of a chess queen (*left*) and a solution of the 8-queens problem (*right*)

on their fitness (with a higher fitness meaning a higher chance of getting selected). Then genetic operators (like mutation and crossover) are applied to the selected individuals. Next, the modified population (or at least the new individuals in it, which have been created by the genetic operators) is evaluated and the cycle starts over. This process continues until the chosen termination criterion is fulfilled.

11.4 The n-Queens Problem

The n-queens problem consists in the task to place n queens (a piece in the game of chess) of the same color onto an $n \times n$ chessboard in such a way that no rank (chess term for row), no file (chess term for column) and no diagonal contains more than one queen. Drawing on the rules of how a queen may move in the game of chess (namely horizontally, vertically or diagonally by any number of squares, but not onto a square that is occupied by a piece of the same color or beyond a square that is occupied by a piece of either color, see Fig. 11.1), we may say that the queens must be placed in such a way that none of them obstructs the possible moves of any other. As an example, Fig. 11.1 shows a solution of the 8-queens problem.

A well-known approach to solve the n-queens problem is a backtracking algorithm, which can be seen as an essentially exhaustive exploration of the space of candidate solutions with a depth-first search. Such an algorithm exploits the obvious fact that each rank (row) of the chessboard must contain exactly one queen. Hence it proceeds by placing the queens rank by rank. For each rank, it is checked for each possible square whether a queen placed on it obstructs the moves of any queens placed earlier (that is, whether there is already a queen on the same file (column) or the same diagonal). If this is not the case, the algorithm proceeds recursively to the next rank. However, if the newly placed queen obstructs any of the queens placed earlier or if the recursion to the next rank returns with the result that no solution could be found because obstructions could not be avoided, the queen is removed again and the algorithm continues with the next square. More formally, this backtracking algorithm can be described, for example, by the following function:

Algorithm 11.2 (Backtracking Solution of the n-Queens Problem)

function queens (n: int, k: int, *board*: array of array of boolean) : boolean;
begin (∗ recursively solve n-queens problem ∗)
  if $k \geq n$ **then return true**; (∗ if all ranks filled, abort with success ∗)
  for $i = 0$ **up to** $n - 1$ **do begin** (∗ traverse the squares of rank k ∗)
    $board[i][k] \leftarrow$ **true**; (∗ place a queen on square (i, k) ∗)
    if **not** $board[i][j]\ \forall j : 0 < j < k$ (∗ if no other queen is obstructed ∗)
    and not $board[i-j][k-j]\ \forall j : 0 < j \leq \min(k, i)$
    and not $board[i+j][k-j]\ \forall j : 0 < j \leq \min(k, n-i-1)$
    and queens (n, $k+1$, *board*) (∗ and the recursion succeeds, ∗)
    then return true; (∗ a solution has been found ∗)
    $board[i][k] \leftarrow$ **false**; (∗ remove queen from square (i, k) ∗)
  end (∗ for $i = 0 \ldots$ ∗)
  return false; (∗ if no queen could be placed, ∗)
end (∗ abort with failure ∗)

This function is called with the number n of queens that defines the problem size, $k = 0$ indicating that the board should be filled starting from rank 0, and *board* being an $n \times n$ Boolean matrix that is initialized to *false* in all elements. If the function returns *true*, the problem can be solved. In this case, one possible placement of the queens is indicated by the *true* entries in *board*. If the function returns *false*, the problem cannot be solved (the 3-queens problem, for example, has no solution). In this case the variable *board* is in its initial state of all *false* entries.

Note that the above algorithm can easily be modified to yield all possible solutions of an n-queens problem. In this case, the first if-statement, which checks whether all ranks have been filled, must be extended by a procedure that reports the found solution. In addition, the recursion must not be terminated if the recursion succeeds (and thus a solution has been found), but the loop over the squares of the current rank must be continued to find possibly existing other solutions.

Although a backtracking approach is very effective for sufficiently small n (up to, say, $n \approx 30$), it can take a long time to find a solution if n is larger. If we are interested in only one solution (i.e., one placement of the queens), there exists a better method, namely an analytical solution (which is slightly less well known than the backtracking approach). We compute the positions of the queens as follows:

Algorithm 11.3 (Analytical Solution of the n-Queens Problem)

- If $n = 2$ or $n = 3$, the n-queens problem does not have a solution.
- If n is odd (that is, if $n \bmod 2 = 1$),
  then we place a queen onto the square $(n-1, n-1)$ and decrement n by 1.
- If $n \bmod 6 \neq 2$, then we place
  the queens in the rows $y = 0, \ldots, \frac{n}{2} - 1$ in the columns $x = 2y + 1$ and
  the queens in the rows $y = \frac{n}{2}, \ldots, n - 1$ in the columns $x = 2y - n$.
- If $n \bmod 6 = 2$, then we place
  the queens in the rows $y = 0, \ldots, \frac{n}{2} - 1$ in the columns $x = (2y + \frac{n}{2}) \bmod n$ and
  the queens in the rows $y = \frac{n}{2}, \ldots, n - 1$ in the columns $x = (2y - \frac{n}{2} + 2) \bmod n$.

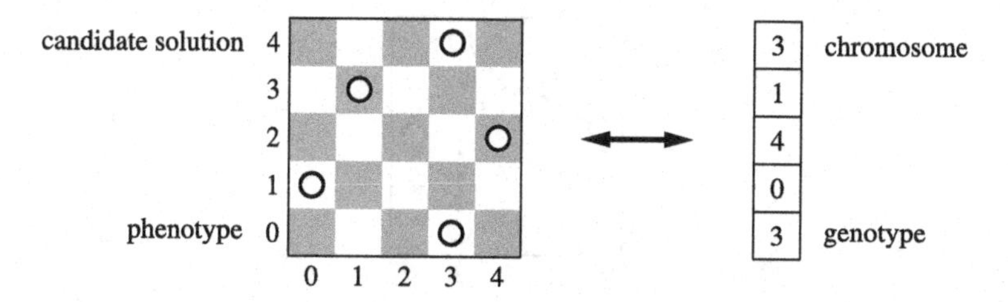

Fig. 11.2 Encoding of candidate solutions in the n-queens problem (here: $n = 5$)

Due to this analytical solution, it is not quite appropriate to approach the n-queens problem with an evolutionary algorithm. Here we do so nevertheless, because this problem allows us to illustrate certain aspects of evolutionary algorithms very well.

In order to solve the n-queens problem with an evolutionary algorithm, we first need an encoding of the solution candidates. For this, we draw on the same obvious fact that was already exploited for the backtracking algorithm, namely that each rank (row) of the chessboard must contain exactly one queen. Therefore, we describe a candidate solution by a chromosome with n genes, each of which refers to one rank of the chessboard and has n possible alleles, namely the possible file (column) numbers 0 to $n - 1$. Such a chromosome is interpreted as demonstrated in Fig. 11.2 (for $n = 5$): the allele of each gene indicates the file in which the queen is placed in the rank to which the gene refers. Note that with this encoding we can clearly distinguish between the genotype, which is an array of numbers, and the phenotype, which is the actual placement of the queens on the chessboard.

Note that this way of encoding the solution candidates has the advantage that we already exclude candidate solutions with more than one queen per rank. As a consequence, the search space becomes much smaller and thus can be explored more quickly and more effectively by an evolutionary algorithm. In order to reduce the search space even further, we may even consider restricting the chromosomes to permutations of the file (column) numbers. That is, each file number must occur for exactly one gene. However, although this clearly shrinks the search space even further, it introduces complications w.r.t. the genetic operators and thus we refrain from introducing this requirement here (however, cf. the discussion in Sect. 12.3).

In order to create an initial population, we simple generate a random sequence of n numbers in $\{0, 1, \ldots, n - 1\}$ for each individual, because there are no special conditions that such a sequence has to satisfy to be an element of the search space.

The fitness function is derived directly from the defining characteristics of a solution: we compute for each queen the number of obstructions, that is, the number of other queens that obstruct its moves. Then we sum these numbers over the queens, divide by 2 and negate the result (see Fig. 11.3 for an example). Clearly, for an actual solution the fitness computed in this manner is zero, whereas it is negative for all other candidates. Note that we divide by two, because each obstruction is counted twice with the above procedure as obstruction is symmetric: if queen 1 obstructs queen 2, then queen 2 also obstructs queen 1. Note also that we negate the result, because we want a fitness function that has to be maximized. For the example shown

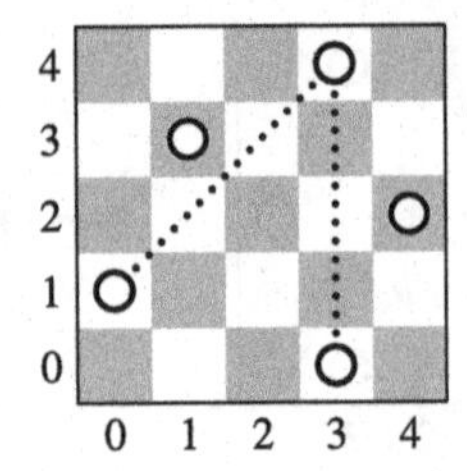

Fig. 11.3 A solution candidate for the 5-queen problem with four obstructions and thus fitness value -2

in Fig. 11.3, we have four (pairwise symmetric) obstructions (we may also say: two collisions between queens) and thus the fitness value is -2.

The fitness function immediately fixes the termination criterion: since a solution has the (maximally possible) fitness value of 0, we stop the algorithm as soon as a solution candidate with fitness 0 has been generated. However, to be on the safe side, we should also introduce a limit for the number of generations, so that the algorithm is guaranteed to terminate. Note, though, that with such an additional criterion the evolutionary algorithm may stop without having found a solution.

For the selection operation we choose a simple, but often very effective form of so-called **tournament selection**. That is, from the individuals of the current population a (small) sample of individuals is drawn that carry out a tournament with each other. This tournament is won by the individual with the highest fitness (ties are broken randomly). A copy of the winning individual is then added to the next generation and the participants of the tournament are replaced into the current population. The process is repeated until the next generation is complete, which usually means that it has reached the same size as the current population. Alternative selection methods as well as variants of tournament selection are discussed in Sect. 12.2.

In order to alter the selected individuals, we need genetic operators for recombination and variation. For the former we use so-called **one-point crossover**, which chooses a random cut point on the chromosomes of two parent individuals and exchanges the part on one side of the cut point between these individuals to create two children. An example for $n = 5$ is shown in Fig. 11.4: the genes below the randomly chosen cut point (the second out of the four possible ones) are exchanged. This example demonstrates what one hopes a genetic recombination may achieve: by combining partial solutions that are present in two deficient individuals (that is, both parent individuals have a negative fitness) a complete solution is obtained (the left child has a fitness of 0 and thus is a solution of the 5-queens problem). Alternative crossover operators are discussed in Sect. 12.3.

As a variation operation, we use a random replacement of the alleles of randomly selected genes (so-called **standard mutation**). An example is shown in Fig. 11.5, in which two genes of a chromosome representing a candidate solution for the 5-queens problem receive new values. This example is fairly typical in the respect that most mutations reduce the fitness of the individual affected by them. However, mutation is nevertheless important, because alleles that are not present in the initial population cannot be created by recombination, which only reorganizes existing al-

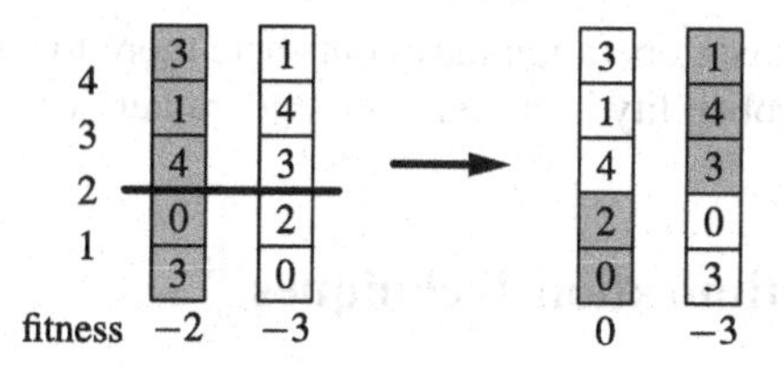

Fig. 11.4 One-point crossover of two chromosomes

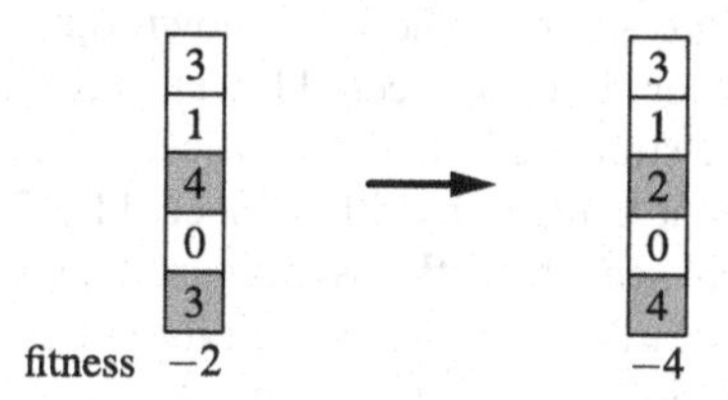

Fig. 11.5 Mutation of two genes of a chromosome

leles. Generally, mutation can more easily introduce new alleles into chromosomes. Alternative mutation operators are discussed in Sect. 12.3.

Finally, we have to choose the values of several parameters. These include the size of the population to evolve (for example, $\mu = 1000$ individuals), the maximum number of generations to compute (for example, $g = 100$ generations), the size of the tournaments to carry out (for example, $\mu_t = 5$ individuals), the fraction of individuals that are subject to crossover (for example, $p_c = 0.5$), and the probability that a gene is subject to a mutation (for example, $p_m = 0.01$). Once all parameters have been chosen, the evolutionary algorithm is fully specified and can be executed according to the general scheme presented in Algorithm 11.1 on page 180.

An implementation of this evolutionary algorithm, which can be found (as a command line program) on the web site for this book, shows that one can find solutions to the n-queens problem even for somewhat larger n than backtracking allows (also available as a command line program on the web site for this book), at least if a sufficiently large population is used and a sufficient number of generations is computed. However, as it is not guaranteed that a solution can be found, the program sometimes ends with a candidate solution that has a high fitness (like -1 or -2), but does not really solve the problem since obstructions remain.

By experimenting with the parameters, especially the fraction of individuals that are subject to crossover or the probability that a gene gets mutated, one can discover some interesting properties. For example, it turns out that mutations seem to be more important than crossover, since the speed with which a solution is found or the quality of solutions that are found in a given number of generations is not reduced if the fraction of individuals that are subject to crossover is reduced to zero. On the other hand, disallowing mutations causes the (average) solution quality to degrade significantly. Note, however, that these are not general characteristics of evolutionary algorithms, but are exhibited only for this particular problem and chosen encoding, selection and genetic operators etc. It cannot be transferred directly to other applications, where the crossover operation may contribute more to a solution being found and mutation, if its probability is chosen too large, rather degrades performance.

11.5 Related Optimization Techniques

In classical optimization (for example, in operations research) many techniques and algorithms have been developed that are fairly closely related to evolutionary algorithms. Among these are some of the best-known optimization techniques, for instance, gradient ascent or descent (Sect. 11.5.1), hill climbing (random ascent or descent, Sect. 11.5.2), simulated annealing (Sect. 11.5.3), threshold accepting (Sect. 11.5.4) and the great deluge algorithm (Sect. 11.5.5). All of these methods are sometimes called **local search methods**, because they make only small steps in the search space and thus carry out a local search for better solutions.

Like evolutionary algorithms, these techniques are based on the assumption that for similar solution candidates s_1 and s_2 the values of the function to optimize—that is, the values $f(s_1)$ and $f(s_2)$—are also similar. The main difference to evolutionary algorithms consists in the fact that the mentioned approaches focus on a single solution candidate rather than a whole population of solution candidates. These methods are relevant in the context of evolutionary algorithms, because one may see them, to some degree, as evolutionary algorithms with a population of size 1. In addition, they are often employed to improve solutions candidates locally or as a final optimization step for the output of an evolutionary algorithm. Finally, they illustrate some of the fundamental ideas underlying evolutionary algorithm techniques.

11.5.1 Gradient Ascent or Descent

While all methods discussed in the following sections only assume similarity of the quality of similar solution canditates, gradient ascent (or descent), requires in addition that the following two conditions hold:

- The search space is a subset of the n-dimensional space of real numbers: $\Omega \subseteq \mathbb{R}^n$.
- The function $f : \Omega \to \mathbb{R}$ to optimize is differentiable (everywhere).

Technically, the **gradient** is a differential operation that creates a vector field. That is, the gradient assigns to each point in the domain of the function of which the gradient is computed a vector that points into the direction of the steepest ascent of the function in that point. An illustration of the gradient of a two-dimensional function $z = f(x, y)$ at a point $\mathbf{p} = (x_0, y_0)$ is shown in Fig. 11.6 (see also Sect. 5.4 in Part I). Formally, the gradient at a point (x_0, y_0) is defined as

$$\nabla z|_{(x_0,y_0)} = \left(\frac{\partial z}{\partial x}\bigg|_{(x_0,y_0)}, \frac{\partial z}{\partial y}\bigg|_{(x_0,y_0)} \right).$$

The basic idea of gradient ascent (or descent) is to start at a randomly chosen point and then to make small steps in the search space Ω in (or against) the direction of the

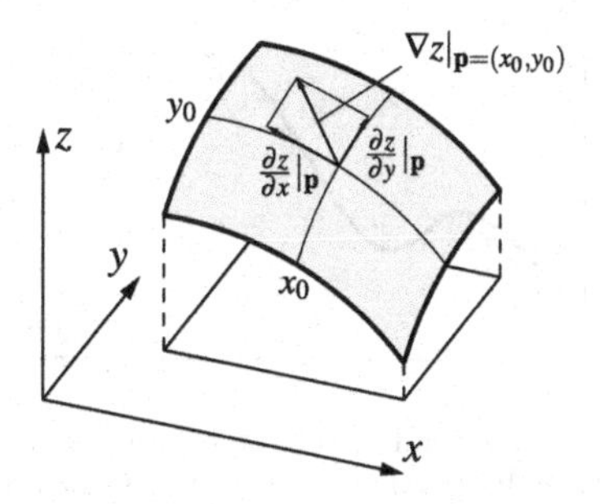

Fig. 11.6 Intuitive interpretation of the gradient of a real-valued function $z = f(x, y)$ at a point $\mathbf{p} = (x_0, y_0)$. It is $\nabla z|_{(x_0,y_0)} = (\frac{\partial z}{\partial x}|_{(x_0,y_0)}, \frac{\partial z}{\partial y}|_{(x_0,y_0)})$

steepest slope of the function f until a (local) optimum is reached. More formally, gradient ascent or descent works according to the following general scheme:

Algorithm 11.4 (Gradient Ascent or Descent)

1. Choose a (random) starting point $\mathbf{x}^{(0)} = (x_1^{(0)}, \dots, x_n^{(0)})$.
2. Compute the gradient at the current point $\mathbf{x}^{(t)}$

$$\nabla_{\mathbf{x}} f(\mathbf{x}^{(t)}) = \left(\frac{\partial}{\partial x_1} f(\mathbf{x}^{(t)}), \dots, \frac{\partial}{\partial x_n} f(\mathbf{x}^{(t)})\right).$$

3. Make a small step in the direction of the gradient (for gradient ascent, positive sign) or against the direction of the gradient (for gradient descent, negative sign):

$$\mathbf{x}^{(t+1)} = \mathbf{x}^{(t)} \pm \eta \nabla_{\mathbf{x}} f(\mathbf{x}^{(t)})$$

   where η is a parameter that controls the step width (also called "learning rate" in the training of (artificial) neural networks, see Sect. 5.4).
4. Repeat steps 2 and 3 until some termination criterion is fulfilled (for example, until a user-specified number of steps has been executed or until the gradient is smaller than a user-specified threshold).

Although simple and often very effective, this optimization technique is not without drawbacks. As discussed in more detail in Sect. 5.6 in Part I, the choice of the step width η is critical. If its value is too small, it may take very long until a (local) optimum is reached, because only tiny steps are executed, especially if the gradient is small. On the other hand, if the step width is too large, the optimization process may oscillate (jump back and forth in the search space), never converging to a (local) optimum. Some approaches to mitigate this problem (like a using a momentum term or an adaptive step width parameter) are discussed in Sect. 5.7 in Part I.

A fundamental problem of gradient ascent or descent is the choice of the starting point, since a bad choice can make it essentially impossible to find the global or even just a good local optimum. Unfortunately, there is little that can be done to improve this situation. Thus, the best option is to execute the procedure multiple

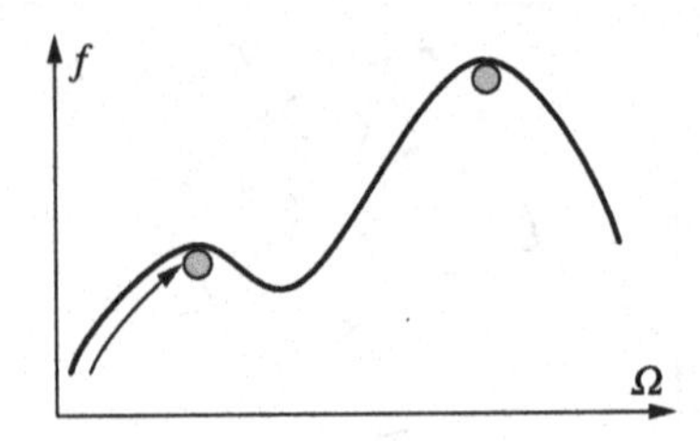

Fig. 11.7 Getting stuck in local maxima is a core problem of gradient ascent as well as of hill climbing. In order to reach the global maximum (or at least a better local optimum) a temporary worsening of the solution quality has to be accepted

times with different starting points and finally to choose the best result obtained. This provides an argument for looking at *populations* of solution candidates (as in evolutionary algorithms). Such an approach offers, in addition, the possibility to exchange information between solution candidates to improve the optimization process. In evolutionary algorithm this is achieved by recombination operators.

11.5.2 Hill Climbing

If the function f is not differentiable, gradient ascent or descent is not an option. However, we may try to determine a direction in which f increases by evaluating random points in the vicinity of the current point. The result is known as *hill climbing*, which is probably the simplest local search method. It works as follows:

Algorithm 11.5 (Hill Climbing)

1. Choose a (random) starting point $s_0 \in \Omega$.
2. Choose a point $s \in \Omega$ "in the vicinity" of s_t (for example, by a small random variation of s_t).
3. Set

$$s_{t+1} = \begin{cases} s & \text{if } f(s) > f(s_t), \\ s_t & \text{otherwise.} \end{cases}$$

4. Repeat steps 2 and 3 until a termination criterion is fulfilled.

As for gradient ascent or descent, the biggest problem of this fairly naïve approach is that it has a strong tendency to get stuck in local optima. This is illustrated in Fig. 11.7, where a solution candidate (indicated by the gray ball) cannot rise any further by merely local modifications and thus gets stuck in a local maximum. All subsequent methods try to mitigate this fundamental problem by accepting, under certain conditions, solution candidates s that are worse than the current solution s_t. The idea is to enable solutions candidates to cross regions of lower solution quality between the local maximum and a global one (or at least a better local one), see Fig. 11.7. The different methods we study below differ mainly in the exact conditions under which they accept solution candidates that are worse.

11.5.3 Simulated Annealing

Simulated annealing (Metropolis et al. 1953; Kirkpatrick et al. 1983) can be seen as extension of both hill climbing and gradient descent. The fundamental idea of this method, which is also discussed in another context in Sect. 8.5 in Part I, is that transitions from lower to higher (local) maxima should be more probable than transitions in the opposite direction. This is achieved as follows: we randomly create variants of the current solution—in exactly the same way as for hill climbing. Better solution candidates are always accepted. Solution candidates that are worse are accepted with a certain probability which depends on both the quality difference between the current and the new solution candidate and a temperature parameter that decreases over time. The guiding principles are that small reductions of the solution quality are more readily accepted than large ones, and that reductions of the solution quality are more easily accepted in early steps of the algorithm than in later ones.

Algorithm 11.6 (Simulated Annealing)

1. Choose a (random) starting point $s_0 \in \Omega$.
2. Choose a point $s \in \Omega$ "in the vicinity" of s_t
   (for example, by a small random variation of s_t).
3. Set

$$s_{t+1} = \begin{cases} s & \text{if } f(s) \geq f(s_t), \\ s & \text{with probability } p = e^{-\frac{\Delta f}{kT}} \text{ and} \\ s_t & \text{with probability } 1 - p \text{ otherwise.} \end{cases}$$

   where $\Delta f = f(s_t) - f(s)$ is the quality reduction of the solution, $k = \Delta f_{\max}$ an estimate of the range of quality values and T is a temperature parameter that is (slowly) decreased over time.
4. Repeat steps 2 and 3 until a termination criterion is fulfilled.

For (very) small T this method is almost identical to hill climbing, because the probability of accepting a worse solution candidate is (very) small. For larger T, quality reductions are accepted with a non-negligible probability, thus allowing the solution candidates to cross regions of reduced quality. Nevertheless the algorithm cannot guarantee, of course, that the global optimum is reached. However, the risk of getting stuck in a local optimum is reduced and thus one obtains better chances of reaching the global optimum or at least a good local one.

11.5.4 Threshold Accepting

The idea of threshold accepting (Dueck and Scheuer 1990) is very similar to that of simulated annealing. Again worse solutions are sometimes accepted, however, with an upper bound for the quality degradation. Threshold accepting works as follows:

Algorithm 11.7 (Threshold Accepting)

1. Choose a (random) starting point $s_0 \in \Omega$.
2. Choose a point $s \in \Omega$ "in the vicinity" of s_t
   (for example, by a small random variation of s_t).
3. Set

$$s_{t+1} = \begin{cases} s & \text{if } f(s) \geq f(s_t) - \theta, \\ s_t & \text{otherwise,} \end{cases}$$

   where $\theta > 0$ is a threshold for accepting worse solution candidates that is (slowly) decreased over time. ($\theta = 0$ is equivalent to standard hill climbing.)
4. Repeat steps 2 and 3 until a termination criterion is fulfilled.

11.5.5 Great Deluge Algorithm

Similar to simulated annealing and threshold accepting, the great deluge algorithm (Dueck 1993) also accepts worse solutions. The difference is that an absolute lower bound is used, which is slowly increased over time. One may imagine the procedure as if the "landscape" formed by the quality function to optimize is slowly "flooded" (in a great deluge, hence the name of the algorithm) and only solution candidates that "sit on dry land" are acceptable. The higher the water level rises, the more strongly the tendency becomes to accept only better solution candidates.

Algorithm 11.8 (Great Deluge Algorithm)

1. Choose a (random) starting point $s_0 \in \Omega$.
2. Choose a point $s \in \Omega$ "in the vicinity" of s_t
   (e.g., by a small random variation of s_t).
3. Set

$$s_{t+1} = \begin{cases} s & \text{if } f(s) \geq \theta_0 + t \cdot \eta, \\ s_t & \text{otherwise,} \end{cases}$$

   where θ_0 is a lower bound for the quality of the candidate solutions at $t = 0$ (that is, an initial "water level") and η is a step width parameter that can be seen as corresponding to the "speed of the rain" causing the flood.
4. Repeat steps 2 and 3 until a termination criterion is fulfilled.

11.5.6 Record-to-Record Travel

Similar to the great deluge algorithm, record-to-record travel uses a rising water level (Dueck 1993). However, it is linked to the fitness of the best individual that has been found so far. Thus, a possible degradation is always seen in relation to the best found individual. Only if there is an improvement, then the current individual is important. Similar to threshold accepting, a monotonously increasing sequence θ of real numbers controls the selection of poor individuals. More formally:

Algorithm 11.9 (Record-to-Record Travel)

1. Choose a (random) starting point $s_0 \in \Omega$ and set $s_{\text{best}} = s_0$.
2. Choose a point $s \in \Omega$ "in the vicinity" of s_t (for example, by a small random variation of s_t).
3. Set

$$s_{t+1} = \begin{cases} s & \text{if } f(s) \geq f(s_{\text{best}}) - \theta, \\ s_t & \text{otherwise,} \end{cases}$$

and

$$s_{\text{best}} = \begin{cases} s & \text{if } f(s) > f(s_{\text{best}}), \\ s_{\text{best}} & \text{otherwise,} \end{cases}$$

where $\theta > 0$ is a threshold for accepting solution candidates worse than the currently best solution that is (slowly) decreased over time.
4. Repeat steps 2 and 3 until a termination criterion is fulfilled.

11.6 The Traveling Salesman Problem

To illustrate the application of the local search methods discussed above to an optimization problem, we take a look at the famous **traveling salesman problem** (TSP): we are given both a set of n cities (idealized as points on a plane) and the distances or costs of the routes between the cities. A traveling salesman has business to conduct in each of the cities, but, of course, wants to travel as cheaply (or as shortly) as possible. Hence, we desire to find a minimum cost (or minimum distance) round trip through the cities, so that each city is visited exactly once.

Mathematically, the traveling salesman problem is defined as follows: we are given a graph with weighted edges and we desire to find a so-called **Hamiltonian cycle** in this weighted graph, that is, an ordering of the vertices of the graph in which neighbors as well as the last and the first vertex are connected by an edge and which minimizes the total weight of the connecting edges. In the form of our definition of an optimization problem (as we gave it in Definition 11.1 on page 173), a traveling salesman problem can be stated as follows:

Definition 11.2 (Traveling Salesman Problem) Let $G = (V, E, w)$ be a weighted graph with the vertex set $V = \{v_1, \dots, v_n\}$ (each v_i represents a city), the edge set $E \subseteq V \times V - \{(v, v) \mid v \in V\}$ (each edge represents a connection between two cities) and the edge weight function $w : E \to \mathbb{R}_+$ (which represents the distances or costs of the connections). The traveling salesman problem is the optimization problem $(\Omega_{\text{TSP}}, f_{\text{TSP}})$ where Ω_{TSP} contains all permutations π of the numbers $\{1, \dots, n\}$ that satisfy $\forall k; 1 \leq k \leq n : (v_{\pi(k)}, v_{(\pi(k) \bmod n)+1}) \in E$ and the function f_{TSP} is defined as

$$f_{\text{TSP}}(\pi) = -\sum_{k=1}^{n} w\big((v_{\pi(k)}, v_{\pi((k \bmod n)+1)})\big).$$

A traveling salesman problem is called **symmetric** if

$$\forall i, j \in \{1, \dots, n\}, i \neq j:$$
$$(v_i, v_j) \in E \Rightarrow (v_j, v_i) \in E \wedge w\big((v_i, v_j)\big) = w\big((v_j, v_i)\big),$$

that is, if all connections can be traversed in both directions and the directions have the same costs. Otherwise the traveling salesman problem is called **asymmetric**.

No algorithm is known that solves this problem in polynomial time unless a non-deterministic formalism (like a non-deterministic Turing machine) is used (which is essentially irrelevant for practical purposes). More technically, one says that this problem is non-deterministic polynomial-time-complete (or *NP-complete* for short). Intuitively, this means that in order to guarantee that one obtains the optimal solution one cannot do fundamentally better than trying all possibilities (exhaustive exploration of the search space). As a consequence, for large n only approximate solutions can be computed in reasonable time, because an exhaustive exploration is exponential in n. Of course, we may actually find the best solution with a guided (random) search like evolutionary algorithms, but this is not guaranteed. Here we consider, as an illustration of the methods discussed above, how the traveling salesman problem can be tackled with hill climbing and simulated annealing.

We work with the following algorithm, which captures hill climbing for $T \equiv 0$:

Algorithm 11.10 (Simulated Annealing for the Traveling Salesman Problem)

1. Order the cities randomly (that is, create a random round trip).
2. Randomly choose two pairs of cities such that each pair consists of cities that are neighbors in the current round trip and such that all four cities are distinct. Split the round trip between the cities of each pair and reverse the interjacent part.
3. If this new round trip is better (that is, shorter or cheaper) than the old, then replace the old round trip with the new one. Otherwise replace the old round trip with probability $p = \exp\big(-\frac{\Delta Q}{kT}\big)$ where ΔQ is the quality difference between the old and the new round trip, k is an estimate of the range of round trip qualities and T is a temperature parameter that is reduced over time, e.g. $T = \frac{1}{t}$.
4. Repeat steps 2 and 3 until a termination criterion is met.

Since we cannot know the range k of quality values, we estimate it, for example, by $k_t = \frac{t+1}{t}(\max_{i=1}^{t} Q_i - \min_{i=1}^{t} Q_i)$ where Q_i is the quality of the ith solution candidate and t is the current time step. The employed variation operator is introduced more generally as a mutation operator in Sect. 12.3.1 under the name **inversion**.

As an illustration, we consider the simple traveling salesman problem with only five cities that is shown in Fig. 11.8 on the left. We simply use the Euclidean distance of the points as the cost function, that is, we search for a round trip of minimum length. A possible starting point (chosen in step 1) is the round trip that is shown in Fig. 11.8 on the right, which has a length of $2\sqrt{2} + 2\sqrt{5} + 4 \approx 11.30$.

All five possible splits of this round trip that could be chosen in step 2 of the above algorithm are shown in Fig. 11.9: the two edges that are cut by a dashed line identify the cities of each pair. The new round trips that result from inverting the interjacent part (that is, traversing it in the opposite direction) are shown in Fig. 11.10 on the left. The new edges of the round trip are shown in gray.

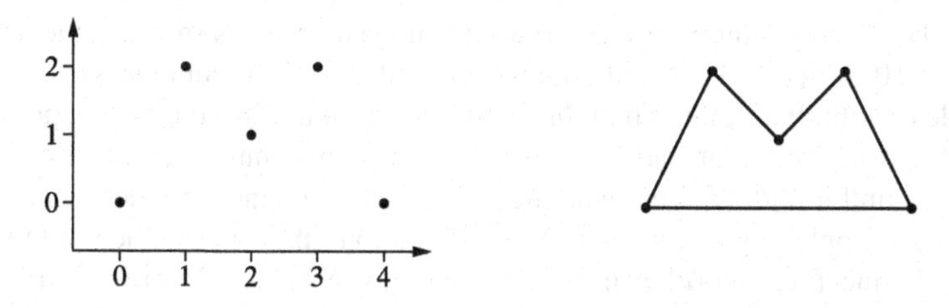

Fig. 11.8 Example of a traveling salesman problem (*left*) with initially created round trip (*right*)

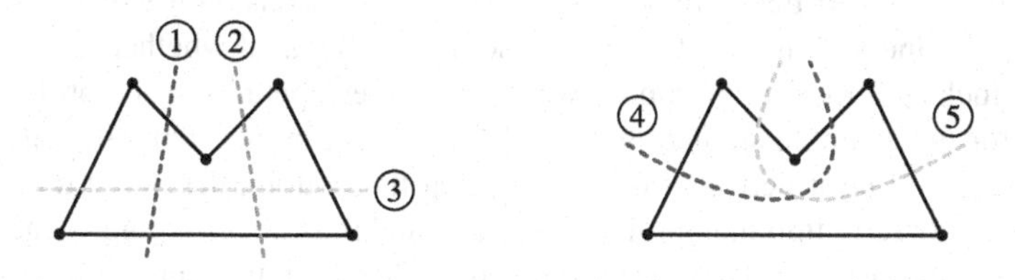

Fig. 11.9 Possible separations of the exemplary round trip

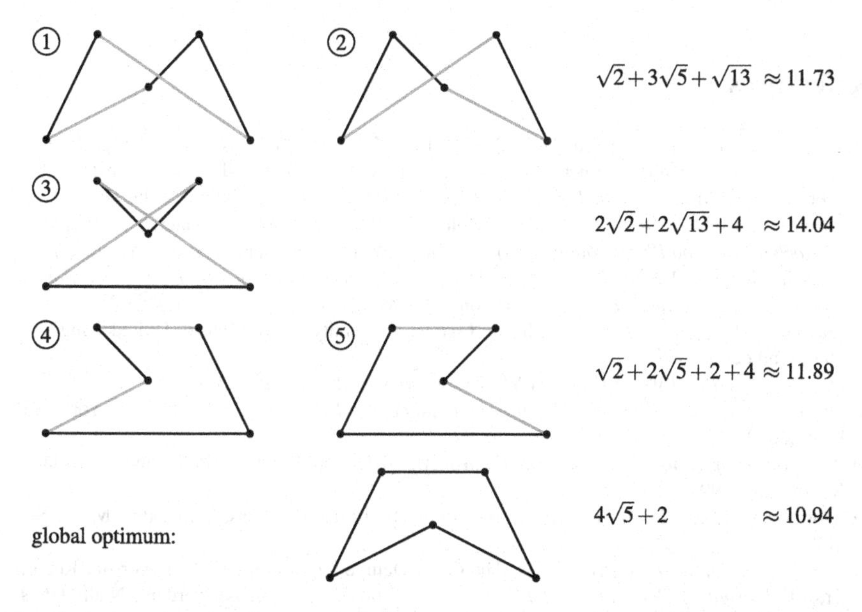

Fig. 11.10 Modifications of the initial round trip and the global optimum with the corresponding fitness values. Compared to the initial round trip, all its possible variations are worse and therefore hill climbing cannot reach the global optimum

Figure 11.10 also demonstrates that all five splits lead to a round trip that is worse than the initial tour (see the length computations on the right). As a consequence, a hill climbing approach (which accepts only better modifications) is unable to improve the initial round trip and thus returns it as the solution. However, this round trip is only a local optimum, as can be seen from the tour that is shown at the bottom of Fig. 11.10, which is the global optimum (round trip of minimum length).

While hill climbing cannot find this optimum (from the chosen starting point—it can find it from a different starting point), simulated annealing offers at least certain chances to find it (although no guarantee). For instance, since the modifications 1, 2, 4 and 5 are only slightly worse than the initial tour, they may be accepted in the simulated annealing algorithm in the random choice in step 3. At least from the modifications 1 and 2 the optimal solution can be obtained with a single additional modification that reverses the right or the left edge, respectively.

It should be noted that it may depend on the operations with which a solution candidate "in the vicinity" of the current solution is created whether hill climbing can get stuck in a local optimum. If we permit other operations, the problem disappears for the example studied in Fig. 11.8 to 11.10: by removing the city in the center and inserting it between the two cities at the bottom, the initial tour can be transformed directly into the optimal round trip and thus hill climbing is enabled to find the optimal solution. However, for this modified operation set another example can be found, in which hill climbing can get stuck in a local optimum.

References

M. Aidoo, D.J. Terlouw, M.S. Kolczak, P.D. McElroy, F.O. ter Kuile, S. Kariuki, B.L. Nahlen, A.A. Lal, and V. Udhayakumar. Protective Effects of the Sickle Cell Gene Against Malaria Morbidity and Mortality. *The Lancet* 359:1311–1312. Elsevier, Amsterdam, Netherlands, 2002

M. Brenner and V.J. Hearing. The Protective Role of Melanin Against UV Damage in Human Skin. *Photochemistry and Photobiology* 84(3):539–549. J. Wiley & Sons, New York, NY, USA, 2008

C. Darwin. *On the Origin of Species by Means of Natural Selection, or the Preservation of Favoured Races in the Struggle for Life*. John Murray, London, United Kingdom, 1859

R. Dawkins. *The Selfish Gene*, 2nd edition. Oxford University Press, Oxford, United Kingdom, 1976; 2nd edition 1989

R. Dawkins. *The Blind Watchmaker*. W.W. Norton, New York, NY, USA, 1986

R. Dawkins. *The Greatest Show on Earth: The Evidence for Evolution*. Free Press, New York, NY, USA, 2009

M. Dorigo. *Optimization, Learning and Natural Algorithms*. PhD Thesis, Politecnico di Milano, Milan, Italy, 1992

M. Dorigo and T. Stützle. *Ant Colony Optimization*. Bradford/MIT Press, Cambridge, MA, USA, 2004

G. Dueck. New Optimization Heuristics: The Great Deluge Algorithm and the Record-to-Record Travel. *Journal of Computational Physics* 104(1):86–92. Elsevier, Amsterdam, Netherlands, 1993

G. Dueck and T. Scheuer. Threshold Accepting: A General Purpose Optimization Algorithm appearing Superior to Simulated Annealing. *Journal of Computational Physics* 90(1):161–175. Elsevier, Amsterdam, Netherlands, 1990

J.D. Farmer, N. Packard and A. Perelson. The Immune System, Adaptation and Machine Learning. *Physica D: Nonlinear Phenomena* 2:187–204. Elsevier, Amsterdam, Netherlands, 1986

S.S. Harris. Vitamin D and African Americans. *Journal of Nutrition* 136(4):1126–1129. American Society for Nutrition, Bethesda, MD, USA, 2006

D.L. Hartl and A.G. Clark. *Principles of Population Genetics*, 4th edition. Sinauer Associates, Sunderland, MA, USA, 2007

J. Kennedy and R. Eberhart. Particle Swarm Optimization. *Proc. IEEE Int. Conf. on Neural Networks*, vol. 4:1942–1948. IEEE Press, Piscataway, NJ, USA, 1995

S. Kirkpatrick, C.D. Gelatt, and M.P. Vercchi. Optimization by Simulated Annealing. *Science* 220:671–680. High Wire Press, Stanford, CA, USA, 1983

T. Krink and F. Vollrath. Analysing Spider Web-building Behaviour with Rule-based Simulations and Genetic Algorithms. *Journal of Theoretical Biology* 185(3):321–331. Elsevier, Amsterdam, Netherlands, 1997

J.-B. Lamarck. *Philosophie zoologique, ou, Exposition des considérations relative à l'histoire naturelle des animaux.* Paris, France, 1809

N. Metropolis, N. Rosenblut, A. Teller, and E. Teller. Equation of State Calculations for Fast Computing Machines. *Journal of Chemical Physics* 21:1087–1092. American Institute of Physics, Melville, NY, USA, 1953

S. Nakrani and S. Tovey. On Honey Bees and Dynamic Server Allocation in Internet Hosting Centers. *Adaptive Behavior* 12:223–240. SAGE Publications, New York, NY, USA, 2004

G. Vollmer. Der wissenschaftstheoretische Status der Evolutionstheorie: Einwände und Gegenargumente. In: G. Vollmer (ed.) *Biophilosophie*, 92–106. Reclam, Stuttgart, Germany, 1995

Chapter 12
Elements of Evolutionary Algorithms

Evolutionary algorithms are not fixed procedures, but contain several elements that must be adapted to the optimization problem to be solved. In particular, the encoding of the candidate solution needs to be chosen with care. Although there is no generally valid rule or recipe, we discuss in Sect. 12.1 some important properties a good encoding should have. In Sect. 12.2, we turn to the fitness function and review the most common selection techniques as well as how certain undesired effects can be avoided by adapting the fitness function or the selection method. Section 12.3 is devoted to genetic operators, which serve as tools to explore the search space, and covers sexual and asexual recombination and other variation techniques.

12.1 Encoding of Solution Candidates

The way in which candidate solutions to the given optimization problem are encoded can have a considerable impact on how easily an evolutionary algorithm finds a (good) solution. With a bad or unfavorable encoding, it may not even find a useful solution at all. As a consequence, one should spend a lot of effort and care on designing a good encoding and the corresponding genetic operators.

For example, in the n-queens problem we discussed in Sect. 11.4, the encoding we chose, namely using an array of n integer numbers with values in $\{0, \dots, n-1\}$, is much better than representing the placement of the queens by a binary array with n^2 elements, in which each element refers to a square of the $n \times n$ chessboard and encodes whether this square is occupied by a queen (bit is 1) or not (bit is 0). The reason is that our encoding already rules out candidate solutions with more than one queen in the same rank (row) and thus considerably reduces the search space. In addition, it ensures that there are always exactly n queens on the board, while with a binary encoding the genetic operators we employed (standard mutation and one-point crossover) may produce a candidate solution with more or less queens.

An even better encoding than the one we chose is to restrict the candidate solutions to permutations of the numbers $\{0, \dots, n-1\}$. Such an encoding not only

R. Kruse et al., *Computational Intelligence*, Texts in Computer Science,
DOI 10.1007/978-1-4471-5013-8_12,

guarantees that each rank contains exactly one queen, but at the same time that each file (column) contains exactly one queen. Hence, it reduces the search space even further, making the task much easier for the evolutionary algorithm. We merely refrained from using this encoding in Sect. 11.4, because it would have caused problems with the genetic operators, since standard mutation and one-point crossover do not maintain that a candidate solution is a permutation. However, these problems can be solved with specialized operators as we demonstrate in Sect. 12.3.

Generally, it is important to pay attention to the interplay between the chosen encoding and the genetic operators. If the encoding reduces the search space, but it turns out to be difficult to find genetic operators that guarantee that the result of their application is in this reduced search space, additional efforts may be needed to handle such cases. These efforts may very well be so costly that it can turn out to be better to use an encoding that defines a larger search space (incorporating fewer constraints), but allows for simpler choices of genetic operators.

This brief discussion already shows that there are no general "cookbook recipes" with which one can find a (good) encoding. Nevertheless, we can identify the following desirable properties that an encoding should have:

- Similar phenotypes should be represented by similar genotypes.
- Similarly encoded candidate solutions should have a similar fitness.
- If possible, the search space Ω (i.e. the set of all possible candidate solutions) should be closed under the used genetic operators.

Of course, these are not absolute rules, but rather guidelines. Depending on the problem at hand one may consider breaking them in order to gain other advantages. However, it is usually a good idea to refrain from using an encoding that does not have these properties unless there are very good reasons for doing so.

12.1.1 Hamming Cliffs

Similar phenotypes should be represented by similar genotypes

Two genotypes are clearly similar if they differ in few genes, because then few mutations, for example, are necessary to transform the one genotype into the other. That is, the similarity of genotypes is defined as how easy it is to transform one into the other with the available genetic operators. It may be measured, for example, as the needed minimum number of applications of genetic operators (like mutation).

However, what is subject to the fitness evaluation is the phenotype, that is, the actual solution candidate, and we only presuppose that similar solution candidates have similar fitness (so that fitness information can be exploited to guide the search for better solution candidates). As we only modify the chromosomes (and thus the genotype), this can clearly lead to problems if similar phenotypes and thus similar solution candidates are described by (very) dissimilar genotypes, because then it may be impossible to obtain a similar phenotype by (small) genetic modifications.

As an example we consider a parameter optimization problem: we are given an n-ary real-valued function $y = f(x_1, \dots, x_n)$ and we desire to find an argument vector $(x_1, \dots, x_n)$ that optimizes the value of y. We represent the real-valued arguments by encoding them as binary numbers, which are then concatenated to obtain a binary chromosome. Unfortunately, such an encoding can create a serious problem in an evolutionary algorithm, because it may introduce so-called **Hamming cliffs**.

To make the example precise, we briefly consider how such a binary encoding of real numbers is computed. Let a real interval $[a, b]$ and an encoding precision ε be given. We look for an encoding rule for any number $x \in [a, b]$ as a binary number z such that the encoded number z differs by less than ε from its actual value x. The basic idea is to divide the interval $[a, b]$ into equally sized segments of a length smaller than or equal to ε. That is, we create 2^k segments with $k = \lceil \log_2 \frac{b-a}{\varepsilon} \rceil$ which are mapped to the numbers $0, \dots, 2^k - 1$. Thus, we obtain the encoding

$$z = \left\lfloor \frac{x-a}{b-a} (2^k - 1) \right\rfloor.$$

Alternatively, we may use the encoding

$$z = \left\lfloor \frac{x-a}{b-a} (2^k - 1) + \frac{1}{2} \right\rfloor,$$

which allows us to reduce the number of intervals to half, that is, we only need $k = \lceil \log_2 \frac{b-a}{2\varepsilon} \rceil$ segments. Decoding is performed by

$$x' = a + z \cdot \frac{b-a}{2^k - 1}.$$

Note that the decoded x' and the original x may differ, although by at most ε.

As an example, suppose that we want to encode the number $x = 0.637197$ in the interval $[-1, 2]$ with precision $\varepsilon = 10^{-6}$. Then k and z are computed as follows:

$$k = \left\lceil \log_2 \frac{2-(-1)}{10^{-6}} \right\rceil = \lceil \log_2 3 \cdot 10^6 \rceil = 22 \quad \text{and}$$

$$z = \left\lfloor \frac{0.637197 - (-1)}{2-(-1)} (2^{22} - 1) \right\rfloor = 2288966_{10}$$
$$= 1000101110110101000110_2.$$

Studying such an encoding in more detail reveals that it may encode numbers from adjacent segments very differently. That is, although the numbers are close (they differ by no more than 2ε if they come from adjacent segments), their encodings may have a large Hamming distance, where the Hamming distance of two bit strings is simply the number of different bits. Large Hamming distances can be overcome by mutation and crossover only with great difficulties, simply because they require many bits to be modified. As a consequence, they have been called "Hamming cliffs" to express that they are difficult to "climb" in a solution improvement process.

Table 12.1 Gray code of 4-bit numbers

binary	Gray	binary	Gray	binary	Gray	binary	Gray
0000	0000	0100	0110	1000	1100	1100	1010
0001	0001	0101	0111	1001	1101	1101	1011
0010	0011	0110	0101	1010	1111	1110	1001
0011	0010	0111	0100	1011	1110	1111	1000

In order to make the problem perfectly clear, we consider the number range from 0 to 1 encoded by 4 bits. That is, we map the real numbers $\frac{k}{15}$ onto the integer numbers k, thus ensuring an ε of $\frac{1}{30}$. In this case, the encoding of $\frac{7}{15}$ is 0111 the encoding of $\frac{8}{15}$ is 1000. Therefore they have a Hamming distance of 4 because every bit is different. Although the phenotypical distance between the two numbers is comparatively small (only $2\varepsilon = \frac{1}{15}$), the genotypical distance is maximal. As a consequence, if the actual optimum of a parameter optimization problem is at $\frac{8}{15}$, it is of no help to the algorithm to discover that $\frac{7}{15}$ has a high fitness, because it cannot generate the encoding for $\frac{8}{15}$ from the encoding of $\frac{7}{15}$ with few genetic operations.

This problem can be solved by introducing so-called **Gray codes**, which are defined as any binary representation of integer numbers in which adjacent numbers differ in only one bit. For 4-bit numbers, a possible Gray code is shown in Table 12.1. Note that a Gray code is not unique as can already be seen from the simple fact that any cyclic permutation of the codes in Table 12.1 is again a Gray code.

The most common form of Gray encoding and decoding, respectively, is

$$g = z \oplus \left\lfloor \frac{z}{2} \right\rfloor \quad \text{and} \quad z = \bigoplus_{i=0}^{k-1} \left\lfloor \frac{g}{2^i} \right\rfloor,$$

where $\oplus$ is the *exclusive or* (XOR gate) of the binary representation.

As an illustration, we reconsider the example we discussed above, namely representing the real number $x = 0.637197$ in the interval $[-1, 2]$ with precision $\varepsilon = 10^{-6}$. Drawing on the binary number encoding derived above we obtain the Gray code

$$\begin{aligned} g &= 1000101110110101000110_2 \\ &\oplus 0100010111011010100011_2 \\ &= 1100111001101111100101_2. \end{aligned}$$

12.1.2 Epistasis

Similarly encoded candidate solutions should have a similar fitness

In biology, **epistasis** means the phenomenon that one allele of a gene (the so-called epistatic gene) suppresses the effect of all possible alleles of another gene. It might

Table 12.2 Impact of a mutation in the traveling salesman problem (second encoding option)

	chromosome	remaining cities	trip
before the mutation	5	1, 2, 3, 4, **5**, 6	5
	3	1, 2, **3**, 4, 6	3
	3	1, 2, **4**, 6	4
	2	1, **2**, 6	2
	2	1, **6**	6
	1	**1**	1

	chromosome	remaining cities	trip
after the mutation	1	**1**, 2, 3, 4, 5, 6	1
	3	2, 3, **4**, 5, 6	4
	3	2, 3, **5**, 6	5
	2	2, **3**, 6	3
	2	2, **6**	6
	1	**2**	2

even be that several other genes are suppressed by one epistatic gene. In evolutionary algorithms, epistasis describes the interaction between the genes of a chromosome. That is, how much the fitness of a solution candidate changes if a gene is modified, in extreme cases even which trait of a solution candidate is represented by some genes, strongly depends on the value(s) of (an)other gene(s). This is undesirable.

As a side remark, we mention that in biology epistasis explains certain deviations from the Mendelian laws. For instance, if homozygous black and white seed beans are crossed, then black, white and brown seed beans are obtained in the second offspring generation in a ratio of 12:1:3, which contradicts the Mendelian laws.

In order to illustrate the occurrence and effects of possible epistasis in evolutionary algorithms, we draw on the **traveling salesman problem** as an example (see Sect. 11.6). We consider two possible encodings of round trips through the cities (Hamiltonian cycles) in order to illustrate the epistasis problem:

1. A round trip is represented by a permutation of the cities (as in Definition 11.2 on page 191). This means that the city at the kth position is visited in the kth step. Such an encoding exhibits low epistasis. For instance, swapping two cities alters the fitness (that is, the cost of the round trip) by comparable amounts, regardless of which two cities are swapped. Such a swap also changes the tour only locally, because the part of the tour through the other cities is not affected.
2. A round trip is specified by a list of numbers that state the position of the next city to be visited in a (sorted) list from which all already visited cities have been deleted. An example of how a chromosome is interpreted in this encoding is shown in the top part of Table 12.2. In contrast, this encoding exhibits high epistasis. Modifying a single gene may alter a large part of the trip, and the more so, the closer to the front (top) the gene is located. The bottom part of Table 12.2 shows an extreme case in which the trip is changed entirely, even though the genetic modification is minimal: only a single gene is mutated.

Of course, the reason for the difference in the two encodings is that in the latter the interpretation of the values of any gene depends on the values of the preceding genes, thus introducing a strong dependence between genes, that is, high epistasis.

If the chosen encoding exhibits high epistasis, the optimization problem is often difficult to solve for an evolutionary algorithm. One reason is that epistasis in the extreme form we studied in the above example destroys the assumption that small changes to the chromosomes (which produce similar genotypes) also effect only small changes in the represented candidate solutions (the phenotypes). More generally, if small changes of the genotype can lead to large changes of the fitness, a fundamental assumption underlying evolutionary algorithms, namely that **gradual improvement** is possible (see Sect. 11.3.1), is no longer valid.

It has been tried to classify optimization problems as "easy or hard to solve by an evolutionary algorithm" based on the notion of epistasis (Davidor 1990). However, this does not work, because epistasis is a property of the encoding and *not of the problem itself.* This can already be seen from the example of the traveling salesman problem, for which we presented an encoding with high and an encoding with low epistasis. In addition, there are problems that can be encoded with low epistasis and that are nevertheless difficult to solve with the help of an evolutionary algorithm.

12.1.3 Closedness of the Search Space

If possible, the search space Ω should be closed under the used genetic operators

In particular, if the solution candidates have to satisfy certain constraints, it is not necessarily the case that the genetic operators, if applied to elements of the search space Ω, always yield other (valid) elements of this search space. Generally, we say that an (operator created) individual lies outside of the search space if

- its chromosome cannot be meaningfully interpreted or decoded,
- the represented candidate solution does not fulfill certain basic requirements,
- the represented candidate solution is evaluated incorrectly by the fitness function.

Clearly, such individuals are undesirable and require some action to treat them, so they should be avoided if possible at all. If one faces a situation nevertheless in which the chosen encoding of candidate solutions together with the preferred genetic operators can produce individuals outside of the search space, one has essentially the following options to tackle the problem:

- Choose or design a **different encoding**, which does not suffer from this problem. Note that this may require enlarging the search space.
- Choose or design **encoding-specific genetic operators** under which the search space is closed. That is, find genetic operators which ensure that only elements of the search space can be produced.
- Use **repair mechanisms**, with which an individual outside the search space is modified in such a way that it is brought back into the search space.

Fig. 12.1 One-point crossover of a permutation

Fig. 12.2 Repair mechanism for permutations

- Accept individuals outside the search space in the evolutionary algorithm, but introduce a **penalty term** that reduces the fitness of such individuals, so that the selection process is endowed with a clear tendency to eliminate them.

In order to illustrate the problem and its solutions, recall the n-queens problem as we discussed it in Sect. 11.4. We already mentioned at the beginning of Sect. 12.1 that a better encoding for this problem (than the one used in Sect. 11.4) are chromosomes that are permutations of the numbers $\{1, \ldots, n-1\}$, because this reduces the search space and thus makes it potentially easier for an evolutionary algorithm to find a solution. However, if we apply the same genetic operators, namely one-point crossover and standard mutation (see Figs. 11.4 and 11.5 on page 185) with this encoding, we may produce chromosomes that are not permutations. This is immediately clear for mutation if it is applied to a single gene, because changing a single number in a permutation cannot possibly yield a permutation: for this at least two numbers have to be changed. A crossover example is shown in Fig. 12.1, in which two permutations are processed with one-point crossover, yielding two children, both of which are not permutations: the top one contains number 5 twice and lacks number 1 while it is the other way round for the bottom child. A closer inspection quickly reveals that it is actually fairly unlikely that a random choice of the parents and the cut point creates a situation in which both children are permutations.

According to the above list of options, we can solve this problem as follows:

- **Different encoding:** The problem vanishes if we allow any sequence of numbers in $\{1, \ldots, n-1\}$, not just permutations. We chose this solution in Sect. 11.4 in order to be able to employ the one-point crossover and standard mutation. However, this has the disadvantage that the search space is enlarged, thus making it possibly more difficult for an evolutionary algorithm to find a solution.
- **Encoding-specific genetic operators:** Instead of standard mutation, one may use gene pair swaps as a mutation operation (cf. Sect. 12.3.1). Likewise, a permutation preserving crossover operation can be designed (see Sect. 12.3.2). These choices ensure that the search space (i.e., the set of permutations of $\{1, \ldots, n-1\}$) becomes closed under the genetic operators.
- **Repair mechanisms:** If a genetic operator produced a chromosome that is not a permutation, repair it, that is, modify it in such a way that it becomes a permutation. For example, find and remove duplicate occurrences of the same (file, i.e., column) numbers and append the missing numbers (see Fig. 12.2).

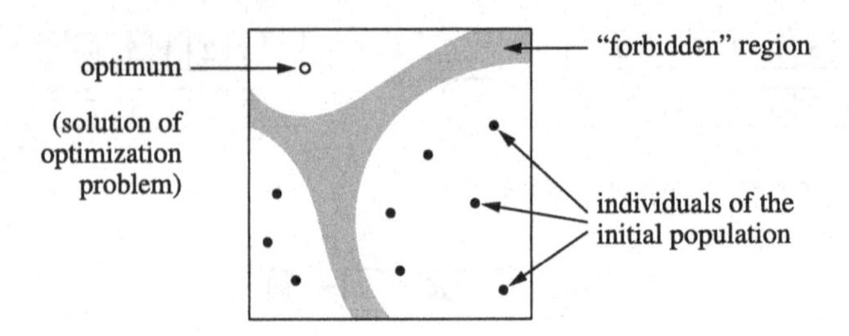

Fig. 12.3 Disconnected areas of a search spacecomplicate the search for an evolutionary algorithm, especially when repair mechanisms are used

- **Penalty term:** Allow the population to contain chromosomes that are not permutations, but add a penalty term to the fitness function, which reduces the fitness of such non-permutations. For example, the fitness could be reduced by the number of missing (file, i.e., column) numbers, possibly multiplied by a weighting factor.

For the n-queens problem the best solution is clearly to employ permutation preserving genetic operators, since they are not much more costly than simple one-point crossover or standard mutation and the evolutionary algorithm certainly benefits from the reduced search space. This is also the solution one would choose for the traveling salesman problem, where essentially the same situation occurs: a solution candidate is best described by a permutation of the cities to visit.

Repair mechanisms are actually closely related to encoding-specific genetic operators, because the application of a genetic operator that may produce an individual outside of the search space followed by the application of a repair mechanism may be seen as one operation w.r.t. which the search space is closed.

However, in certain cases, encoding-specific genetic operators or repair mechanism may complicate the search. If the search space is disconnected, encoding-specific genetic operators never produce individuals in the "forbidden" regions and repair mechanisms immediately restore such individuals to a permissible area. The possibly resulting problem is illustrated in the sketch in Fig. 12.3: if the optimum is in a region of the search space in which no individuals of the initial population are located, it may be very difficult for an evolutionary algorithm to find the optimum, because the "forbidden" regions are difficult to cross. In such cases a penalty term may be preferable, which reduces the fitness of candidate solutions in "forbidden" regions, but it does not eliminate them altogether. As a consequence, the individuals are left with at least some chance to transgress "forbidden" regions.

12.2 Fitness and Selection

The basic principle of **selection** is that better individuals (that is, individuals with a higher fitness) have better chances to procreate. We referred to this principle before as **differential reproduction** (see Sect. 11.2), because it states that individuals differ in their (expected) reproduction depending on their fitness. How strongly individuals with higher fitness are preferred for producing offspring is called **selective pressure**. High selective pressure means that even small fitness differences can cause considerably differing chances of procreation, while a low selective pressure means that reproduction chances depend only a little on fitness differences.

It should be clear that we need at least some selective pressure for evolutionary algorithms to work, because without any selective pressure the search for a solution is essentially random and thus unlikely to be successful (cf. Sect. 11.2, especially the car parts example). On the other hand, if the selective pressure is very high, the search may focus too quickly on individuals that happened to be the best in the initial population, trying to optimize them further. In such a case, other regions of the search space, which were under-represented in the initial population, may never be explored and thus good solutions contained in them may never be found. This effect is related to the principle of **opportunism** mentioned in Sect. 11.2, namely that better and optimal solutions are not found if the intermediary stages exhibit certain fitness handicaps. (We consider this problem in more detail below.)

More generally, when trying to adjust the selective pressure by choosing a selection mechanism or its parameters, we face the problem that there is a trade-off between **exploration of the search space** and **exploitation of good individuals**. With a low selective pressure we favor search space exploration, because the fitness has only little influence on the reproductive chances. Thus, even individuals in regions of the search space with a generally lower fitness have good chances to multiply and hence (due to random modifications and recombinations) to spread over the corresponding region of the search space. On the other hand, with a high selective pressure we favor the exploitation of good individuals, because in this case only highly fit individuals have good reproduction chances and thus their vicinity in the search space will be focused on to find even better modifications of them.

The best strategy is usually a time-dependent selective pressure: in early generations the selective pressure is kept fairly low, so that the search space is well explored, with the hope that we obtain subpopulations in all promising regions of the search space. In later generations, the selective pressure is then increased in order to find the best (local) modifications in the promising regions and thus the best solution candidates. The selective pressure can be controlled by either adapting (in particular, scaling) the fitness function (see Sect. 12.2.4) or by choosing a selection method and/or adapting its parameters (see Sects. 12.2.6 and 12.2.7).

12.2.1 Fitness Proportionate Selection

Roulette-wheel selection is certainly among the best-known selection methods. It takes the (absolute) fitness of each individual and computes their relative fitness as

$$f_{\mathrm{rel}}(s) = \frac{f_{\mathrm{abs}}(s)}{\sum_{s' \in \mathrm{pop}(t)} f_{\mathrm{abs}}(s')}.$$

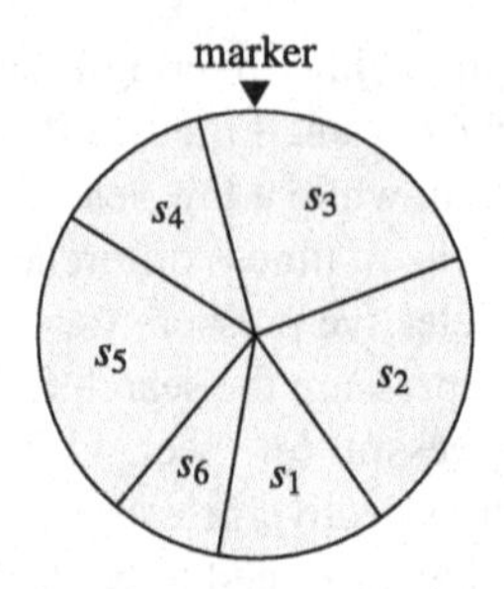

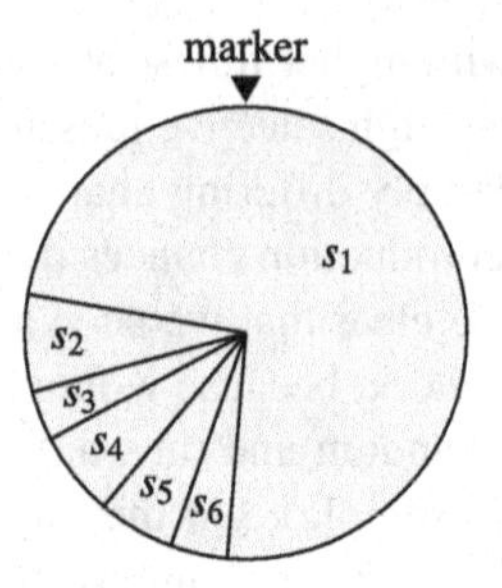

Fig. 12.4 Fitness proportionate selection and the dominance problem

This value is then interpreted as the probability with which the corresponding individual may be selected (to become a member of the next generation). Since with this method the selection probability of an individual is directly proportional to its fitness, this method is also called **fitness proportionate selection**.

Note that the absolute fitness $f_{\text{abs}}(s)$ must not be negative in order for this method to be applicable. This is no real constraint, though. If necessary, we may add a suitable (positive) number to all fitness values and/or we may set all (remaining) negative fitness values equal to zero (provided there are only few). Furthermore, the fitness function must be such that it is to be maximized. Otherwise this methods clearly selects bad individuals with high probability rather than good ones.

The name "roulette-wheel selection" for this method stems from the fact that a roulette wheel is a good illustration of how this method works. Each individual is represented by a sector of a roulette wheel (see Fig. 12.4 on the left, which shows a roulette wheel for six individuals s_1 to s_6). The angle (or equivalently, the area) of the sector for an individual s represents its relative fitness value $f_{\text{rel}}(s)$. In addition, there is a marker at the top of the roulette wheel. An individual is selected by setting the roulette wheel into motion and waiting until it stops. Then the individual is chosen that corresponds to the sector the marker points to. For example, in Fig. 12.4 on the left, individual s_3 is selected. It should be clear that with this procedure, assuming that all angles are equally likely as stopping positions, the probability that a certain individual gets selected equals its relative fitness.

12.2.2 The Dominance Problem

A drawback of fitness proportionate selection is that an individual with a very high fitness may **dominate** the selection, (almost) suppressing all other individuals. This is illustrated in Fig. 12.4 on the right: the individual s_1 has a much higher fitness than all other individuals (even than all other individuals together). Thus, it is extremely likely to get selected, while any other individual has only very low chances of getting into the next generation. In subsequent generations, this dominance may become even stronger, since many copies of the dominating individual and many individuals very similar to it will be present. As a consequence, we may observe what is often

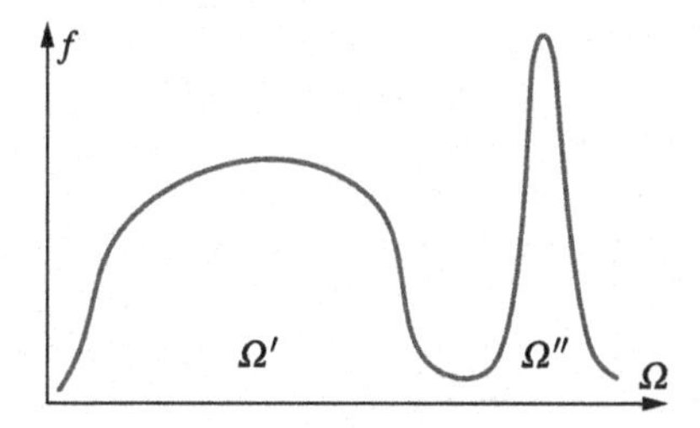

Fig. 12.5 The problem of premature convergence: individuals close to the transition area between Ω' and Ω'' have very low reproduction chances

called **crowding**, that is, the population becomes very dense in one or only few regions of the search space (that is, these regions become "crowded"), while the rest of the search space becomes (almost) void of individuals.

Crowding usually results in very fast convergence to a (local) optimum in the crowded region(s). That is, only one or very few good individuals are exploited (i.e., optimized locally), while a wider exploration of the search space may (almost) cease. As already mentioned, this may be desirable in later generations to actually find the best solutions. In early generations, however, it should be avoided in favor of a thorough exploration of the search space in order to increase the chances that all promising regions are sufficiently covered. Otherwise, **premature convergence** may occur, that is, the search focuses too quickly on those regions of the search space in which the best individuals of the initial population happened to be located.

Figure 12.5 illustrates the drawbacks of such a behavior. Since the region Ω' is much larger than the region Ω'', it may easily happen that the (randomly created) initial population contains only individuals from Ω'. With crowding, the best individuals in this region may be found, but it is highly unlikely that, by random modification and recombination, individuals can reach the region Ω''. The reason is the opportunism of evolution processes: since the transition region between Ω' and Ω'' exhibits severe fitness handicaps, individuals in this region have low chances of surviving the selection process, let alone multiplying themselves. As a consequence, the better solutions in region Ω'' may never be found.

12.2.3 Vanishing Selective Pressure

While the dominance problem demonstrates the disadvantages of large fitness differences between individuals, (very) small differences are also undesirable as they can lead to **vanishing selective pressure**. That is, if the (relative) fitness differences between individuals are too small, fitness proportionate selection does not prefer good individuals enough to push solution candidates towards better and better fitness values. As an illustration, consider the fitness function g in Fig. 12.6. Since the (relative) differences between the fitness values are fairly small, the reproductive chances of all individuals are roughly the same, regardless of where they are located, thus rendering the search almost equivalent to a random search.

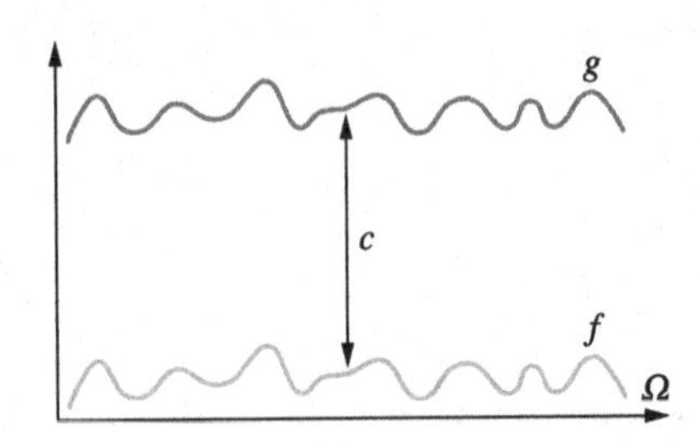

Fig. 12.6 Vanishing selective pressure: although the maxima are located at the same points, they are differently easy to find with an evolutionary algorithm. With the function g there are only (too) small differences in the relative fitness of individuals

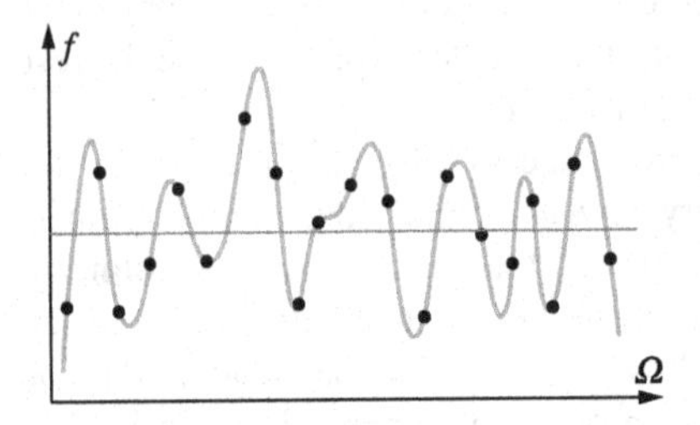

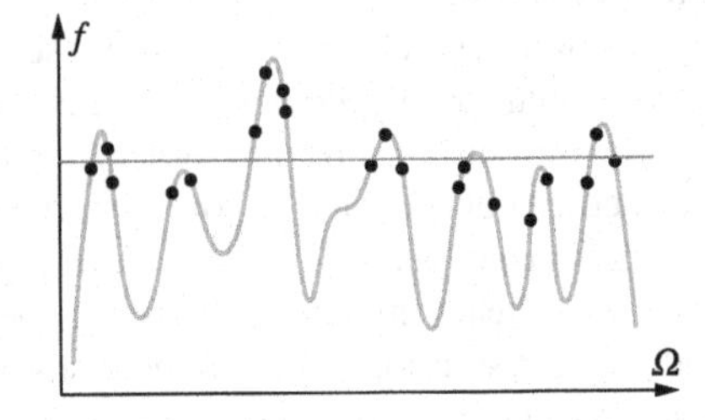

Fig. 12.7 Example of vanishing selective pressure. The *dots* represent individuals in a very early stage of evolution (*left*) and in a later generation (*right*). The average fitness is shown as a *line*

More generally, we have to consider how the absolute fitness values relate to their range over the population or even the whole search space. For example, compare the functions f and g depicted in Fig. 12.6. Both have the same shape and thus the same maxima. More precisely, it is $g \equiv f + c$ for some constant $c \in \mathbb{R}$. An evolutionary algorithm may work well for f, because the relative difference of the fitness values provided by f allows for sufficient selective pressure to find the maxima of f. However, since $c \gg \sup_{s \in \Omega} f(s)$, we have $\forall s \in \text{pop}(t) : g_{\text{rel}}(s) \approx \frac{1}{\mu}$ where μ is the population size and thus almost no selective pressure to find the maxima of g.

Since an evolutionary algorithm has a tendency to increase the (average) fitness of individuals from one generation to the next (as better individuals are selected with higher probability), it may even create the problem of vanishing selective pressure itself. Due to a rather random distribution of the initial fitness values, there may be a comparatively high selective pressure in early generations. In later generations, however, the selection of better individuals may have increased the average fitness. As a consequence, the range of fitness values and thus the selective pressure as it results from fitness proportionate selection may be reduced. This is illustrated in Fig. 12.7, which shows a possible situation in an early (or the initial) generation on the left and a possible situation in a later generation on the right. The gray curve is the fitness function, the black dots represent individuals (candidate solutions).

Note that the desirable behavior is exactly opposite: low selective pressure in early generations (to favor exploration) and higher selective pressure in later generations (to favor exploitation)—see the beginning of this section. As a consequence, this is a particularly unfortunate behavior, which should be counteracted with one of the methods that are presented in the following sections.

12.2.4 Adapting the Fitness Function

A popular possibility to overcome both the dominance problem and the problem of vanishing selective pressure is to **scale the fitness function**. A very simple approach is **linear dynamic scaling**, which scales the fitness values according to

$$f_{\mathrm{lds}}(s) = \alpha \cdot f(s) - \min\{f(s') \mid s' \in \mathrm{pop}(t)\}, \quad \alpha > 0.$$

The factor α determines the strength of the scaling. Instead of the minimum of $\mathrm{pop}(t)$, one may also use the minimum of the last k generations.

An alternative to linear dynamic scaling is the popular **σ-scaling**, which transforms the fitness of all individuals according to

$$f_\sigma(s) = f(s) - \big(\mu_f(t) - \beta \cdot \sigma_f(t)\big), \quad \beta > 0,$$

where $\mu_f(t) = \frac{1}{\mu}\sum_{s\in\mathrm{pop}(t)} f(s)$ and $\sigma_f(t) = \sqrt{\frac{1}{(\mu-1)}\sum_{s\in\mathrm{pop}(t)}(f(s) - \mu_f(t))^2}$ are the mean value and the standard deviation, respectively, of the fitness values of the individuals in the current population and β is a parameter.

Obviously, a problem of both approaches is how to choose the parameters α and β. To solve this problem one considers the so-called **coefficient of variation** v of the fitness function (in the current population), which is defined as

$$v = \frac{\sigma_f}{\mu_f} = \frac{\sqrt{\frac{1}{|\Omega|-1}\sum_{s'\in\Omega}(f(s') - \frac{1}{|\Omega|}\sum_{s\in\Omega} f(s))^2}}{\frac{1}{|\Omega|}\sum_{s\in\Omega} f(s)}, \quad \text{or} \quad v(t) = \frac{\sigma_f(t)}{\mu_f(t)}.$$

Note that v is defined on the whole search space Ω (left formula). However, in practice (where we certainly cannot compute it for the complete search space Ω, as then we could also carry out a complete enumeration search), it is estimated from its value for the current population $\mathrm{pop}(t)$ (right formula, $\mathrm{pop}(t)$ instead of Ω).

In order to illustrate the coefficient of variation, Fig. 12.8 shows the fitness function of Fig. 12.6 lifted to different levels, so that the coefficient of variation becomes roughly 0.2 (top left), 0.1 (top right) and 0.05 (bottom left).

Empirically it was found that a value of $v \approx 0.1$ yields a good rapport of exploration and exploitation. As a consequence, if v deviates (significantly) from this (optimal) value, one should try to adapt the fitness function f, for example, by scaling or exponentiation, so that $v \approx 0.1$ is obtained. That is, one computes $v(t)$ for the current populations and then adapts the fitness values in such a way that $v(t)$ becomes 0.1. This is particularly easy for σ-scaling, since β and v are directly related. To be precise, one should choose $\beta = \frac{1}{v^*}$ with $v^* = 0.1$.

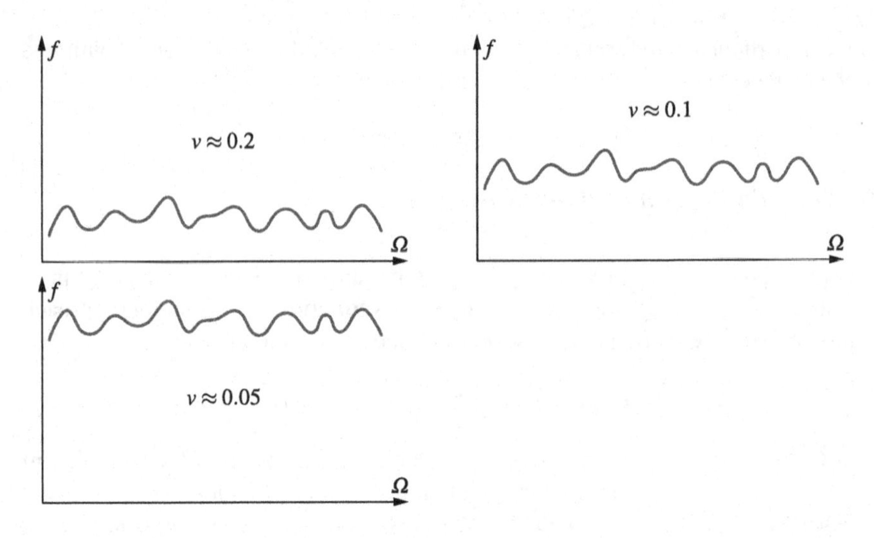

Fig. 12.8 Illustrations of the coefficient of variation. High coefficients of variation can cause premature convergence, low values cause vanishing selective pressure. Empirically it was found that $v \approx 0.1$ yields a good compromise

Another advisable adaptation of the fitness function is to introduce a **dependence on time**. That is, we do not compute the relative fitness values directly from the function $f(s)$ to be optimized, but from $g(s) \equiv (f(s))^{k(t)}$. The time-dependent exponent $k(t)$ controls the selective pressure and should be chosen in such a way that the coefficient of variation v stays close to $v^* \approx 0.1$. Michalewicz (1996) proposes

$$k(t) = \left(\frac{v^*}{v}\right)^{\beta_1} \left(\tan\left(\frac{t}{T+1} \cdot \frac{\pi}{2}\right)\right)^{\beta_2 \left(\frac{v}{v^*}\right)^{\alpha}},$$

where v^*, β_1, β_2, α are parameters, v is the coefficient of variation (estimated, for instance, from the initial population), T is the maximum number of generations to be computed and t is the current point in time (that is, the generation index). For this function, Michalewicz (1996) recommends $v^* = 0.1$, $\beta_1 = 0.05$, $\beta_2 = 0.1$, $\alpha = 0.1$.

An alternative time-dependent fitness function is **Boltzmann selection**. It determines the relative fitness from $g(s) \equiv \exp(\frac{f(s)}{kT})$. The time-dependent temperature parameter T controls the selective pressure and k is a normalization constant that is analogous to the Boltzmann constant. The temperature may decrease (usually linearly) until a predefined maximum number of generations is reached. The idea of this selection method resembles the idea underlying **simulated annealing** (see Sect. 11.5.3): in early generations, the temperature parameter is high and the relative differences between the fitness values are therefore small. In later generations, the temperature parameter is decreased, which causes increasing fitness differences. As a consequence, the selective pressure increases in the course of the generations.

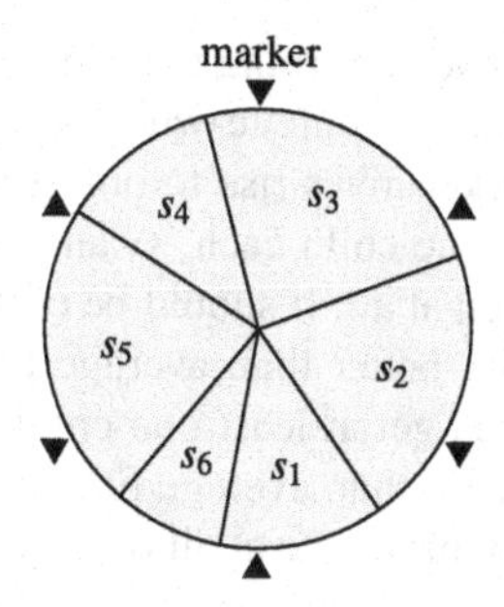

Fig. 12.9 Stochastic universal sampling. This method implements the expected value model, according to which each individual is selected at least as often as the expected number of times rounded to the next lower integer

12.2.5 The Variance Problem

Although roulette-wheel selection strives to select individuals in proportion to their fitness, deviations from an exactly proportional behavior have to be expected, since the selection process is random. As a consequence, there is no guarantee that good individuals (say, individuals that are better than average) enter the next generation. Even the best individual of the population, although it certainly has the best *chances* to be selected and thus to procreate, may not get selected. More generally, the number of offspring individuals that are produced per individual of the current generation may differ considerably from the expected value. This phenomenon is also known as the **variance problem** of roulette-wheel or fitness proportionate selection.

A very simple, though not really advisable solution to this problem is to discretize the range of fitness values. Based on the mean $\mu_f(t)$ and the standard deviation $\sigma_f(t)$ of the fitness values in the population, offspring is created according to the following rule: If $f(s) < \mu_f(t) - \sigma_f(t)$, then s is discarded and does not create a descendant. If $\mu_f(t) - \sigma_f(t) \leq f(s) \leq \mu_f(t) + \sigma_f(t)$, then s is allowed to have one descendant. If $f(s) > \mu_f(t) + \sigma_f(t)$, then two descendants of s are created.

An alternative way to solve the variance problem is to employ the **expected value model**, with which one strives to give each individual a number of children that is close to the expected value. More precisely, in this model $\lfloor f_{\text{rel}}(s) \cdot |\text{pop}(t)|\rfloor$ offspring individuals are created for every individual of the current population (where $|\text{pop}(t)|$ is the size of the population). However, this usually produces too few individuals, since generally $\sum_{s \in \text{pop}(t)} \lfloor f_{\text{rel}}(s) \cdot |\text{pop}(t)|\rfloor < |\text{pop}(t)|$. In order to fill the remaining spots in the next generation, different methods may be drawn on, with roulette-wheel selection being among the most straightforward choices. Alternatives include techniques that are known from voting evaluation for the apportionment of political mandates or seats in a parliament. Among these are the *largest remainder method*, the *Haré–Niemeyer method*, the *D'Hondt method* etc.

A very elegant way of implementing the expected value model is so-called **stochastic universal sampling**, which can be seen as a variant of roulette-wheel selection. As shown in Fig. 12.9, it uses a roulette wheel with as many markers as there are individuals in the population. These markers are located at equal distances around the roulette wheel. Instead of turning the roulette wheel once for each individual to be selected (as in standard roulette-wheel selection), the roulette wheel is turned only once and each marker gives rise to one selected individual. For example, in Fig. 12.9, s_1 and s_2 get one child each, s_3 and s_5 get two children, whereas s_4 and s_6 do not get any children at all. It should be clear that this selection method guarantees that individuals with better than average fitness (and thus an expected value of children greater than 1) get at least one child and they may receive more than one. Individuals with a lower than average fitness (and thus an expected value of children less than 1) may not procreate at all or may get at most one child.

An alternative approach (which, however, does not guarantee each individual as many children as the integer part of the expected value indicates) is the following procedure: individuals are chosen by roulette-wheel selection. After each selection of an individual, its fitness is reduced by a certain amount Δf (and the relative fitness of all individuals is recomputed). If an individual's fitness becomes negative, it is discarded and cannot produce offspring anymore. Methods for choosing Δf include

$$\Delta f = \frac{\sum_{s \in \text{pop}(t)} f(s)}{|\text{pop}(t)|},$$

which renders this model very similar to the expected value model, and

$$\Delta f = \frac{1}{k} \max\{f(s) \mid s \in \text{pop}(t)\},$$

which restricts the best individual to at most k children. The latter specifically targets the dominance problem, which can be limited with this choice.

12.2.6 Rank-Based Selection

In **rank-based selection** the individuals are sorted w.r.t. their fitness and thus a rank is assigned to each individual of the population. The idea underlying this method stems from statistics, especially from distribution-free techniques like rank correlation. Each rank is assigned a probability, with higher ranks (and thus better individuals) receiving higher probabilities. The actual selection is then be carried out with roulette-wheel selection or any of its variants (like the expected value model), with the rank probability taking the place of the relative fitness.

The advantage of rank-based selection is that it decouples the fitness value and the selection probability (which are proportional in standard roulette-wheel selection). Only the order of the fitness values, but not their absolute value determines the selection probability. As a consequence, one can easily avoid the dominance problem, namely by choosing the rank probabilities in such a way that the higher ranks, although endowed with higher probabilities, do not completely dominate the lower ones. In addition, the progress of an evolutionary algorithm no longer produces a vanishing selection pressure (cf. Sect. 12.2.3), since the distribution of the rank probabilities does not depend on the distribution of the fitness values. By adapting the rank probabilities over time, one even has a convenient way of controlling how the selective pressure develops. By moving probability mass from the lower to the higher ranks, one can easily install a drift from exploration to exploitation.

The only disadvantage of rank-based selection is the fact that the individuals need to be sorted which causes an computational effort of $O(|\,\text{pop}\,| \cdot \log_2 |\,\text{pop}\,|)$. In contrast, fitness proportionate selection, at least in the form of stochastic universal sampling, is linear in the number of individuals, because it only has to sum the fitness values and then to select $|\,\text{pop}\,|$ individuals.

12.2.7 Tournament Selection

In **tournament selection** a certain number k of individuals are drawn at random from the current population. These individuals carry out a tournament, which is won by the individual with the highest fitness (ties are broken arbitrarily). As a prize, the winning individual receives a descendant in the next generation. After the tournament, all participants are returned to the current population (including the winner). Since each tournament selects one individual, $|\,\text{pop}\,|$ tournaments have to be carried out to select all individuals of the next generation.

Note that all individuals have the same probability to be chosen to participate in a tournament. Their fitness does not influence how likely it is that they *participate* in a tournament, but only how likely it is that they *win* a tournament in which they participate. Clearly, individuals with a high fitness are more likely to win the tournaments in which they participate. However, individuals with a low fitness may still produce offspring, namely if they happen to participate in a tournament in which all other participants have an even lower fitness.

The individuals that are to participate in a tournament may be drawn with or without replacement, which usually does not make much of a difference given the fact that typical population sizes in an evolutionary algorithm are in the thousands.

The number k, the **tournament size**, is a parameter of this selection method that has to be chosen by a user, $k \in \{2, 3, \ldots, |\,\text{pop}\,|\}$. With this parameter, the selective pressure can be controlled: the larger the tournament, the higher the selective pressure. If tournaments are small, individuals with a low fitness have a higher chance of finding themselves in a tournament in which all of their opponents have even lower fitness. Only the $k - 1$ worst individuals do not have any chance of reproducing at all (assuming that all fitness values are different). On the other hand, the larger the tournaments, the higher the chance that at least one participant has a high fitness, wins the tournament, and thus deprives less fit participants of potential offspring.

Tournament selection is an excellent method to tackle the dominance problem, since the fitness value does not directly influence the selection probability. For example, even for the best individual the expected number of offspring is only the expected number of tournaments in which this individual participates. This number is controlled by the tournament size k, but not by the individual's fitness.

A modification of tournament selection is to replace the deterministic rule that the best participating individual wins the tournament by a fitness proportionate selection. That is, for each individual the tournament-specific relative fitness is computed and the winner is determined by roulette-wheel selection. This modification allows worse individuals to produce offspring even if they participate only in tournaments in which at least one other participant has a better fitness.

An important advantage of tournament selection is that it lends itself perfectly to parallelization. While fitness-proportionate selection requires a central agent that collects and normalizes the fitness values (computes the relative fitness values), arbitrarily many tournaments can be carried out in parallel without any need for centralized computations or communication between the tournament hosts. As a consequence, parallelized implementations often use tournament selection.

12.2.8 Elitism

Due to the randomness of the selection procedures we discussed up to now, only the expected value model (and some of its variants) ensures that the best individual enters the next generation. However, even if this is the case, the best individual is not protected from modifications by genetic operators (mutation or recombination with another individual). As a consequence, it is not guaranteed that the quality of the best solution candidate never worsens from one generation to the next.

As this possibility is clearly undesirable, evolutionary algorithms often employ a technique known as **elitism**. That is, the best individual (or, alternatively, the k best individuals with k to be chosen by a user) are transferred without modification to the next generation. This ensures that the best solution(s) that have been found up to now—that is, the *elite* of the population—never gets lost or destroyed. Note, though, that (other copies of) these elite individuals still enter the normal selection and modification process in the hope that they may be improved by genetic operators.

A closely related technique is **local elitism**, which refers to the treatment of individuals that undergo modification by genetic operators. In a standard evolutionary algorithm, the products of an application of a genetic operator (mutation or crossover) enter the new generation, while the originals are discarded. We may also say that the products (children) replace the originals (parents). With local elitism, however, the fitness of the involved individuals decides which actually enter the next generation. For example, a mutated individual replaces its original only if it has a better fitness. Of the four individuals involved in a crossover (two parents, two children), the best two are determined and passed on to the next generation (which may mean that the two parents are maintained and the children are discarded).

Evolutionary algorithms employing (global or local) elitism usually exhibit better convergence characteristics, since local optima are approached more consistently. However, especially local elitism bears a certain danger of converging prematurely and getting stuck in local optima, because as no (local) degradations are possible.

12.2.9 Niche Techniques

The objective of **niche techniques** is to prevent crowding as it was discussed in Sect. 12.2.1, that is, a lack of diversity many similar individuals being formed or selected. Here we consider *deterministic crowding* and *sharing*.

The idea of **deterministic crowding** is that generated offspring should always replace those individuals in the population that are most similar. As a consequence, the local density of individuals in the search space cannot grow so easily. Of course, in order to be able to apply this idea, we need a similarity or distance measure for the individuals. If chromosomes are binary coded, the *Hamming distance* may be a viable choice. In other cases specialized similarity or distance measures may be needed that take the concrete encoding of solutions candidates into account. As a consequence, it is not possible to provide generally applicable measures.

A variant of deterministic crowding, which includes ideas of elitism (see the preceding section) is the following approach: in a crossover, the two parents and two children are grouped into two pairs, each consisting of one parent and one child. The guiding principle is that a child is assigned to the parent to which it is more similar. If both children happen to be assigned to the same parent, the child that is less similar is reassigned to the other parent. Ties are broken arbitrarily. From each pair, the better individual is selected and passed on into the next generation. The advantage of this variant is that much fewer similarity computations are needed than in a global approach that finds the most similar individuals in the population as a whole.

The idea of **sharing** is to reduce the fitness of an individual if there are other individuals in its neighborhood in the search space (and hence we need again a similarity or distance measure for individuals). Intuitively, the individuals share the resources of a niche, that is, a region in the search space, which has a negative effect on their fitness. A possible choice for the fitness reduction is

$$f_{\text{share}}(s) = \frac{f(s)}{\sum_{s' \in \text{pop}(t)} g(d(s, s'))},$$

where d is a distance measure for individuals and g defines both shape and size of the niche. A concrete example is so-called **power law sharing**, which employs

$$g(x) = \begin{cases} 1 - \left(\frac{x}{\rho}\right)^\alpha & \text{if } x < \rho, \\ 0, & \text{otherwise,} \end{cases}$$

where ρ is the radius of the niche and α controls the strength of the influence that individuals in the niche have on each other.

12.2.10 *Characterization of Selection Methods*

Selection methods are often characterized by certain terms, which describe their properties. Some of the most important ones are collected in Table 12.3.

The distinction of "static" and "dynamic" mainly refers to whether the selective pressure changes over time (preferably from exploration—low selective pressure—to exploitation—high selective pressure), which is governed by the (relative) selection probabilities for individuals with different fitness. Characterizing selection

Table 12.3 Characterization of selection methods

term	meaning
static dynamic	The probability of selection remains constant. The probability of selection changes.
extinguishing preservative	Probabilities of selection may be 0. All probabilities of selection must be greater than 0.
pure-bred under-bred	Individuals can only have offspring in one generation. Individuals are allowed to have offspring in more than one generation.
right left	All individuals of a population may reproduce. The best individuals of a population may *not* reproduce.
generational on the fly	The set of parents is fixed until all offspring are created. Created offspring directly replace their parents.

methods as "extinguishing" versus "preservative" based on whether selection probabilities may be zero or not may be somewhat misleading, since even preservative methods allow for solution candidates to go extinct. The reason is simply that due to the randomness of the selection procedures a positive probability does not guarantee survival. The purpose of allowing individuals to have offspring in only one generation ("pure-bred") rather than in multiple generations ("under-bred") is to reduce the danger of crowding, since offspring tends to be similar to its parents. "Left" selection methods, in which the best individuals are not allowed to reproduce, are meant to prevent premature convergence by forcing offspring to result from worse individuals, thus explicitly favoring exploration. In contrast to this, "right" selection methods do not introduce such a guidance. Finally, "on the fly" methods are constantly modifying the population, as it is also the case in nature, which "generational" methods employ a strict discretization of time.

12.3 Genetic Operators

Genetic operators are applied to a certain part of the individuals in a generation to create modifications and recombinations of existing solution candidates. Although the majority of these modifications can be expected to be harmful, there is reasonable hope that a few of these modifications result in (slightly) better individuals.

Genetic operators are commonly categorized according to the number of parents into **mutation** or **variation** operators (only one parent, see Sect. 12.3.1), **crossover** operators (two parents, see Sect. 12.3.2) and multi-parent operators (see Sect. 12.3.3). The latter two categories (that is, with more than one parents) are sometimes generally called **recombination** operators.

An important aspect of genetic operators is whether the search space is closed under their application (see Sect. 12.1.3). For example, if candidate solutions are encoded as permutations (see, for instance, the traveling salesman problem, Sect. 11.6), then the genetic operators should preserve this property. That is, if they are applied to permutations, the results should also be permutations.

12.3.1 Mutation Operators

Genetic one-parent operators are generally referred to as **mutation** or **variation** operators. Such operators mainly serve the purpose to introduce an element of **local search**, that is, to produce a solution candidate that is (very) similar to its parent.

If solution candidates are encoded as bit strings (that is, the chromosomes are composed of zeroes and ones), then **bit mutation** (also called *bit flipping*) is commonly chosen. It consists in flipping randomly chosen alleles, that is, turning a 1 into a 0 and *vice versa*. The following algorithm formalizes this operation:

Algorithm 12.1 (Bit Mutation)

procedure mutate_bits (**var s**: array of bit, p_m: real);

begin (* mutation rate p_m *)

  for $i \in \{1, \ldots, \text{length}(s)\}$ **do begin**

    $u \leftarrow$ randomly choose according to $U([0, 1))$;

    if $u \leq p_m$ **then** $s_i \leftarrow 1 - s_i$; **end**

  end

end

Empirically it was found that choosing $p_m = 1/\text{length}(s)$ is often close to optimal.

While the number of bits that are flipped in standard bit mutation may vary (to be precise: it follows a binomial distribution with the parameter p_m), a variant fixes the number of randomly chosen bits to be flipped. In this case, the mutation rate p_m is replaced by a number n, $1 \leq n < \text{length}(s)$, of bits to be flipped, or a fraction p_b, $1 < p_b < 1$, of bits, which translates into the number of bits by $n = \lfloor p_b \cdot \text{length}(s) \rfloor$. We refer to this variant as **n-bit mutation**. The special case $n = 1$, which flips exactly one randomly chosen bit of the chromosome, is called **one-bit mutation**.

For chromosomes that are arrays of real-valued numbers so-called **Gaussian mutation** is most commonly used. It adds a random number that is sampled from a normal or Gaussian distribution to every gene, as shown in the following algorithm:

Algorithm 12.2 (Gaussian Mutation)

procedure mutate_gauss (**var x**: array of real, σ: real);

begin (* standard deviation σ *)

  for $i \in \{1, \ldots, \text{length}(x)\}$ **do begin**

    $u \leftarrow$ sample randomly from $N(0, \sigma)$;

    $x_i \leftarrow x_i + u$;

    $x_i \leftarrow \max\{x_i, l_i\}$; (* lower limit l_i of range of x_i *)

    $x_i \leftarrow \min\{x_i, u_i\}$; (* upper limit u_i of range of x_i *)

  end

end

The parameter σ controls the spread of the random numbers and corresponds to the standard deviation of the normal or Gaussian distribution. It may be used to control to some degree whether exploration of the search space should be favored (large σ) or whether a local optimization should be performed (exploitation, small σ).

An extension of Gaussian mutation is so-called **self-adaptive Gaussian mutation**. Instead of a single standard deviation σ, which is the same for all chromosomes, self-adaptive Gaussian mutation endows each chromosome x with its own standard deviation σ_x, which is used to mutate its genes. In addition, not only the genes of the chromosome x, but also its standard deviation σ_x is adapted in a mutation operation. The following algorithm formalizes the operation:

Algorithm 12.3 (Self-adaptive Gaussian Mutation)
procedure mutate_gsa (**var x**: array of real, **var** σ_x: real);
begin (* chromosome-specific standard deviation σ_x *)
  $u \leftarrow$ sample randomly from $N(0, 1)$;
  $\sigma_x \leftarrow \sigma_x \cdot \exp(u/\sqrt{\text{length}(x)})$;
  for $i \in \{1, \ldots, \text{length}(x)\}$ **do begin**
    $u \leftarrow$ randomly choose according to $N(0, \sigma_x)$;
    $x_i \leftarrow x_i + u$;
    $x_i \leftarrow \max\{x_i, l_i\}$; (* lower limit $l[i]$ of range of x_i *)
    $x_i \leftarrow \min\{x_i, u_i\}$; (* upper limit $u[i]$ of range of x_i *)
  end
end

The idea of self-adaptive Gaussian mutation is to exploit the process of evolutionary adaptations not only to find good solution candidates, but at the same time to optimize the mutation step widths (evolution-strategic principle, see Sect. 11.2). Intuitively, we may say that chromosomes with a "suitable" standard deviation—that is, a standard deviation that causes steps of "suitable width" in the region of the search space in which the chromosome is located—are more likely to produce good offspring. As a consequence, this standard deviation will spread in the population, at least among individuals located in the same region of the search space.

A generalization of (one-)bit mutation to chromosomes consisting of (more or less) arbitrary computational objects is so-called **standard mutation**. It simple replaces the current allele of a randomly chosen gene with another randomly chosen allele. An example for a chromosome consisting of integer numbers (as we used it, for example, in Sect. 11.4) is shown in Fig. 12.10: in the third gene the allele 4 is replaced with the allele 6. If desired, multiple genes may be mutated (cf. the n-queens problem in Sect. 11.4, especially Fig. 11.5 on page 185).

Like bit mutation, standard mutation receives as a parameter either a mutation rate p_m, which indicates the probability that a gene gets mutated, or a number n of genes that are to be mutated. The new allele is simply chosen from the other possible alleles using a uniform distribution (equal probabilities).

The mutation operator **transposition** or **pair swap** exchanges the alleles of two randomly chosen genes in a chromosome with each other (see Fig. 12.11). Of course, this operator can only be applied is the two affected genes have the same set of possible alleles. Otherwise a chromosome describing an individual outside of

Fig. 12.10 Example of standard mutation: an allele of a gene is replaced by another allele

Fig. 12.11 Example of transposition: two genes exchange their alleles in a chromosome

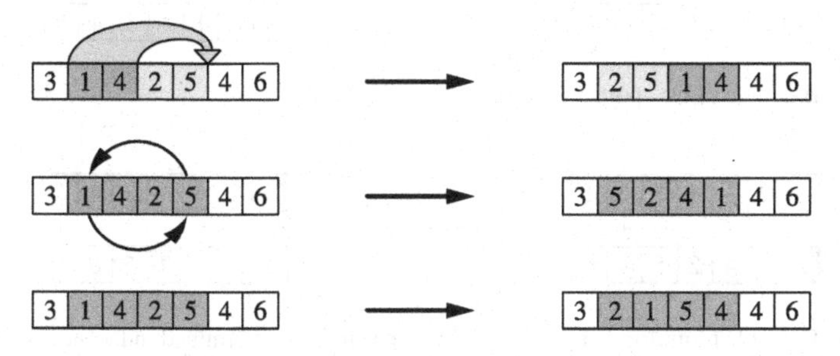

Fig. 12.12 Mutation operators on subsequences: shift, inversion and arbitrary permutation

the search space may result, which needs special treatment (cf. Sect. 12.1.3). Transposition is an excellent mutation operator in case the chromosomes are permutations of a set of integer numbers (like for traveling salesman problem, see Sect. 11.6), because the set of permutations is obviously closed under it.

Generalizations of transposition are different forms of (constrained) permutations of the alleles of a group of genes, usually forming a subsequence of $3, 4, \ldots, k$ genes in the chromosome. Among these are the **shift** of a subsequence to a new location (which may also be seen as a **cyclic permutation**), the **inversion** of a subsequence (that is, reversing the order of the alleles) and finally applying an arbitrary **permutation** to the alleles in a subsequence. Figure 12.12 illustrates these operators for chromosomes consisting of integer arrays. Clearly, all of these operators require that the sets of alleles of the genes in the affected subsequence are equal. They may be parametrized with the length of the subsequence or with a probability distribution over such lengths. Since all of these operators merely permute the alleles in a subsequence, they clearly preserve the property that a chromosome is a permutation of a set of numbers and thus are nicely applicable to cases where this is required to make the search space closed under the mutation operators (like the traveling salesman problem, see Sects. 11.6 and 12.1.3).

12.3.2 Crossover Operators

Genetic operators that involve two parent individuals are generally referred to as *crossover* operators. The best-known crossover operator is so-called **one-point**

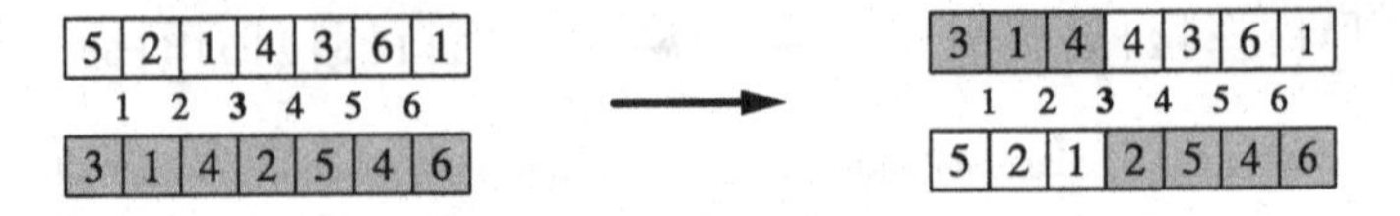

Fig. 12.13 Example of one-point crossover

Fig. 12.14 Example of two-point crossover

Fig. 12.15 Example of uniform crossover. For every gene it is determined independently whether it is exchanged (+) or not (−)

crossover (see Fig. 12.13): a cut point is chosen at random and the gene sequences on one side of the cut point are exchanged between the two (parent) chromosomes.

A straightforward extension of one-point crossover is **two-point crossover** (see Fig. 12.14). In this case, two cut points are chosen at random and the section between the two cut points is exchanged between the (parent) chromosomes.

A natural generalization of these forms is n**-point crossover**, for which n cut points are chosen. Offspring is created by alternately exchanging and maintaining the parental gene sequences between two consecutive cut points.

As an example and because we draw on it in later algorithms (cf. Sect. 13.1), we formulate one-point crossover as an algorithm (note that the value of "allele" depends on the encoding; it may be a bit, an integer, a real-valued number etc.):

Algorithm 12.4 (One-point Crossover)
procedure crossover_1point (**var r**, **s**: array of allele);
begin (∗ exchange part of two chromosomes ∗)
  $c \leftarrow$ random element of $\{1, \dots, \text{length}(\mathbf{s}) - 1\}$; (∗ choose cut point ∗)
  for $i \in \{0, \dots, c - 1\}$ **do begin** (∗ traverse the section to exchange ∗)
    $t \leftarrow r_i$; $r_i \leftarrow s_i$; $s_i \leftarrow t$; **end** (∗ swap genes up to the cut point ∗)
end

Instead of choosing a certain number of cut points, **uniform crossover** determines for each gene independently whether it is exchanged or not based on an exchange probability p_x (see Fig. 12.15; a "+" means that the genes are exchanged, a "−" that they are kept). Note that uniform crossover is *not* equivalent to $(L - 1)$-point crossover (where $L = \text{length}(s)$), as $(L - 1)$-point crossover alternately exchanges

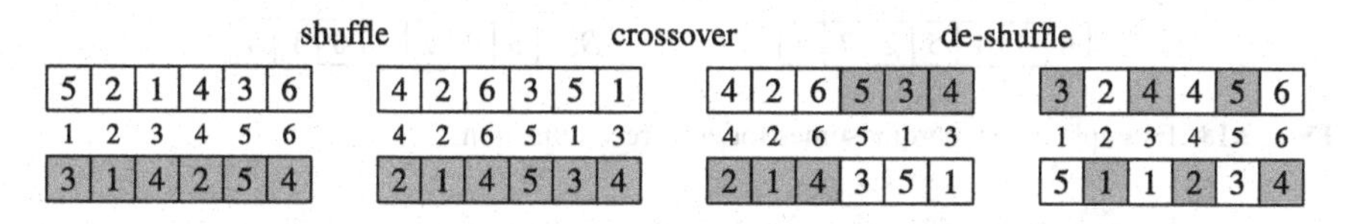

Fig. 12.16 Example of shuffle crossover (which is a modified one-point crossover)

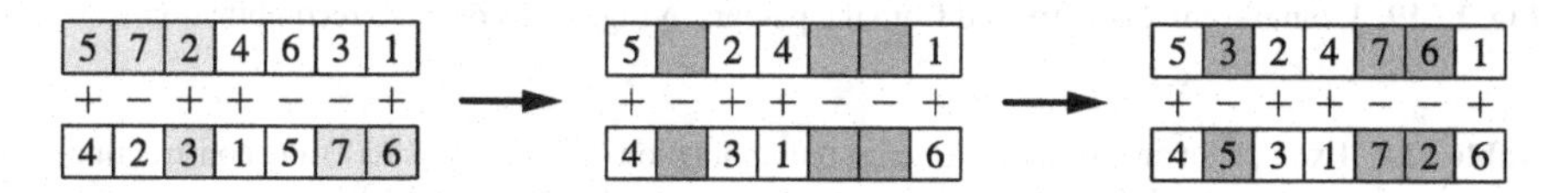

Fig. 12.17 Example of uniform order-based crossover

and keeps genes. Uniform crossover rather chooses the number of cut points randomly. (Note, however, that the different numbers are *not* equally likely: numbers close to $L/2$ are more likely than small or large numbers of cut points.)

Shuffle crossover shuffles the genes randomly before applying an arbitrary two-parent operator and then de-shuffles the genes again. Most commonly, shuffle crossover is implemented with one-point crossover, as shown in Fig. 12.16. It should be noted that shuffle crossover is *not* equivalent to uniform crossover. With uniform crossover the number of exchanged genes is binomially distributed with the parameter p_x. With shuffle crossover in combination with one-point crossover, however, every number of gene exchanges between the chromosomes is equally likely. This makes shuffle crossover one of the most recommendable crossover operators.

If the chromosomes are permutations, none of the crossover operators we studied up to now is suitable, because none guarantees that the children are permutations. For permutations, special so-called permutation-preserving crossover operators should be used, of which we discuss two variants here.

Uniform order-based crossover resembles uniform crossover, since it also decides for each gene independently whether it should be kept or not (based on an keep probability p_k). It differs from uniform crossover in how the genes are handled that are not to be kept, that is, in how the gaps between the genes to keep are filled. These gaps cannot be filled by simply taking the corresponding genes from the other parent (as in uniform crossover), as this could create a chromosome that is not a permutation. Rather the missing numbers are found and inserted into the gaps in the order (hence "order-based") in which they occur in the other parent.

An example is shown in Fig. 12.17: the plus signs mark the genes that are to be kept. This leaves three gaps to be filled. In the top chromosomes the numbers 3, 6 and 7 are missing, which occur in the bottom parent in the order 3, 7, 6. In this order they are entered into the gaps of the top chromosome. In the bottom chromosome the numbers 2, 5, and 7 are missing, which occur in the top parent in the order 5, 7, 2. In this order they are entered into the gaps of the bottom chromosome.

Note that not only the space of permutations is closed under this operator, but that this operator also preserves order information. Clearly, the kept genes are in the

A: | 6 | 3 | 1 | 5 | 2 | 7 | 4 | B: | 3 | 7 | 2 | 5 | 6 | 1 | 4 |

Fig. 12.18 Example parent chromosomes for edge recombination

| 6 | 5 | 2 | 7 | 4 | 3 | 1 |

Fig. 12.19 Example child constructed from the parents in Fig. 12.18 by edge recombination

Table 12.4 Example of an edge table for edge recombination (left, cf. Fig. 12.18) and constructing offspring from such an edge table by edge recombination (right, cf. Fig. 12.19)

allele	neighbor in *A*	in *B*	aggregated
1	3, 5	6, 4	3, 4, 5, 6
2	5, 7	7, 5	5*, 7*
3	6, 1	4, 7	1, 4, 6, 7
4	7, 6	1, 3	1, 3, 6, 7
5	1, 2	2, 6	1, 2*, 6
6	4, 3	5, 1	1, 3, 4, 5
7	2, 4	3, 2	2*, 3, 4

allele	neighbors	choice: 6	5	2	7	4	3	1
1	3, 4, 5, 6	3, 4, 5	3, 4	3, 4	3, 4	3		
2	5*, 7*	5*, 7*	7*	**7***	—	—	—	—
3	1, 4, 6, 7	1, 4, 7	1, 4, 7	1, 4, 7	1, 4	1	**1**	—
4	1, 3, 6, 7	1, 3, 7	1, 3, 7	1, 3, 7	1, 3	**1, 3**	—	—
5	1, 2*, 6	1, 2*	**1, 2***	—	—	—	—	—
6	1, 3, 4, 5	**1, 3, 4, 5**	—	—	—	—	—	—
7	2*, 3, 4	2*, 3, 4	2*, 3, 4	3, 4	**3, 4**	—	—	—

same order, while the genes with which the gaps are filled are in the order in which they occur in the other parent. This property can be useful for certain problems and corresponding encodings of the solution candidates.

An alternative permutation-preserving crossover operator is so-called **edge recombination**. It has been developed specifically for the traveling salesman problem (see Sect. 11.6), in which the round trips are encoded as permutations of the cities. In this method the chromosome is seen as a graph or, more precisely, as a chain or ring of edges: each allele is connected by edges to its neighbors in the chromosome. In addition, the first and the last allele are connected by an edge. The crossover operation consists in recombining the edges of two parent rings (hence the name of the method). It preserves neighborhood information rather than order information.

Edge recombination is a rather complex method and proceeds in two steps. In the first step, an **edge table** is constructed as follows: for every allele its neighbors (in both parents) are listed (including the last allele as a neighbor of the first and *vice versa*). If an allele has the same neighbor in both parents (where the side is irrelevant), this neighbor is listed only once, but it is marked to indicate that it has to be treated specially. As an example, Table 12.4 shows on the left an edge table for the two parent chromosomes depicted in Fig. 12.18. In the column "aggregated" duplicate neighbors are listed only once and marked with a star.

In the second step, a child is constructed, as demonstrated in Table 12.4 on the right. For the first allele of the child, the first allele of a randomly chosen parent is taken. That is, with the example parents shown in Fig. 12.18, we may start either with 6 (first allele of A) or with 3 (first allele of B). For this example we choose 6. The chosen allele is deleted from all neighbor lists in the edge table and its own list of neighbors is retrieved (see the third column of the table on the right of Table 12.4; the list of neighbors is shown in bold print). From this neighbor list, an allele is chosen respecting the following precedences:

1. marked neighbors (i.e., neighbors that occur in both parents),
2. neighbors with the shortest neighborhood list (marked neighbors count once),
3. any neighbor,
4. any allele not yet in the child.

If there are multiple choices in the highest applicable precedence class, a random allele from this class is chosen (i.e., ties are broken arbitrarily). In Table 12.4, allele 6 has the neighbors 1, 3, 4 and 5. None of these is marked, so we move on to the second precedence class. Here the neighbor lists of 1, 3 and 4 all contain four elements, while the neighbor list of 5 contains only three (due to the fact that allele 2 is marked in this list). Therefore, we have to choose allele 5, which thus becomes the second allele of the child. The process is then repeated with this allele: it is deleted from all entries in the edge table and a neighbor from its list of neighbors is chosen according to the above precedences (see the fourth column of the table on the right of Table 12.4; the list of neighbors is again shown in bold print). Since allele 2 is marked in the neighbor list, it becomes the third allele of the child. The neighbor selection process is then iterated until the child is complete, as shown in the remaining columns of the table on the right of Table 12.4. It finally yields the child chromosome depicted in Fig. 12.19.

In analogy to this process, a second child may be constructed from the first allele of the other parent (here: 3, since we started with 6). However, this is rarely done; rather only a single child is constructed from every pair of parent chromosomes.

Note that the precedence rules prefer marked alleles to non-marked in order to increase the chances that edges that are present in both parents are also present in the child. Alleles with short neighbor lists are preferred in order to reduce the chances that in a later step the fallback class (any allele not yet in the child) has to be invoked, because this introduces a new edge that is not present in either parent. The rationale is very simple: short neighbor lists run a higher risk of becoming empty due to allele selections, so one should choose from them earlier than from longer lists.

Note also that edge recombination may be employed just as well if the first and the last allele of a chromosome are *not* to be regarded as neighbors. In this case, the corresponding edges (that close the ring or cycle) are simply excluded from the edge table. On the other hand, if we regard the first and last allele as neighbors, then we may, in principle, select an arbitrary starting allele, and not just one of the first alleles of the parents. This constraint (or its equivalent, namely having to choose the *last* allele of one of the parents as a starting point) is needed only if the first and the last allele are *not* regarded as neighbors, so that the child is constructed from one of the end points of the chains that are represented by the parents.

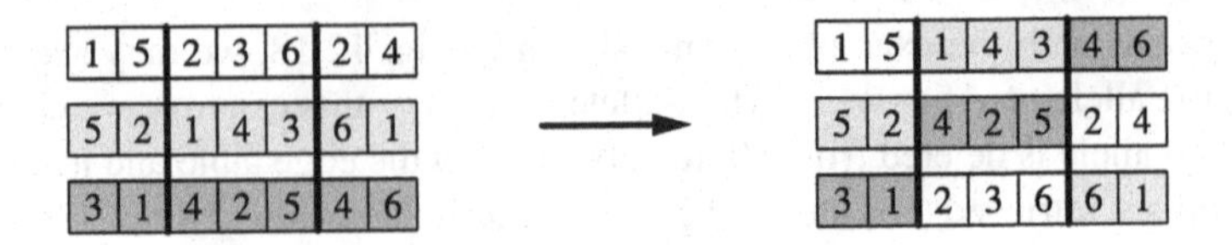

Fig. 12.20 Example of diagonal crossover

12.3.3 Multi-parent Operators

Diagonal crossover is a recombination operator that can be applied to three or more parents and that can be seen as a generalization of one-point crossover. For k parent chromosomes, one randomly chooses $k-1$ distinct cut points in $\{1,\dots,L-1\}$, where L is the length of the chromosomes. For $i=2,\dots,k$ the ith section (between the $(i-1)$th and the ith cut point, where the kth cut point is the end of the chromosomes) is then shifted cyclically $(i-1)$ steps across the k chromosomes. As an example, Fig. 12.20 shows diagonal crossover for three parent chromosomes and hence two cut points. Diagonal crossover is said to lead to a very good exploration of the search space, especially for a large number of parents (around 10–15).

12.3.4 Characteristics of Recombination Operators

Recombination operators are often categorized based on whether they have certain properties, which can aid in selecting the best operators for a given problem. Common properties that are used to characterize recombination operators with this purpose include whether they exhibit positional bias and/or distributional bias.

A recombination operator exhibits **positional bias** if the probability that two genes are jointly inherited from the same parent depends on the (relative) position of these genes in the chromosome. Positional bias is generally an undesirable property, because it can make the exact arrangement of the different genes in a chromosome crucial for the success or failure of an evolutionary algorithm. If genes for certain traits are located in an unfortunate way w.r.t. the positional bias of a crossover operator, it may be very difficult to optimize the allele combination of these genes.

A simple example of a recombination operator that exhibits positional bias is one-point crossover: the probability that two genes are jointly inherited is the higher, the closer the genes are together in the chromosome. The reason is that genes that are close together have only few possible cut points between them. Only if one of these cut points is chosen, they are separated from each other. Since all cut points are equally likely, the probability that two genes are jointly inherited is inversely proportional to the number of cut points between the genes and thus to their distance in the chromosome. As an extreme case consider the first and the last gene of a chromosome. They can never be jointly inherited under one-point crossover, because *any* choice of a cut point separates them from each other. On the other hand, two genes that are neighbors in a chromosome are separated by one-point crossover only with a probability of $1/(L-1)$, where L is the length of the chromosome.

A recombination operator exhibits **distributional bias** if the probability that a certain number of genes is exchanged between the parent chromosomes is not the same for all possible numbers of genes. Distributional bias is often undesirable, because it causes partial solutions of different lengths to have different chances of progressing to the next generation. However, distributional bias is usually less critical (that is, more easily tolerable) than positional bias.

A simple example of a recombination operator that exhibits distributional bias is uniform crossover. Since for every gene it is decided with probability p_x and independently of all other genes whether it is exchanged or not, the number K of exchanged genes is binomially distributed with the parameter p_x. That is, we have

$$P(K=k) = \binom{L}{k} p_x^k \, (1-p_x)^{L-k},$$

where L is the total number of genes. Consequently, very small and very large numbers are less likely than numbers close to Lp_x. In contrast to this, one-point crossover exhibits no distributional bias: all cut points are equally likely and since the genes on one side of the cut point are exchanged, all numbers of exchanged genes are obviously equally likely. An example of a crossover operator that exhibits neither positional nor distributional bias is *shuffle crossover* based on one-point crossover.

12.3.5 Interpolating and Extrapolating Recombination

All recombination operators that we discussed up to now merely recombine alleles that already exist in the parent chromosomes, but do not create any new alleles. As a consequence, their effectiveness depends crucially on the diversity of the population. If there is only little variation in the population, recombination operators cannot create sufficiently different offspring and thus the search may be confined to certain limited regions of the search space that can be reached with the individuals of the (initial) population. On the other hand, if a population is very diverse, such recombination operators can explore the search space well.

Especially in the realm of numeric parameter optimization, however, a different kind of recombination operator becomes possible, which can blend the traits of the parents in such a way that offspring with new traits is created, encoded by alleles that are not present (at least in exactly this form) in the parents. An example of such an operator is interpolating recombination, which blends alleles of the parents with a randomly chosen mixing parameter. A more concrete example for chromosomes that are real-valued arrays is **arithmetic crossover**, which can be seen as interpolating between the points that are represented by the parent chromosomes:

Algorithm 12.5 (Arithmetic Crossover)
function crossover_arith ($\mathbf{r}, \mathbf{s}$: array of real) : array of real;
begin
  $\mathbf{s}' \leftarrow$ new array of real with length($\mathbf{r}$) elements;
  $u \leftarrow$ choose randomly from $U([0, 1])$;
  for $i \in \{1, \ldots, \text{length}(\mathbf{r})\}$ **do**
    $s_i' \leftarrow u \cdot r_i + (1 - u) \cdot s_i$;
  return $\mathbf{s}'$;
end

It should be noted, though, that an exclusive application of such a blending operator can cause the so-called **Jenkins nightmare**, that is, the complete disappearance of all variation in a population, because the blending operation introduces a strong tendency to average all parameters (that is, all genes) that are present in the population. Therefore, arithmetic crossover should be combined—at least in the early generations of an evolutionary algorithm, in which exploration is vital—with a strongly random-based, diversity-preserving mutation operator.

An alternative are extrapolating recombination operators, which try to infer information from several individuals. Intuitively, they create a prognosis in what direction from the examined parent individuals one can expect fitness improvements. Hence extrapolating recombination may leave the region of the search space in which the individuals are located, from which the prognosis is derived. Extrapolating recombination is one of only few recombination methods that may take fitness values into account. A simple example of an extrapolating operator is arithmetic crossover with $u \in U([-1, 2])$. It should be noted, though, that an extrapolating recombination operator usually cannot compensate a lack of diversity in a population.

References

Y. Davidor. Lamarckian Sub-Goal Reward in Genetic Algorithm. *Proc. Euro. Conf. on Artificial Intelligence (ECAI, Stockholm, Sweden)*, 189–194. Pitman, London/Boston, United Kingdom/USA, 1990

Z. Michalewicz. *Genetic Algorithms + Data Structures = Evolution Programs*, 3rd (extended) edition. Springer-Verlag, New York, NY, USA, 1996

Chapter 13
Fundamental Evolutionary Algorithms

The preceding chapter presented all relevant elements of evolutionary algorithms, namely guidelines of how to choose an encoding for the solution candidates, procedures how to select individuals based on their fitness, and genetic operators with which modified solution candidates can be obtained. Equipped with these ingredients we proceed in this chapter to introducing basic forms of evolutionary algorithms, including classical genetic algorithms (in which solution candidates are encoded as bit strings, see Sect. 13.1), evolution strategies (which focus on numerical optimization, see Sect. 13.2) and genetic programming (which tries to derive function expressions or even (simple) program structures with evolutionary principles, see Sect. 13.3). Finally, we take a look at related population-based approaches (like ant colony and particle swarm optimization, see Sect. 13.4).

13.1 Genetic Algorithms

In nature, all genetic information is described in an essentially quaternary code, based on the four *nucleotides* adenine, cytosine, guanine and thymine, which are stringed together in a DNA sequence on a backbone of phosphate-deoxyribose (so-called *primary structure* of a nucleic acid). Although there are also higher level structures (like the fact that the nucleotides are organized in so-called *codons*, which are triplets of nucleotides), this is the basis of the genetic code. Transferring this structure to computer science, it seems natural to base all encoding on the ultimately binary structure of information in a computer. That is, we use chromosomes that are bit strings. This is the distinctive feature of so-called **genetic algorithms** (GA).

In principle, of course, all evolutionary algorithms running on a computer can be seen as genetic algorithms in this sense, simply because a computer ultimately encodes all information in bits, that is, in zeros and ones. For example, the chromosomes we considered for the n-queens problem, which are arrays of integers, are stored using a binary representation of these numbers. To be concrete, the chromosome $(4, 2, 0, 6, 1, 7, 5, 3)$, which describes a solution of the 8-queens problem,

R. Kruse et al., *Computational Intelligence*, Texts in Computer Science,
DOI 10.1007/978-1-4471-5013-8_13, © Springer-Verlag London 2013

could be stored (using 3 bits for each number) as 100 010 000 110 001 111 101 011. The difference between a genetic algorithm and the evolutionary algorithm we considered for the n-queens problem consists in the fact that in a genetic algorithm we consider merely the bit string that a chromosome is and ignore any higher level structure. (In the above concrete example: the fact that bit triplets are interpreted together, each indicating a file (column) in which a queen is placed.) That is, we completely separate the encoding from the genetic mechanisms, while in an evolutionary algorithm certain aspects of the encoding are considered, for instance, to restrict the genetic operators. (In the above example: in the evolutionary algorithm for the 8-queens problem, we allow as cut points for a crossover operator only the points between bit triplets, because these are the interpretable information units, while in a genetic algorithm we allow cuts between arbitrary bits.)

A typical genetic algorithm works like this:

Algorithm 13.1 (Genetic Algorithm)
function genalg (f: function, μ: int, p_x: real, p_m: real) : array of bit;
begin
  $t \leftarrow 0$; (∗ initialize the generation counter ∗)
  pop(t) ← create a population of μ random bit strings; (∗ μ must be even ∗)
  evaluate pop(t) with the fitness function f; (∗ compute initial fitness ∗)
  while termination criterion is not fulfilled **do begin**
    $t \leftarrow t + 1$; (∗ count the created generation ∗)
    pop(t) ← ∅; (∗ build the next generation ∗)
    pop′ ← select μ individuals $\mathbf{s}_1, \ldots, \mathbf{s}_\mu$ from pop(t)
      with *roulette wheel selection* ;
    for $i \leftarrow 1, \ldots, \mu/2$ **do begin** (∗ process individuals in pairs ∗)
      u ← choose random number from $U([0, 1))$;
      if $u \leq p_x$ **then** crossover_1point($\mathbf{s}_{2i-1}, \mathbf{s}_{2i}$); **end**
      mutate_bits($\mathbf{s}_{2i-1}$, p_m); (∗ crossover rate p_x and ∗)
      mutate_bits($\mathbf{s}_{2i}$, p_m); (∗ mutation rate p_m ∗)
      pop(t) ← pop(t) ∪ $\{\mathbf{s}_{2i-1}, \mathbf{s}_{2i}\}$; (∗ add (modified) individuals ∗)
    end (∗ to the next generation ∗)
    evaluate pop(t) with the fitness function f;
  end (∗ compute new fitness ∗)
  return best individual from pop(t); (∗ return the solution ∗)
end

That is, a genetic algorithm follows essentially the scheme of a general evolutionary algorithm (see Algorithm 11.1 on page 180), only that it uses bit strings as its chromosomes and applies genetic operators to these chromosomes that ignore any higher level structure of the encoding. A genetic algorithm requires mainly three parameters: the population size μ, for which an even number is chosen in order to simplify the implementation of a random application of the crossover operator, the crossover probability p_x, with which it is decided for a pair of chromosomes whether they undergo crossover or not, and the mutation probability p_m, with which it is decided for each bit of a chromosome whether it is flipped or not.

13.1.1 The Schema Theorem

So far we answered the question why evolutionary algorithms work only by providing plausibility and intuition-based arguments. Due to their restriction to bit strings, genetic algorithms, however, allow for a more formal investigation, which was first proposed in Holland (1975). This leads to the famous **schema theorem**.

Since genetic algorithms work on bit strings only, they allow us to confine our considerations to binary chromosomes. More precisely, we consider **schemata**, that is, partly specified binary chromosomes. We then investigate how the number of chromosomes matching a schema evolve over several generations of a genetic algorithm. The objective of this investigation is to derive a rough stochastic statement that describes how a genetic algorithm explores the search space.

In order to keep things simple, we confine ourselves (following Holland 1975) to bit strings of a fixed length L. In addition, we generally assume the specific form of a genetic algorithm as it was presented in pseudo-code in the preceding section. That is, we assume that chromosomes enter the intermediate population pop′ by fitness-proportionate selection (to be precise, by roulette-wheel selection as it was introduced in Sect. 12.2.1) and that one-point crossover (see Sect. 12.3.2), applied with probability p_x to chromosome pairs, and bit mutation (see Sect. 12.3.1), using the mutation probability p_m, are employed as the genetic operators.

We start with the necessary technical definitions of *schema* and *matching*.

Definition 13.1 (Schema) A *schema* h is a character string of length L over the alphabet $\{0, 1, *\}$, that is, $h \in \{0, 1, *\}^L$. The character $*$ is called *wildcard character* or *don't-care symbol.*

Definition 13.2 (Matching) A (binary) chromosome $c \in \{0, 1\}^L$ *matches* a schema $h \in \{0, 1, *\}^L$, written as $c \lhd h$, if and only if it coincides with h at all positions where h is 0 or 1. Positions at which h is $*$ are not taken into account (which explains the names "don't care symbol" or "wildcard character" for the character $*$).

As an illustration consider the following simple example: let h be a schema of length $L = 10$ and c_1, c_2 two different chromosomes of this length, which look like this:

$$h = \texttt{**0*11*10*}$$
$$c_1 = \texttt{1100111100}$$
$$c_2 = \texttt{1111111111}$$

Clearly, chromosome c_1 matches schema h, that is, $c_1 \lhd h$, because c_1 differs from h only at positions where h is $*$. On the other hand, chromosome c_2 does not match h, that is, $c_2 \not\lhd h$, because c_2 contains 1s at positions where h is 0 (positions 3 and 9).

Generally, there are 2^L possible chromosomes and 3^L different schemata. Every chromosome matches $\sum_{i=0}^{L} \binom{L}{i} = 2^L$ schemata (because any choice of i positions, at which the bit in the chromosome is replaced with a $*$, yields a schema that the chromosome matches). Based on this fact, Holland (1975) started from the idea

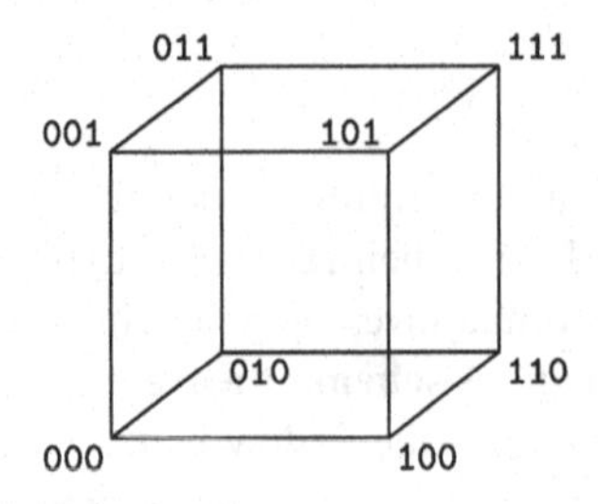

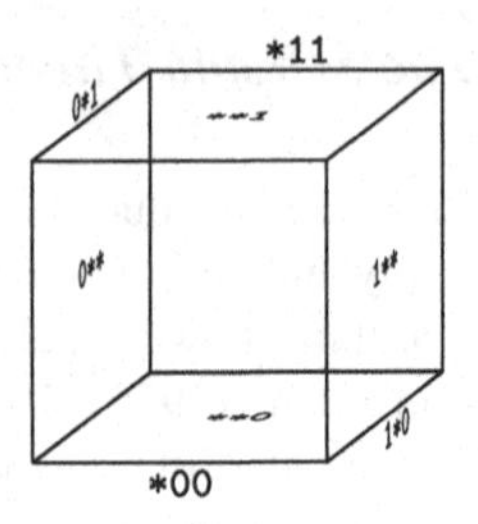

Fig. 13.1 Geometric representation of schemata as hyperplanes in a unit hypercube

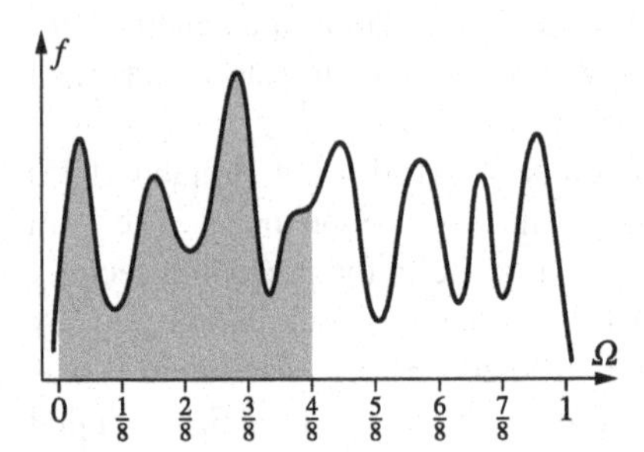

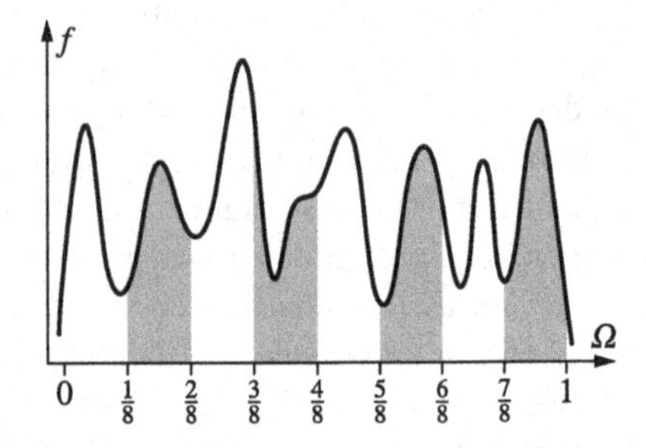

Fig. 13.2 Representation of schemata as domain of a function. On the left the schema $0 * * \cdots *$ is shown, on the right the schema $* * 1 * \cdots *$

that the observation of one chromosome corresponds to the observation of many schemata at the same time. This is what Holland called **implicit parallelism**.

Note that a population of size μ can, in principle, match close to $\mu 2^L$ different schemata. However, the number of actually matched schemata is usually a lot smaller, especially in later generations of a genetic algorithm, because the selective pressure produces similar chromosomes. (Assuming that similar chromosomes have similar fitness, selecting individuals with a high fitness has a tendency to select similar chromosomes, cf. Fig. 12.7 on page 208 as an illustration.)

In order to understand schemata better, let us take a look at two interpretations. Geometrically, a schema can be seen as describing a hyperplane in a unit hypercube. Not general hyperplanes, though, but only hyperplanes that are parallel or orthogonal to the sides of the hypercube. This is illustrated in Fig. 13.1 for three dimensions: the schema $*00$ describes the edge connecting the corners 000 and 100 (bottom front). The left face of the cube is captured by the schema $0**$. The schema $***$, which consists of wildcard characters only, describes the whole cube.

An alternative interpretation considers the domain of the fitness function. Suppose we are given a unary fitness function $f : [0, 1] \to \mathbb{R}$ and that the argument of this function is encoded as a binary number (in the usual way—for reasons of simplicity, we disregard here that such a code introduces **Hamming cliffs** (see Sect. 12.1.1), which can be avoided by using a Gray code). In this case, a schema corresponds to a "strip pattern" in the domain of the function f. This is illustrated in Fig. 13.2 for the two schemata $0 * * \cdots *$ (left diagram) and $* * 1 * \cdots *$ (right diagram).

In order to carry out our plan of tracking the evolution of chromosomes that match a schema, we have to examine how selection and applying genetic operators (that is, one-point crossover and bit mutation) influence these chromosomes. We do so in three steps. In the first step, we consider the effect of selection, in the second step the effect of one-point crossover, and in the third step the effect of bit mutation. We distinguish the populations in these steps (and quantities referring to them) by splitting the transition from time t to time $t+1$ into the steps t (original population), $t+\Delta t_s$ (population after selection, that is, intermediate population pop′), $t+\Delta t_s+\Delta t_x$ (population after selection and crossover), $t+\Delta t_s+\Delta t_x+\Delta t_m=t+1$ (population after selection, crossover and mutation, which is identical to the new population at time $t+1$). In these steps, we are mainly interested in the (expected) number of chromosomes that match a schema h. We denote these numbers by $N(h,t')$, where $t' \in \{t, t+\Delta t_s, t+\Delta t_s+\Delta t_x, t+1\}$. Our objective is to derive a (stochastic) relationship between $N(h,t)$ and $N(h,t+1)$.

In order to capture the effect of selection, we have to consider what fitness the chromosomes have that match a schema h. With fitness proportionate selection, the expected number of offspring of a chromosome s is $\mu \cdot f_{\text{rel}}(s)$. Hence, the expected number of chromosomes that match schema h after selection is

$$N(h, t+\Delta t_s) = \sum_{s \in \text{pop}(t), s \lhd h} \mu \cdot f_{\text{rel}}^{(t)}(s).$$

In order to express this number without referring to individual chromosomes s, it is convenient to define the *mean relative fitness* of chromosomes that match a schema.

Definition 13.3 (Mean Relative Fitness) The *mean relative fitness* of chromosomes that match schema h in the population pop(t) is

$$f_{\text{rel}}^{(t)}(h) = \frac{\sum_{s \in \text{pop}(t), s \lhd h} f_{\text{rel}}^{(t)}(s)}{N(h,t)}.$$

With this definition, we can write $N(h, t+\Delta t_s)$ as

$$N(h, t+\Delta t_s) = N(h,t) \cdot \mu \cdot f_{\text{rel}}^{(t)}(h).$$

By inserting the definition of relative fitness (see page 205), we can transform the expression $\mu \cdot f_{\text{rel}}(h)$ on the right-hand side according to

$$\mu \cdot f_{\text{rel}}(h) = \frac{\sum_{s \in \text{pop}(t), s \lhd h} f_{\text{rel}}(s)}{N(h,t)} \cdot \mu = \frac{\frac{\sum_{s \in \text{pop}(t), s \lhd h} f(s)}{\sum_{s \in \text{pop}(t)} f(s)}}{N(h,t)} \cdot \mu$$

$$= \frac{\frac{\sum_{s \in \text{pop}(t), s \lhd h} f(s)}{N(h,t)}}{\frac{\sum_{s \in \text{pop}(t)} f(s)}{\mu}} = \frac{\overline{f_t(h)}}{\overline{f_t}},$$

where $\overline{f_t(h)}$ is the mean fitness of the chromosomes that match schema h in generation t and $\overline{f_t}$ is the mean fitness of all chromosomes in generation t. Therefore, we can write the expected number of chromosomes that match schema h after selection as

$$N(h, t + \Delta t_s) = N(h, t) \cdot \frac{\overline{f_t(h)}}{\overline{f_t}}.$$

In order to incorporate the effect of the genetic operators, we need measures with which we can compute probabilities that the match to a schema is preserved (or destroyed). For one-point crossover, we have to consider how likely it is that the involved parent chromosomes are cut in such a way that all fixed elements of a schema under consideration are inherited from the same parent. In this case, the created offspring matches the schema if the corresponding parent does. Otherwise the match to the schema can get lost as the following example demonstrates:

$$\begin{array}{lll} h \quad = \texttt{***0*|1*1**} & & \texttt{***0*1*1**} = h \\ h \rhd c_1 = \texttt{00000|11111} & \rightarrow & \texttt{1111111111} = c_1' \not\lhd h \\ h \not\rhd c_2 = \texttt{11111|00000} & \rightarrow & \texttt{0000000000} = c_2' \not\lhd h \end{array}$$

Chromosome c_1 matches the schema h, but c_2 does not. A one-point crossover at the point marked by | creates two offspring chromosomes, both of which do not match the schema. With a different cut point, however, offspring c_1' matches the schema:

$$\begin{array}{lll} h \quad = \texttt{***|0*1*1**} & & \texttt{***0*1*1**} = h \\ h \rhd c_1 = \texttt{000|0011111} & \rightarrow & \texttt{1110011111} = c_1' \lhd h \\ h \not\rhd c_2 = \texttt{111|1100000} & \rightarrow & \texttt{0001100000} = c_2' \not\lhd h \end{array}$$

Obviously, whether an offspring chromosome matches the schema can depend crucially on the location of the cut point relative to the fixed characters of the schema. This gives rise to the notion of the *defining length* of a schema:

Definition 13.4 (Defining Length of a Schema) The *defining length* $\text{deflen}(h)$ of a schema h is the difference between the position of the last and the first non-$*$ in h.

For instance, $\text{deflen}(\texttt{**0*11*10*}) = 9 - 3 = 6$, because the first character that is not a wildcard, here a 0, is at position 3 and the last character that is not a wildcard, here also a 0, is at position 9. The defining length is the difference of these numbers.

In one-point crossover, all possible cut points are equally likely. Therefore, the probability that the cut point splits a chromosome in such a way that some of the fixed characters of a schema lie on one side of the cut and some on the other (and therefore are *not* all inherited from the same parent chromosome) is $\frac{\text{deflen}(h)}{L-1}$, because there are $\text{deflen}(h)$ cut points between the first and the last fixed character and $L - 1$ possible cut points in total (as L is the length of the chromosomes). In contrast, the probability that all fixed characters of the schema lie on the same side of the cut (and therefore *are* all inherited from the same parent, thus ensuring that the corresponding child matches the schema if the parent does) is $1 - \frac{\text{deflen}(h)}{L-1}$.

In order to derive an expression for $N(h, t + \Delta t_s + \Delta t_x)$, we have to consider whether a chromosome is subject to crossover (otherwise its matching status w.r.t. the schema obviously remains unchanged) and if it is, whether the cut point lies in such a way that the fixed characters of the schema are inherited from the same parent (then the matching status of this parent is transferred to the child). W.r.t. the latter alternative we also have to take into account that a cut point that lets a child inherit the fixed characters of a schema partially from a parent matching the schema and partially from the other may still match the schema, since it may happen that the characters inherited from the other parent match the schema. In particular, this situation occurs if both parents happen to match the schema. Finally, offspring chromosomes that match a schema can be created from parents, both of which do not match the schema, as can be seen from the following example:

$$\begin{array}{llcl} h & = \texttt{***0*|1*1**} & & \texttt{***0*1*1**} = h \\ h \not\triangleright c_1 & = \texttt{00010|11111} & \to & \texttt{1110111111} = c_1' \triangleleft h \\ h \not\triangleright c_2 & = \texttt{11101|00100} & \to & \texttt{0001000100} = c_2' \ntriangleleft h \end{array}$$

Clearly the reason why a match to the schema is created is that both of the parent chromosomes match the schema *partially* and that these partial matches are combined. Note, however, that at most one child can match the schema in such a case.

As a result of the above considerations we write

$$\begin{aligned} &N(h, t + \Delta t_s + \Delta t_x) \\ &= \underbrace{(1 - p_x) \cdot N(h, t + \Delta t_s)}_{A} + \underbrace{p_x \cdot N(h, t + \Delta t_s) \cdot (1 - p_{\text{loss}})}_{B} + C, \end{aligned}$$

where p_x is the probability that a chromosome is subject to crossover and p_{loss} is the probability that after applying one-point crossover the offspring does not match schema h anymore. A is the expected number of chromosomes that match schema h and *are not* subject to one-point crossover (and therefore still match the schema h). B is the expected number of chromosomes that *are* subject to one-point crossover and still match the schema h afterward. Finally, C is the expected number of chromosomes matching schema h that are gained by favorable recombinations of chromosomes that do not match schema h themselves.

Since the term C is almost impossible to estimate properly, we simply neglect it. As a consequence, we obtain only a lower bound for $N(h, t + \Delta t_s + \Delta t_x)$, which, however, is sufficient for our purposes. In order to derive an expression for p_{loss}, we draw on the fact that a loss of match is possible only if the randomly chosen cut point falls in such a way that the fixed characters of the schema are *not* all inherited from the same parent. As argued above, this probability is $\frac{\text{deflen}(h)}{L-1}$. Even in these cases, however, the result may still match the schema (see above). As it is difficult to obtain an expression for *all* possible cases in which a match is preserved, we confine ourselves to those cases in which both parents match the schema and therefore the location of the cut point is irrelevant. This provides us with the expression

$$p_{\text{loss}} \leq \frac{\text{deflen}(h)}{L-1} \cdot \left(1 - \frac{N(h, t + \Delta t_s)}{\mu}\right).$$

The first factor captures the probability that the choice of the cut point is potentially harmful and the second factor captures the probability that the other parent does *not* match the schema and hence there is a certain chance that the result does not match the schema. Clearly, however, this product yields only an upper bound for p_{loss}, because it assumes that any one-point crossover with a potentially harmful cut point and another parent that does not match the schema destroys the match to the schema. This is certainly not the case, as the following example demonstrates:

$$\begin{array}{llcl} h & = \texttt{***0*|1*1**} & & \texttt{***0*1*1**} = h \\ h \rhd c_1 & = \texttt{00000|11111} & \rightarrow & \texttt{1110111111} = c_1' \lhd h \\ h \not\rhd c_2 & = \texttt{11101|00100} & \rightarrow & \texttt{0000000100} = c_2' \not\lhd h \end{array}$$

Even though chromosome c_2 does not match the schema h and the cut point of one-point crossover falls in such a way that some of the fixed characters of the schema h are inherited from c_1 and some from c_2, the offspring chromosome c_1' matches the schema h. The reason is, of course, that c_2 matches the schema h *partially*. However, capturing partial matches in this analysis is extremely difficult and therefore we confine ourselves to the upper bound for p_{loss} stated above.

Plugging the upper bound for p_{loss} into the formula for $N(h, t + \Delta t_s + \Delta t_x)$ yields

$$\begin{aligned} & N(h, t + \Delta t_s + \Delta t_x) \\ & \quad \geq (1 - p_x) \cdot N(h, t + \Delta t_s) \\ & \quad\quad + p_x \cdot N(h, t + \Delta t_s) \cdot \left(1 - \frac{\text{deflen}(h)}{L - 1} \cdot \left(1 - \frac{N(h, t + \Delta t_s)}{\mu}\right)\right) \\ & \quad = N(h, t + \Delta t_s)\left(1 - p_x + p_x \cdot \left(1 - \frac{\text{deflen}(h)}{L - 1} \cdot \left(1 - \frac{N(h, t + \Delta t_s)}{\mu}\right)\right)\right) \\ & \quad = N(h, t + \Delta t_s) \cdot \left(1 - p_x \frac{\text{deflen}(h)}{L - 1} \cdot \left(1 - \frac{N(h, t + \Delta t_s)}{\mu}\right)\right) \\ & \quad = N(h, t) \cdot \frac{\overline{f_t(h)}}{\overline{f_t}} \cdot \left(1 - p_x \frac{\text{deflen}(h)}{L - 1} \cdot \left(1 - N(h, t) \cdot f_{\text{rel}}(h)\right)\right), \end{aligned}$$

where we exploited in the last step the relationship $N(h, t + \Delta t_s) = N(h, t) \cdot \mu \cdot f_{\text{rel}}(h)$. Note that we obtain only an inequality, because we used an upper bound for p_{loss} and because we neglected potential gains from recombinations of chromosomes that do not match the schema h (captured above by the term C, which is missing here).

Having incorporated the effect of crossover, we now turn to mutation. The effect of bit mutation can easily be captured by the *order* of a schema:

Definition 13.5 (Order of a Schema) The *order* $\text{ord}(h)$ of a schema h is the number of zeroes and ones in h, that is, $\text{ord}(h) = \#(h, 0) + \#(h, 1) = L - \#(h, *)$ where the operator # counts the number of occurrences of its second argument in its first.

For instance, $\text{ord}(\texttt{**0*11*10*}) = 5$, because the chromosome contains 2 zeros and 3 ones and thus a total of 5 fixed (that is, not *don't care*) characters.

With the notion of the order of a schema, we can express the probability that a match to a schema does *not* get lost due to a bit mutation of a chromosome as $(1 - p_m)^{\text{ord}(h)}$. The reason is that a single bit gets flipped with the probability p_m and thus remains unchanged with probability $1 - p_m$ (see Algorithm 12.1 on page 217). If any of the fixed characters in the schema h, of which there are $\text{ord}(h)$, is flipped in the chromosome, the chromosome does not match the schema anymore. We do not care about the $L - \mathrm{ord}(h)$ remaining bits, because the chromosome matches the schema regardless of their value. Since the bit flips are decided independently, the probability that none of the fixed bits in the schema is flipped is $(1 - p_m)^{\mathrm{ord}(h)}$.

As a consequence, we can express the effect of bit mutation as

$$\begin{aligned} N(h, t+1) &= N(h, t + \Delta t_s + \Delta t_x + \Delta t_m) \\ &= N(h, t + \Delta t_s + \Delta t_x) \cdot (1 - p_m)^{\mathrm{ord}(h)}. \end{aligned}$$

Note that alternative mutation models are easy to handle as well. For example, if at most one bit is flipped in a chromosome (so-called one-bit mutation, see Sect. 12.3.1), then the effect can be described by

$$\begin{aligned} N(h, t+1) &= N(h, t + \Delta t_s + \Delta t_x + \Delta t_m) \\ &= N(h, t + \Delta t_s + \Delta t_x) \cdot \left(1 - \frac{\mathrm{ord}(h)}{L}\right), \end{aligned}$$

where $\mathrm{ord}(h)/L$ is the probability that a fixed character in the schema h is flipped in the chromosome (assuming that all bits are equally likely).

Plugging in the result for $N(h, t + \Delta t_s + \Delta t_x)$ that we derived above, we finally obtain **schema theorem** (for bit mutation):

$$N(h, t+1) \geq N(h, t) \cdot \frac{\overline{f_t(h)}}{\overline{f_t}} \left(1 - p_x \frac{\mathrm{deflen}(h)}{L-1} \left(1 - \frac{N(h, t)}{\mu} \cdot \frac{\overline{f_t(h)}}{\overline{f_t}}\right)\right) (1 - p_m)^{\mathrm{ord}(h)}.$$

The general form of this relationship between $N(h, t+1)$ and $N(h, t)$ is clearly

$$N(h, t+1) \geq N(h, t) \cdot g(h, t).$$

Simplifying, we may therefore say that the number of chromosomes that match a schema h is multiplied in each generation by some factor and thus develops exponentially in the course of several generations. If $g(h, t) > 1$, the number of matching chromosomes grows exponentially, if $g(h, t) < 1$, it decreases exponentially. Since the number of matching chromosomes cannot decrease for *all* schemata (simply because the population size is constant and the contained chromosomes must match some schemata), there must be schemata for which the number of matching chromosomes grows (unless the number of matching chromosomes stays the same for *all* schemata, which, however, implies that the population is essentially constant).

By considering the factors of $g(h, t)$, we can therefore try to derive properties of schemata for which the number of matching chromosomes grows particularly quickly (that is, for which $g(h, t)$ is large). Since $g(h, t)$ is a product, every factor should be as large as possible. Therefore, such schemata should have

- high mean fitness (due to the factor $\overline{f_t(h)}/\overline{f_t}$),
- small defining length (due to the factor $1 - p_x \,\mathrm{deflen}(h)/(L-1) \ldots$), and
- low order (due to the factor $(1 - p_m)^{\mathrm{ord}(h)}$).

Such schemata are also called **building blocks**, due to which the schema theorem is sometimes also referred to as the **building block hypothesis**: the evolutionary search focuses on promising building blocks of solution candidates.

It should be kept in mind, though, that the schema theorem or the building block hypothesis applies in the derived form only to bit strings, fitness proportionate selection, one-point crossover and bit mutation. If we use different genetic operators, building blocks may be characterized by other properties than order or defining length. However, a high mean fitness is always among the characteristic features, since all selection methods favor such chromosomes, although differently strongly and not always in direct proportion to the fitness values.

It should also be noted that the schema theorem is open to many different lines of criticism. It widely neglects the interplay of different schemata, as well as the possibility of *epistasis* (the whole derivation implicitly assumes (very) low epistasis). It works with expected values that are strictly valid only for an infinite population size (which obviously cannot be achieved in practice, where effects of stochastic drift need to be taken into account). The factor $g(h,t)$ is clearly not constant, but changes over time due to its dependence on the population at time t, so that the claim of an exponential behavior over several generations is slightly dubious (especially, since saturation effects can be expected) etc.

13.1.2 The Two-Armed Bandit Argument

The schema theorem implies that a genetic algorithm achieves a near-optimal balance between exploration of the search space and exploitation of good solution candidates. As an argument for this claim, Holland used the **two-armed bandit** model as an analogy, that is, a slot machine with two independent arms (Holland 1975; Michell 1998). The two arms have different expected (per trial) payoffs μ_1 and μ_2 with variances σ_1^2 and σ_2^2, respectively, all of which are unknown. Without loss of generality we may assume $\mu_1 > \mu_2$, though. It is also unknown, however, which of the two arms of the slot machine has the higher payoff (that is, it is unknown whether μ_1 is assigned to the left or to the right arm). Suppose we may play N games with such a slot machine. What is the best strategy to maximize our winnings?

If we knew which arm has the greater payoff, we would simply use that arm for all N games, thus clearly maximizing our expected winnings. However, since we do not know which arm is better, we must invest some trials into gathering information about which arm might be the one with the higher payoff. For example, we may choose to use $2n$ trials, $2n < N$, for this task, in which we play both arms equally often (that is, we use each arm n times). Afterward, we evaluate which arm has given us the higher average payoff per trial (exploration). In the remaining $N - 2n$ trials, we then exclusively play the arm that has the higher observed payoff (exploitation). Our original question can thus be reformulated as: how should we choose n relative to N in order to maximize our (expected) winnings or, equivalently, to minimize our (expected) loss relative to always having chosen the better arm? In other words, how should we balance exploration (initial $2n$ trials) and exploitation (final $N - 2n$ trials)?

Clearly, there are two types of losses involved here: (1) inevitable loss in the information gathering phase, in which we play the worse arm n times (regardless of which arm is the worse one, since we play both arms equally often), and (2) loss due to the fact that we determine the better arm only based on an empirical payoff estimate, which may point us to the wrong arm as the better one. The former refers to the fact that in the $2n$ trials we devote to exploration we necessarily lose

$$L_1(N,n) = n(\mu_1 - \mu_2),$$

because we do n trials with the arm with lower payoff μ_2 instead of using the arm with the higher payoff μ_1. The loss from the remaining $N - 2n$ trials can only be given in stochastic terms. Let p_n be the probability that the average payoffs per trial, as determined empirically in the first $2n$ trials, actually identify the correct arm. (The index n of this probability is due to the fact that it obviously depends on the choice of $2n$: the larger $2n$, the higher the probability that the empirical payoff estimate from the $2n$ exploration trials identifies the correct arm.) That is, with probability p_n we use the arm actually having the higher payoff μ_1 for the remaining $N - 2n$ trials, while with probability $1 - p_n$ we use the arm actually having the lower payoff μ_2 in these trials. In the former case, there is no additional loss (beyond what is incurred in the exploration phase), while in the latter case we lose

$$L_2(N,n) = (N - 2n)(\mu_1 - \mu_2)$$

in the exploitation phase, because we choose the wrong arm (that is, the one actually having the lower payoff μ_2). Therefore, the expected total loss is

$$\begin{aligned} L(N,n) &= \underbrace{L_1(N,n)}_{\text{exploration loss}} + (1 - p_n) \underbrace{L_2(N,n)}_{\text{incorrect exploitation loss}} \\ &= n(\mu_1 - \mu_2) + (1 - p_n)(N - 2n)(\mu_1 - \mu_2) \\ &= (\mu_1 - \mu_2)\big(np_n + (1 - p_n)(N - n)\big). \end{aligned}$$

The final form nicely captures that in case the better arm is correctly identified (probability p_n), we lose winnings from n times using the worse arm in the exploration phase, while in case the better arm is incorrectly identified (probability $1 - p_n$), we lose winnings from $N - n$ trials with the worse arm (n of which happen in the exploration phase and $N - 2n$ happen in the exploitation phase).

We now have to minimize the loss function $L(N,n)$ w.r.t. n. The main problem here is to express the probability p_n in terms of n (since p_n clearly depends on n: the longer the exploration phase, the higher the chance that it yields a correct decision). Going into details is beyond the scope of this book, so we only present the final result (Holland 1975; Michell 1998): n should be chosen according to

$$n \approx c_1 \ln\left(\frac{c_2 N^2}{\ln(c_3 N^2)}\right),$$

where c_1, c_2 and c_3 are certain positive constants. By rewriting this expression, we can turn it into (Holland 1975; Michell 1998)

$$N - n \approx e^{n/2c_1} \sqrt{\frac{\ln(c_3 N^2)}{c_2}} - n.$$

Since with growing n the term $e^{n/2c_1}$ dominates the expression on the right hand side, this equation can be simplified (accepting further approximation) by

$$N - n \approx e^{cn}.$$

In other words, the total number of trials $N - n$ that are executed with the arm that is observed to be better should increase exponentially compared to the number of trials n that are executed with the arm that is observed to be worse.

This result, though obtained for a two-armed bandit, can be transferred to multi-armed bandits. In this more general form, it is then applied to the schemata that are considered in the schema theorem: the arms of the bandit correspond to different schemata, their payoff to the (average) fitness of chromosomes matching them. A chromosome in a population that matches a schema is seen as a trial of the corresponding bandit arm. Recall, however, that a chromosome matches many schemata, thus exhibiting an inherently parallel exploration of the space of schemata.

As we saw in the preceding section, the schema theorem states that the number of chromosomes matching schemata with better than average fitness grows essentially exponentially over several generations. The two- or multi-armed bandit argument now says that this is an optimal strategy to balance exploration of schemata (playing all arms of the bandit) and their exploitation (playing the arm or arms that have been observed to be better than the others).

13.1.3 The Principle of Minimal Alphabets

The **principle of minimal alphabets** is sometimes invoked to claim that binary encodings, as used by genetic algorithms, are "optimal" in a certain sense. The core idea is that the number of possible schemata should be maximized relative to the size of the search space (or the population size), so that the parallelism inherent in the search for schemata is maximally effective. That is, with the chromosomes of the population a number of schemata should be covered that is as large as possible.

If chromosomes are defined as strings of length L over an alphabet $\mathcal{A}$, then the ratio of the number of schemata to the size of the search space is $(|\mathcal{A}| + 1)^L / |\mathcal{A}|^L$. Clearly, this ratio is maximized if the size $|\mathcal{A}|$ of the alphabet is minimized. Since the smallest usable alphabet has $|\mathcal{A}| = 2$, binary codings optimize this ratio.

A more intuitive form of this argument was put forward by Goldberg (1989): the larger the size of the alphabet, the more difficult it is to find meaningful schemata, because a schema is matched by a larger number of chromosomes. Since a schema averages over the fitness of the matching chromosomes, the quality of a schema may be tainted by some bad chromosomes (low fitness) that happen to match the same schema. Therefore one should strive to minimize the number of chromosomes that are matched by a schema. Since a schema h matches $(L - \mathrm{ord}(h))^{|\mathcal{A}|}$ chromosomes, we should use an alphabet of minimal size. Again, since the smallest usable alphabet has $|\mathcal{A}| = 2$, binary codings can be expected to be optimal.

Whether these arguments are convincing is debatable. At least one has to admit that there is a tradeoff between maximizing the number of schemata relative to the search space and expressing the problem in a more natural manner with the help of larger alphabets (Goldberg 1991). Furthermore, a strong argument *in favor* of larger alphabets has been put forward by Antonisse (1989).

13.2 Evolution Strategies

Evolution strategies (ES) (Rechenberg 1973) are the oldest form of an evolutionary algorithm. They focus on numerical optimization problems and therefore work exclusively with chromosomes that are arrays of real-valued numbers. Their name points to the evolution-strategic principles we mentioned in Sect. 11.2: in (natural) evolution not only the organisms are optimized, but also the mechanisms of evolution. These include parameters like reproduction and mortality rates, life spans, vulnerability to mutations, mutation step sizes etc. Apart from the focus on numerical optimization problems, it is a distinctive feature of evolution strategies that in many forms they adapt mutation step sizes as well as their direction.

To be more precise, we are given a function $f : \mathbb{R}^n \to \mathbb{R}$, for which we want to find an optimal argument vector $\mathbf{x} = (x_1, \dots, x_n)$, that is, an argument vector that yields a maximum or minimum of the function f. Chromosomes are therefore such vectors of real-valued numbers. Evolution strategies often (though not always) abandon **crossover**, that is, there may be no recombination of chromosomes. Rather they focus on **mutation** as the core variation operator. Mutation in evolution strategies consists generally in adding a random vector $\mathbf{r}$, each element r_i, $i = 1, \dots, n$, of which is the realization of a normally distributed random variable with mean zero (independent of the element index i) and variances σ_i^2 or standard deviations σ_i. The variances σ_i^2 may or may not depend on the element index i (one variance for the whole vector or a specific variance for each element) and may or may not depend on the generation counter t (time dependent or independent variance).

As we study in more detail below, the variances may also be coupled to the chromosome and may be subject to mutation themselves. In this way, an adaptation of the mutation step sizes and step directions can be realized. Intuitively, we may say that chromosomes with a "suitable" mutation variance—that is, a variance that causes steps of "suitable width" in the region of the search space in which the chromosome is located—are more likely to produce good offspring. As a consequence, this variance can be expected to spread in the population, at least among individuals located in the same region of the search space. It should be noted, though, that the adaptation of the mutation parameters is thus indirect and therefore cannot be expected to be as effective and efficient as the optimization of the function arguments itself. Nevertheless it can help the search process considerably.

13.2.1 Selection

Selection in evolution strategies follows a strict **elite principle**: only the best individuals enter the next generation. There is no random element involved, like, for instance, in the various forms of fitness proportionate selection (see Sect. 12.2.1), which gives better individuals only better chances to enter the next generation, but does not rule out entirely that individuals with a low fitness have offspring.

Even though the elite principle is fixed, there are two different forms of selection, which are distinguished by whether only offspring or parents and offspring together are considered in the selection process. Let μ be the number of individuals in the parent generation and λ the number of offspring individuals that were created by mutation. In the so-called **plus strategy** the parents and children are pooled for the selection process, that is, the selection works on $\mu + \lambda$ individuals (hence an evolution strategy working with this scheme is also called $(\mu + \lambda)$-ES). In contrast to this, in the so-called **comma strategy** (also called (μ, λ)-ES) selection considers only the offspring individuals. In both cases, the μ best individuals (either from the $\mu + \lambda$ pooled individuals for the plus strategy or only from the $\lambda > \mu$ offspring individuals in the comma strategy) are selected for the next generation.

In both the plus and the comma strategy, the number λ of offspring individuals (usually) exceeds the number μ of parent individuals (considerably). This approach implements the principle of birth surplus or overproduction, see Sect. 11.2. It is motivated by the fact that usually the majority of mutations are harmful.

Note that the comma strategy actually requires $\lambda \gg \mu$ or at least $\lambda > \mu$, so that there are enough individuals to choose from. (Clearly, $\lambda < \mu$ leads to shrinking populations and $\lambda = \mu$ ignores the fitness of the individuals, since all offspring individuals have to be chosen regardless of their fitness.) However, for the plus strategy it is usually also advisable to create many more individuals than there are parents, in order to reduce the risk of getting stuck in a local optimum, which is caused by the strict elite principle. A typical choice (for both strategies) is $\mu : \lambda = 1 : 7$.

In order to counteract that the plus strategy gets stuck in a local optimum, it is sometimes replaced for some generations with the comma strategy, which increases the diversity of the population again. On the other hand, since in the comma strategy all parent individuals are definitely lost, it is advisable to keep track of the best individual encountered so far in this strategy, while in the plus strategy the best individual is automatically preserved by the strict elite principle.

As a simple example, we consider the special case of the $(1 + 1)$-ES. The initial "population" $\mathbf{x}_0 \in \mathbb{R}^n$ consists of a single random vector of real numbers. The created offspring individual is $\mathbf{x}_t^* = \mathbf{x}_t + \mathbf{r}_t$, where $\mathbf{r}_t \in \mathbb{R}^n$ is random vector of real numbers sampled from a normal distribution. Selection consists in setting

$$\mathbf{x}_{t+1} = \begin{cases} \mathbf{x}_t^* & \text{if } f(\mathbf{x}_t^*) \geq f(\mathbf{x}), \\ \mathbf{x}_t & \text{otherwise.} \end{cases}$$

Further generations are created until a termination criterion is met. Clearly, this procedure is identical to hill climbing, as we discussed it in Sect. 11.5.2.

As a consequence, we may interpret a more general plus strategy (with $\mu > 1$, but still with $\lambda = \mu$) as a kind of parallel hill climbing. Instead of considering only one current point, the search is performed simultaneously at several places in the search space, always pursuing the μ most promising paths. Note, however, the difference to executing μ hill climbing processes in parallel: in an evolution strategy, both parent and child of a hill climbing pair may enter the next generation (thus creating a fork in the search), for which another pair is extinguished completely (neither parent nor child enter the next generation). As a consequence, there is an exchange of information (about fitness values) between the hill climbing processes, which focuses the search more strongly on promising regions. With $\lambda > \mu$ the search is even more efficient, because several paths are explored from the same parent. However, this may also have the effect of increasing the risk to get stuck in a local optimum.

13.2.2 Global Variance Adaptation

As we mentioned at the beginning of this section, a distinctive feature of evolution strategies is that they try to adapt mutation step sizes. In the simplest form, there is only one global variance σ^2 (or standard deviation σ) that controls the mutations of all chromosomes. This variance is adapted in the course of the generations, so that the mean convergence rate is (approximately) optimized.

In order to obtain a rule how to adapt a global variance, Rechenberg (1973) determined the optimal variance σ^2 for the two functions $f_1(x_1, \ldots, x_n) = a + bx_1$ and $f_2(x_1, \ldots, x_n) = \sum_{i=1}^{n} x_i^2$ by determining the probabilities for a successful (that is, improving) mutation. These probabilities are $p_1 \approx 0.184$ for f_1 and $p_2 \approx 0.270$ for f_2. From this result Rechenberg (1973) heuristically inferred the so-called $\frac{1}{5}$ success rule: under the plus strategy the mutation step size (as expressed in σ or σ^2) is appropriate if approximately $\frac{1}{5}$ of the offspring are better than the parents.

With the $\frac{1}{5}$ success rule, the standard deviation σ is adapted as follows: if more than $\frac{1}{5}$ of the children are better than the parents, the variance should be increased. On the other hand, if less than $\frac{1}{5}$ of the children are better than the parents, the variance should be reduced. The rationale is that under the assumption that similar individuals have similar fitness, smaller modifications are more likely to create better individuals than larger modifications. In order to obtain a simple rule, the standard deviation σ is increased by multiplying it with a user-specified factor $\alpha > 1$, and reduced by dividing it by the same factor. We thus obtain the following procedure:

Algorithm 13.2 (Global-Variance-Adaption)
function varadapt_global (σ, p_s, θ, α: real) : real;
begin (* standard deviation σ, success rate p_s *)
  if $p_s > \theta$ **then return** $\sigma \cdot \alpha$; (* threshold $\theta = \frac{1}{5}$, modification factor $\alpha > 1$ *)
  if $p_s < \theta$ **then return** σ / α;
  return σ;
end

Here p_s stands for the fraction of the children that are better than the parents, that is, the success rate of the mutations. Note that this rate may also be seen as a measure for the balance of exploration and exploitation. If the success rate is too large, exploitation of good individuals dominates, which can lead to effects of premature convergence (cf. Sect. 12.2.2). On the other hand, if the success rate is too low, exploration dominates, which can lead to slow convergence (though never to vanishing selection pressure, due to the strict elite principle employed in evolution strategies).

Note that for larger populations, the $\frac{1}{5}$ success rule is sometimes too optimistic. In addition, in analogy to simulated annealing (see Sect. 11.5.3), one may define a function that increases the threshold over time (and thus introduces a tendency to reduce the variance). This is the reason for making θ a parameter of the above function instead of fixing it at the value $\frac{1}{5}$ (even though this is frequently chosen).

The complete algorithm for a comma strategy that works with global variance adaption every k generations is shown below. For a plus strategy, one merely has to replace the statement "pop′ ← ∅" with "pop′ ← pop(t − 1)."

Algorithm 13.3 (Adaptive-ES)

```
function evostrat_global (f: function, μ, λ, k: int, θ, α: real) : object;
begin                        (* objective function f, population size μ *)
  t ← 0;                     (* number of offspring λ, modification frequency k *)
  s ← 0;                     (* threshold θ = 1/5, modification factor α > 1 *)
  σ ← value for the initial step size;
  pop(t) ← create a population with μ individuals;
  evaluate pop(t) with the function f;
  while termination criterion is not fulfilled do begin
    t ← t + 1;               (* count the created generation *)
    pop′ ← ∅;                (* for plus strategy pop′ ← pop(t − 1) *)
    for i = 1, ..., λ do begin
      x ← select random parent uniformly from pop(t);
      y ← copy of x;         (* create a mutated child *)
      mutate_gauss(y, σ);
      if f(y) > f(x) then s ← s + 1; end
      pop′ ← pop′ ∪ {y};     (* count the successful mutations *)
    end
    pop(t) ← select best μ individuals from pop′;
    if t mod k = 0 then      (* every k generations *)
      σ ← varadapt_global(σ, s/kλ, θ, α);
      s ← 0;                 (* adapt the variance and *)
    end                      (* reinitialize the success counter *)
  end
  return best individual in pop(t);
end
```

Note that for a comma strategy it may be advisable to keep track of the best solution found in the process (since parents are discarded), while a plus strategy automatically does so due to the strict elite principle employed for selection.

13.2.3 Local Variance Adaptation

In contrast to its global counterpart (see above), **local variance adaptation** employs chromosome-specific variances. That is, a chromosome not only consists of an array of real-valued numbers that are the arguments of the function f to optimize, but includes a standard deviation or even an array of standard deviations (one for each function argument), which prescribe chromosome-specific (and gene-specific) mutation step sizes. In the evolution process, not only the function arguments are modified, but the standard deviations are adapted as well. It is plausible that chromosomes with "bad" standard deviation(s) create a lot of "bad" offspring, which is filtered out in the selection process. Chromosomes with "good" standard deviations(s), on the other hand, can be expected to create a larger number of "good" offspring. As a consequence, "good" standard deviation(s) should spread in the population, even though they do not influence the fitness of an individual directly.

A complete self-adaptive algorithm with chromosome-specific mutation step widths, which employs the self-adaptive Gaussian mutation (as it was introduced in Algorithm 12.3 on page 218), is shown below. For a plus strategy, one merely has to replace the statement "pop$' \leftarrow \emptyset$" with "pop$' \leftarrow$ pop$(t-1)$."

Algorithm 13.4 (Self-Adaptive-ES)
function adaptive_evostr (f: function, μ, λ: int) : object;
begin (∗ objective function f, population size μ ∗)
  $t \leftarrow 0$; (∗ number of offspring λ ∗)
  pop$(t) \leftarrow$ create population with μ individuals;
  evaluate pop(t) with the function f;
  while termination criterion not fulfilled **do begin**
    $t \leftarrow t+1$; (∗ count the created generation ∗)
    pop$' \leftarrow \emptyset$; (∗ for plus selection: pop$' \leftarrow$ pop(t) ∗)
    for $i = 1, \dots, \lambda$ **do begin**
      $(\mathbf{x}, \sigma_x) \leftarrow$ randomly select parent uniformly from pop(t);
      $(\mathbf{y}, \sigma_y) \leftarrow$ copy of $(\mathbf{x}, \sigma_x)$;
      mutate_gsa$(\mathbf{y}, \sigma_y)$; (∗ mutate individual and variance ∗)
      pop$' \leftarrow$ pop$' \cup \{(\mathbf{y}, \sigma_y)\}$;
    end
    evaluate pop$'$ with the function f;
    pop$(t) \leftarrow$ select best μ individuals from pop$'$;
  end
  return best individual in pop(t);
end

Note that for element-specific mutation step sizes (that is, a vector of standard deviations per chromosome instead of merely a single standard deviation per chromosome), a somewhat more complex mutation operator is needed.

A commonly used rule for adapting element-specific mutation step sizes (that is, there is one standard deviation σ_i for each function argument) is:

$$\sigma_i' = \sigma_i \cdot \exp(r_1 \cdot u_0 + r_2 \cdot u_i).$$

Here $\mathbf{u} = (u_0, u_1, \ldots, u_n)$ is a vector, each element of which is sampled from a standard normal distribution with mean 0 and variance 1 (that is, from $N(0, 1)$) and n is the number of arguments of the function f to optimize (and thus the length of the chromosome $\mathbf{x}$). Bäck and Schwefel (1993) recommend to choose $r_1 = 1/\sqrt{2n}$ and $r_2 = 1/\sqrt{2\sqrt{n}}$, while Nissen (1997) proposes to use $r_1 = 0.1$ and $r_2 = 0.2$. In addition, a lower bound (greater than the mandatory lower bound 0 as variances cannot be negative) is specified for the mutation step widths.

13.2.4 Covariances

In the standard form of variance adaption, the variances of different vector elements are independent of each other. That is, the **covariance matrix** of the mutation operator is a diagonal matrix. As a consequence, the mutation operator is only able to prefer directions in the search space that are parallel to the coordinate axes of the search space. An oblique direction cannot be represented, even though it may be better than an axes-parallel direction. As a simple example consider a two-dimensional search space. If the best mutations change the two arguments about equally strongly and in the *same* direction, this cannot be represented by merely using independent variances for the arguments. The best approximation are equal variances for both arguments, which, however, allow with the same probability that both arguments are changed by about the same amount, but in *opposite* directions.

This problem can be solved by introducing not only element-specific variances, but covariances for them as well. For instance, in the simple two-dimensional search space example, we may use a covariance matrix like

$$\Sigma = \begin{pmatrix} 1 & 0.9 \\ 0.9 & 1 \end{pmatrix}.$$

With this covariance matrix changes of the two arguments in the same direction (either both are increased or both diminished) are much more likely than changes in opposite directions (increasing one and reducing the other).

As an illustration of the effect of a covariance matrix, which introduces **correlations** between the changes of the arguments, Fig. 13.3 shows uncorrelated, weakly positively, strongly positively and strongly negatively correlated mutations in a two-dimensional space. Note that in these diagrams the variances in the coordinate directions are the same (which may be a result of normalizing the dimensions by their variance, which is exactly what distinguishes correlation from covariance). If one allows for different variances in the directions of the axes, it becomes clear that correlated mutations allow us to prefer mutation directions in the search space that are (arbitrarily) oblique to the coordinate axes.

In order to understand the meaning of a covariance matrix better, recall the one dimensional case. A variance is often a bit difficult to interpret: due to its quadratic

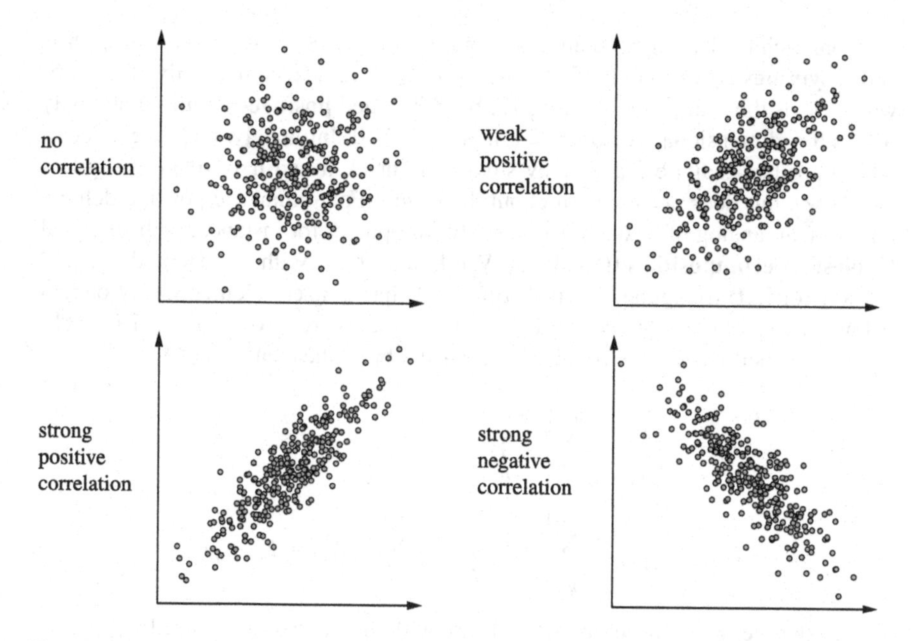

Fig. 13.3 Illustration of covariance and correlation

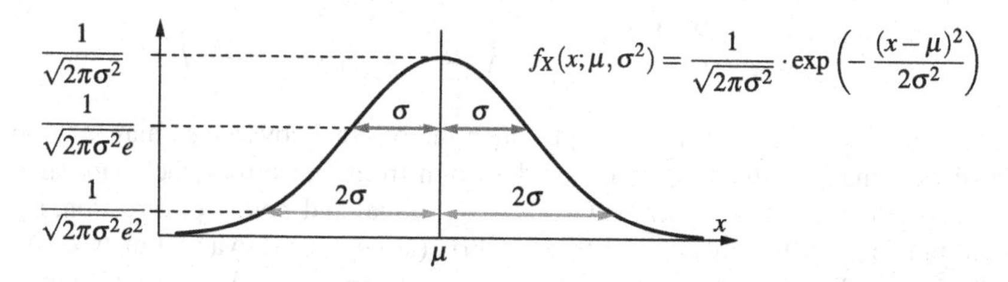

Fig. 13.4 Interpretation of the standard deviation σ for a one-dimensional normal distribution

nature, it is not so easily related to the underlying quantity. This problem is usually handled by computing the standard deviation, which is the square root of the variance and which has the same unit as the underlying quantity. Assuming that the described random process is governed by a normal (or Gaussian) distribution, the standard deviation is easily interpreted, as demonstrated in Fig. 13.4.

This convenient interpretation in the one-dimensional case naturally raises the question whether it can be transferred to two or more dimensions in order to obtain an intuitive interpretation of a covariance matrix. The problem is essentially the same: due to its quadratic nature, a covariance matrix is difficult to interpret. If we could form a "square root" of it, we may obtain an analog of standard deviation, which may be easier to interpret. Thus, we face the problem to compute the "square root" of a matrix. Not a general one, though, but a covariance matrix, which has the convenient properties of being symmetric and positive definite.

More technically, let $\mathbf{S}$ be an $n \times n$ matrix, that is, $\mathbf{S} = (s_{ij})_{1 \le i \le m, 1 \le j \le n}$. $\mathbf{S}$ is called **symmetric** if and only if $\forall 1 \le i, j \le n : s_{ij} = s_{ji}$ (or, equivalently, if $\mathbf{S}^\top = \mathbf{S}$, where $\mathbf{S}^\top$ is the transpose of the matrix $\mathbf{S}$). $\mathbf{S}$ is called **positive-definite** if and only if for all n-dimensional vectors $\mathbf{v} \neq \mathbf{0}$ it is $\mathbf{v}^\top \mathbf{S} \mathbf{v} > 0$. Intuitively, mapping a vector with a positive-definite matrix may stretch or shrink it and may rotate it (by less than $\pi/2$), but does not reflect it about the origin. For symmetric positive-definite matrices, an analog of a square root may be computed, for instance, with so-called **Cholesky decomposition** (Golub and Van Loan 1996). With this method, we find a lower (or left) triangular matrix $\mathbf{L}$ (that is, $\mathbf{L}$ has nonzero elements only on and below (or left) of the diagonal, while all other elements are zero) such that $\mathbf{L}\mathbf{L}^\top = \mathbf{S}$. By simply spelling out this equation, we easily obtain the elements of $\mathbf{L}$ as

$$l_{ii} = \left(s_{ii} - \sum_{k=1}^{i-1} l_{ik}^2 \right)^{\frac{1}{2}},$$

$$l_{ji} = \frac{1}{l_{ii}} \left(s_{ij} - \sum_{k=1}^{i-1} l_{ik} l_{jk} \right), \quad j = i+1, i+2, \dots, n.$$

As an example, we consider the special case with only two dimensions. In this case, an analog of a square root can be computed for a covariance matrix

$$\Sigma = \begin{pmatrix} \sigma_x^2 & \sigma_{xy} \\ \sigma_{xy} & \sigma_y^2 \end{pmatrix} \quad \text{as} \quad \mathbf{L} = \begin{pmatrix} \sigma_x & 0 \\ \frac{\sigma_{xy}}{\sigma_x} & \frac{1}{\sigma_x}\sqrt{\sigma_x^2\sigma_y^2 - \sigma_{xy}^2} \end{pmatrix}.$$

The matrix $\mathbf{L}$ is much easier to interpret than the original covariance matrix Σ: it is a (linear) mapping that describes the deviation from an isotropic behavior (that is, direction independent, from the Greek ἴσος: equal and τρόπος: direction, rotation) that is described by the covariance matrix (as compared to a unit matrix). As an illustration, Fig. 13.5 shows how a unit circle is mapped with the lower triangular matrix resulting for the covariance matrix

$$\Sigma = \begin{pmatrix} 1.5 & 0.8 \\ 0.8 & 0.9 \end{pmatrix} \quad \text{and thus} \quad \mathbf{L} \approx \begin{pmatrix} 1.2248 & 0 \\ 0.6532 & 0.6880 \end{pmatrix}.$$

The mapping with $\mathbf{L}$ has a meaning for an n-dimensional normal (or Gaussian) distribution that is analogous to the meaning of standard deviation for a one-dimensional normal (or Gaussian) distribution. To understand this, let us consider the probability density of an n-dimensional normal distribution, that is,

$$f_{\mathbf{X}}(\mathbf{x}; \boldsymbol{\mu}, \Sigma) = \frac{1}{\sqrt{(2\pi)^m |\Sigma|}} \cdot \exp\left(-\frac{1}{2} (\mathbf{x} - \boldsymbol{\mu})^\top \Sigma^{-1} (\mathbf{x} - \boldsymbol{\mu}) \right),$$

where μ is an n-dimensional expected value vector and Σ is an $n \times n$ covariance matrix. This probability density is illustrated for $n = 2$ in Fig. 13.6: the left diagram shows a standard normal distribution (unit covariance matrix), while the right diagram shows a general normal distribution (with the above covariance matrix).

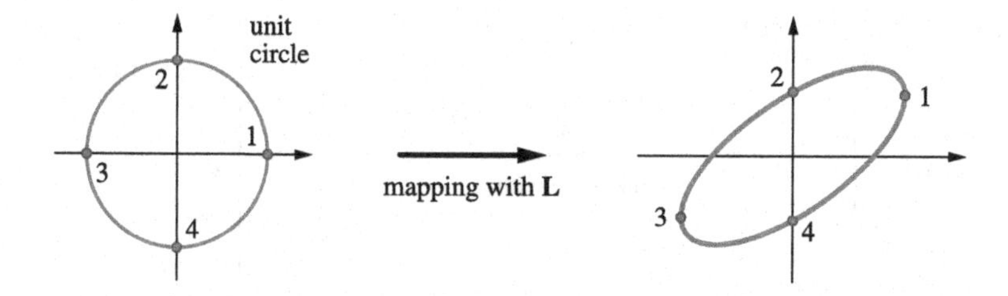

Fig. 13.5 Mapping of the unit circle with the help of Cholesky decomposition

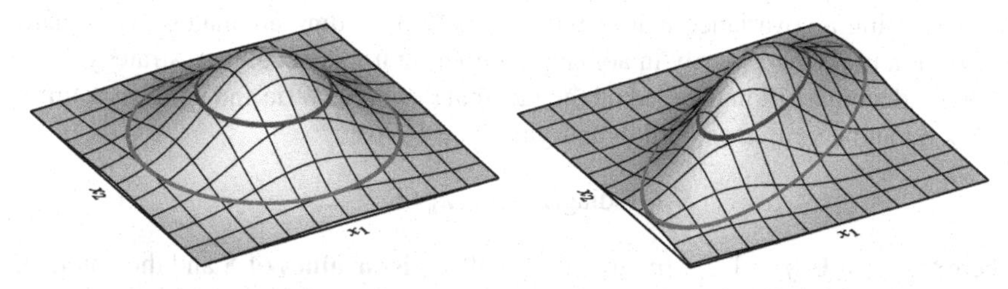

Fig. 13.6 Lines of equal probability density for a standard normal (or Gaussian) distribution (*left*, variance 1 for both axes, no covariance) and a general normal distribution with covariances (*right*)

For a one-dimensional normal distribution, the standard deviation measures the spread (or dispersion) of the probability density by stating the distance between the mean and the points having specific probability densities relative to the height of the mode of the density (see Fig. 13.4). If, in an analogous manner, we mark points with the same probability density for a two-dimensional distribution, we obtain closed curves, which are indicated by gray lines in Fig. 13.6: the darker gray corresponds to σ, the lighter gray to 2σ (in analogy to Fig. 13.4).

As shown in Fig. 13.6, these curves are circles for a standard normal distribution (left), while they are ellipses for a normal distribution with covariances (right). These ellipses are obtained by mapping the circles of the standard normal distribution with the matrix $\mathbf{L}$ that results from a Cholesky decomposition of the covariance matrix of the general normal distribution. Since the diagrams in Figs. 13.5 and 13.6 are computed with the same covariance matrix, this is particularly obvious.

If we do not require the "square root" of a covariance matrix to be a lower (or left) triangular matrix (as in Cholesky decomposition), the "square root" is, in general, not unique. That is, there may be multiple matrices which, if multiplied with their own transpose, produce the given covariance matrix. Thus, it is not surprising that Cholesky decomposition is not the only method with which an analog of standard deviation can be computed. A particularly nice alternative is **eigen decomposition** (Golub and Van Loan 1996), because it yields a decomposition that consist of elements that make it even better interpretable. Its disadvantage is, however, that it is more costly to compute than Cholesky decomposition (Press et al. 1992). Fortunately, though, this is irrelevant for our purposes, because we are mainly interested

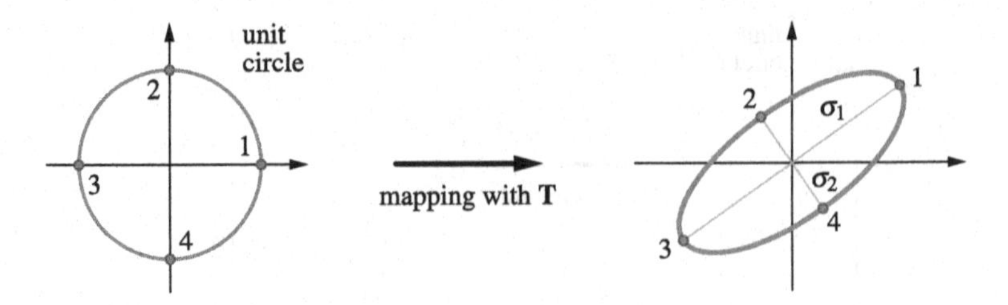

Fig. 13.7 Mapping of the unit circle with the help of eigen decomposition

in interpreting a covariance matrix with the help of finding an analog of a square root of it, and not necessarily in actually computing it in an evolution strategy.

Eigen decomposition is based on the fact that any symmetric and positive definite matrix $\mathbf{S}$ can be written as (Golub and Van Loan 1996)

$$\mathbf{S} = \mathbf{R}\,\mathrm{diag}(\lambda_1, \ldots, \lambda_n)\,\mathbf{R}^{-1},$$

where the $\lambda_j \geq 0$, $j = 1, \ldots, n$, are the so-called **eigenvalues** of $\mathbf{S}$ and the columns of $\mathbf{R}$ are the (normalized, that is, length 1) **eigenvectors** of $\mathbf{S}$. The eigenvectors are pairwise orthogonal to each other. As a consequence, $\mathbf{R}^{-1} = \mathbf{R}^{\top}$, or, in other words, $\mathbf{R}$ is a rotation matrix. With such a decomposition, we can write $\mathbf{S} = \mathbf{T}\mathbf{T}^{\top}$ with

$$\mathbf{T} = \mathbf{R}\,\mathrm{diag}(\sqrt{\lambda_1}, \ldots, \sqrt{\lambda_m}).$$

Eigen decomposition has the clear advantage that the resulting mapping can be interpreted as a scaling of the axes (namely by the roots of the eigenvalues, denoted by $\sqrt{\lambda_i}$, $i = 1, \ldots, n$) and a rotation (encoded in the rotation matrix $\mathbf{R}$). For the special case with only two dimensions ($n = 2$), eigen decomposition leads to

$$\mathbf{T} = \begin{pmatrix} c & -s \\ s & c \end{pmatrix} \begin{pmatrix} \sigma_1 & 0 \\ 0 & \sigma_2 \end{pmatrix}, \qquad \begin{aligned} \sigma_1 &= \sqrt{c^2\sigma_x^2 + s^2\sigma_y^2 + 2sc\sigma_{xy}}, \\ \sigma_2 &= \sqrt{s^2\sigma_x^2 + c^2\sigma_y^2 - 2sc\sigma_{xy}}, \end{aligned}$$

where $s = \sin\phi$, $c = \cos\phi$ and $\phi = \frac{1}{2}\arctan\frac{2\sigma_{xy}}{\sigma_x^2 - \sigma_y^2}$. How a unit circle is mapped with $\mathbf{T}$ is shown (in complete analogy to Fig. 13.5) in Fig. 13.7 for

$$\Sigma = \begin{pmatrix} 1.5 & 0.8 \\ 0.8 & 0.9 \end{pmatrix} \quad \text{and thus} \quad \mathbf{T} \approx \begin{pmatrix} 1.1781 & -0.3348 \\ 0.8164 & 0.4832 \end{pmatrix},$$

computed from $\phi \approx 0.606 \approx 34.7°$, $\sigma_1 = \sqrt{\lambda_1} \approx 1.4333$ and $\sigma_2 = \sqrt{\lambda_2} \approx 0.5879$. Note that the mapping differs from the one shown in Fig. 13.5—even though the ellipse as a whole is the same—as can be seen from the images of the points labeled 1, 2, 3 and 4, which are in different positions.

A mutation operator employing a covariance matrix is used, for example, in the **covariance matrix adaptation evolution strategy** (CMA-ES) (Hansen 2006), which is among the most highly recommended variants of evolution strategies.

Without going into mathematical details and neglecting many subtleties, this algorithm works roughly as follows: an initial population is created by sampling from a standard normal distribution with a chosen mean and variance. In the course of the algorithm, the mean vector, a covariance matrix (initially a unit matrix), and two scaling factors for the mutation are updated. Essentially, the latter three quantities describe a probability distribution from which the mutation vectors are sampled, while the mean vector is mainly used in their update. As a consequence, this approach can be seen as closely related to so-called **estimation of distribution algorithms** (Larrañaga and Lozano 2002; Lozano et al. 2006), which are genetic algorithms that build and sample probabilistic models of promising solution candidates.

Subsequent populations are generated by applying a mutation operator that combines an isotropic and an anisotropic component (where the Greek prefix ἄν- expresses negation, that is, "anisotropic" means *not* direction independent and thus direction dependent), which are governed by the two scaling factors. The mean vector, the covariance matrix and the two scaling factors are updated in each iteration in such a way that the likelihood that the mutation operator repeats beneficial mutations is increased. Essentially, this is achieved by updating the current covariance matrix with a (properly normalized) covariance matrix that is computed from the current mean vector and the individuals of the new population. Note that the covariance matrix only captures the direction dependence of the mutations, while the actual mutation step width is controlled by the two scaling factors. Technical details as well as a comparison to other methods can be found in Hansen (2006).

Note that the approach can be seen as related to estimating the inverse Hessian matrix of the function to optimize, as it is used, for instance, in the Quasi-Newton method in classical optimization (Press et al. 1992). As such, the covariance matrix adaptation evolution strategy uses the population mainly to explore the neighborhood of the mean vector it maintains (and which can be seen as a representative of the currently preferred solution candidate) in order to optimize the mutation direction and step width. Therefore it is more closely related to local search methods as we discussed them in Sect. 11.5 than to approaches that use a population rather to achieve a broader exploration of the search space, looking for optima (and not just improvement directions) in many places at the same time.

Although such an approach may also be considered, covariance matrices do not lend themselves well to a local variance adaptation scheme, because this requires to incorporate $\frac{n(n+1)}{2}$ mutation parameters into the chromosomes, thus increasing the size of the chromosomes inordinately. In addition, it is debatable whether the indirect adaptation of so many parameters works effectively and efficiently.

13.2.5 Recombination Operators

Evolution strategies are often executed without a crossover operator. If a crossover operator is employed, it is commonly defined as a random selection of components from two parents, that is, analogous to uniform crossover (see Sect. 12.3.2):

$$\begin{matrix} (\mathbf{x_1}, x_2, x_3, \ldots, \mathbf{x_{n-1}}, x_n) \\ (y_1, \mathbf{y_2}, \mathbf{y_3}, \ldots, y_{n-1}, \mathbf{y_n}) \end{matrix} \quad \Rightarrow \quad (x_1, y_2, y_3, \ldots, x_{n-1}, y_n).$$

In principle, any other crossover operator discussed in Sect. 12.3.2, like 1-, 2- or n-point crossover etc. is applicable as well. An alternative is **blending**, as implemented, for example, by arithmetic crossover (see Sect. 12.3.5):

$$\begin{matrix} (x_1, \ldots, x_n) \\ (y_1, \ldots, y_n) \end{matrix} \quad \Rightarrow \quad \tfrac{1}{2}(x_1 + y_1, \ldots, x_n + y_n).$$

In case this crossover operator is used, one should bear in mind, though, that it carries the danger of **Jenkins Nightmare**, that is, a total disappearance of any diversity in a population due to the averaging effect of blending.

13.3 Genetic Programming

With **genetic programming** (GP) it is tried to evolve symbolic expression or even computer programs with certain properties. The purpose of these expressions or programs is usually to associate certain inputs with certain outputs in order to solve a given problem. Genetic programming can be seen as a very general way to learn or to create computer programs, even as complex as programs playing checkers (Fogel 2001). Its application areas are huge, because many problems can be seen as a search for a program, for example, controller development, scheduling, knowledge representation, symbolic regression, decision tree induction etc. (Nilsson 1998).

Up to now we considered only chromosomes that were arrays of a fixed length, for example, arrays of bits for genetic algorithms or arrays of real-valued numbers for evolution strategies. For genetic programming, we abandon the restriction to a fixed length and allow chromosomes that differ in their length. To be more precise, the chromosomes of genetic programming are functional terms and programs, which are commonly called **genetic programs** (also often abbreviated by "GP").

The formal basis of genetic programming is a grammar that describe the language of the genetic programs. Following the standard approach in formal languages, we define two sets, namely the set $\mathcal{F}$ of **function symbols and operators** and the set $\mathcal{T}$ of **terminal symbols** (constants and variables). These sets are problem-specific and thus comparable to the encoding that we considered w.r.t. other types of chromosomes (see Sect. 12.1). $\mathcal{F}$ and $\mathcal{T}$ should be limited in size in order to restrict the search space to a feasible size, but "rich" enough to enable a problem solution.

As an illustration, we take a look at two examples of symbol sets. Suppose we want to learn a Boolean function that maps n binary inputs to associated binary outputs. In this case, the following symbol sets are a natural choice:

$$\mathcal{F} = \{\text{and}, \text{or}, \text{not}, \text{if} \ldots \text{then} \ldots \text{else} \ldots, \ldots\},$$

$$\mathcal{T} = \{x_1, \ldots, x_n, 1, 0\} \quad \text{or} \quad \mathcal{T} = \{x_1, \ldots, x_n, \text{true}, \text{false}\}.$$

If the task is symbolic regression,[1] the following sets appear to be appropriate:

$$\mathcal{F} = \{+, -, *, /, \sqrt{\ }, \sin, \cos, \log, \exp, \ldots\},$$
$$\mathcal{T} = \{x_1, \ldots, x_m\} \cup \mathbb{R}.$$

A desirable property of $\mathcal{F}$ is that all functions in it are domain complete, that is, the functions in $\mathcal{F}$ should accept any possible input value. If this is not the case, the genetic program may cause an error and thus may not be able to complete its execution, which causes problems in the evolution program. Simply examples of functions that are not domain complete are division, which causes an error if the divisor is zero, or a logarithm, which usually accepts only positive arguments.

If the function set $\mathcal{F}$ contains functions that are not domain complete, we have to solve an optimization problem with constraints: we must ensure that the expressions and programs are structured in such a way that a function is never applied to an unacceptable input. For this, we could employ repair mechanisms or may introduce a penalty term for the fitness function (cf. Sect. 12.1.3).

As an alternative, we may render all functions in $\mathcal{F}$ domain complete by implementing **protected versions** of error-prone functions. For example, we may define a protected division that returns zero or the (appropriately signed) maximally representable value if the divisor is zero, or a protected nth root that operates on the absolute value of its argument, or a protected logarithm that yields $\log(x) = 0$ for all $x \le 0$. Along the same lines, we may also reinterpret data types and thus render functions that are defined for one data type applicable to the other. For example, if the chosen function set $\mathcal{F}$ contains Boolean as well as numeric functions, we may define 0 as "false" and any value not equal to 0 as "true" (this is the convention, for example, in the programming languages C and C++), thus making it possible to apply Boolean functions to numeric arguments. By fixing a value for a "true" result of a Boolean function, for example 1 or -1, numeric functions become applicable to Boolean arguments. If necessary, we may also define a conditional operator (if ... then ... else ...) that executes the else-part unless the condition is a proper Boolean "true," in which case only the then-part is executed.

Another important property is the completeness of the function sets $\mathcal{F}$ and $\mathcal{T}$ w.r.t. the functions (mapping inputs to outputs) they can represent. Genetic programming can only effectively solve a given problem if $\mathcal{F}$ and $\mathcal{T}$ are sufficient to find an appropriate program. For example, in Boolean propositional logic $\mathcal{F} = \{\land, \neg\}$ and $\mathcal{F} = \{\rightarrow, \neg\}$ are complete sets of operators, because any Boolean function with any number of arguments can be represented by appropriate combinations of the operators in these sets. However, $\mathcal{F} = \{\land\}$ is not a complete set, because even the simple negation of an argument cannot be represented. Finding the smallest complete set of operators for a given set of functions to represent is (usually) NP-hard. As a consequence, $\mathcal{F}$ usually contains more functions than are actually necessary.

[1] Regression finds a function from a given class to given data by minimizing the sum of squared deviations and is also called the *method of least squares*, see Sect. 10.2.

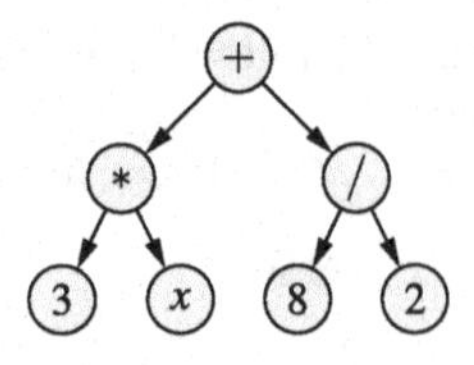

Fig. 13.8 Parse tree of the symbolic expression (+ (∗ 3 x) (/ 8 2))

However, this is not necessarily a disadvantage, since richer function sets may allow for simpler and thus more easily interpretable solutions.

Given the sets $\mathcal{F}$ and $\mathcal{T}$ genetic programs—that is, the chromosomes of genetic programming—can be created. They consist of elements from $\mathcal{C} = \mathcal{F} \cup \mathcal{T}$ and possibly opening and closing parenthesis in order to clarify precedence of operators (if necessary) or to group a function symbol and its arguments. However, the genetic programs are not arbitrary sequences of these symbols. Rather we confine ourselves to so-called "well-defined" expressions, that is, expressions that can be created by following certain rules, so that the expressions are interpretable. These rules specify a grammar, which thus defines the language of genetic programs.

Most commonly, this grammar uses a **recursive definition** based on a prefix notation of the functions and operators, similar to the style of functional programming languages like Lisp or Scheme. In these languages, all expressions are either atoms or nested lists. The first list element corresponds to the function symbol or the operator and the subsequent elements are the function arguments or the operands.

More formally, valid symbolic expressions are defined as follows:

- Constants and variables (i.e., the elements of $\mathcal{T}$) are symbolic expressions.
- If $t_1, \dots, t_n$ are symbolic expressions and $f \in \mathcal{F}$ is an n-ary function symbol, then $(f t_1 \dots t_n)$ is a symbolic expression.
- Any other character sequence is not a symbolic expression.

For example, the character sequence "(+ (∗ 3 x) (/ 8 2))" is a valid symbolic expression, which, in more conventional terms, means $3 \cdot x + \frac{8}{2}$. In contrast to this, the character sequence "2 7 ∗ (3 /)" is not a valid or "well-defined" symbolic expression.

For the following discussion, it is convenient to represent a symbolic expression by its so-called **parse tree**. Parse trees are commonly used, for example, in compilers, especially for arithmetic expressions. Such a parse tree for the symbolic regression which we used as an example above is shown in Fig. 13.8.

To find good symbolic expressions for a given problem, genetic programming follows the same general procedure of an evolutionary algorithm as we presented it in Algorithm 11.1 on page 180. That is, first we create an **initial population** of random symbolic expressions, which are *evaluated* by computing their fitness values. Here the fitness is a measure how well a genetic program maps certain vectors of input values to their corresponding output values. For example, suppose we want to find a symbolic expression that computes a Boolean function, which is given as a set of pairs of input and output vectors, all elements of which are either *true* or *false*. In this case the fitness function could simply be the number of correctly computed outputs summed over the input–output pairs. If the task is symbolic regression, the fitness function may simply be the sum of squared deviations from the desired output values. For example, if we are given a data set consisting of pairs (x_i, y_i), $i = 1, \dots, n$, where x is the input and y the output variable, the fitness may be computed a $f(c) = \sum_{i=1}^{n}(g(x_i) - y_i)^2$, where g stands for the genetic program that maps a input x to an output value. **Selection** is implemented by any of the methods that have been studied in Sect. 12.2. Finally **genetic operators**, usually only crossover, are applied to the selected individuals and the resulting individuals are evaluated again. The procedure of selecting individuals, applying genetic operators and evaluating the (new) individuals is repeated until a termination criterion is met, for example, that the an expression exceeding a user-specified fitness threshold has been found. In the following, we discuss the individual steps in more detail.

13.3.1 Initialization

Creating an initial population is somewhat more complex in genetic programming that in the algorithms we studied up to now, because we cannot simple create random sequences of function symbols, constants, variables and parentheses. We have to respect the recursive definition of a valid or "well-defined" expression. The most natural approach to create a random genetic program for the initial population is a recursive procedure that simply follows the recursive definition. In addition, since it is also convenient for the later evaluation of genetic programs, we actually do not create sequences of symbols, but directly their corresponding parse trees.

In order to limit the size of the parse trees (and to ensure that the recursive procedure terminates), we may specify a maximum tree height or a maximum number of nodes. If this maximum tree height or maximum number of nodes is reached, all unfilled function arguments are chosen only from the terminal symbols to avoid further recursion. A simple form of this procedure looks like this (Koza 1992):

Algorithm 13.5 (Initialize-Grow)

```
function init_grow (d, d_max: int) : node;
begin                                  (* current depth d, maximal depth d_max *)
  if    d = 0                          (* avoid mere constants or variables *)
  then n ← draw from F using a uniform distribution;
  elseif d ≥ d_max                     (* stop at the maximal tree height *)
  then n ← draw from T using a uniform distribution;
  else  n ← draw from F ∪ T using a uniform distribution; end
  forall c ∈ arguments of n do         (* if arguments/operands are needed, *)
     c ← init_grow(d + 1, d_max);      (* create sub-expressions recursively *)
  return n;                            (* if n ∈ T, n has no arguments *)
end                                    (* finally return the created node *)
```

Instead of simply pooling $\mathcal{F}$ and $\mathcal{T}$ if $d < d_{\max}$, we may explicitly specify the probabilities with which a function symbol or a terminal symbol is chosen. With such a parameter the size and complexity of the trees can be controlled to some degree: it is the higher, the more likely it is that a function symbol is chosen.

An common alternative to the above procedure is a "full" initialization, which always grows the parse trees to the specified maximum tree height (Koza 1992):

Algorithm 13.6 (Initialize-Full)

```
function init_full (d, d_max: int) : node;
begin                                  (* current depth d, maximal depth d_max *)
  if d ≥ d_max then                    (* stop at the maximal tree height *)
    n ← draw from T using a uniform distribution;
  else                                 (* below the maximal tree height *)
    n ← draw from F using a uniform distribution;
    for c ∈ arguments of n do          (* always choose a function symbol *)
      c ← init_full(d + 1, d_max);     (* and create its arguments recursively *)
  end
  return n;                            (* finally return the created node *)
end
```

Each of the two methods *grow* and *full* presented above may be used as the only parse tree generation method, or they may be combined, creating half of the initial population with the method *grow* and the other half with the method *full*. Furthermore, it is advisable to vary the maximum tree depth between 1 and some user-specified maximum, so that trees of different height are created. This approach commonly called *ramped half-and-half* initialization (Koza 1992), where the term "ramped" refers to the varying maximum tree height, which is implemented by increasing this parameter by one with each created tree (pair).

Algorithm 13.7 (Initialize-Ramped-Half-and-Half)

```
function init_halfhalf (μ, d_max: int) : set of node;
begin                           (* maximal depth d_max *)
  P ← ∅;                        (* population size μ (even multiple of d_max) *)
  for i ← 1 ... d_max do begin
    for j ← 1 ... μ/(2 · d_max) do begin
      P ← P ∪ init_grow(0, i);
      P ← P ∪ init_full(0, i);
    end                         (* initialize half the trees with grow *)
  end                           (* and the other half with full *)
  return P;                     (* return the created population *)
end
```

This ramped half-and-half method has the advantage that it ensures a good diversity of the population, with trees of different height and complexity.

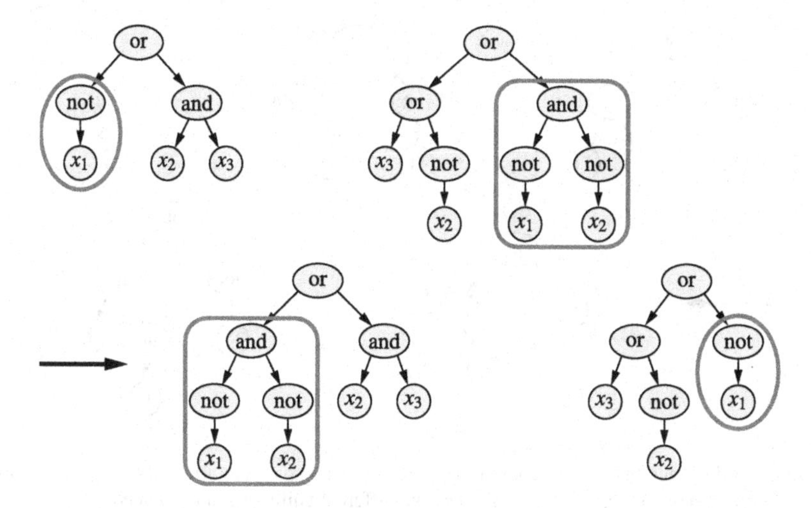

Fig. 13.9 Crossover of two sub-expressions or sub-trees

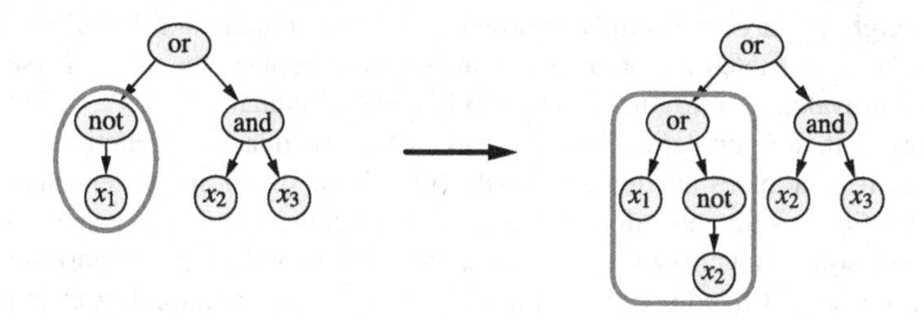

Fig. 13.10 Mutation of a sub-expression or sub-tree

13.3.2 Genetic Operators

The initial population of genetic programming (as of any evolutionary algorithm) usually has a very low fitness, because it is highly unlikely that a random generation of parse trees yields even a remotely adequate solution of any non-trivial problem. In order to obtain better solution candidates, genetic operators are applied. For convenience, we describe these genetic operators (crossover and mutation) by showing how they modify the parse trees underlying the genetic programs.

In genetic programming, **crossover** consists in an exchange of two sub-expressions (and thus sub-trees of the parse trees). A simple example for a Boolean function problem is shown in Fig. 13.9, with the parents at the top and the produced offspring at the bottom. The exchanged sub-trees are encircled.

The mutation operator replaces a sub-expression (that is, a sub-tree of the parse tree) by a randomly created sub-expression. To create a new sub-expression, we may simply draw on the initialization method init_grow presented above, for which we may choose the maximum tree height, for example, as a (small) modification of the old sub-tree height. A simple example, again for a Boolean function problem, is

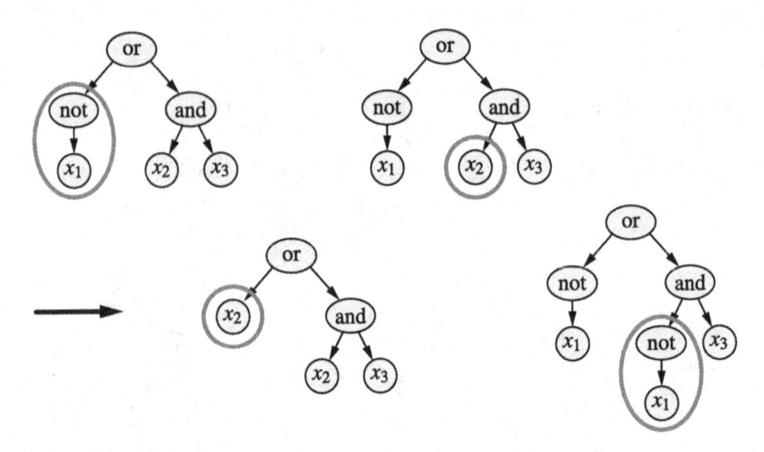

Fig. 13.11 Advantage of genetic programming crossover compared to array-based crossover operators: applied to two identical chromosomes, two different children can be created

shown in Fig. 13.10. It is common to restrict mutation to replacing small sub-trees (with a height no larger than three or four), so that an actually similar individual is created. An unrestricted mutation could, in principle, replace the whole parse tree, which is equivalent to introducing an entirely new individual.

However, if the population is sufficiently large, so that the "genetic material" present in it guarantees adequate diversity (mainly of function and terminal symbols), mutation is often abandoned and crossover becomes the only genetic operator.[2] The reason is that crossover is—compared to other evolutionary algorithms that work with arrays of fixed length—a much more powerful operator. For example, if we apply it to two identical individuals, the simple fact that different sub-trees may be chosen creates the possibility that two different individuals result. An example demonstrating this is shown in Fig. 13.11: even though the two (identical) parents are fairly simple parse trees, a crossover can create considerable modifications. In contrast to this, any of the operators discussed in Sect. 12.3.2 cannot create any variation if they are applied to two identical chromosomes.

13.3.3 Application Examples

As an illustration of genetic programming, we briefly discuss two applications: learning the Boolean function of a multiplexer and symbolic regression.

13.3.3.1 The 11-Multiplexer Problem

A classical example of genetic programming is to learn a Boolean 11-bit multiplexer (Koza 1992), that is, a multiplexer with 8 data and 3 address lines. A sketch of such a

[2]Note that this is exactly opposite to evolution strategies (see Sect. 13.2), in which crossover is often abandoned and mutation is the only genetic operator.

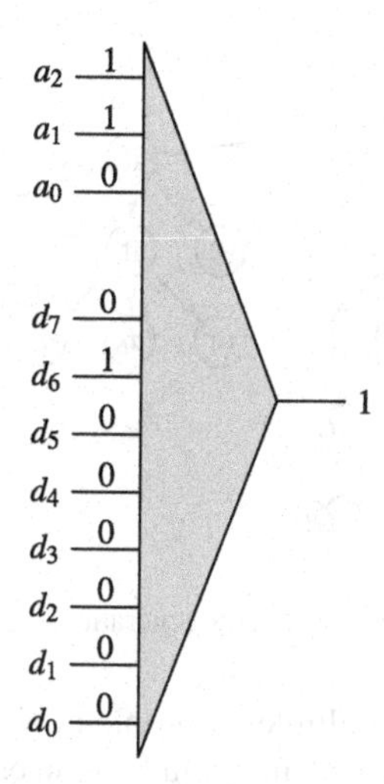

Fig. 13.12 Configuration of a Boolean 11-bit multiplexer. Since the address lines a_0 to a_2, interpreted as a binary number, have value 6, the data line d_6 is passed through to the output

multiplexer, together with one valid input vector and corresponding output, is shown in Fig. 13.12. The purpose of a multiplexer is to pass the value of the data line to the output that is indicated by the address lines. In the example of Fig. 13.12, the values of the address lines represent the number 6 (in a binary encoding) and thus the value of data line d_6 is passed through to the output. In total, there are $2^{11} = 2048$ possible input configurations of an 11-bit multiplexer. Each of these configurations is associated with a single bit output.

In order to learn a symbolic expression that describes the function of an 11-bit multiplexer with genetic programming, we choose the following symbol sets:

$$\mathcal{T} = \{a_0, a_1, a_2, d_0, \ldots, d_7\} \quad \text{and} \quad \mathcal{F} = \{\text{not}/1, \text{and}/2, \text{or}/2, \text{if}/3\}.$$

That is, the set $\mathcal{T}$ of terminal symbols contains the 11 input variables and the function set $\mathcal{F}$ contains four simple Boolean operators, with which these variables can be combined to compute the output. The numbers after the "/" indicate the arity of these operators. That is, the negation takes a single argument, "and" and "or" take two arguments, and "if" takes three arguments, which correspond to the condition, the then-part, and the else-part. Obviously, since we are dealing merely with Boolean values, all functions are domain complete. In addition, the function set is complete in the sense that one can represent any Boolean function with it.

As the fitness function we choose $f(s) = 2048 - \sum_{i=1}^{2048} e_i$ where e_i is the error for the ith input configuration. That is, $e_i = 0$ if the computed output for configuration i coincides with the desired output for this configuration and $e_i = 1$ otherwise. As the termination criterion we may choose that we obtain an individual with fitness 2048. Since such an individual produces the correct output for all 2048 configurations (that is, it makes no mistakes), it is clearly a solution of the problem.

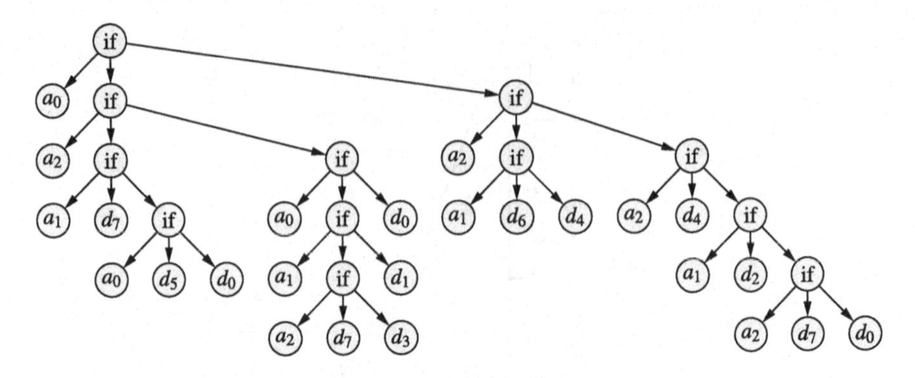

Fig. 13.13 Solution of the 11-multiplexer after 9 generations

In order to solve the 11-bit multiplexer problem, Koza (1992) used a population size of $\mu = 4{,}000$. This number of individuals is needed as with a lower number it becomes unlikely that a good program can be found in a search space as large as the one encountered here. (Note that the expressions may, in principle, become arbitrarily complex unless measures are taken to limit their size.) The initial tree depth (maximum depth in the initial population) was set to 6 and the maximum tree depth to 17 (no parse trees with a depth larger than 17 can be generated).

In the experiment reported by Koza (1992), the fitness values of the initial population ranged between 768 and 1280, with an average fitness of 1063. Note that the expected value is 1024 since a random output is correct for about 50 % of all configurations. Hence, this is a plausible result for an initial population of genetic programs, which produces essentially random output as it has not yet been geared towards the task at hand. 23 expressions of the initial population had a fitness of 1280, with one of them corresponding to a 3-multiplexer, namely (if a_0 d_1 d_2). This is important, because such individuals provide building blocks for the whole 11-bit multiplexer.

In his experiment, Koza (1992) used fitness-proportionate selection method. 90 % of the individuals (that is, 3,600) were modified by crossover and the remaining 10 % (that is, 400) were left unchanged. According to Koza (1992), after only 9 generations the solution depicted in Fig. 13.13 was found, which actually has the (maximally possible) fitness of 2048. However, doe to its complexity, this solution is somewhat difficult to interpret for human beings (although a little effort shows that it actually computes the Boolean function of an 11-bit multiplexer).

In order to obtain a better solution, the expression is *edited*. **Editing** can be seen as an asexual (genetic) operation that is applied to single individuals. It serves the purpose to simplify an expression and is divided into general and special editing.

General editing evaluates a sub-tree and replaces it with its result if the sub-tree yields a constant result, for example, because its leaves contain only constants or because its structure makes it independent of its variable arguments. An example of the latter is the expression "(or x_5 (not x_5))" which is a tautology and thus may be replaced by the constant *true*, despite the fact that x_5 is a variable.

Special editing exploits certain equivalences that do not change the value of an expression, but may make it possible to express the same function in a simpler way.

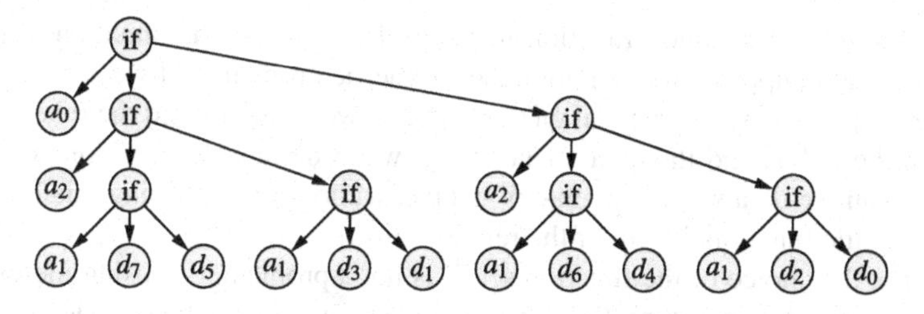

Fig. 13.14 The 11-bit multiplexer solution of Fig. 13.13 after special editing

Simple examples of such identities in Boolean logic are $A \wedge A \equiv A$, $B \vee B \equiv B$, where A and B are arbitrary Boolean expressions, or De Morgan's laws, that is, $\neg(A \vee B) \equiv (\neg A \vee \neg B)$ and its dual. An example for a numerical expression is $\sqrt{x^4} \equiv x^2$.

In principle, editing may already be performed during the evolutionary search, namely by using it as a genetic operator (alongside crossover and mutation). In this case it reduces the number of unnecessarily inflated individuals at the expense of diversity in the population. However, more commonly editing is used to simplify the solutions found by genetic programming in order to obtain a result that is easier to interpret. For the 11-bit multiplexer example, an edited solution (derived from the one shown in Fig. 13.13 after several editing steps) is shown in Fig. 13.14.

Note that this (edited) solution has a hierarchical structure. That is, it divides the 11-multiplexer problem into two smaller problems, namely two 6-multiplexers with two address bits and 4 data bits, which in turn are divided into two 3-multiplexers with one address bit and 2 data bits. At the top level, the address variable a_0 is used to distinguish between the odd data lines d_7, d_5, d_3, and d_1 (the-part of the top-level if) and the even data lines d_6, d_4, d_2, and d_0 (else-part of the top-level if). On the next level the address variable a_2 distinguishes (in both branches) between the higher pair in each quadruple and the lower pair. At the bottom level, the address variable a_1 makes the final decision between the remaining two data lines.

Although the population size was large enough with 4,000 generations, we do not want to conceal our doubts that finding a solution to the 11-bit multiplexer problem in only 9 generations is a typical result. Judging from own experiments with genetic programming (for a related problem), we feel that it usually takes longer to find a solution and that solutions are much more complex than the one shown in Fig. 13.13. Possibly this is due to the fact that we did not use editing, neither in the search nor to simplify our solutions, or that our methods of restricting the complexity of the parse trees were not effective enough. However, we certainly feel that genetic programming is not quite as straightforward as we may have presented it here, but may involve a lot of tweaking to obtain satisfactory results.

13.3.3.2 Symbolic Regression with Constants

Finding expressions that compute Boolean functions with genetic programming is comparatively simple (though certainly not a trivial task), because there are at most (if one chooses to use them) two constants: *true* and *false*. For symbolic regression that is, fitting a real-valued function to given data points—the situation is much more complex. Suppose, for example, the geometric problem that we are given the radius r and the area A of several circles and that we desire to find the relationship between the radius and the area. (That is, we want to learn the function $A = \pi r^2$ from the sample data we are given—of course, without presupposing this function or any specific functional form of the relationship.)

Clearly, the choice of the function set $\mathcal{F}$ is not a problem. Choosing merely the basic mathematical operators (that is, $+$, $-$, $*$, $/$) may already be sufficient and we may easily add functions like square root, logarithm, some trigonometric functions without creating any serious problems: $\mathcal{F}$ can always be kept finite. However, the set of terminal symbols is less easy to choose. For this particular problem (finding an expression for the relationship of the area of a circle on its radius), we do not lose the possibility of finding the solution by using a finite set $\mathcal{T}$, because basic mathematical constants like π and e can (and should) be made elements of $\mathcal{T}$. However, more general cases, already starting with finding a mere regression line $y = ax + b$ for given pairs (x_i, y_i) causes problems: clearly we cannot make all numbers in $\mathbb{R}$ elements of $\mathcal{T}$, so that we have any values we may need for a and b available, simply because genetic programming, as we described it up to now, requires $\mathcal{T}$ to be finite. Because $\mathcal{T}$ needs to be finite, even $\mathbb{N}$ or $\mathbb{Q}$ already cannot be added to $\mathcal{T}$. However, how can we then be sure that we have the constants necessary to properly describe the relationship in question? That is, how can we make sure that $\mathcal{F}$ and $\mathcal{T}$ together are rich enough to cover all functions we may want to explore in the search?

The solution to this problem is to introduce so-called **ephemeral random constants** (Koza 1992). That is, instead of adding all elements of $\mathbb{R}$ to $\mathcal{T}$, we add only an special terminal symbol, for instance, the *symbol* "$\mathbb{R}$" (not the *set* of real numbers). If in the process of initializing a genetic program, we choose this special symbol as a constant, it is not used literally, but we rather sample a random value from a meaningful interval, thus creating a constant from a potentially infinite set without having these constants all literally contained in $\mathcal{T}$.

If ephemeral random constants are employed, it may be advisable to extend the mutation operator, so that it can also add a random value (sampled, for instance, from a normal distribution with a certain standard deviation) to a constant in a genetic program. The reason is that the standard mutation can only replace a constant with constant that is newly created from scratch (new sub-tree of height 1) and thus may differ completely from the constant that was present before. However, in order to achieve a gradual improvement of solution candidates in symbolic regression, it is more appropriate to modify already existing constants only slightly.

13.3.4 The Problem of Introns

Unless explicit countermeasures are taken, genetic programming exhibits a tendency to produce larger and more complex individuals with every generation. The main reason for this phenomenon are so-called **introns**. In biology, introns are parts of the

DNA sequence that do not carry any information in the sense that they do not code for any phenotypical trait. Introns may be either inactive (they lack a reading trigger and thus are possibly obsolete) or actually functionless nucleotide sequences beyond a gene or in-between genes (also known as *junk DNA*). In genetic programming, introns can occur, for example, if sub-expressions like "(if $2 > 1$ then ... else ...)" are created. Here the else-part of the conditional statement is an intron, because it can never be executed (unless the tested condition is changed).

Note that a mutation or a crossover operation that only affects an intron is fitness-neutral, because the intron is never executed and the fitness only depends on the active program code (that is, the sub-expressions that are actually evaluated for some input data). As a consequence, introns can grow arbitrarily in size and complexity, since this complexity does not carry any fitness penalty. In nature, the situation is somewhat different, because a larger genome usually means a larger metabolic overhead for copying it etc. As a consequence, some organisms, for which the metabolic overhead is a serious fitness aspect (especially certain forms of bacteria), have very "streamlined" genomes that contain no or almost no introns.

It is generally advisable to prevent the creation of introns as much as possible, because they inflate the chromosomes unnecessarily, increase the processing time and make it more difficult to interpret found solutions. These disadvantages can be seen as analogies to the metabolic overhead that introns carry in nature. This analogy also suggests the idea to introduce a fitness penalty for large and complex chromosomes, which may, for example, be a function of the height or the number of nodes of the parse tree. An alternative countermeasure is **editing**, which we already discussed above in the example of the 11-bit multiplexer. If it is used as a genetic operator, it serves the purpose to keep the chromosomes simple, although at the price of reducing the variety of "genetic material" in a population.

Other methods that have been suggested to reduce the creation of introns are modified genetic operators like, for instance, **brood recombination**. This operator creates many children from the same two parents by applying a crossover operator with different parameters. Only the best child of the brood enters the next generation. This method is particularly useful if combined with a fitness penalty, because then it favors children that achieve the same result with simpler means (that is, with a less complex chromosome). **Intelligent recombination** chooses crossover points purposefully, which can help to prevent the creation of introns.

13.3.5 Extensions

In the course of the further development of genetic programming, many extensions and improvements have been suggested. For example, in order to automatically define new functions, for instance, one may introduce **encapsulation**: potentially good sub-expressions are protected from getting destroyed by crossover and mutation. One way to achieve this is to define new functions for certain sub-expressions (of a chromosome with high fitness) and to add corresponding new symbols to the function set $\mathcal{F}$. The arity of the new function is the number of (different) variables in the leaves of its sub-tree or the number of all different symbols in the leaves (including constants), so that occurring constants can be replaced by different values in an instantiation. Other extensions include **iterations** and **recursion**, which introduce more powerful programming constructs. However, a detailed discussion of these forms and other extensions is beyond the scope of this book. An interested reader can find more information, for example, in Banzhaf et al. (1998).

13.4 Other Population-Based Approaches

Swarm intelligence is a research area in artificial intelligence in which the collective behavior of populations of simple agents is studied. Swarm intelligence systems are often inspired by the behavior of certain species of animals, especially social insects (like ants, termites, bees etc.) and animals that live and search for food in swarms, schools, flocks, herds or packs (like fish, birds, deer, wolves etc.). By cooperating with each other, these animals are often able to solve fairly complex problems. For example, they can find (shortest paths to) food sources (for example, ants, see below), build nests (like bee hives), hunt for prey (for example, packs of wolves), protect themselves against predators (for example, fish and deer) etc.

Swarm intelligence systems consist of agents that are usually very simple individuals with limited abilities, which interact with each other and the environment only locally and operate without any central control. Nevertheless, if observed from a higher level, these systems exhibit an "intelligent" global behavior, which emerges from **self-organization**. Core features of swarm intelligence systems are how individuals relate to solution candidates (at least if they are employed for optimization tasks, which is our main topic here) and how information is exchanged between the individuals. As a consequence, we categorize different approaches based on these features. A fairly complete, though brief overview of nature-inspired algorithms and their implementations can be found in Brownlee (2011).

Not surprisingly, evolutionary algorithms can also be seen as swarm intelligence systems. In basically all of their variants, individuals represent solution candidates and information is communicated by exchanging genetic material (encoding phenotypical traits) between individuals. An alternative approach that is also inspired by the biological evolution of organisms is **population-based incremental learning** (see Sect. 13.4.1). Although individuals still represent solution candidates, there is no (explicit) population and thus no inheritance or recombination. Rather only population statistics are updated, from which new individuals are created.

Ant colony optimization (see Sect. 13.4.2) models the ability of certain ant species to find the shortest path to food sources. Information is exchanged through modifications of the environment, so-called **stigmergy**, in particular the deposit of secreted chemical substances called **pheromones**. Stigmergy is closely related to the concept of an extended phenotype according to Dawkins (1982), which includes in a phenotype all effects that a gene has, regardless of whether they reside inside or outside the body of an organism. In ant colony optimization, individuals *construct* candidate solutions, mainly by finding (shortest) paths in a graph.

Finally, particle swarm optimization (see Sect. 13.4.3) models how swarms (or flocks or schools) of fish or birds search for food. Individuals represent solution candidates and are endowed with a memory that influences their movement. Information is exchanged by aggregating individual solutions and exploiting this aggregate to influence the movement of the individuals.

13.4.1 Population-Based Incremental Learning

Population-based incremental learning (PBIL) (Baluja 1994) is a special form of a genetic algorithm, that is, it works with a binary encoding of solution candidates (cf. Sect. 13.1). In contrast to a normal genetic algorithm, it does not maintain an explicit population of solution candidates. Rather only population statistics are stored and updated, which capture certain aspects of a population of (good) individuals (without actually representing these individuals). In particular, it is recorded and updated for each bit of the chosen encoding for what fraction of (good) individuals the bit has value 1. Population-based incremental learning works as follows:

Algorithm 13.8 (PBIL)
function pbil (f: function, μ, α, β, p_m: real) : object;
begin (∗ f: fitness function, μ: population size ∗)
  $t \leftarrow 0$; (∗ α: learning rate, β, p_m: mutation strength ∗)
  $s_{\text{best}} \leftarrow$ randomly create individual from $\{0,1\}^L$;
  $P^{(t)} \leftarrow (0.5, \ldots, 0.5) \in [0,1]^L$; (∗ initialize the population statistics ∗)
  while termination criterion not fulfilled **do begin**
    $t \leftarrow t+1$; (∗ count the processed generation ∗)
    $s^* \leftarrow$ create individual from $\{0,1\}^L$ according to $P^{(t-1)}$;
    for $i \leftarrow 2, \ldots, \mu$ **do begin** (∗ create μ individuals ∗)
      $s \leftarrow$ create individual from $\{0,1\}^L$ according to $P^{(t-1)}$;
      if $f(s) > f(s^*)$ **then** $s^* \leftarrow s$; **end**
    end (∗ find the best created individual ∗)
    for $k \in \{1, \ldots, L\}$ **do** (∗ update population statistics ∗)
      $P_k^{(t)} \leftarrow \alpha \cdot s_k^* + (1-\alpha) \cdot P_k^{(t-1)}$;
    if $f(s^*) > f(s_{\text{best}})$ **then** $s_{\text{best}} \leftarrow s^*$; **end** (∗ update best individual ∗)
    for $k \in \{1, \ldots, L\}$ **do begin** (∗ traverse the bits of the encoding ∗)
      $u \leftarrow$ choose random number from $U((0,1])$;
      if $u < p_m$ **then** (∗ for randomly selected bits ∗)
        $v \leftarrow$ choose random number from $U(\{0,1\})$;
        $P_k^{(t)} \leftarrow v \cdot \beta + P_k^{(t)}(1-\beta)$;
      end (∗ mutate population statistics ∗)
    end
  end
  return s_{best}; (∗ return the best individual found ∗)
end

Table 13.1 Same population statistics for different populations (Weicker 2007)

Population 1					Population 2			
1	1	0	0	individual 1	1	0	1	0
1	1	0	0	individual 2	0	1	1	0
0	0	1	1	individual 3	0	1	0	1
0	0	1	1	individual 4	1	0	0	1
0.5	0.5	0.5	0.5	population statistics	0.5	0.5	0.5	0.5

Note that the generation of new chromosomes can be seen as related to uniform crossover (cf. Sect. 12.3.2), because the value of each bit is decided independently, according to the statistics $P^{(t)}$. Of the created individuals (which are never kept to form an actual population) only the best individual s^* (strict elite principle, cf. Sect. 12.2.8) is used to update the population statistics according to

$$P_k^{(t)} \leftarrow s_k^* \cdot \alpha + P_k^{(t-1)}(1 - \alpha).$$

Here α is a kind of "learning rate" that determines how strongly the best individual of a generation changes the population statistics. A low learning rate is typically used to explore the search space (because it keeps the population statistics close to a vector with only 0.5 entries, and thus does not prefer specific bit values). A high learning rate, on the other hand, causes an exploitation of the aggregated population statistics. Therefore one may consider a time-dependent α, which prefers exploration in early generations and exploitation in later ones (cf. Sect. 12.2).

The parameter p_m (mutation probability) determines the probability with which the statistics for a bit are subject to a random change, while β controls the strength of this change. A common choice of the parameters of population-based incremental learning is: population size μ between 20 and 100, learning rate α between 0.05 and 0.2, mutation rate p_m between 0.001 and 0.02, mutation strength $\beta = 0.05$.

Note that population-based incremental learning treats the bits of the chosen encoding independently of each other. This is equivalent to assuming vanishing *epistasis* (cf. Sect. 12.1.2). Looking from the point of view of the schema theorem (cf. Sect. 13.1.1), we observe that population-based incremental learning does *not* create any implicit parallelism w.r.t. the exploration of schemata, but rather uses a single probabilistic schema (namely the population statistics). Furthermore, it should be kept in mind that very different populations can have the same population statistics, as is demonstrated by the simple example shown in Table 13.1. As a consequence, population-based incremental learning is not necessarily well suited for problems in which specific combinations of bits need to be found.

In order to overcome this drawback, improved approaches try to model the dependences between bits. A convenient tool for this purpose are Bayesian networks, which are graphical models that represent high-dimensional probability distributions effectively by capturing the obtaining dependences and independences. Bayesian networks are discussed extensively in Part IV. An example of using them for population-based incremental learning is the so-called **Bayesian optimization**

algorithm (BOA) (Pelikan et al. 2000). This algorithm starts with a Bayesian network consisting of unconnected nodes, that is, with independent bits (equivalent to PBIL). It samples a population from the current network and evaluates its individuals with the fitness function. Then a new Bayesian network is created by learning it (cf. Chap. 25) from the best individuals of the population or by endowing all individuals with case weights that are proportional to their fitness. This process is repeated until some termination criterion is met. More details about the Bayesian optimization algorithm can be found in Pelikan et al. (2000) or Brownlee (2011).

13.4.2 Ant Colony Optimization

Ant colony optimization (ACO) (Dorigo 1992; Dorigo and Stützle 2004) is inspired by the path-finding behavior of certain species of ants. Since food from a discovered source has to be transported to the nest, many species of ants form "transport trails," which are marked by "odor signatures" of secreted chemical substances called **pheromones** (from the Greek *φέρειν*: to bear, to carry and *ὁρμή*: impetus). Since most ants are practically blind, pheromones are (besides sound and touch) their main means of communication. By following pheromone traces left by their fellows, ants find their way to a discovered food source. The amount of secreted pheromone signals both the quality and the amount of the discovered food.

The process of leaving traces in the environment, which trigger actions of other individuals, is commonly called **stigmergy** (from the Greek *στίγμα*: mark, sign and *ἔργον*: work, action). Stigmergy enables ants to discover and follow **shortest paths** without having any global overview of the situation. All they need to adapt their behavior to the global requirements is local information. Intuitively, starting from a random exploration, the shortest path receives more pheromone than other paths, because it is traversed by more individuals in the same amount of time.

A striking illustration of this phenomenon is the so-called **double-bridge experiment** (Goss et al. 1989). In this experiment the nest of an ant colony of the species *Iridomyrmex humilis* was connected to a food source by a double bridge, the two branches of which had different length (see Fig. 13.15 for a sketch). Since the ants are practically blind, they are unable to see which side of the bridge is shorter. Due to the construction of the bridge they also cannot derive any information from the angle at which the two branches fork off the initial path (see Fig. 13.15: both branches start with 45°; the longer branch changes its direction only later).

In most of the trials Goss et al. (1989) conducted, almost all ants took the shorter branch of the bridge after only a few minutes. This phenomenon can be explained as follows (see Fig. 13.15, in which the amount of deposited pheromone is indicated by shades of gray): at the beginning, both branches are chosen by the same number of ants (that is, the branches are chosen with the same probability), because there is no pheromone on either of them (steps 1 and 2 in Fig. 13.15). However, the ants following the shorter branch reach the food source earlier (simply because a shorter path means less travel time, step 3). Ants returning from the food source observe

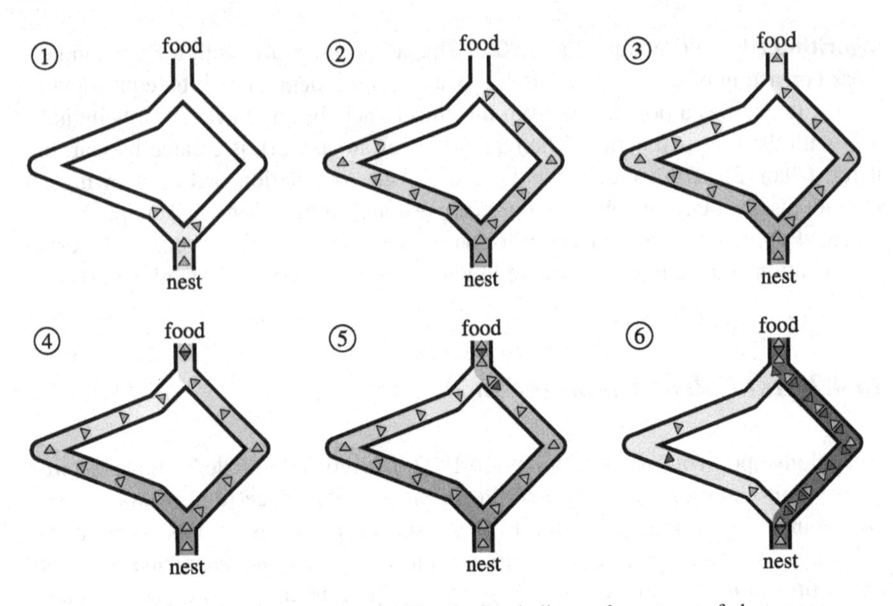

Fig. 13.15 The double-bridge experiment. The *shading* indicates the amount of pheromone

more pheromone on the shorter branch, because more ants have already reached the food source on this branch (and more ants mean more secreted pheromone, steps 4 and 5). This leads to an increasing preference of the shorter branch, so that after some time the shorter branch is chosen almost exclusively (step 6). The core principle here is that the shorter path is systematically reinforced, which is also called **auto-catalysis**: The more pheromone is on a path, the more ants choose this path; more ants traveling a path deposit more pheromone on it and so on.

Note that the shortest path can be found only if the ants deposit pheromone in *both directions*, that is, on the way from the nest to the food *and* on the way from the food to the nest. Suppose, for instance, that they deposited pheromone only on the way to the food source. Although the first ants returning from the food source choose the shorter path (because there is more pheromone on this path, see the discussion above), the amount of pheromone on this path is not systematically increased, because according to our assumption the ants do not deposit pheromone on their way back from the food source. The initial pheromone difference is rather eventually equalized by the ants that arrive (though somewhat later) on the longer path. The same argument applies for the opposite direction, at least if we assume that the ants cannot remember which path they came on and thus choose the return path randomly based on the amount of pheromone: in this case the pheromone difference would have to be brought about by ants returning on the shorter path and thus arriving earlier at the nest. Although initially ants starting out from the nest after the first ants have returned from the food source observe a pheromone difference, this difference is eventually equalized by the ants returning on the longer branch. As a consequence, no preference for the shorter path can develop.

Of course, due to random fluctuations in selecting a path (the ants choose essentially randomly with probabilities corresponding to the amount of pheromone), the search may still converge to one of the branches under such conditions. That is, eventually almost all ants may choose the same branch of the bridge. However, whether this is the shorter or the longer branch is entirely random.

Furthermore, note that (under the original conditions, that is, pheromone is deposited in both directions) the shortest path is found only if both branches exist from the beginning and neither contains any pheromone: a preference for an established path (marked with pheromone) is maintained. This plausible claim is supported by a second bridge experiment (Goss et al. 1989): in an initial setup the nest and the food source were connected only by the longer branch of the bridge. The second, shorter branch was added only after some time. In this setup the majority of the ants kept using the longer branch, which they had established in the early stage of the experiment. Only in very rare cases the ants switched to the shorter path (likely caused by a strong random fluctuation in the path selections).

The described natural principle can be transferred to computer-aided optimization by considering the problem of finding shortest paths in weighted graphs, for example, the shortest path between two given vertices. Each ant constructs a candidate solution. It starts at one of the given vertices and then moves from vertex to vertex, choosing the edge to follow according to a probability distribution that is proportional to the amount of pheromone it observes on the edges.

Unfortunately, a fundamental problem of such a straightforward approach are cycles traversed by the ants, because they introduce a tendency to reinforce themselves: if an ant traverses a cycle, the pheromone deposited by it makes it likely that the ant traverses the same cycle again. This drawback is counteracted by depositing pheromone only after the complete path has been constructed. In addition, before pheromone is deposited, any cycles that a path may contain are removed.

Another potential problem is that the search may focus on solution candidates that are constructed early in the process. Since these solution candidates receive pheromone, there is a tendency to stick to them (or minor variations of them). This can lead to **premature convergence**, similar to what we studied in Sect. 12.2.2 for evolutionary algorithms. In order to handle this problem, it is common to introduce **pheromone evaporation** (which plays only a minor role in nature). Other beneficial extensions and improvements include making the amount of pheromone that is deposited dependent on the quality of the constructed candidate solution and the introduction of heuristics to improve the edge selection, for example, by considering not only the pheromone, but also the weight of the edge.

With these considerations it should be clear that ant colony optimization is well suited to tackle the traveling salesman problem (cf. Sect. 11.6). We represent this problem by an $n \times n$ matrix $\mathbf{D} = (d_{ij})_{1 \leq i,j \leq n}$, where n is the number of cities and the d_{ij} are the distances between the cities i and j. Obviously, $\forall i \in \{1, \dots, n\} : d_{ii} = 0$. However, $\mathbf{D}$ need not be symmetric, that is, it may be $d_{ij} \neq d_{ji}$ for some i and j. We desire to find a round trip through all cities that has minimum length. Formally, we try to find a permutation π of the numbers $\{1, \dots, n\}$ that minimizes the sum of the edge weights of a tour that visits the cities in the order given by π.

Corresponding to the distance matrix **D**, pheromone deposits are represented by an $n \times n$ matrix $\Phi = (\phi_{ij})_{1 \leq i,j \leq n}$. Intuitively, a matrix element ϕ_{ij}, $i \neq j$, states how desirable it is to visit city j directly after city i, while the ϕ_{ii} are not used (formally, we may set them to 0 for all i). Like **D**, the matrix Φ need not be symmetric. The matrix elements ϕ_{ij}, $1 \leq i, j \leq n$, $i \neq j$, are initialized with the same arbitrary small value ε. That is, at the beginning every edge carries the same amount of pheromone and thus there is no preference between the edges. As an alternative, the pheromone deposits may be initialized with values that are inversely proportional to the edge weights. Following the general description given above, the pheromone values are used to determine the probability with which the next city to visit is chosen in the tour construction process. They are updated based on the quality of constructed round trips by depositing pheromone on the edges of good trips.

More formally, we proceed as follows: each ant constructs a solution by traversing a (random) Hamiltonian cycle. In order to avoid that an already visited city is revisited, we endow each ant with a "memory," consisting of the set C of indices of the cities that have not been visited yet. (Note that this deviates from the biological prototype!) An ant constructs a round trip (Hamiltonian cycle) as follows:

1. The ant starts in an arbitrary city (randomly chosen).
2. The ant observes the pheromones on the connections from its current city i to any city j it has not yet visited. Then it chooses to move to city j with the probability
$$p_{ij} = \frac{\phi_{ij}}{\sum_{k \in C} \phi_{ik}},$$
where C is the set of indices of cities that have not been visited and ϕ_{ij} is the amount of pheromone on the connection from city i to city j.
3. The ant repeats step 2 until it has visited all cities.

After a round trip has been constructed, represented as a permutation π of the city indices, the pheromone matrix Φ is updated according to

$$\forall i \in \{1, \ldots, n\}: \quad \phi_{\pi(i)\pi((i \bmod n)+1)} \leftarrow \phi_{\pi(i)\pi((i \bmod n)+1)} + Q(\pi),$$

where Q is a function that measures the quality of the solution. A natural choice for Q for the traveling salesman problem is the (scaled) inverse trip length

$$Q(\pi) = c \cdot \left(\sum_{i=i}^{n} d_{\pi(i)\pi((i \bmod n)+1)} \right)^{-1},$$

where c is a user-specified constant that controls the strength of the pheromone changes. Intuitively, this means that the shorter the trip (and thus the better the solution), the more pheromone is deposited on its edges.

In addition, after μ ants the pheromone is updated by **pheromone evaporation**:

$$\forall i, j \in \{1, \ldots, n\}: \quad \phi_{ij} \leftarrow (1 - \eta) \cdot \phi_{ij},$$

where η is an evaporation factor (fraction of pheromone that evaporates).

The complete algorithm for the traveling salesman problem looks like this:

Algorithm 13.9 (Ant-Colony-Optimization-TSP)

```
function aco_tsp (D: matrix of real, μ: int, η: real) : list of int;
begin                                       (* D = (d_ij)_{1≤i,j≤n}: distance matrix *)
  Φ ← (φ_ij)_{1≤i,j≤n}                      (* μ: number of ants, η: evaporation *)
        with φ_ij = ε for all i, j;         (* initialize the pheromone matrix *)
  π_best ← random permutation of {1, ..., n};
  repeat                                    (* search loop *)
    for a ∈ {1, ..., μ} do begin            (* construct candidate solution *)
      C ← {1, ..., n};                      (* set of cities to visit *)
      i ← randomly choose start city from C;
      π ← (i);                              (* start tour at the chosen city and *)
      C ← C \ {i};                          (* remove it from the unvisited cities *)
      while C ≠ ∅ do begin                  (* not all cities have been visited *)
        j ← choose next city of the trip from C
              with probability p_ij = φ_ij / Σ_{k∈C} φ_ik;
        π.append(j);                        (* add the chosen city to the tour and *)
        C ← C \ {j};                        (* remove it from the unvisited cities *)
        i ← j;                              (* finally go to the selected city *)
      end
      update pheromone matrix Φ with π and Q(π);
      if Q(π) > Q(π_best) then π_best ← π; end
    end                                     (* update the best tour *)
    update pheromone matrix Φ with evaporation η;
  until termination criterion is fulfilled;
  return π_best;                            (* return the best tour found *)
end
```

This basic algorithm may be extended in several ways. For instance, we may introduce a preference of cities that are close, in analogy to a nearest neighbor heuristics: choose to move from city i to city j with probability

$$p_{ij} = \frac{\phi_{ij}^{\alpha} \tau_{ij}^{\beta}}{\sum_{k \in C} \phi_{ik}^{\alpha} \tau_{ik}^{\beta}},$$

where C contains the indices of cities that have not been visited yet and $\tau_{ij} = d_{ij}^{-1}$.

A "greedy" approach introduces a stronger tendency to choose the best edge (than is already present due to the fact that is has the highest probability) with a user-specified probability p_{exploit}. That is, the ant moves from city i to city j_{best} with

$$j_{\text{best}} = \arg\max_{j \in C} \phi_{ij} \quad \text{or} \quad j_{\text{best}} = \arg\max_{j \in C} \phi_{ij}^{\alpha} \tau_{ij}^{\beta},$$

with probability p_{exploit}, while it chooses the next city according to the rule given above (that is, based on probabilities p_{ij}) with probability $1 - p_{\text{exploit}}$.

Furthermore, we may draw on an **elite principle** by reinforcing the best known round trip: after every iteration of the search (that is, after every μ ants), additional pheromone is deposited on the edges of the best round trip that has been found so far. The amount of pheromone can conveniently be specified as the number of ants that traverse this tour (in addition to the tours constructed in the normal process).

Further variants include **rank-based updating**, in which pheromone is deposited only on the edges of the best m solution candidates of the last iteration (consisting of the runs of μ ants), and maybe also on the best solution candidate found so far. This approach can be seen as analogous to rank-based selection (cf. Sect. 12.2.6), whereas the standard approach is analogous to fitness proportionate selection (cf. Sect. 12.2.1). **Strict elite principles** are extreme forms of rank-based updating: pheromone is deposited only on the best solution candidate of the last iteration or even only on the best solution found so far. However, this approach carries the risk of premature convergence and thus of getting stuck in a local optimum.

In order to avoid extreme values of the pheromone deposits, it can be advisable to introduce **lower and upper bounds** for the amount of pheromone on an edge. They correspond to lower and upper bounds for the probability of selecting an edge and thus help to enforce a better exploration of the search space (though at the price of slower convergence). A similar effect can be achieved by **restricted evaporation**: pheromone evaporates only from edges that have been traversed in the last iteration. This reduces the risk of pheromone deposits becoming very small.

Improvements of the standard approach, which are meant to lead to better solution candidates, are local improvements of the round trip (like removing edge crossings, which obviously cannot be optimal). More generally, we may consider simple operations as they could be used in a hill climbing approach (cf. Sect. 11.5.2) and thus try (in a limited number of steps) to optimize solution candidates locally. Among such operations are: exchange of cities that are visited in consecutive steps, permutation of adjacent triplets, "inverting" a part of a round trip (cf. Sect. 11.6) etc. More costly local optimization should only be considered to improve the best solution candidate before it is returned from the search procedure.

A program illustrating ant colony optimization for the traveling salesman problem (containing several of the mentioned improvements) can be found at:

http://www.borgelt.net/acopt.html

In order to apply ant colony optimization to other optimization problems, the problem has to be formulated as a search in a graph. In particular, it must be possible to describe a solution candidate as a set of edges. However, these edges need not form a path. As long as there is an iterative procedure with which the edges of the set can be chosen (based on pheromone-described probabilities), ant colony optimization is applicable. Even more generally, ant colony optimization is applicable if solution candidates are constructed with the help of a series of (random) decisions, where every decision extends a (partial) solution. The reason is that the sequence of decisions can be interpreted as a path in a **decision graph** (also called **construction graph**). The ants explore paths in this decision graph and try to find the best (shortest, cheapest) path, which yields a best set or sequence of decisions.

As a final remark we consider the convergence properties of ant colony optimization. Referring to a "standard procedure," in which (1) pheromone evaporates with a constant factor from all edges, (2) new pheromone is deposited only on the edges of the best found candidate solution (strict elite principle), and (3) the pheromone values are bounded from below by $\phi_{\min}$, the search converges in probability to the optimal solution (Dorigo and Stützle 2004). That is, if we let the number of computation steps go to infinity, then the probability that the optimal solution is found approaches 1. If the lower bound $\phi_{\min}$ goes to 0 "sufficiently slowly" (for instance, $\phi_{\min} = \frac{c}{\ln(t+1)}$ where t is the iteration step and c is some constant), one can even show that with the number of iterations going to infinity, every ant in the colony will construct the optimal solution with a probability approaching 1.

13.4.3 Particle Swarm Optimization

Particle swarm optimization (PSO) (Kennedy and Eberhart 1995) is inspired by the behavior of animal species that search for food in swarms, schools, or flocks, for example, bird or fish. This search is characterized by individuals exploring the environment in the vicinity of the swarm up to some distance, but always returning to the swarm as well. In this way information about discovered food sources is conveyed to the other members of the swarm, namely when the discovering member returns to the food source and other members follow it. Eventually the whole swarm may move toward the food source (provided it is attractive enough).

Particle swarm optimization can be seen as a method that combines gradient-based search (for example, gradient descent, cf. Sect. 11.5.1, and hill climbing, cf. Sect. 11.5.2) with population-based search (like evolutionary algorithms). Like gradient descent, it requires that the search space Ω is real-valued, that is, $\Omega \subseteq \mathbb{R}^n$. Hence the function to be optimized must be of the form $f : \mathbb{R}^n \to \mathbb{R}$.

The rationale of the search is to use a "swarm" of m candidate solutions instead of a single one and to aggregate information from the individual solution to guide the search better. Every candidate solution corresponds to a "particle" having a position $\mathbf{x}_i$ in the search space and a velocity $\mathbf{v}_i$, $i = 1, \ldots, m$. In each step the position and the velocity of the ith particle are updated according to

$$\begin{aligned}
\mathbf{v}_i(t+1) &= \alpha \mathbf{v}_i(t) + \beta_1\big(\mathbf{x}_i^{(\text{local})}(t) - \mathbf{x}_i(t)\big) + \beta_2\big(\mathbf{x}^{(\text{global})}(t) - \mathbf{x}_i(t)\big),\\
\mathbf{x}_i(t+1) &= \mathbf{x}_i(t) + \mathbf{v}_i(t).
\end{aligned}$$

Here $\mathbf{x}_i^{(\text{local})}$ is the **local memory** of the individual (i.e. particle). It is the best point in the search space that this particle has visited up to time step t. That is,

$$\mathbf{x}_i^{(\text{local})} = \mathbf{x}_i\big(\arg\max_{u=1}^{t} f\big(\mathbf{x}_i(u)\big)\big).$$

Similarly, $\mathbf{x}^{(\text{global})}$ is the **global memory** of the swarm as a whole. It is the best point in the search space that an individual of the swarm has visited up to step t. That is,

$$\mathbf{x}^{(\text{global})}(t) = \mathbf{x}_j^{(\text{local})}(t) \quad \text{with } j = \arg\max_{i=1}^{m} f\big(\mathbf{x}_i^{(\text{local})}\big).$$

The parameters β_1 and β_2, with which the strength of "attraction" of the points in the local and the global memory are controlled, are randomly chosen in every step. The parameter α is chosen in such a way that it decreases over time. That is, as time advances, the influence of the (current) velocity of the particle decreases and thus the (relative) influence of the attraction of the local and the global memory increases. Generally, it can be observed that after starting with fairly high speed (caused by the attraction of the global memory) the particles become slower and slower and finally come (almost) to a halt at the best point discovered by the swarm.

The general procedure of particle swarm optimization looks like this:

Algorithm 13.10 (Particle-Swarm-Optimization)
function pso (m: int, a, b, c: real) : array of real;
begin (* m: number of particles *)
  $t \leftarrow 0$; $q \leftarrow -\infty$; (* a, b, c: update parameters *)
  for $i \in \{1, \dots, m\}$ **do begin** (* initialize the particles *)
    $\mathbf{v}_i \leftarrow 0$; (* initialize velocity and position *)
    $\mathbf{x}_i \leftarrow$ choose a random point of $\Omega = \mathbb{R}^n$;
    $\mathbf{x}_i^{(\text{local})} \leftarrow \mathbf{x}_i$; (* initialize the local memory *)
    if $f(\mathbf{x}_i) \geq q$ **then begin** $\mathbf{x}^{(\text{global})} \leftarrow \mathbf{x}_i$; $q \leftarrow f(\mathbf{x}_i)$; **end**
  end (* compute initial global memory *)
  repeat (* update the swarm *)
    $t \leftarrow t + 1$; (* count the update step *)
    for $i \in \{1, \dots, m\}$ **do begin** (* update the local and global memory *)
      if $f(\mathbf{x}_i) \geq f(\mathbf{x}_i^{(\text{local})})$ **then** $\mathbf{x}_i^{(\text{local})} \leftarrow \mathbf{x}_i$; **end**
      if $f(\mathbf{x}_i) \geq f(\mathbf{x}^{(\text{global})})$ **then** $\mathbf{x}^{(\text{global})} \leftarrow \mathbf{x}_i$; **end**
    end
    for $i \in \{1, \dots, m\}$ **do begin** (* update the particles *)
      $\beta_1 \leftarrow$ sample random number uniformly from $[0, a)$;
      $\beta_2 \leftarrow$ sample random number uniformly from $[0, a)$;
      $\alpha \leftarrow b/t^c$; (* example of a time dependent α *)
      $\mathbf{v}_i(t+1) \leftarrow \alpha \cdot \mathbf{v}_i(t) + \beta_1(\mathbf{x}_i^{(\text{local})}(t) - \mathbf{x}_i(t)) + \beta_2(\mathbf{x}^{(\text{global})}(t) - \mathbf{x}_i(t))$;
      $\mathbf{x}_i(t+1) \leftarrow \mathbf{x}_i(t) + \mathbf{v}_i(t)$; (* update velocity and position *)
    end
  until termination criterion is fulfilled;
  return $\mathbf{x}^{(\text{global})}$; (* return the best point found *)
end

A demonstration program illustrating particle swarm optimization with the above algorithm for a two-dimensional function with many local maxima can be found at:

http://www.borgelt.net/psopt.html

Since this program visualizes each particle with a "tail" of previous positions, the movement of the particles can be observed well.

Extensions of the basic particle swarm optimization algorithm are manifold. If the search space is not unbounded, but a proper subset of $\mathbb{R}^n$ (for example, a hypercube $[a, b]^n$), we may let particles get reflected at the boundaries of the subset instead of merely clamping their position to the admissible region. Furthermore, one may focus the update more strongly on the **local environment** of a particle. In this variant, the global swarm memory is replaced by the best local memory of particles that reside in the vicinity of the particle to update. This improves the exploration of the search space. One may also introduce **automatic parameter adaptation**, for example, methods that adapt the size of the swarm (for instance: a particle whose local memory is significantly worse than that of particles in its vicinity is removed). Finally, **diversity control mechanisms** are meant to prevent a premature convergence to suboptimal solutions. One way of achieving this is to introduce an additional random component into the velocity update, which increases the diversity by letting particles explore the environment of the swarm in a more versatile fashion.

References

J. Antonisse. A New Interpretation of Schema Notation that Overturns the Binary Encoding Constraint. *Proc. 3rd Int. Conf. on Genetic Algorithms*, 86–97. Morgan Kaufmann, San Francisco, CA, USA, 1989

T. Bäck and H.-P. Schwefel. An Overview of Evolutionary Algorithms for Parameter Optimization. *Evolutionary Computation* 1(1):1–23. MIT Press, Cambridge, MA, USA, 1993

S. Baluja. Population-based Incremental Learning: A Method for Integrating Genetic Search Based Function Optimization and Competitive Learning. Technical Report CMU-CS-94-163, School of Computer Science, Carnegie Mellon University, Pittsburgh, PA, USA, 1994

W. Banzhaf, P. Nordin, R.E. Keller, and F.D. Francone. *Genetic Programming—An Introduction: On the Automatic Evolution of Computer Programs and Its Applications*. Morgan Kaufmann, San Francisco, CA, USA, 1998

J. Brownlee. *Clever Algorithms: Nature-Inspired Programming Recipes*. Lulu Press, Raleigh, NC, USA, 2011

R. Dawkins. *The Extended Phenotype: The Long Reach of the Gene*. Oxford University Press, Oxford, United Kingdom, 1982; new edition 1999

M. Dorigo. *Optimization, Learning and Natural Algorithms*. PhD Thesis, Politecnico di Milano, Milan, Italy, 1992

M. Dorigo and T. Stützle. *Ant Colony Optimization*. MIT Press, Cambridge, MA, USA, 2004

D.B. Fogel. *Blondie24: Playing at the Edge of AI*. Morgan Kaufmann, San Francisco, CA, USA, 2001

D.E. Goldberg. *Genetic Algorithms in Search Optimization and Machine Learning*. Addison Wesley, Reading, MA, USA, 1989

D.E. Goldberg. The Theory of Virtual Alphabets. *Proc. 1st Workshop on Parallel Problem Solving in Nature (PPSN 1991, Dortmund, Germany)*, LNCS 496:13–22. Springer-Verlag, Heidelberg, Germany, 1991

G.H. Golub and C.F. Van Loan. *Matrix Computations*, 3rd edition. The Johns Hopkins University Press, Baltimore, MD, USA, 1996

S. Goss, S. Aron, J.-L. Deneubourg, and J.M. Pasteels. Self-organized Shortcuts in the Argentine Ant. *Naturwissenschaften* 76:579–581. Springer-Verlag, Heidelberg, Germany, 1989

N. Hansen. The CMA Evolution Strategy: A Comparing Review. In: Lozano et al. (2006), 1769–1776

J.H. Holland. *Adaptation in Natural and Artificial Systems: An Introductory Analysis with Applications to Biology, Control, and Artificial Intelligence*. University of Michigan Press, Ann Arbor, MI, USA, 1975. 2nd edition: MIT Press, Cambridge, MA, USA, 1992

J. Kennedy and R. Eberhart. Particle Swarm Optimization. *Proc. IEEE Int. Conf. on Neural Networks*, vol. 4:1942–1948. IEEE Press, Piscataway, NJ, USA, 1995

J.R. Koza. *Genetic Programming: On the Programming of Computers by Means of Natural Selection*. MIT Press, Boston, MA, USA, 1992

P. Larrañaga and J.A. Lozano (eds.) *Estimation of Distribution Algorithms: A New Tool for Evolutionary Computation*. Kluwer Academic Publishers, Boston, USA, 2002

J.A. Lozano, T. Larrañaga, O. Inza, and E. Bengoetxea (eds.) *Towards a New Evolutionary Computation. Advances on Estimation of Distribution Algorithms*. Springer-Verlag, Berlin/Heidelberg, Germany, 2006

M. Michell. *An Introduction to Genetic Algorithms*. MIT Press, Cambridge, MA, USA, 1998

N.J. Nilsson. *Artificial Intelligence: A New Synthesis*. Morgan Kaufmann, San Francisco, CA, USA, 1998

V. Nissen. *Einführung in evolutionäre Algorithmen: Optimierung nach dem Vorbild der Evolution*. Vieweg, Braunscweig/Wiesbaden, Germany, 1997

M. Pelikan, D.E. Goldberg, and E.E. Cantú-Paz. Linkage Problem, Distribution Estimation, and Bayesian Networks. *Evolutionary Computation* 8:311–340. MIT Press, Cambridge, MA, USA, 2000

W.H. Press, S.A. Teukolsky, W.T. Vetterling, and B.P. Flannery. *Numerical Recipes in C: The Art of Scientific Computing*, 2nd edition. Cambridge University Press, Cambridge, United Kingdom, 1992

I. Rechenberg. *Evolutionstrategie: Optimierung technischer Systeme nach Prinzipien der biologischen Evolution*. Fromman-Holzboog, Stuttgart, Germany, 1973

K. Weicker. *Evolutionäre Algorithmen*, 2nd edition. Teubner, Stuttgart, Germany, 2007

Chapter 14
Special Applications and Techniques

With this chapter, we close our discussion of evolutionary algorithms by giving an overview of an application of and two special techniques for this kind of metaheuristics. In Sect. 14.1, we consider behavioral simulation for the iterated prisoners dilemma with an evolutionary algorithm. In Sect. 14.2, we study evolutionary algorithms for multi-criteria optimization, especially in the presence of conflicting criteria, which instead of returning a single solution try to map out the so-called **Pareto-frontier** with several solution candidates. Finally, we take a look at parallelized versions of evolutionary algorithms in Sect. 14.3.

14.1 Behavioral Simulation

Up to now we applied evolutionary algorithms only to numerical or discrete optimization problems. In this section, however, we employ them for behavioral simulation, in order to study strategies for social interaction as well as so-called population dynamics. The basis for the presented approach is **game theory**, which has proven to be highly useful to analyze social and economic situations and is one of the most important theoretical foundation of economics. The rationale of game theory is to model agents and their actions as game moves in a formally specified framework. Here we consider how the behavioral strategy underlying the moves of an agent in a specific situation can be encoded in a chromosome, so that an evolutionary algorithm can be used to study the properties of successful strategies as well as the influence of the distribution of strategies in a population.

14.1.1 The Prisoner's Dilemma

The best-known and most thoroughly studied problem of the game theory is the so-called **prisoner's dilemma**. This delicate problem is commonly described as

R. Kruse et al., *Computational Intelligence*, Texts in Computer Science,
DOI 10.1007/978-1-4471-5013-8_14, © Springer-Verlag London 2013

Table 14.1 Special payoff matrix of the prisoner's dilemma

A \ *B*	keeps silent	confesses
keeps silent	−1, −1	−10, 0
confesses	0, −10	−5, −5

Table 14.2 General payoff matrix of the prisoner's dilemma

A \ *B*	cooperate	defect
cooperate	R, R	S, T
defect	T, S	P, P

follows: suppose two people robbed a bank and were arrested. Unfortunately, the available circumstantial evidence is not sufficient for a conviction because of the bank robbery. There is, however, sufficient evidence for a conviction because of a lesser criminal offense (say, illegal possession of firearms). Suppose this lesser criminal offense carries a sentence of one year in prison. In order to attain a sentence for the bank robbery, the prosecutor offers both prisoners to become a key witness: if one of them confesses to the bank robbery (and thus incriminates the other), he/she is exempted from punishment, while the other prisoner will be punished with the full force of the law, which means 10 years imprisonment. The problem resides in the fact that both prisoners are offered this possibility and thus both may be tempted to confess. However, if both confess, the key witness rule is inapplicable and thus both will be punished. Since they both pleaded guilty, though, they receive a mitigated sentence, meaning that both of them have to spend 5 years in prison.

A popular tool to analyze situations like the prisoner's dilemma formally is a **payoff matrix**. Such a matrix states for all possible pairs of action choices the payoff each of the agents receives. For the constellation of the prisoner's dilemma, we may use the payoff matrix shown in Table 14.1, in which the number of years in prison is used as the payoff—stated as a negative number to indicate that the larger the (absolute) value, the worse the outcome for the two agents.

Seen from a global point of view, it is best if both prisoners keep silent, because this minimizes the total number of years they have to spend in prison. However, there is a temptation to confess, because a confession, provided the other prisoner keeps silent, leads to a better outcome for the confessing prisoner. Unfortunately, this carries the risk to end up in a situation in which both have to spend 5 years in prison. As a consequence, there is a tendency that only a suboptimal payoff result is achieved. More technically, a double confession is the so-called **Nash equilibrium** (Nash 1951) of this payoff matrix: neither agent can improve its payoff by changing its action, while the other agent maintains the same action. An improvement is only possible if *both* agents change their action choice. Nash (1951) showed that every payoff matrix has at least one Nash equilibrium.

Generalizing from the special situation stated above, the prisoner's dilemma can be defined with a general payoff matrix as it is shown in Table 14.2. The letters mean: **R** reward for mutual cooperation, **P** punishment for mutual defection, **T** temptation to defect, **S** sucker's payoff. Note that the exact values of **R**, **P**, **T** and **S** are not important as long as the following two inequalities hold

$$\mathbf{T} > \mathbf{R} > \mathbf{P} > \mathbf{S} \quad \text{and} \quad 2 \cdot \mathbf{R} > \mathbf{T} + \mathbf{S}. \tag{14.1}$$

The first condition states that the payoff for a defect in case the other agent cooperates must be better than the reward for cooperation, so that there is actually a temptation to defect. In addition, cooperation must lead to a better outcome than defection. Finally, mutual defection should still be preferable to being exploited (sucker's payoff), so that there is an incentive to avoid getting exploited. The second condition is needed to make ongoing cooperation preferable to alternating exploitation. With these conditions, mutual defection is a Nash equilibrium of the payoff matrix.

Many situations in everyday life are analogous to the prisoner's dilemma: we often face the choice to exploit an interaction partner and thus to improve our payoff or to cooperate with the chance to achieve an globally better result, but at the risk of getting exploited ourselves. It is often argued that the rational choice in the prisoner's dilemma is to defect in order to ensure that one is not exploited. However, we observe cooperating behavior on a large scale, not only among humans, but among animals as well. Therefore, Axelrod (1980) raised the question under what conditions cooperation emerges in a world of egoists without any central authority. Several centuries earlier, in the book *Leviathan* (Hobbes 1651), a famous classic of political philosophy, it had already been claimed: under no conditions whatsoever! Before governmental order and thus a directing central authority existed, the state of nature was dominated by egoistic individuals that competed against each other in such a reckless way that life was "solitary, poor, nasty, brutish, and short" (Hobbes 1651). However, on an international level we observe that countries cooperate (economically as well as politically), even though there is *de facto* no central authority (unless one sees the United Nations as such, despite their lack of directive power). In addition, animals show cooperative behavior without any form of government.

14.1.2 The Iterated Prisoner's Dilemma

Axelrod (1980, 1984) proposed to find an explanation why we observe cooperation despite the unfavorable Nash equilibrium of the prisoner's dilemma by looking at the **iterated version** of this game. The reason is that in real life, interactions are rarely isolated, but take place in a social context that makes it likely that we will interact with the same agent again in the future. The idea is that the expectation of future iterations of the game may reduce the temptation to defect, since this opens up the possibility of retaliation. This could introduce an incentive to cooperate.

These considerations lead to the following two questions:

- Is cooperation created in the iterated prisoner's dilemma?
- What is the best strategy in the iterated prisoner's dilemma?

Table 14.3 Axelrod's payoff matrix for the iterated prisoner's dilemma

A \ *B*	cooperate	defect
cooperate	3 \ 3	0 \ 5
defect	5 \ 0	1 \ 1

In order to obtain a concrete framework (Axelrod 1980, 1984) specified the **payoff matrix** shown in Table 14.3. The chosen values for the quantities **R**, **P**, **T** and **S** are the smallest non-negative integer numbers that satisfy the two conditions stated in Eq. (14.1). R. Axelrod then invited scientists from diverse disciplines (psychology, social and political sciences, economics, mathematics) to encode what they believed to be an optimal strategy for the iterated prisoner's dilemma with this payoff matrix. The programs were to have access to all games already played against the same agent, that is, they could try to exploit information gained about the behavioral strategy of their opponents from the result of earlier games. Each program was to be evaluated by the total payoff it achieved in a round robin tournament (that is, every participant plays against every other participant).

With this framework, R. Axelrod conducted two tournaments. In the *first tournament*, 14 programs and one random player[1] competed against each other in a round-robin tournament with 200 matches per pairing. The winner was a surprising simple program submitted by the mathematician A. Rapoport, which implemented the common sense strategy of *tit for tat*: cooperate in the first move; in subsequent moves simply copy the move of the opponent from the preceding match. The name "tit for tat" for this strategy derives from the simple fact that it retaliates in case its opponent does not cooperate by defecting itself in the next match.

In order to substantiate the result, R. Axelrod published the program code of all participants of this tournament together with the payoff results and invited a second tournament. The idea was that by analyzing the results of the first tournament, insights about what constitutes a good strategy may be gained, so that better programs could be designed. R. Axelrod then conducted a *second tournament* in which 62 programs and one random player participated.[2] The tournament conditions were identical, that is, a round-robin tournament with 200 matches per pairing. Surprisingly, the winner was also the same, namely *tit for tat.*

The result may be surprising, because *tit for tat* does *not* win generally against any other strategy. For example, against a strategy that defects in all moves (and thus behaves "rationally" as seen from the point of view of the uniterated prisoner's dilemma) it loses due to the exploitation it suffers in the first match. However, if there are agents in the population with whom it can cooperate, it can gain an overall advantage. Another problem of *tit for tat* is that it may react inadequately to mistakes. Suppose two instances of *tit for tat* play against each other, cooperating nicely

[1] All programs were written in the programming language *Fortran*.

[2] All programs were written in the programming languages *Fortran* and *Basic*.

Table 14.4 Encoding of strategies for the prisoner's dilemma. The first element of every pair corresponds to the own move, the second one to the opposing move

		1st match	2nd match	3rd match	next move
1st bit:	response to	(C,C),	(C,C),	(C,C):	C
2nd bit:	response to	(C,C),	(C,C),	(C,D):	D
3rd bit:	response to	(C,C),	(C,C),	(D,C):	C
⋮		⋮		⋮	⋮
64th bit:	response to	(D,D),	(D,D),	(D,D):	D

at first, but in some move one instance accidentally plays defect. Clearly, the *tit for tat* strategy is unable to recover from the ensuing train of mutual retaliations.

As a consequence, it may be worthwhile to consider *tit for two tat* as an alternative: this strategy starts retaliating only after having been exploited twice. This behavior maintains cooperation even after an accidental defect. On the other hand, if the *tit for two tat* strategy is fixed and known to the opponent, it can be exploited: an opponent that defects in every other move gains a very clear advantage.

14.1.3 A Genetic Algorithm Approach

Although the two tournaments saw participants written by capable and renowned researchers, one may still harbor doubts whether the space of possible strategies was sufficiently well explored. In addition, the winner may depend on the selection of participants. In order to obtain better even substantiated results, Axelrod (1987) approached the problem by simulating and evolving populations of strategies for the prisoner's dilemma with a genetic algorithm.

In this approach, a strategy is encoded by considering all possible sequences of three consecutive games. Since three games mean six moves (three moves by either opponent), each of which can either be cooperate or defect, there are $2^6 = 64$ possible sequences. Thus the chromosome consists of 64 bits, which simply state what move should be played in the next game if the corresponding sequence of three matches was observed (see Table 14.4). In addition, each chromosome is endowed with 6 bits that encode the course of the game "before" the first move. This is necessary, so that a uniform strategy is specified that also covers the first three moves, in which the history is shorter than three matches. In total, every chromosome consists of 70 binary genes that are either C (cooperate) or D (defect).

The genetic algorithm works as follows: the initial population is created by randomly sampling bit sequences of length 70. The individuals of a population are evaluated by pairing each individual with sufficiently many opponents that are randomly selected from the population. (Depending on the size of the population a full round robin tournament may be prohibitively costly, since its cost is quadratic in the population size. Note also that the experiment was conducted in 1987 and thus at a time were computing power was much more limited than today.) In each pairing, 200 matches were played (like in the tournaments reported about above). The fitness of an individual is the average payoff it gained per pairing. Individuals are selected for the next generation according to the simplified expectation value model mentioned in Sect. 12.2.5: let $\mu_f(t)$ be the average fitness of the individuals in the population at time t and $\sigma_f(t)$ its standard deviation. Individuals s with $f(s,t) < \mu_f(t) - \sigma_f(t)$ do not receive offspring; individuals s with $\mu_f(t) - \sigma_f(t) \leq f(s,t) \leq \mu_f(t) + \sigma_f(t)$ have one child and individuals s $f(s,t) > \mu_f(t) + \sigma_f(t)$ have two children. As genetic operators, standard mutation and one-point crossover were applied. The algorithm was than run for a certain number of generations and the best individuals of the final population were examined.

From the result, Axelrod (1987) identified the following general patterns:

- **Don't rock the boat.** Cooperate after three times cooperate.
  (C, C), (C, C), (C, C) → C
- **Be provokable.** Play defect after a sudden defect of the opponent.
  (C, C), (C, C), (C, D) → D
- **Accept an apology.** Cooperate after mutual exploitation.
  (C, C), (C, D), (D, C) → C
- **Forget.** (Do not be resentful.) Cooperate after cooperation has been restored after one defect (also without retaliation).
  (C, C), (C, D), (C, C) → C
- **Accept a rut.** Play defect after three times defect of the opponent.
  (D, D), (D, D), (D, D) → D

Clearly the *tit for tat* strategy, which won the two tournaments with human-designed programs, has all of these properties. The *tit for two tat* strategy only lacks the "be provokable" property as it reacts only after two exploitation. As already mentioned above, this makes it vulnerable if the strategy is fixed and known to the opponent.

Note that this result should still not be taken as an argument that *tit for tat* is generally the best strategy. A lone individual playing *tit for tat* in a population of individuals that constantly play defect cannot strive. In order to gain the upper hand, *tit for tat* needs a (sufficiently large) sub-population of individuals with which it can cooperate. Its growth may be facilitated, though, if individuals can choose whom to play with based on their experience, so that they can prefer for future pairings individuals with whom they cooperated successfully in the past. Clearly, this is an important aspect that is also decisive in the real world.

14.1.4 Extensions

In order to generalize and possibly improve the results obtained by Axelrod (1987) longer match histories may be considered (exploiting the greater computing power we have available nowadays). The strategies may also be described by Moore machines (a specific type of finite state machine) or even general programs that are evolved with the principles of genetic programming (cf. Sect. 13.3).

The prisoner's dilemma may also be extended in many different ways to render its setup more realistic and to capture more and more general situations. We already mentioned above that individuals usually choose with whom to play again and are not forced to repeatedly play against opponents that exploited them. In addition, in the real world, we often find ourselves in situations in which more than two agents are involved, thus leading to a multiplayer version of the prisoner's dilemma. Furthermore, in the real world the consequences of actions are not always perfectly observable. That is, it may not always be perfectly clear whether the last move of our opponent was actually a defect, even though it appeared to be (as judged, for example, by the payoff we received). In order to avoid getting stuck in a train of mutual retaliations, variations of the *tit for tat* strategy may be needed. A very simple method to achieve this is the introduction of a random component, at least if a pairing is stuck in a rut of mutual defect: after a random number of steps try to restore cooperation by cooperating even though the last moves of the opponent were defects. Note that this may require a random component (in order to make the moment unpredictable at which cooperation is played), which cannot be encoded with the scheme we described in the preceding section.

14.2 Multi-criteria Optimization

In everyday life, we frequently encounter situations that cannot be described in the simple form of the optimization problems as we defined them in Definition 11.1 on page 173. In particular, we often face the task to select from a set of options that satisfy several criteria to different degrees. Often enough, these criteria are even conflicting, that is, trying to improve one causes another criterion to be less well satisfied. Consider, for example, the task of finding an apartment. You may desire:

- large floorspace,
- a certain number of bedrooms, bathrooms etc., a garage, a balcony,
- low rent (excluding service charges),
- low service charges,
- short distance to work,
- short distance to shopping centers and public facilities,
- good environment (low noise, air pollution etc.),
- good neighborhood etc.

Clearly, several of these criteria are conflicting. Usually, the larger the floorspace and the quality of the facilities, the higher the rent and the service charges. In addition, the rent usually also reflects the quality of the location. A short distance to work or to shopping centers may conflict with the objective of living in a quiet place etc.

Similar situations occur when you are considering to buy almost any consumer good. Generally, quality and price are conflicting criteria (getting high quality at a low price is rarely possible). Design and usability are also often conflicting (what looks better is often less convenient to use).

14.2.1 Weight Combination of Criteria

Formally, multi-criteria optimization can be described by k objective functions

$$f_i : \Omega \to \mathbb{R}, \quad i = 1, \dots, k.$$

Our objective is to find an element of the search space for which all functions yield a value that is as high as possible. The simplest approach to this problem is to combine all k objective functions into a single function and thus to reduce it to a standard optimization problem. For example, we may compute a weighted sum

$$f(s) = \sum_{i=1}^{k} w_i \cdot f_i(s),$$

where the absolute values of the weights specify the relative importance we attach to the different criteria (taking their range of values into account), and their signs may be used to account for both criteria to maximize and to minimize.

Unfortunately, an approach based on combining several criteria in this way has severe drawbacks: apart from the fact that it may not be easy to choose proper weights, we thus lose the possibility to adapt our relative preferences based on the properties of potential solutions we obtain (which is something we definitely do in decision making processes like the search for an apartment as we considered it above). However, the problem is even more fundamental: in general, we face here the problem of having to **aggregate preferences**: each criterion defines a preference order of the solution candidates and we have to aggregate these preference orders over the different criteria to obtain an ordering of the solution candidates.

Note that the same problem generally occurs in elections: each voter has a preference order for the candidates, and these preference orders must be aggregated in order to determine the result of an election. Unfortunately, as shown by Arrow (1951), there is no aggregation function that has all desirable properties. This result is also known as **Arrow's paradox**. Although the consequences of Arrow's impossibility theorem (Arrow 1951) can, in principle, be avoided by using **scaled preference assignments**, the needed scaling functions introduce an additional degree of freedom that makes the aggregation specification even more difficult.

14.2.2 Pareto-Optimal Solutions

An alternative approach to combining multiple criteria in optimization is to try to find all or at least many *Pareto-optimal* solutions.

Definition 14.1 (Pareto optimality) An element $s \in \Omega$ is called *Pareto-optimal* w.r.t. the objective functions f_i, $i = 1, \dots, k$ if there does not exist any element $s' \in \Omega$ for which the following two properties hold:

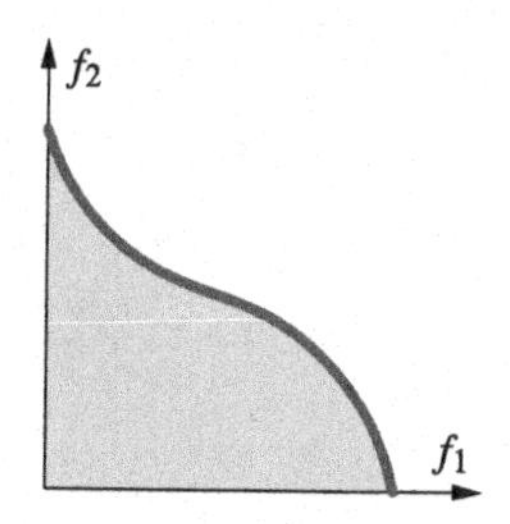

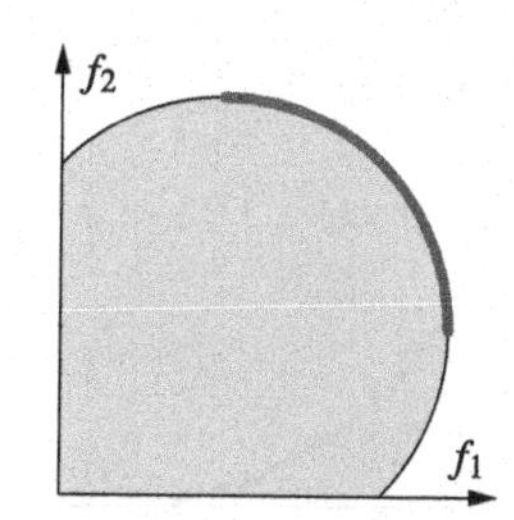

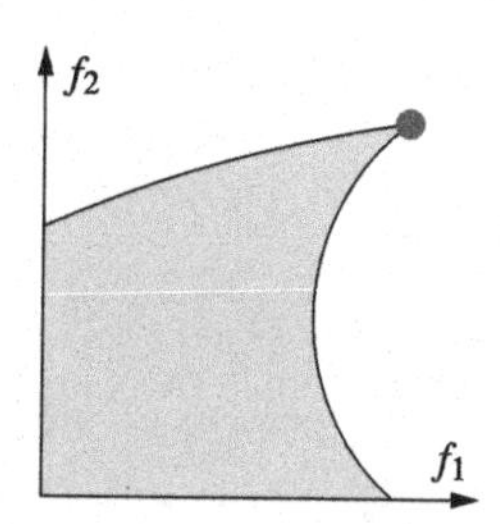

Fig. 14.1 Illustration of Pareto-optimal solutions, i.e. the so-called *Pareto frontier*. All points of the search space are situated in the gray area (with the functions f_1 and f_2 providing the coordinates). Pareto-optimal solutions are located on the part of the border that is drawn in *dark gray*. With the exception of the right diagram, there are multiple Pareto-optimal solutions

$$\forall i, 1 \le i \le k: \quad f_i(s') \ge f_i(s) \quad \text{and}$$
$$\exists i, 1 \le i \le k: \quad f_i(s') > f_i(s).$$

Intuitively, Pareto optimality (named after the Italian economist V. Pareto) means that the satisfaction of any criterion cannot be improved without harming another.

The notion of Pareto optimality may also be defined in two steps as follows: an element $s_1 \in \Omega$ **dominates** an element $s_2 \in \Omega$ if and only if

$$\forall i, 1 \le i \le k: \quad f_i(s_1) \ge f_i(s_2).$$

An element $s_1 \in \Omega$ **strictly dominates** an element $s_2 \in \Omega$ if s_1 dominates s_2 and

$$\exists i, 1 \le i \le k: \quad f_i(s_1) > f_i(s_2).$$

An element $s_1 \in S$ is called **Pareto-optimal** if it is *not* strictly dominated by any element $s_2 \in \Omega$. The notions of "dominates" and "strictly dominates" introduced here will be very useful below to describe the procedure of some algorithms.

Clearly, an advantage of searching for Pareto-optimal solutions is that the objective functions need not be combined (and thus there is no need to specify any weights or any aggregation function). In addition, we preserve the possibility to adapt our view of how important a criterion is relative to the others based on the solutions we obtain. However, the disadvantage is that there is rarely only just one Pareto-optimal solution and thus that there is no unique solution of the optimization problem. This is demonstrated in Fig. 14.1, which shows three different forms of the so-called **Pareto frontier**, which is the set of Pareto-optimal solutions, for two criteria. The gray areas contain all solution candidates, which are located according to the values that the functions f_1 and f_2 assign to them (that is, the gray area is the search space). The dark gray lines in the two diagrams on the left and the dark gray dot in the diagram on the right are the Pareto frontiers. Clearly, the points marked in dark gray are exactly the solution candidates that are not strictly dominated by any other solution candidate. Only in the diagram on the right we have a unique solution. In the other two diagrams, several candidate solutions are Pareto optimal.

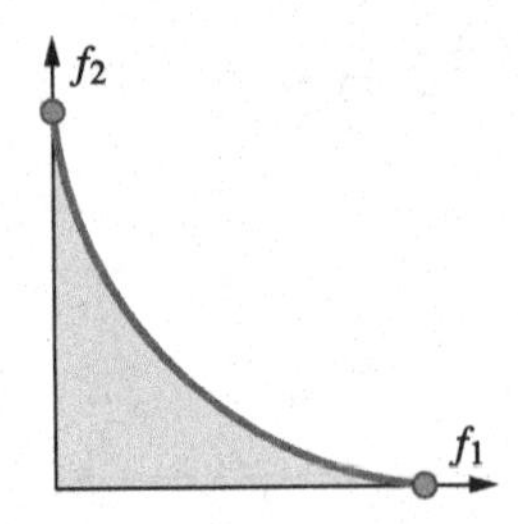

Fig. 14.2 Problem of the VEGA approach: the search focuses on the "corners," while solution candidates that satisfy all criteria moderately well are neglected

14.2.3 Finding Pareto-Frontiers with Evolutionary Algorithms

Although we always assumed up to now that we are given a single function to optimize, evolutionary algorithms may be applied to multi-criteria optimization problems. In this case, the goal is to find a selection of solution candidates on or at least close to the Pareto frontier, which covers this frontier sufficiently well. That is, we desire that not all solutions candidates are located in the same part of the Pareto frontier, while other parts are not covered by any solutions candidates.

The goal to cover the Pareto frontier sufficiently well rules out an approach that combines the objective functions by weights, so that we may simply apply one of the methods we already discussed. Although this approach will certainly yield a solution that is on or at least close to the Pareto frontier, it covers only a single point. Even if we return not just one, but, say, the best r individuals of the final population, it is unlikely that they are distributed over the Pareto frontier, because the weights prefer a specific point on the Pareto frontier to any other. As an illustration consider the diagrams in Fig. 14.1 again: assigned weights can be visualized as a straight line through the origin with a slope that is given by the ratio of the weights. The solutions found with a weighted combination of the objective functions are located close to the intersection of this straight line with the Pareto frontier. Provided, of course, that there is an intersection, which may not be the case in the two diagrams on the right.

An obvious alternative is the so-called **Vector Evaluated Genetic Algorithm** (VEGA) (Schaffer 1985), which works as follows: for every $i \in \{1, \dots, k\}$, where k is the number of criteria, $\frac{|\text{pop}|}{k}$ individuals are chosen based on the fitness function f_i. Intuitively, this can be seen as pooling sub-populations, each of which evolves w.r.t. a different objective function. The clear benefit of this approach is its simplicity and low computational costs. Its major drawback is that solutions that meet all criteria moderately well, but none of them particularly well have a significant selection disadvantage. As a consequence, with a Pareto frontier that is shaped a shown in Fig. 14.2, the search focuses on the "corners." Taking genetic drift into account, it may even happen that eventually the evolutionary algorithm converges to a randomly chosen corner, which is clearly undesirable.

A better approach consists in exploiting the notion that some solution candidates *dominate* others as we introduced it above. The basic idea is to divide the individuals in the population into ranked sets by iteratively removing non-dominated individuals. To be more specific, the division process works as follows:

1. Find all non-dominated candidate solutions of the population.
2. Assign the highest rank to these candidate solutions and remove them from the population.
3. Repeatedly determine and remove non-dominant candidate solutions, assigning them progressively lower ranks, until the population is empty.

Based on the assigned ranks, we may then perform rank-based selection to choose the individuals for the next generation (cf. Sect. 12.2.6). This approach is usually combined with niche techniques (cf. Sect. 12.2.9) to distinguish individuals having the same rank. Their purpose is, of course, to ensure a proper distribution of the individuals over the area close to the Pareto frontier. For example, we may employ **power law sharing**: the fitness assigned to an individual is the lower, the more individuals in the population have similar function values. Note that in contrast to Sect. 12.2.9 the similarity measure for the individuals is based here on the objective function values and not on the structure of the chromosome.

An alternative is the **non-dominated sorting genetic algorithm** (NSGA) (Srinivas and Deb 1994). Instead of rank-based selection, this algorithm relies on a scheme that is closely related to *tournament selection* (cf. Sect. 12.2.7), where the winner of the tournament is determined by the dominance term. In addition, a niche technique is drawn on. To be more specific, the selection of individuals of the next generation works as follows:

Algorithm 14.1 (NSGA-Selection)
function nsga_sel (A: array of array of real, n: int, ε: real) : list of int;
begin (* $A = (a_{ij})_{1 \le i \le r, 1 \le j \le k}$: fitness values *)
  $I \leftarrow ()$; (* n: tournament size, ε: neighborhood size *)
  for $t \in \{1, \ldots, r\}$ **do begin**
    $a \leftarrow U(\{1, \ldots, r\})$;
    $b \leftarrow U(\{1, \ldots, r\})$;
    $Q \leftarrow$ subset of $\{1, \ldots, r\}$ of size n;
    $d_a \leftarrow \exists i \in Q : A_i >_{\text{dom}} A_a$
    $d_b \leftarrow \exists i \in Q : A_i >_{\text{dom}} A_b$
    if d_a **and not** d_b **then** I.append(b);
    elseif d_b **and not** d_a **then** I.append(a);
    else
      $n_a \leftarrow \left|\{1 \le i \le r \mid d(A^{(i)}, A^{(a)}) < \varepsilon\}\right|$;
      $n_b \leftarrow \left|\{1 \le i \le r \mid d(A^{(i)}, A^{(b)}) < \varepsilon\}\right|$;
      if $n_a < n_b$ **then** I.append(a);
      else I.append(b); **end**
    end
  end
  return I;
end

That is, the algorithm repeatedly chooses two (indices of) individuals randomly and compares them to a sample of a user-specified size n (tournament participants). If only one of the chosen individuals is not dominated by any of the tournament participants, it is selected for the next generation. If neither or both are non-dominated, the individual is chosen that has fewer neighbors in a (fitness) neighborhood of a user-specified size ε. The algorithm returns a list of indices of chosen individuals.

Unfortunately, it may happen that the above algorithm approximates the Pareto frontier badly. On the one hand, this is due to the parameter setting of ε, which can be crucial for making good choices. On the other hand, this algorithm uses the population for two entirely different purposes, namely as a memory for non-dominated individuals (Pareto frontier) and as living population to explore the search space, and these two purpose do not necessarily harmonize with each other.

The mentioned problem can be fixed by removing non-dominated individuals from the population and keeping them separate in an archive. This archive usually has a finite size, so that new individuals can only be added if (dominated) individuals are deleted from it. If it cannot be filled completely, because not enough non-dominated individuals are known, it is filled with dominated individuals.

This approach is called **strength Pareto evolutionary algorithm 2** (SPEA2) (Zitzler et al. 2001) and works as follows:

Algorithm 14.2 (SPEA2)
function spea2 ($F_1, \dots, F_k$: function, μ, $\tilde{\mu}$: int) : set of object;
begin (∗ objective functions $F_1, \dots, F_k$ ∗)
  $t \leftarrow 0$; (∗ population size μ, archive size $\tilde{\mu}$ ∗)
  $P(t) \leftarrow$ create population with μ individuals;
  $R(t) \leftarrow \emptyset$;
  while termination criterion not fulfilled **do begin**
    evaluate $P(t)$ with the functions $F_1, \dots, F_k$;
    for $A \in P(t) \cup R(t)$ **do**
      $\text{noDom}(A) \leftarrow |\{B \in P(t) \cup R(t) \mid A >_{\text{dom}} B\}|$;
    for $A \in P(t) \cup R(t)$ **do begin**
      $d \leftarrow$ distance of A and its $\sqrt{\mu + \tilde{\mu}}$ nearest individuals in $P(t) \cup R(t)$;
      $A.F \leftarrow \frac{1}{d+2} + \sum_{B \in P(t) \cup R(t), B >_{\text{dom}} A} \text{noDom}(B)$;
    end
    $R(t+1) \leftarrow \{A \in P(t) \cup R(t) \mid A \text{ is non-dominated}\}$;
    while $|R(t+1)| > \tilde{\mu}$ **do**
      remove from $R(t+1)$ the individual A that has the smallest value $A.F$;
    if $|R(t+1)| < \tilde{\mu}$ **then**
      fill $R(t+1)$ with the best dominated individuals from $P(t) \cup R(t)$; **end**
    $t \leftarrow t+1$;
    if termination criterion not fulfilled **then**
      $P(t) \leftarrow$ select from $P(t-1)$ with tournament selection;
      apply recombination and mutation to $P(t)$;
    end
  end
  return non-dominated individuals from $R(t+1)$;
end

SPEA2 is a fairly ordinary evolutionary algorithm with a combined evaluation function. The archive, which is always kept at a user-specified size, may contain non-dominated individuals and is also used to compute the fitness. A niche technique is employed that evaluates at a certain number of nearest neighbors.

As a final example of evolutionary algorithms for multi-criteria optimization we mention here the **Pareto-archived evolutionary strategy** (PAES) (Knowles and Corne 1999). This approach is based on a $(1 + 1)$-ES (cf. Sect. 13.2), also employs an archive of non-dominated solution candidates, and works as follows:

Algorithm 14.3 (PAES)
function paes ($F_1, \dots, F_k$: function, $\tilde{\mu}$: int) : set of object;
begin (∗ objective functions $F_1, \dots, F_k$ ∗)
  $t \leftarrow 0$; (∗ archive size $\tilde{\mu}$ ∗)
  $A \leftarrow$ create random individual;
  $R(t) \leftarrow \{A\}$ organized as a multidimensional hash table;
  while termination criterion not fulfilled **do begin**
    $B \leftarrow$ mutation of A;
    evaluate B with the functions $F_1, \dots, F_k$;
    if $\forall C \in R(t) \cup \{A\}$: **not** $C >_{\text{dom}} B$ **then**
      if $\exists C \in R(t)$: $B >_{\text{dom}} C$ **then**
        remove all individuals from $R(t)$ that are dominated by B;
        $R(t) \leftarrow R(t) \cup \{B\}$;
        $A \leftarrow B$;
      elseif $|R(t)| = \tilde{\mu}$ **then**
        $g^* \leftarrow$ hash entry with the most entries;
        $g \leftarrow$ hash entry for B;
        if entries in $g <$ entries in g^* **then**
          remove one entry from g^*;
          $R(t) \leftarrow$ add B to $R(t)$;
        end
      else
        $R(t) \leftarrow$ add B to $R(t)$;
        $g_A \leftarrow$ hash entry for A;
        $g_B \leftarrow$ hash entry for B;
        if entries in $g_B <$ entries in g_A **then** $A \leftarrow B$; **end**
      end
    end
    $t \leftarrow t + 1$;
  end
  return non-dominated individuals from $R(t + 1)$;
end

Unless the archive is full, new solution candidates are added to it. If it is full, all dominated solution candidates are removed from it. If there are no dominated solution candidates, one of the individuals in the hash entry with the most members is removed (that is, the hash code represents a niche).

Although a wide variety of evolutionary algorithms for multi-criteria optimization exist, most of them start to have trouble to approximate the Pareto frontier well if more than three criteria are given. One of the reasons is, naturally, that the size of the Pareto frontier grows with the number of criteria, thus making it more difficult for algorithms to cover it sufficiently well or even to find solution candidates sufficiently close to it. This problem may be mitigated by presenting solutions to a user during the search and letting the user choose the direction the search space in which the search should continue (semi-automatic search).

14.3 Parallelization

Evolutionary algorithms are fairly expensive optimization methods. In order to obtain sufficiently good solutions, one often has to work with both a large population (a few thousand up to several tens of thousands of individuals) and a large number of generations (a couple of hundreds). Although this drawback is often compensated by a slightly better solution quality compared to other approaches, the execution time of an evolutionary algorithm can be unpleasantly long. One way to improve this situation is **parallelization**, that is, to distribute the necessary operations on several processors (exploiting that essentially all modern processors have multiple cores and multi-processor machines are also becoming more frequent). In this section, we discuss which steps can be parallelized (sufficiently easily) and what additional, specialized techniques are inspired by a parallel organization of the algorithm.

14.3.1 Parallelizable Operations

Creating an initial population is often easy to parallelize, because usually the chromosomes of the initial population are created randomly and independently of each other. The attempt to prevent duplicate individuals may, however, pose obstacles to a parallel execution. Parallelizing this step of fairly little importance overall, though, because the initial population is created just once.

The **evaluation** of chromosomes is also easily parallelizable because usually an individual is evaluated independently of any other individual. Since for many important problems the evaluation of the chromosomes is the most costly task, this is a decisive advantage. Even in the evolutionary algorithm used to study the prisoner's dilemma (cf. Sect. 14.1.1), we can process pairings in parallel. In order to **compute (relative) fitness values** or a ranking of the individuals, however, we need a central agent that collects and processes evaluations. As a consequence, whether the **selection** of the individuals for the next generation is parallelizable, depends heavily on the chosen selection method: the *expected value model* and *elitism* all require to consider the population as a whole, thus need a central agent, and therefore are difficult to parallelize. *Roulette-wheel and rank-based selection* can be parallelized after the initial step of computing the relative fitness values or sorting the individuals according to their fitness has been carried out. The initial step, however, needs a central agent that collects and processes all fitness values. *Tournament selection* is usually best suited for a parallel execution, especially for small tournament sizes, because all tournaments are independent and thus can be held in parallel.

Genetic operators can easily be applied in parallel, since they affect only one (mutation) or two chromosomes (crossover), and are independent of any other chromosomes. Even if multi-parent operators (like diagonal crossover) are used and thus more chromosomes are affected at the same time, different crossover procedures can be executed in parallel. If combined with tournament selection, a steady-state evolutionary algorithm can thus be parallelized very well.

Whether a **termination criterion** can be parallelized depends on the specific form of the termination criterion. The simple test whether a certain number of generations is reached does not cause any problems. However, termination criteria like

- the best individual of the population exceeds a user-specified fitness threshold, or
- the best individual has not changed (a lot) over a certain number of generations

need a central agency that collects this information about the individuals.

14.3.2 Island Model and Migration

Even if we require a selection method that causes some troubles for parallelization (like fitness proportionate selection, which needs a central agency at least to compute the relative fitness values), we may achieve a parallel execution by simply processing several independent populations, each on its own processor. Drawing on an obvious analogy from nature, each population can be seen as inhabiting an island, which explains the name **island model** for such an architecture. A pure island model is equivalent to executing the same evolutionary algorithm multiple times, which may just as well be done in a serial fashion. Usually it yields results that are somewhat worse than those of a single run with a larger population.

However, with a parallel execution, on may consider exchanging individuals between the island populations at certain fixed points in time (doing so in every generation creates too much communication overhead). Again drawing on an obvious analogy from nature, such an approach is commonly called **migration**. The idea underlying this method is that transferring genetic material between the islands improves the exploration properties of the island populations, without needing direct recombinations of chromosomes from different islands.

For the mechanisms that **control the migration between islands**, many different proposals exist. In the *random model* pairs of islands are chosen randomly, which then exchange some of their inhabitants. In this model any two island can, in principle, exchange individuals. This freedom is reduced in the *network model*, in which the islands are arranged into a network or graph. Individuals can migrate only between islands that are connected by an edge in the graph. Typical graph structures include rectangular and hexagonal grids in two or three dimensions. Along which of the edges individuals are exchanged is determined randomly.

Instead of merely exchanging individuals and thus genetic material, island populations may also be seen as competing with each other (*contest model*). In this case the evolutionary algorithms that are applied on the islands differ in approaches and/or parameters. The effect of the contest is that the population size of an island is increased or decreased according to the average fitness of its individuals. Usually a lower bound for the population size is set, so that islands cannot become empty.

14.3.3 Cellular Evolutionary Algorithms

Cellular evolutionary algorithms are a form of parallelization that is also called "*isolation by distance*." They work with a large number of (virtual) processors, each of which handles a single individual (or only a small number of individuals). The processors are arranged in an rectangular grid, usually in the shape of a torus in order to avoid boundary effects. Selection and crossover are restricted to adjacent processor, that is, to processors that are connected by an edge of the grid. *Selection* means that a processor chooses the best chromosome of the (four) processors adjacent to it (or one of these chromosomes randomly based on their fitness). The processor then performs *crossover* of the selected chromosome with its own. The better child resulting from such a crossover replaces the chromosome of the processor (local elite principle). A processor may also *mutate* its chromosome, the result of which, however, replaces the old chromosome only if it is better (local elite principle again). In such an architecture, groups of adjacent processors are created that maintain similar chromosomes. This mitigates the usually destructive effect of the crossover.

14.3.4 Combination with Local Search Methods

The approach of Mühlenbein (1989) is a combination of an evolutionary algorithm and hill climbing. After an individual has been created by a mutation or a crossover operation, it is optimized with hill climbing: random mutations are applied and kept if they are advantageous. Otherwise they are retracted and a different mutation is tried. Obviously, the local hill climbing optimization can easily be parallelized, because it is executed independently by the individuals.

Furthermore, individuals search for a crossover partner not in the whole population, but only in their (local) neighborhood, thus easing a parallel execution. Note that this requires a distance measure for the individuals and relates the approach to niche techniques (cf. Sect. 12.2.9). The offspring (the crossover products) perform local hill climbing. The individuals of the next generation are selected with a *local elite principle* (cf. Sect. 12.2.8), that is, the best two individuals among the four involved individuals (two parents and two children) replace the parents.

References

K.J. Arrow. *Social Choice and Individual Values*. J. Wiley & Sons, New York, NY, USA, 1951

R. Axelrod. More Effective Choice in the Prisoner's Dilemma. *Journal of Conflict Resolution* 24:379–403. SAGE Publications, New York, NY, USA, 1980

R. Axelrod. *The Evolution of Cooperation*. Basic Books, New York, NY, USA, 1984

R. Axelrod. The Evolution of Strategies in the Iterated Prisoner'S Dilemma. In: L. Davis (ed.) *Genetic Algorithms and Simulated Annealing*, 32–41. Morgan Kaufmann, San Francisco, CA, USA, 1987

T. Hobbes. *Leviathan. Or the Matter, Forme and Power of a Commonwealth Ecclesiastical and Civil*, 1651. Reprinted as: Ian Shapiro (ed.) *Leviathan. Or The Matter, Forme, & Power of a Common-Wealth Ecclesiastical and Civil*. Yale University Press, New Haven, CT, USA, 2010

J. Knowles and D. Corne. The Pareto Archived Evolution Strategy: A New Baseline Algorithm for Pareto Multiobjective Optimisation. *Proc. IEEE Congress on Evolutionary Computation (CEC 1999, Washington, DC)*, vol. 1:98–105. IEEE Press, Piscataway, NJ, USA, 1999

H. Mühlenbein. Parallel Genetic Algorithms, Population Genetics and Combinatorial Optimization. *Proc. 3rd Int. Conf. on Genetic Algorithms (CEC 1989, Fairfax, VA)*, 416–421. Morgan Kaufmann, San Francisco, CA, USA, 1989

J.F. Nash. Non-cooperative Games. *Annals of Mathematics* 54(2):286–295. Princeton University, Princeton, NJ, USA, 1951

J.D. Schaffer. Multiple Objective Optimization with Vector Evaluated Genetic Algorithms. *Proc. 1st Int. Conf. on Genetic Algorithms*, 93–100. L. Erlbaum Associates, Hillsdale, NJ, USA, 1985

N. Srinivas and K. Deb. Multiobjective Optimization Using Nondominated Sorting in Genetic Algorithms. *Evolutionary Computing* 2(3):221–248. MIT Press, Cambridge, MA, USA, 1994

E. Zitzler, M. Laumanns and L. Thiele. *SPEA2: Improving the Strength Pareto Evolutionary Algorithm*. Technical Report TIK-Rep. 103. Department Informationstechnologie und Elektrotechnik, Eidgenössische Technische Hochschule Zürich, Switzerland, 2001

Part III
Fuzzy Systems

Chapter 15
Fuzzy Sets and Fuzzy Logic

Many propositions about the real world are not either true or false, rendering classical logic inadequate for reasoning with such propositions. Furthermore, most concepts used in human communication do not have crisp boundaries, rendering classical sets inadequate to represent such concept. The main aim of fuzzy logic and fuzzy sets is to overcome the disadvantages of classical logic and classical sets.

15.1 Natural Languages and Formal Models

Classical logic and mathematics assume that we can assign one of the two values, *true* or *false*, to each logical proposition or statement. If a suitable formal model for a certain problem or task can be specified, conventional mathematics provides powerful tools which help us to solve the problem. When we describe such a formal model, we use a terminology which has much more stringent rules than natural language. This specification often requires more work and effort, but by using it we can avoid misinterpretations. Furthermore, based on such models we can prove or reject hypotheses or derive unknown correlations.

However, in our everyday life formal models do not concern the interhuman communication. Human beings are able to assimilate easily linguistic information without thinking in any type of formalization of the specific situation. For example, a person will have no problems to accelerate slowly while starting a car, if he is asked to do so. If we want to automate this action, it will not be clear at all, how to translate this advice into a well-defined control action. It is necessary to determine a concrete statement based on an unambiguous value, that is, step on the gas at the velocity of half a centimeter per second. On the other hand, this kind of information will not be adequate or very helpful for a person.

Therefore, automated control is usually not based on a linguistic description of heuristic knowledge or knowledge from one's own experience, but it is based on a formal model of the technical or physical system. This method is definitely a suitable approach, especially if there is a good model to be determined.

R. Kruse et al., *Computational Intelligence*, Texts in Computer Science,
DOI 10.1007/978-1-4471-5013-8_15,

However, a completely different technique is to use knowledge formulated in natural language directly for the design of the control strategy. In this case, a main problem will be the translation of the verbal description into concrete values, i.e. assigning "step on the gas slowly" into "step on the gas at the velocity of a centimeter per second" as in the above mentioned example.

When describing an object or an action, we usually use uncertain or vague concepts. In natural language we hardly ever find exactly defined concepts like supersonic speed for the velocity of a passing airplane. Supersonic speed characterizes an unambiguous set of velocities, because the speed of sound is a fixed entity and therefore it is unambiguously clear whether an airplane flies faster than sound or not. Frequently used vague concepts, like *fast*, *very big*, *small* and so on, make it impossible to decide unambiguously whether a given value satisfies such a vague concept or not. One of the reasons for this is that vague concepts are usually context dependent. Talking about airplanes *fast* has a different meaning than using this characteristic while referring to cars. But also if we agree that we are talking about cars it is not easy to distinguish clearly between fast and non-fast cars. The difficulty here is not to find a value telling us whether a car (or its top speed) is fast or not, but we had to presuppose that such a value does exist. It is more likely that we will be reluctant to fix such a value because there are velocities, we can classify as fast for a car and there are some we can classify as not fast, and in between there is a wide range of velocities which are considered as more or less fast.

15.2 Fuzziness versus Uncertainty

Any notation is said to be fuzzy when its **meaning** is not fixed by sharp boundaries. This means that the statement can be applied fully, to a certain degree, or not at all. The gradual degrees of this membership is also called **fuzziness**. A proposition is fuzzy if it contains gradual predicates. Such propositions may be neither *true* nor *false*, but anything in-between, for example, *to a certain degree* or *partially true*. Forms of such degrees can be found in natural languages, like *very*, *rather*, or *almost*.

Without these fuzzy statements, we can get stuck in different problems such as the **Sorites Paradox**. If a heap of sand is small, adding one grain of sand to it leaves it small. Obviously, a heap of sand with one single grain is small. Hence, all heaps of sand are small. This paradox dissolves if the degree of "heap of sand is small" slowly decreases by adding one grain after another.

Why is there a need for vagueness in natural languages? Any language is discrete, but the real world is continuous. To close this gap, we use fuzzy statements. Consider the word *young* for human beings. The more fine-grained the scale of age, for example, going from years to months, weeks, days, etc., the more difficult is it to fix the threshold below which *young* fully applies and above which *young* does not at all. So there is a conflict between the linguistic finite term set {young, mature, old} and the numerical representation by a real-valued interval, e.g. [0, 120] years for human beings.

Fuzzy set theory does not assume any threshold. The consequences for the logic behind fuzzy set theory will be discussed later in this book.

In contrast, uncertainty describes the probability of a well defined proposition. For example, rolling a die will either lead to exactly 6 or not, but not to something around 6. Uncertainty also comes from conflicting but precisely observed pieces of information and is usually found in statistics.

Here are some examples for a better distinction of fuzziness and uncertainty: "This car is about 10 and 15 years old." This proposition is fuzzy, because we have a lack of information or a lack of ability to measure or evaluate the numerical feature exactly. "This car was probably built in Germany." This sentence expresses uncertainty about the well-defined proposition *made in Germany*. Perhaps this uncertainty is coming from statistics like a random experiment. "The car I chose randomly is perhaps very big." In this example there are both kind of concepts, that is, the uncertain statement *perhaps* and the fuzzy description *very big*.

15.3 Fuzzy Sets

The idea of fuzzy sets is to solve this problem by avoiding the sharp separation of conventional sets into two values—complete membership or complete nonmembership. Instead, fuzzy sets can handle partial membership. So in fuzzy sets we have to determine to what degree or extend an element is a member of this fuzzy set. Therefore, we define:

Definition 15.1 A *fuzzy subset* or simply a *fuzzy set* μ of a set X (the universe of discourse) is a mapping $\mu : X \to [0, 1]$, which assigns to each element $x \in X$ a *degree of membership* $\mu(x)$ to the fuzzy (sub)set μ. The set of all fuzzy (sub)sets of X is denoted $\mathcal{F}(X)$.

A conventional set $M \subseteq X$ can be viewed as a special fuzzy set by identifying it with its **characteristic function** or **indicator function**.

$$I_M : X \to \{0, 1\}, \quad x \mapsto \begin{cases} 1 & \text{if } x \in M, \\ 0 & \text{otherwise.} \end{cases}$$

Seen in this way, fuzzy sets can be considered as generalized characteristic functions.

Example 15.1 Figure 15.1 shows the characteristic function of the set of velocities which are higher than 170 km/h. This set does not represent an adequate model of all high velocities. The jump at the value of 170 causes that 169.9 km/h would not be a high velocity but 170.1 km/h would be. Therefore, a fuzzy set (Fig. 15.2) seems to be more adequate to model the concept *high velocity*.

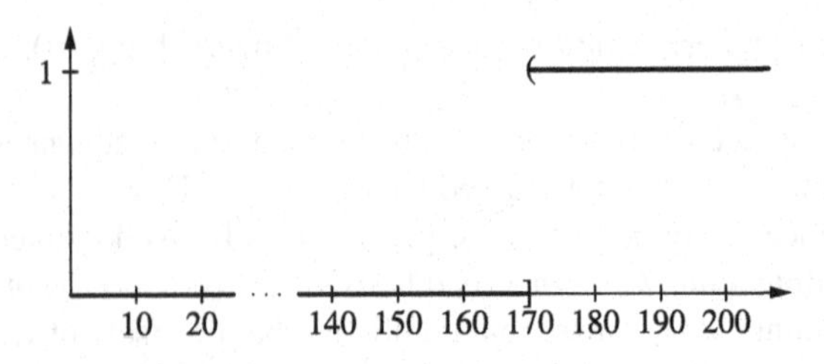

Fig. 15.1 The characteristic function of the set of velocities that are higher than 170 km/h

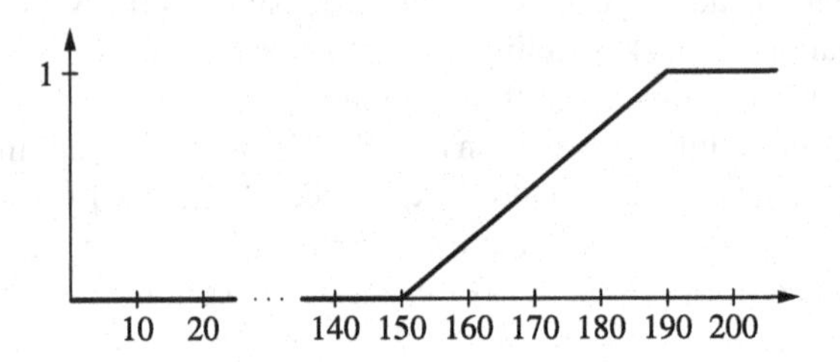

Fig. 15.2 The fuzzy set μ_{hv} of high velocities

Some authors use the term fuzzy set only for a vague concept $\mathcal{A}$ like *high velocity* and call the membership function $\mu_{\mathcal{A}}$, that models the vague concept, a characterizing or membership function of the fuzzy set or the vague concept $\mathcal{A}$. When operating with fuzzy sets, there is no advantage to make this distinction. Only from a philosophical point of view, one might be interested in distinguishing between an abstract vague concept and its concrete model in the form of a fuzzy sets. Since we do not want to initiate a philosophical discussion in this book, we stick to our restricted definition of a fuzzy set as a (membership) function, yielding values in the unit interval.

Besides the formal definition of a fuzzy set as a mapping to the unit interval, there are also other notations which are preferred by some authors, but we will not use them in this book. In some publications, a fuzzy set is written as a set of pairs of the elements of the underlying set and the corresponding degrees of membership in the form of $\{(x, \mu(x)) \mid x \in X\}$ following the fact that in mathematics a function is usually formally defined as a set of pairs, each consisting of one argument of the function and the image of this argument. A little bid more misleading is the notation of a fuzzy set as a formal sum $\sum_{x\in X} x/\mu(x)$ with an at most countable reference set X or as "integral" $\int_{x\in X} x/\mu(x)$ for an uncountable reference set X.

15.3.1 Interpretation of Fuzzy-Sets

We want to emphasize here, that fuzzy sets are formalized in the framework of "conventional" mathematics, just as probability theory is formulated in the framework of "conventional" mathematics. In this sense, fuzzy sets do not open the door to a "new" kind of mathematics, but define merely a new branch of mathematics.

Knowing that a strictly two-valued view is not suitable to model vague concepts adequately, which can be handled by human beings easily, we have introduced the concept of fuzzy sets on a purely intuitive basis. We did not specify precisely, how to interpret degrees of membership. The meanings of 1 as complete membership and 0 as complete non-membership are obvious, but we left open the question, how to interpret a degree of membership of 0.7 or why to prefer 0.7 over 0.8 as degree of membership for a certain element. These questions of semantics are often ignored. Therefore, a consistent interpretation of fuzzy sets is not maintained in some applications and this is one of the reasons, why inconsistencies may occur, when fuzzy sets are applied only on an intuitive basis. Understanding fuzzy sets as generalized characteristic functions, there no is ultimate reason for choosing the unit interval as the canonical extension of the set $\{0, 1\}$. In principle, it is possible that any linearly ordered set or—more generally—a lattice L might be better suited than the unit interval. In this case, the literature refers to L fuzzy sets. However, in real applications they do not play an important role. But even if we agree on the unit interval as the set of possible degrees of membership, we should explain in which sense or as what kind of structure we understand it.

The unit interval can be viewed as an ordinal scale. This means, only the linear ordering of the numbers is considered, that is, for expressing preferences. In this case the interpretation of a number between 0 and 1 as a degree of membership makes sense only, when comparing it to another degree of membership. Thus, we can express that one element belongs more to a fuzzy set than another one. A problem resulting from this purely ordinal view of the unit interval is the incomparability of degrees of membership stated by different persons. The same difficulty appears comparing grades. Two examinees receiving the same grade from different examiners can have shown very different performances. Normally, the scale of grades is not used as a purely ordinal scale. Pointing out which performance or which amount of mistakes leads to which grade is an attempt to make it possible to compare grades given by different examiners.

With the canonical metric quantifying the distance between two numbers and operations like addition and multiplication the unit interval has a considerably richer structure than the linear ordering of the numbers. Therefore, in many cases it is better to understand the unit interval as a metric scale to obtain a more concrete interpretation of the degrees of membership. We will discuss the issue of semantics of degrees of membership and fuzzy sets in Chap. 18. For the moment, we confine ourselves to a naive interpretation of degrees of memberships and say that the property of being an element of a fuzzy set can be satisfied gradually.

We want to emphasize here that gradual membership is a completely different idea than the concept of probability. It is clear that a fuzzy set μ must not be regarded as a probability distribution or density, because, in general, μ does not satisfy the condition

$$\sum_{x \in X} \mu(x) = 1 \quad \text{or} \quad \int_X \mu(x)\, dx = 1.$$

that is required in probability theory for density functions. Also the degree of membership $\mu(x)$ of an element x to the fuzzy set μ should not be interpreted as the probability that x belongs to μ.

To illustrate the difference between a gradual property and probability, we take a look at the example below, following Bezdek (1993).

U denotes the "set" of non-toxic liquids. A person dying of thirst receives two bottles A and B and the information that bottle A belongs to U with a probability of 0.9 and bottle B has a degree of membership of 0.9 to U. From which of the bottles should the person drink? The probability of 0.9 for A could mean that the bottle was selected from among ten bottles in a room where nine were filled with water and one with cyanide. But the degree of membership of 0.9 means that the liquid is "reasonably" drinkable. For instance, B could contain a juice which has already past its best-before date. That is why the thirsty person should choose bottle B.

The liquid in bottle A has the property of being non-toxic either completely (with a probability of 0.9) or not at all (with a probability of 0.1). The liquid in bottle B satisfies the property of being non-toxic in a merely gradual way.

Probability theory and fuzzy sets serve us for modeling completely different phenomena—namely, on the one hand the quantification of the uncertainty whether an event may happen and on the other hand how much a property or statement is satisfied or to what degree a property is fulfilled.

15.4 Representation of Fuzzy Sets

After having introduced fuzzy sets formally as functions from a universe of discourse to the unit interval, we now want to discuss different methods for specifying concrete fuzzy sets and adequate techniques for representing fuzzy sets as well as store them in a computer.

15.4.1 Definition Based on Functions

If the universe of discourse $X = \{x_1, \dots, x_n\}$ is a finite, discrete set of objects x_i, a fuzzy set μ can, in general, only be specified by the degrees of membership $\mu(x)$ for each element $x \in X$, that is, in the form of $\mu \widehat{=} \{(x_1, \mu(x_1)), \dots, (x_n, \mu(x_n))\}$.

In most of the cases, we will consider fuzzy sets here, the universe of discourse X will be the domain of a real-valued variable, that is, a subset of the real line, usually an interval. Then a fuzzy set μ is a real function taking values in the unit interval and can be illustrated by drawing its graph. With a purely graphical definition of fuzzy sets membership degrees of the single elements can only be specified up to a certain, quite rough precision leading to difficulties and errors in further calculations. Thus the graphical representation is only suitable for illustration purposes.

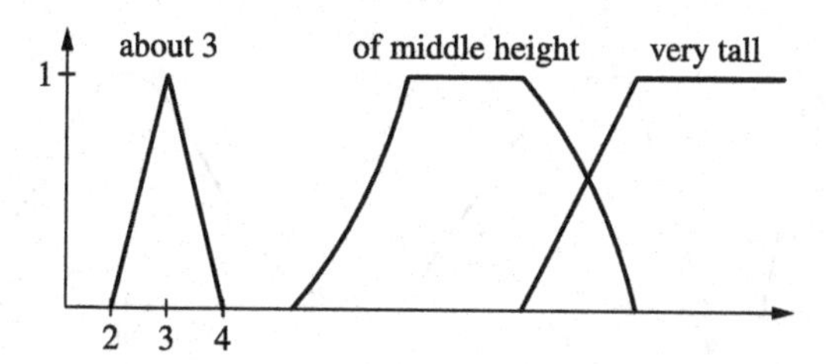

Fig. 15.3 Three convex fuzzy sets

Fig. 15.4 A non-convex fuzzy set

Usually fuzzy sets are used for modeling expressions—sometimes also called *linguistic* expressions in order to emphasize the relation to natural language, for example, 'about 3', 'of middle height' or 'very tall' which describe a vague value or a vague interval. Fuzzy sets associated with such expressions should monotonically increase up to a certain value and monotonically decrease from this value. Such fuzzy sets are called **convex**.

Figure 15.3 shows three convex fuzzy sets which could model the expressions 'about 3', 'of middle height' and 'very tall'. In Fig. 15.4, we see a non-convex fuzzy set. Note that the convexity of a fuzzy set μ does not imply that μ is also convex as real function.

For applications it is very often sufficient to consider only a few basic forms of convex fuzzy sets, so that a fuzzy set can be specified uniquely by a few parameters. Typical examples of such parametric fuzzy sets are **triangular functions** (cf. Fig. 15.5)

$$\Lambda_{a,b,c} : \mathbb{R} \to [0,1], \qquad x \mapsto \begin{cases} \frac{x-a}{b-a} & \text{if } a \le x \le b, \\ \frac{c-x}{c-b} & \text{if } b \le x \le c, \\ 0 & \text{otherwise,} \end{cases}$$

where $a < b < c$ holds.

Triangular functions are special cases of **trapezoidal functions** (cf. Fig. 15.5)

$$\Pi_{a',b',c',d'} : \mathbb{R} \to [0,1], \qquad x \mapsto \begin{cases} \frac{x-a'}{b'-a'} & \text{if } a' \le x \le b', \\ 1 & \text{if } b' \le x \le c', \\ \frac{d'-x}{d'-c'} & \text{if } c' \le x \le d', \\ 0 & \text{otherwise,} \end{cases}$$

where $a' < b' \le c' < d'$ holds. We also permit the following parameter combina-

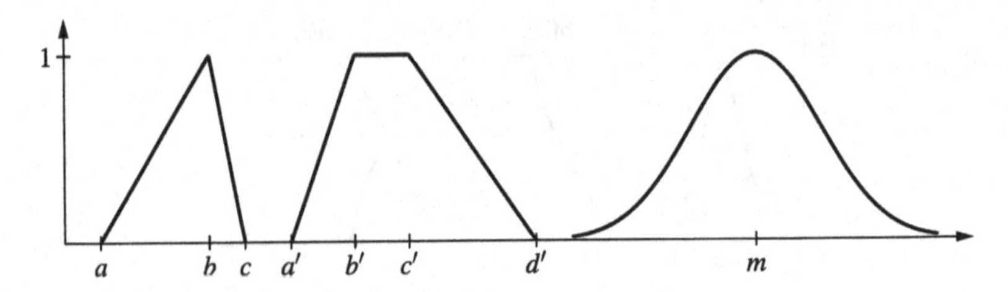

Fig. 15.5 The triangular function $\Lambda_{a,b,c}$, the trapezoidal one $\Pi_{a',b',c',d'}$ and the bell curve $\Omega_{m,s}$

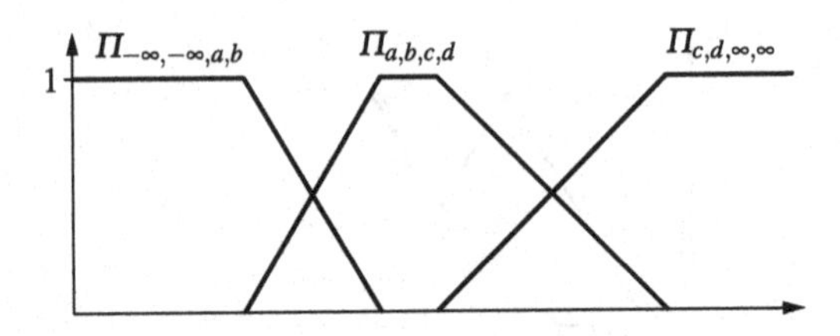

Fig. 15.6 The trapezoidal functions $\Pi_{-\infty,-\infty,a,b}$, $\Pi_{a,b,c,d}$ and $\Pi_{c,d,\infty,\infty}$

tions: $a' = b' = -\infty$ or $c' = d' = \infty$. The resulting trapezoidal functions are shown in Fig. 15.6. For $b' = c'$ we have $\Pi_{a',b',c',d'} = \Lambda_{a',b',d'}$.

If we want to use smooth functions instead of piecewise linear functions like triangular or trapezoidal ones, *bell curves* in the form of

$$\Omega_{m,s} : \mathbb{R} \to [0, 1], \qquad x \mapsto \exp\left(\frac{-(x-m)^2}{s^2}\right)$$

might be a possible choice. We have $\Omega_{m,s}(m) = 1$. The parameter s determines the width of the bell curve.

15.4.2 Level Sets

The representation of a fuzzy set as a function from the universe of discourse to the unit interval, assigning a membership degree to each element is called *vertical view*. Another possibility to describe a fuzzy set is the *horizontal view*. For each value α of the unit interval, we consider the set of elements having a membership degree of at least α to the fuzzy set.

Definition 15.2 Let $\mu \in \mathcal{F}(X)$ be a fuzzy set over the universe of discourse X and let $0 \le \alpha \le 1$. The (usual) set

$$[\mu]_\alpha = \{x \in X \mid \mu(x) \ge \alpha\}$$

is called α*-level set* or α*-cut* of the fuzzy set μ.

Figure 15.7 shows the α-cut $[\mu]_\alpha$ of the fuzzy set μ for the case that μ is a trapezoidal function. In this case, the α-cut is a closed interval. For an arbitrary

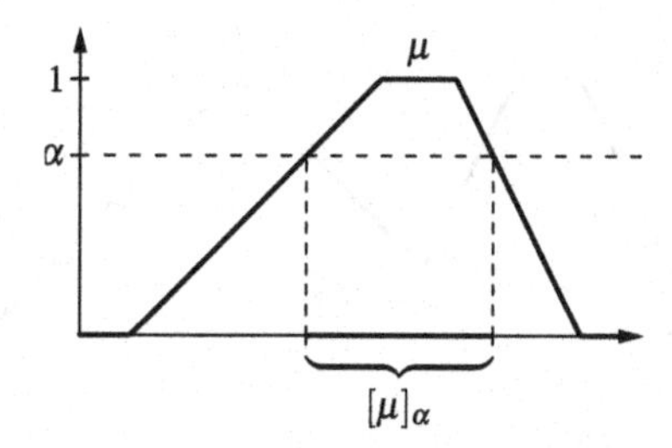

Fig. 15.7 The α-level set or α-cut $[\mu]_\alpha$ of the fuzzy set μ

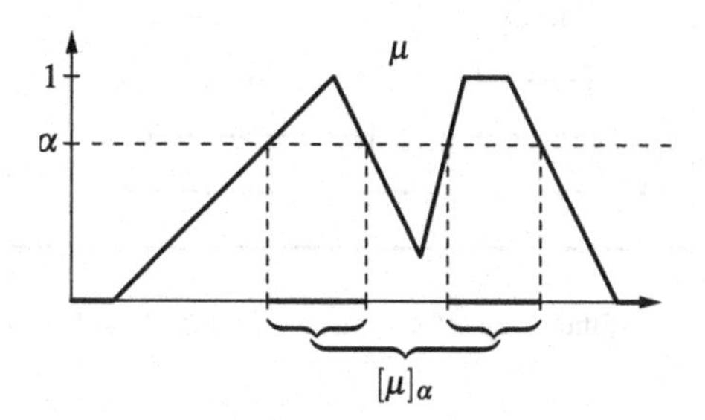

Fig. 15.8 α-cut $[\mu]_\alpha$ of the fuzzy set μ consisting of two disjoint intervals

fuzzy set μ over the real numbers we have that μ is convex as a fuzzy set if all its level sets are intervals. Figure 15.8 shows an α-cut of a non-convex fuzzy set consisting of two disjoint intervals.

The level sets of a fuzzy set have the important property of characterizing the fuzzy set uniquely. When we know the level sets $[\mu]_\alpha$ of a fuzzy set μ for all $\alpha \in [0, 1]$, we can determine the degree of membership $\mu(x)$ of any element x to μ by the equation

$$\mu(x) = \sup\{\alpha \in [0, 1] \mid x \in [\mu]_\alpha\}. \tag{15.1}$$

Geometrically speaking, a fuzzy set is the upper envelope of its level sets.

Characterizing a fuzzy set through its level sets will allow us in Sect. 15.6 to work levelwise with operations on fuzzy sets on the basis of usual sets.

The connection between a fuzzy set and its level sets is frequently used for the internal representation of fuzzy sets in computers. But only the α-cuts for a finite amount of selected values α, for example, 0.25, 0.5, 0.75, 1, are used and the corresponding level sets of the fuzzy set are saved. In order to determine the degree of membership of an element x to the fuzzy set μ, Eq. (15.1) can be used, where the supremum is only taken over a finite number of values for α. Thus, we discretize the degrees of membership and obtain an approximation of the original fuzzy set. Figure 15.10 shows the level sets $[\mu]_{0.25}$, $[\mu]_{0.5}$, $[\mu]_{0.75}$ and $[\mu]_1$ of the fuzzy set μ defined in Fig. 15.9. If we only use these four level sets in order to represent μ, we obtain the fuzzy set

$$\tilde{\mu}(x) = \max\{\alpha \in \{0.25, 0.5, 0.75, 1\} \mid x \in [\mu]_\alpha\}$$

in Fig. 15.11 as an approximation for μ.

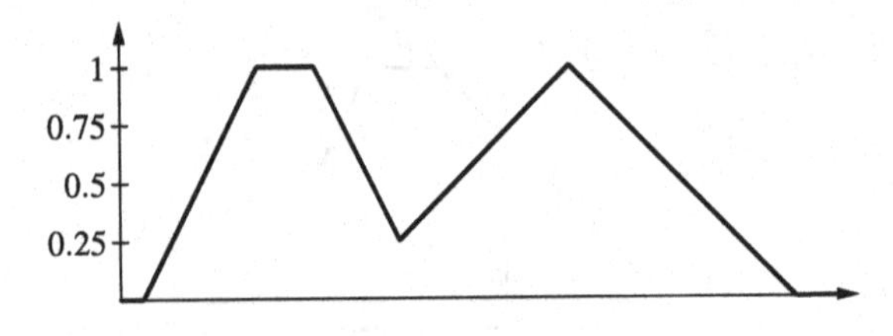

Fig. 15.9 The fuzzy set μ

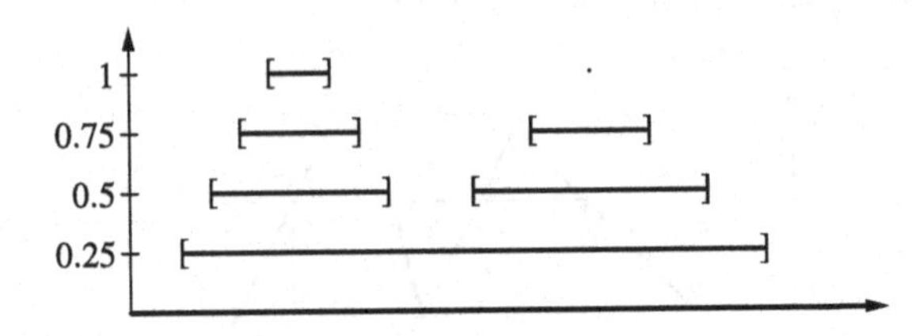

Fig. 15.10 The α-level sets of the fuzzy set μ for $\alpha = 0.25, 0.5, 0.75, 1$

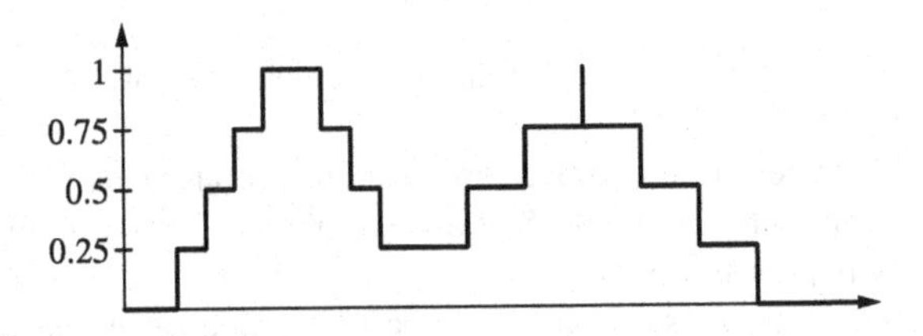

Fig. 15.11 The approximation of the fuzzy set μ resulting from the α-level sets

Confining us to a finite number of level sets in order to represent or save a fuzzy set corresponds to a discretization of the membership degrees. Besides this vertical discretization, we can also discretize the domain (horizontal discretization). Depending on the considered problem, we have to choose how fine the discretization should be chosen in both directions. Therefore, no general rules for discretization can be specified. In general, a refined discretization of the membership degrees seldom leads to significant improvements of a fuzzy system. One reason for this is that the fuzzy sets are usually determined heuristically or can only be specified roughly. Another reason is that human experts tend to use a limited amount of differentiation levels or degrees of acceptance or membership in order to judge a situation.

15.5 Fuzzy Logic

The notion *fuzzy logic* has three different meanings. In most cases, the term fuzzy logic refers to fuzzy logic in the broader sense, including all applications and theories where fuzzy sets or concepts are involved.

Table 15.1 The truth value table for conjunction, disjunction, implication and negation

$[\![\varphi]\!]$	$[\![\psi]\!]$	$[\![\varphi \wedge \psi]\!]$
1	1	1
1	0	0
0	1	0
0	0	0

$[\![\varphi]\!]$	$[\![\psi]\!]$	$[\![\varphi \vee \psi]\!]$
1	1	1
1	0	1
0	1	1
0	0	0

$[\![\varphi]\!]$	$[\![\psi]\!]$	$[\![\varphi \rightarrow \psi]\!]$
1	1	1
1	0	0
0	1	1
0	0	1

$[\![\varphi]\!]$	$[\![\neg\varphi]\!]$
1	0
0	1

On the contrary, the second (and narrower) meaning of the term fuzzy logic focuses on the field of approximative reasoning where fuzzy sets are used and propagated within an inference mechanism as it is for instance common in expert systems.

Finally, fuzzy logic in the narrow sense, which is the topic of this section, considers fuzzy systems from the point of view of multi-valued logic and is devoted to issues connected to logical calculi and the associated deduction mechanisms.

We cannot provide a complete introduction to fuzzy logic as a multi-valued logic. A detailed study of this aspect is found in Gottwald (2003). In this section, we will introduce those notions of fuzzy logic which are necessary or useful to understand fuzzy controllers. In Sect. 19.3 about logic-based fuzzy controllers on page 364, we discuss some further aspects of fuzzy logic in the narrow sense. We mainly need the concepts of (fuzzy) logic to introduce the set theoretical operations for fuzzy sets. The basis for operations like union, intersection and complement are the logical connectives disjunction, conjunction and negation, respectively. Therefore, we briefly repeat some fundamental concepts from classical logic in order to generalize them to the field of fuzzy logic.

15.5.1 Propositions and Truth Values

Classical propositional logic deals with the formal handling of statements (propositions) to which one of the two truth values 1 (for *true*) or 0 (for *false*) can be assigned. We represent these propositions by Greek letters φ, ψ etc. Typical propositions, for which the formal symbols φ_1 and φ_2 may stand are

$$\varphi_1 : \text{Four is an even number.}$$
$$\varphi_2 : 2+5=9.$$

The truth value which is assigned to a proposition φ is denoted by $[\![\varphi]\!]$. For the above propositions we obtain $[\![\varphi_1]\!] = 1$ and $[\![\varphi_2]\!] = 0$. If the truth values of single propositions are known, we can determine the truth values of combined propositions using truth tables that define the interpretation of the corresponding logical connectives. The most important logical connectives are the logical AND $\wedge$ (conjunction), the logic OR $\vee$ (disjunction), the negation NOT $\neg$ and the IMPLICATION $\rightarrow$.

The conjunction $\varphi \wedge \psi$ of two propositions φ and ψ is true, if and only if both φ and ψ are true. The disjunction $\varphi \vee \psi$ of φ and ψ obtains the truth value 1 (true), if and only if at least one of the two propositions is true. The implication $\varphi \rightarrow \psi$ is only false, if the antecedent φ is true and the consequent ψ is false. The negation $\neg\varphi$ of the proposition φ is false, if and only if φ is true. These definitions are shown in the truth value tables for conjunction, disjunction, implication and negation in Table 15.1.

This definition implies that the propositions

$$\text{Four is an even number AND } 2+5=9.$$

and

$$\text{Four is an even number IMPLICATION } 2+5=9.$$

are false, whereas the propositions

$$\text{Four is an even number OR } 2+5=9.$$

and

$$\text{NOT } 2+5=9.$$

are true. Formally expressed, this means that we have $[\![\varphi_1 \wedge \varphi_2]\!] = 0$, $[\![\varphi_1 \rightarrow \varphi_2]\!] = 0$, $[\![\varphi_1 \vee \varphi_2]\!] = 1$ and $[\![\neg\varphi_2]\!] = 1$.

The assumption that a statement is either true or false is suitable for mathematical issues. But for many expressions formulated in natural language such a strict separation between true and false statements would be unrealistic and would lead to counterintuitive consequences. If somebody promises to come to an appointment at 5 o'clock, his statement would have been false, if he came one minute later. Nobody would call him a liar, although, strictly speaking, his statement was not true. Even more complicated is the statement of being at a party at about 5. The greater the difference between the arrival and 5 o'clock the "less true" the statement is. A sharp definition of an interval of time corresponding to "about 5" is impossible.

Humans are able to formulate such "fuzzy" statements, understand them, draw conclusions from them and work with them. If someone starts an approximately four-hour-drive at around 11 o'clock and is going to have lunch for about half an hour, we can use these imprecise pieces of information and conclude at what time more or less the person will arrive. A formalization of this simple issue in a logical calculus, where statements can be either true or false only, is not adequate.

In natural language using fuzzy statements or information is not an exception but normal. In a recipe, nobody would replace the statement "Take a pinch of salt" by "Take 80 grain of salt". A driver will not calculate the distance he will need for stopping his car abruptly on a wet road by using another friction constant in some mathematical formula to calculate this distance. He will consider the rule: the wetter the road, the longer the distance needed for breaking.

In order to model this human information processing in a more appropriate way, we use gradual truth values for statements. This means a statement can not only be true (truth value 1) or false (truth value 0) but also more or less true expressed by a value between 0 and 1.

The connection between fuzzy sets and imprecise or fuzzy statements can be described in the following way. A fuzzy set models a property that elements of the universe of discourse can have more or less. For example, let us consider the fuzzy set μ_{hv} of high velocities from Fig. 15.2 on page 298. The fuzzy set represents the property or the predicate *high velocity*. That means the degree of membership of a specific velocity v to the fuzzy set of high velocities corresponds to the "truth value" which is assigned to the statement "v is a high velocity". In this sense, a fuzzy set determines the corresponding truth values for a set of statements—in our example for all statements we obtain, when we consider in a concrete velocity value for v. In order to understand how to operate with fuzzy sets, it is first of all useful to consider classical crisp propositions.

Dealing with combined propositions like "160 km/h is a high velocity AND the stopping distance is about 110 m" requires the extension of the truth tables for logical connectives like conjunction, disjunction, implication or negation. The truth tables shown in Table 15.1 determine a truth function for each logic connective. For conjunction, disjunction and implication this truth function assigns to each combination of two truth values (the truth value assigned to φ and ψ) one truth value (the truth value of the conjunction, disjunction of φ and ψ or the implication $\varphi \to \psi$). The truth function assigned to the negation has only one truth value as argument. If we denote the truth function by w_* associated with the logical connective $* \in \{\wedge, \vee, \to, \neg\}$, w_* is a binary or unary function. This means

$$w_\wedge, w_\vee, w_\to : \{0, 1\}^2 \to \{0, 1\}, \quad w_\neg : \{0, 1\} \to \{0, 1\}.$$

For fuzzy propositions, where the unit interval $[0, 1]$ replaces the binary set $\{0, 1\}$ as set of possible truth values, we have to assign truth functions to the logic connectives accordingly. These truth functions have to be defined on the unit square or the unit interval.

$$w_\wedge, w_\vee, w_\to : [0, 1]^2 \to [0, 1], \quad w_\to : [0, 1] \to [0, 1].$$

A minimum requirement we demand of these functions is that, limited to the values 0 and 1, they should provide the same values as the corresponding truth function associated with the connectives of classical logic. This requirement says that a combination of fuzzy propositions which are actually crisp (non-fuzzy), because their truth values are 0 or 1 were, coincide with the usual combination of classical crisp propositions.

The most frequently used truth functions for conjunction and disjunction in fuzzy logic are the minimum or maximum. That means $w_\wedge(\alpha, \beta) = \min\{\alpha, \beta\}$, $w_\vee(\alpha, \beta) = \max\{\alpha, \beta\}$. Normally the negation is defined by $w_\neg(\alpha) = 1 - \alpha$. In his seminal work (Zadeh 1965), Lotfi Zadeh introduced the concept of fuzzy sets and used these functions for operating with fuzzy sets.

The implication is often understood in the sense of the **Łukasiewicz implication**

$$w_\to(\alpha, \beta) = \min\{1 - \alpha + \beta, 1\}$$

or the **Gödel implication**

$$w_{\rightarrow}(\alpha,\beta)=\begin{cases}1 & \text{if } \alpha\le\beta,\\ \beta & \text{otherwise.}\end{cases}$$

15.5.2 t-Norms and t-Conorms

Until now we have interpreted the truth values from the unit interval in a purely intuitive way as gradual truths. So, choosing the truth functions for the logical connectives in the above mentioned way seems to be plausible but it is not unique at all. Instead of trying to find more or less arbitrary functions, we might better use an axiomatic approach where we define some reasonable properties a truth function should satisfy and thus confining the possible truth functions. We discuss this axiomatic approach in detail for the conjunction.

Let us consider the function $t:[0,1]^2\rightarrow[0,1]$ as a potential candidate for the truth function of a conjunction for fuzzy propositions. The truth value of a conjunction of several propositions should not depend on the order in which the propositions are considered. In order to guarantee this property t has to be commutative and associative, that means

$$\text{(T1)}\quad t(\alpha,\beta)=t(\beta,\alpha),$$
$$\text{(T2)}\quad t\big(t(\alpha,\beta),\gamma\big)=t\big(\alpha,t(\beta,\gamma)\big)$$

should hold.

The truth value of the conjunction $\varphi\wedge\psi$ should not be less than the truth value of the conjunction $\varphi\wedge\chi$, if χ has a lower truth value than ψ. Therefore, we require some monotonicity condition of t:

$$\text{(T3)}\quad \beta\le\gamma\rightarrow t(\alpha,\beta)\le t(\alpha,\gamma).$$

Because of the commutativity (T1), (T3) implies that t is non-decreasing in both arguments.

Furthermore, we require that the truth value of a proposition φ will be the same as the truth value of the conjunction of φ with any true proposition ψ. For the truth function t this leads to

$$\text{(T4)}\quad t(\alpha,1)=\alpha.$$

Definition 15.3 A function $t:[0,1]^2\rightarrow[0,1]$ is called a *t-norm* (*triangular norm*), if the axioms (T1)–(T4) are satisfied.

In the framework of fuzzy logic, we should always choose a t-norm as the truth function for conjunction. From the property (T4) follows that for every t-norm t we have $t(1,1)=1$ and $t(0,1)=0$. From $t(0,1)=0$ we obtain $t(1,0)=0$ using the commutativity property (T1). Furthermore, because of the monotonic property (T3) and $t(0,1)=0$ we must have $t(0,0)=0$. In this way, every t-norm restricted to the

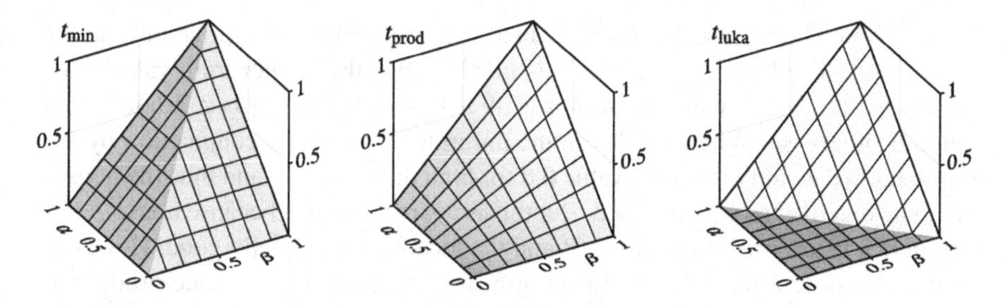

Fig. 15.12 The t-norms t_{min} (minimum), t_{prod} (algebraic product) and t_{luka} (Łukasiewicz)

values 0 and 1 coincides with the truth function given by the truth table of the usual conjunction.

We can verify easily that the discussed truth function $t(\alpha, \beta) = \min\{\alpha, \beta\}$ for the conjunction is a t-norm. Other examples of t-norms are

$$\text{Łukasiewicz t-norm:} \quad t(\alpha, \beta) = \max\{\alpha + \beta - 1, 0\}$$

$$\text{algebraic product:} \quad t(\alpha, \beta) = \alpha \cdot \beta$$

$$\text{drastic product:} \quad t(\alpha, \beta) = \begin{cases} 0 & \text{if } 1 \notin \{\alpha, \beta\}, \\ \min\{\alpha, \beta\} & \text{otherwise.} \end{cases}$$

The minimum, the algebraic product and the Łukasiewicz t-norm are illustrated as three-dimensional function graphs in Fig. 15.12.

These few examples show that the spectrum of the t-norms is very broad. The limits are given by the drastic product, which is the smallest t-norm and which is discontinuous, and the minimum, which is the greatest t-norm. Besides this, the minimum can be considered as a special t-norms, since it is the only idempotent t-norm which means that only the minimum satisfies the property $t(\alpha, \alpha) = \alpha$ for all $\alpha \in [0, 1]$.

Only the idempotence of a t-norm can guarantee that the truth values of the proposition φ and $\varphi \wedge \varphi$ coincide, which at first sight seems to be a canonical requirement, letting the minimum seem to be the only reasonable choice for the truth functions for the conjunction in the context of fuzzy logic. However, the following example shows that the idempotency property is not always desirable.

Example 15.2 A buyer has to decide between the houses A and B. The houses are very similar in most aspects. So, he makes his decision considering the criteria good price and good location. After careful consideration he assigns the following "truth values" to the decisive aspects:

	statement	truth value $[\![\varphi_i]\!]$
φ_1	The price of house A is good.	0.9
φ_2	The location of house A is good.	0.6
φ_3	The price of house B is good.	0.6
φ_4	The location of house B is good.	0.6

He chooses house $x \in \{A, B\}$ for which the proposition "The price of house x is good AND The location of house x is good" yields the greater truth value. This means that the buyer will choose house A if $[\![\varphi_1 \wedge \varphi_2]\!] > [\![\varphi_3 \wedge \varphi_4]\!]$ holds, and house B otherwise. When we determine the truth value of the conjunction by the minimum, we would obtain the value 0.6 for both of the houses and thus the houses would be regarded as equally good. But this is counterintuitive because house A has definitely a better price than house B and the locations are equally good. However, when we choose a non-idempotent t-norm, for example, the algebraic product or the Łukasiewicz t-norm, as truth function for the conjunction, we will always favor house A.

Besides the discussed examples for the t-norms there are many others. In particular, there are whole families of t-norms which can be defined in a parametric way. For example, the Weber family

$$t_\lambda(\alpha, \beta) = \max\left\{\frac{\alpha + \beta - 1 + \lambda\alpha\beta}{1 + \lambda}, 0\right\}$$

which determines a t-norm for each $\lambda \in (-1, \infty)$. For $\lambda = 0$ it results in the Łukasiewicz t-norm.

In most practical applications only the minimum, the algebraic product and the Łukasiewicz t-norm are chosen. Therefore, we will not consider the enormous variety of other t-norms here. For further readings on t-norms, we refer the gentle reader to Butnariu and Klement (1993), Kruse et al. (1994).

In the same way as we have defined t-norms as possible truth functions for the conjunction, we can define candidates for truth functions for the disjunction. Just like the t-norms they should satisfy the properties (T1)–(T3). Instead of (T4) we ask for

$$\text{(T4}') \quad s(\alpha, 0) = \alpha,$$

which means that the truth value of a proposition φ will be the same as the truth value of the disjunction of φ with any false proposition ψ.

Definition 15.4 A function $s : [0, 1]^2 \to [0, 1]$ is called *t-conorm* (*triangular conorm*) if the axioms (T1)–(T3) and (T4′) are satisfied.

t-norms and t-conorms are dual concepts in the following sense. Each t-norm induces a t-conorm s by

$$s(\alpha, \beta) = 1 - t(1 - \alpha, 1 - \beta), \tag{15.2}$$

and vice versa, from a t-conorm s we obtain the corresponding t-norm by

$$t(\alpha, \beta) = 1 - s(1 - \alpha, 1 - \beta). \tag{15.3}$$

Equations (15.2) and (15.3) correspond to De Morgan's Laws

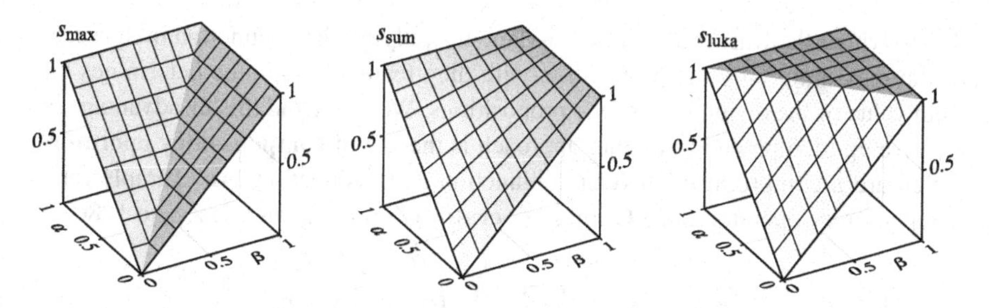

Fig. 15.13 The t-conorms s_{max} (maximum), s_{sum} (algebraic sum) and s_{luka} (Łukasiewicz)

$$[\![\varphi \vee \psi]\!] = [\![\neg(\neg\varphi \wedge \neg\psi)]\!] \quad \text{and} \quad [\![\varphi \wedge \psi]\!] = [\![\neg(\neg\varphi \vee \neg\psi)]\!]$$

if we compute the negation using the truth function $[\![\neg\varphi]\!] = 1 - [\![\varphi]\!]$.

The t-conorms we obtain from the t-norms minimum, Łukasiewicz t-norm, algebraic and drastic product by applying Eq. (15.2) are

maximum: $s(\alpha, \beta) = \max\{\alpha, \beta\}$

Łukasiewicz t-conorm: $s(\alpha, \beta) = \min\{\alpha + \beta, 1\}$

algebraic sum: $s(\alpha, \beta) = \alpha + \beta - \alpha\beta$

drastic sum: $s(\alpha, \beta) = \begin{cases} 1 & \text{if } 0 \notin \{\alpha, \beta\}, \\ \max\{\alpha, \beta\} & \text{otherwise.} \end{cases}$

The minimum, the algebraic sum and the Łukasiewicz t-conorm are illustrated as three-dimensional function graphs in Fig. 15.13.

The duality between t-norms and t-conorms implies immediately that the drastic sum is the greatest, the maximum is the smallest t-conorm, and the maximum is the only idempotent t-conorm. Analogously to t-norms we can define parametric families of t-conorms. Such as

$$s_\lambda(\alpha, \beta) = \min\left\{\alpha + \beta - \frac{\lambda\alpha\beta}{1+\lambda}, 1\right\}$$

which is the Weber family of the t-conorms.

Operating with t-norms and t-conorms we should be aware that not all laws we know for classical conjunction and disjunction hold also for t-norms and t-conorms. For instance, minimum and maximum are not merely the only idempotent t-norms or t-conorms, but also the only pair defined by duality (cf. Eq. (15.2)) which satisfies the distribution laws.

In the example of the man who wanted to buy a house we could see that the idempotency of a t-norm is not always desirable. The same holds for t-conorms. Let us consider the propositions $\varphi_1, \ldots, \varphi_n$ which shall be connected in a conjunctive or disjunctive manner. The significant disadvantage of the idempotency is the following. Applying the conjunction in terms of the minimum, the resulting truth value of the connection of propositions depends only on the truth value of the proposition to which the least truth value is assigned. Applying the disjunction in the sense of the maximum only the proposition with the greatest truth value determines the truth value of the disjunction of the propositions. We can avoid this disadvantage, if we give up idempotency. Another approach is the use of **compensatory operators** which are a compromise between conjunction and disjunction. An example for a compensatory operator is the **Gamma operator** (Zimmermann and Zysno 1980)

$$\Gamma_\gamma(\alpha_1, \dots, \alpha_n) = \left(\prod_{i=1}^{n} \alpha_i\right) \cdot \left(1 - \prod_{i=1}^{n}(1 - \alpha_i)\right)^\gamma$$

with parameter $\gamma \in [0, 1]$. For $\gamma = 0$ this results in the algebraic product, for $\gamma = 1$ we obtain the algebraic sum. Another compensatory operator is the arithmetical mean. Other suggestions for such operators can be found in Mayer et al. (1993). A big disadvantage of these operator is the fact that they do not satisfy associativity. So we do not use them here.

In addition to the connection between t-norms and t-conorms, we can also find connections between t-norms and implications. A continuous t-norm t induces the *residuated implication* $\vec{t}$ by the formula

$$\vec{t}(\alpha, \beta) = \sup\{\gamma \in [0, 1] \mid t(\alpha, \gamma) \le \beta\}.$$

Thus, we obtain by residuation the Łukasiewicz implication from the Łukasiewicz t-norm and the Gödel implication from the minimum.

Later we will need the corresponding **biimplication** $\overleftrightarrow{t}$ which is defined by the formula

$$\begin{aligned} \overleftrightarrow{t}(\alpha, \beta) &= \vec{t}\big(\max\{\alpha, \beta\}, \min\{\alpha, \beta\}\big) \\ &= t\big(\vec{t}(\alpha, \beta), \vec{t}(\beta, \alpha)\big) \\ &= \min\big\{\vec{t}(\alpha, \beta), \vec{t}(\beta, \alpha)\big\}. \end{aligned} \tag{15.4}$$

This formula is motivated by the definition of the biimplication or equivalence in classical logic in terms of

$$[\![\varphi \leftrightarrow \psi]\!] = [\![(\varphi \to \psi) \wedge (\psi \to \varphi)]\!].$$

Besides the logical operators like conjunction, disjunction, implication or negation in (fuzzy) logic, there also exist the quantifiers $\forall$ (all) and $\exists$ (exists).

The universal quantifier $\forall$ and the existential quantifier $\exists$ are closely related to the conjunction and the disjunction, respectively. Let us consider the universe of discourse X and the predicate $P(x)$. For instance, X could be the set $\{2, 4, 6, 8, 10\}$ and $P(x)$ the predicate "x is an even number". If the set X is finite, e.g. $X = \{x_1, \dots, x_n\}$, the statement $(\forall x \in X)(P(x))$ is equivalent to the statement $P(x_1) \wedge \dots \wedge P(x_n)$. Therefore, in this case it is possible to define the truth value of the statement $(\forall x \in X)(P(x))$ on the basis of the conjunction which means

$$[\![\forall x \in X : P(x)]\!] = [\![P(x_1) \wedge \cdots \wedge P(x_n)]\!].$$

If we assign the minimum to the conjunction as truth value function we obtain

$$[\![\forall x \in X : P(x)]\!] = \min\{[\![P(x)]\!] \mid x \in X\}$$

which can be extended to an infinite universe of discourse X by

$$[\![\forall x \in X : P(x)]\!] = \inf\{[\![P(x)]\!] \mid x \in X\}.$$

Other t-norms than the minimum are normally not used for the universal quantifier, since the non-idempotent property leads easily to the truth value zero in the case of an infinite universe of discourse.

The same consideration about the existential quantifier leads to its definition

$$[\![\exists x \in X : P(x)]\!] = \sup\{[\![P(x)]\!] \mid x \in X\}.$$

If the universe of discourse for the existential quantifier is finite, the propositions $\exists x \in X : P(x)$ and $P(x_1) \vee \cdots \vee P(x_n)$ are equivalent.

Example 15.3 As an example, we consider the predicate $P(x)$ with the interpretation "x is a high velocity". Let the truth value $[\![P(x)]\!]$ be given by the fuzzy set of the high velocities from Fig. 15.2 on page 298 which means $[\![P(x)]\!] = \mu_{\text{hv}}(x)$. So, we have for instance $[\![P(150)]\!] = 0$, $[\![P(170)]\!] = 0.5$ and $[\![P(190)]\!] = 1$. Thus, the statement $\forall x \in [170, 200] : P(x)$ ("All velocities between 170 km/h and 200 km/h are high velocities") has the truth value

$$\begin{aligned} [\![\forall x \in [170, 200] : P(x)]\!] &= \inf\{[\![P(x)]\!] \mid x \in [170, 200]\} \\ &= \inf\{\mu_{\text{hv}}(x) \mid x \in [170, 200]\} \\ &= 0.5. \end{aligned}$$

Analogously, we obtain $[\![\exists x \in [100, 180] : P(x)]\!] = 0.75$.

15.5.3 Basic Assumptions and Problems

In this section, about fuzzy logic we have discussed various ways of combining fuzzy propositions. An essential assumption we have used in the section is that of **truth functionality**. This means that the truth value of the combination of several propositions depends only on the truth values of the propositions, but not on the individual propositions. This assumption holds in classical logic but not, for example, in the context of probability theory or probabilistic logic. In probability theory it is not enough to know the probability of two events in order to determine the probability that both events will occur simultaneously or at least one of them will occur. For this we also need information about the dependency of these events. In the case of independence, the probability that both events occur is the product of the single probabilities, and the probability that at least one of the events will occur is the sum of the probabilities if the two events exclude each other. We cannot determine these probabilities without knowing the independence of the events.

We should be aware of the assumption of truth functionality in the framework of fuzzy logic. It is not always satisfied. Coming back to the example of the man buying a house, we gave reasons for using non-idempotent t-norms. If we use these t-norms, like for instance the algebraic product, for propositions like "The price of the house A is good AND ... AND the price of house A is good", this combined proposition can obtain a very small truth value. Depending on how we interpret the conjunction, this effect might be desirable or might lead to inconsistency. If we understand the conjunction in its classical sense, a conjunctive combination of a proposition with itself should be equivalent to itself which is not satisfied for non-idempotent t-norms. Another possibility is to understand the conjunction as a list of pro and con arguments for a thesis or as a proof. The repeated use of the same (fuzzy) argument within a proof might result in a loss of credibility and thus idempotency is not desirable, even for a conjunction of a proposition with itself.

Fortunately, for fuzzy control these consideration are of minor importance, because in this application area fuzzy logic is used in a more restricted context, where we do not have to worry about combining the same proposition with itself. More difficulties will show up, when we apply fuzzy logic in the framework of complex expert systems.

15.6 Operations on Fuzzy Sets

Sections 15.3 and 15.4 described how vague concepts can be modeled using fuzzy sets and how fuzzy sets can be represented. In order to operate with vague concepts or apply some kind of deduction mechanism to them, we need suitable operations for fuzzy sets. Therefore, in this section operations like union, intersection or complement well known from classical set theory will be extended to fuzzy sets.

15.6.1 Intersection

The underlying concept of generalizing fundamental set-theoretic operations to fuzzy sets is explained in detail for the intersection of (fuzzy) sets. For the other operations, a generalization can be carried out in a straight forward manner analogously to intersection. For two ordinary sets M_1 and M_2, we have that an element x belongs to the intersection of the two sets, if and only if it belongs to both M_1 and M_2. Whether x belongs to the intersection depends only on the membership of x to M_1 and M_2 but not on the membership of any other element $y \neq x$ to M_1 and M_2. Formally speaking, this means

$$x \in M_1 \cap M_2 \quad \Longleftrightarrow \quad x \in M_1 \wedge x \in M_2. \tag{15.5}$$

For two fuzzy sets μ_1 and μ_2 we also assume that the degree of membership of an element x to the intersection of the two fuzzy sets depends only on the membership degrees of x to μ_1 and μ_2. We interpret the degree of membership $\mu(x)$ of an element x to the fuzzy set μ as truth value $[\![x \in \mu]\!]$ of the fuzzy proposition "$x \in \mu$", that x is an element of μ. In order to determine the membership degree of an element x to the intersection of the fuzzy sets μ_1 and μ_2, we have to calculate the truth value of the conjunction "x is an element of μ_1 AND x is an element of μ_2" following the equivalence in Eq. (15.5). The previously discussed concepts of fuzzy logic have told us, how we can define the truth value of the conjunction of two fuzzy propositions. Therefore, it is necessary to choose a suitable t-norm t as the truth function for the conjunction. Thus, we define the intersection of two fuzzy sets μ_1 and μ_2 (w.r.t. the t-norm t) as the fuzzy set $\mu_1 \cap_t \mu_2$ with

$$(\mu_1 \cap_t \mu_2)(x) = t\big(\mu_1(x), \mu_2(x)\big).$$

If we interpret the degree of membership $\mu(x)$ of an element x to the fuzzy set μ as truth value $[\![x \in \mu]\!]$ of the fuzzy proposition "$x \in \mu$", that x is an element of μ, the definition of the intersection of two fuzzy sets can be written in the following way

$$\big[\!\big[x \in (\mu_1 \cap_t \mu_2)\big]\!\big] = [\![x \in \mu_1 \wedge x \in \mu_2]\!],$$

where we assign the t-norm t as truth function for the conjunction.

By defining the intersection of fuzzy sets on the basis of a t-norm, the properties of the t-norm are inherited to the intersection operator: the axioms (T1) and (T2) make the intersecting of fuzzy sets commutative and associative, respectively. The monotonicity property (T3) guarantees that replacing a fuzzy set μ_1 by a larger fuzzy set μ_2, which means $\mu_1(x) \leq \mu_2(x)$ for all x, can only lead to a larger intersection:

$$\mu_1 \leq \mu_2 \quad \text{implies} \quad \mu \cap_t \mu_1 \leq \mu \cap_t \mu_2.$$

Axiom (T4) guarantees that the intersection of a fuzzy set with an ordinary set, respectively its characteristic function, results in the original fuzzy set limited to the ordinary set with which we intersected it. If $M \subseteq X$ is an ordinary subset of X and $\mu \in \mathcal{F}(X)$ a fuzzy set of X, we have

$$(\mu \cap_t I_M)(x) = \begin{cases} \mu(x) & \text{if } x \in M, \\ 0 & \text{otherwise.} \end{cases}$$

If not otherwise stated, the intersection of two fuzzy sets will be computed w.r.t. the minimum t-norm. In this case, or when it is clear to which t-norm we refer, we will write $\mu_1 \cap \mu_2$ instead of $\mu_1 \cap_t \mu_2$ for the case of $t = \min$.

Example 15.4 We consider the intersection of the fuzzy set μ_{hv} of high velocities from Fig. 15.2 on page 298 and the fuzzy set $\mu_{170-190}$ of the velocities not much less than 170 km/h and not much greater than 190 km/h from Fig. 15.14. Both of

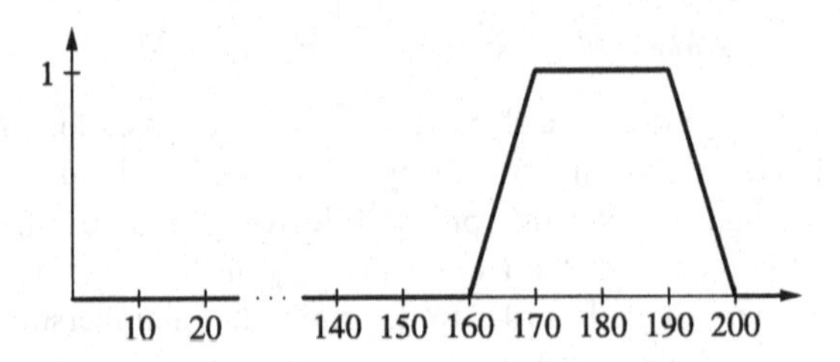

Fig. 15.14 The fuzzy set $\mu_{170-190}$ of the velocities not much less than 170 km/h and not much greater than 190 km/h

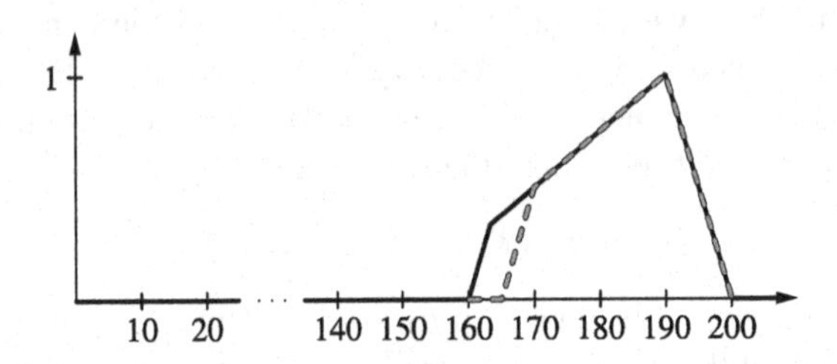

Fig. 15.15 Intersection $\mu_{\text{hv}} \cap_t \mu_{170-190}$ of the fuzzy sets μ_{hv} and $\mu_{170-190}$ calculated with the minimum (*solid black line*) and the Łukasiewicz t-norm (*dashed gray line*)

them are trapezoidal functions:

$$\mu_{\text{hv}} = \Pi_{150,180,\infty,\infty}, \qquad \mu_{170-190} = \Pi_{160,170,190,200}.$$

Figure 15.15 shows the intersection of the two fuzzy sets on the basis of the minimum (solid line) and the Łukasiewicz t-norm (dashed line).

15.6.2 Union

From the representation in Eq. (15.5) we have derived the definition for the intersection of two fuzzy sets. Analogously, we can define the union of two fuzzy sets on the basis of

$$x \in M_1 \cup M_2 \quad \Longleftrightarrow \quad x \in M_1 \vee x \in M_2.$$

This leads to

$$(\mu_1 \cup_s \mu_2)(x) = s\big(\mu_1(x), \mu_2(x)\big),$$

as the union of the two fuzzy sets μ_1 and μ_2 w.r.t. the t-conorm s. In the interpretation of the membership degree $\mu(x)$ of an element x to the fuzzy set μ as truth value $[\![x \in \mu]\!]$ of the fuzzy proposition "$x \in \mu$" we can define the union in the form of

$$\big[\!\!\big[x \in (\mu_1 \cup_t \mu_2)\big]\!\!\big] = [\![x \in \mu_1 \vee x \in \mu_2]\!],$$

where we assign the t-conorm s as the truth function for the disjunction. As in the case of the intersection, we will write $\mu_1 \cup \mu_2$ instead of $\mu_1 \cup_s \mu_2$ when we use the maximum t-conorm or when it is clear which t-conorm we refer to.

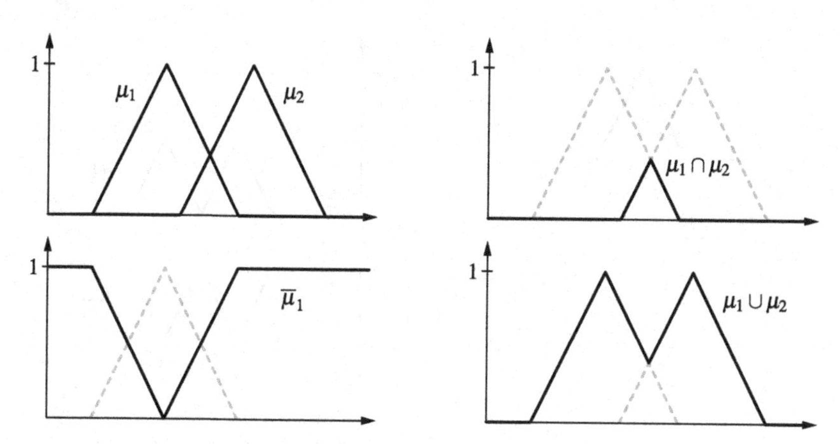

Fig. 15.16 Intersection, union and complement for fuzzy sets

15.6.3 Complement

The complement of a fuzzy set is derived from the formula

$$x \in \overline{M} \quad \Longleftrightarrow \quad \neg(x \in M)$$

for ordinary sets where $\overline{M}$ stands for the complement of the (ordinary) set M. If we assign the truth function $w_{\neg}(\alpha) = 1 - \alpha$ to the negation, we obtain the fuzzy set

$$\overline{\mu_1}(x) = 1 - \mu(x),$$

as the complement $\overline{\mu}$ of the fuzzy set μ. This is also in accordance with

$$[\![x \in \overline{\mu}]\!] = [\![\neg(x \in \mu)]\!].$$

Figure 15.16 illustrates the intersection (top right), union (bottom right) and complement (bottom left) of two fuzzy sets (top left).

Like the complement for ordinary sets, the complementing of fuzzy sets is an involution, which means that $\overline{\overline{\mu}} = \mu$ holds. In classical set theory, we have that the intersection of a set with its complement yields the whole universe of discourse. In the context of fuzzy sets, these two laws are weakened to $(\mu \cap \overline{\mu})(x) \leq 0.5$ and $(\mu \cup \overline{\mu})(x) \geq 0.5$ for all x. Figure 15.17 illustrates this phenomenon.

If the intersection and the union are defined on the basis of minimum or maximum, we can use the representation of fuzzy sets by the level sets introduced in Sect. 15.4 in order to computer the resulting fuzzy set. We have

$$[\mu_1 \cap \mu_2]_\alpha = [\mu_1]_\alpha \cap [\mu_2]_\alpha \quad \text{and} \quad [\mu_1 \cup \mu_2]_\alpha = [\mu_1]_\alpha \cup [\mu_2]_\alpha$$

for all $\alpha \in [0, 1]$. According to these equations the level sets of the intersection and the union of two fuzzy sets are the intersection or the union of the level sets of the single fuzzy sets.

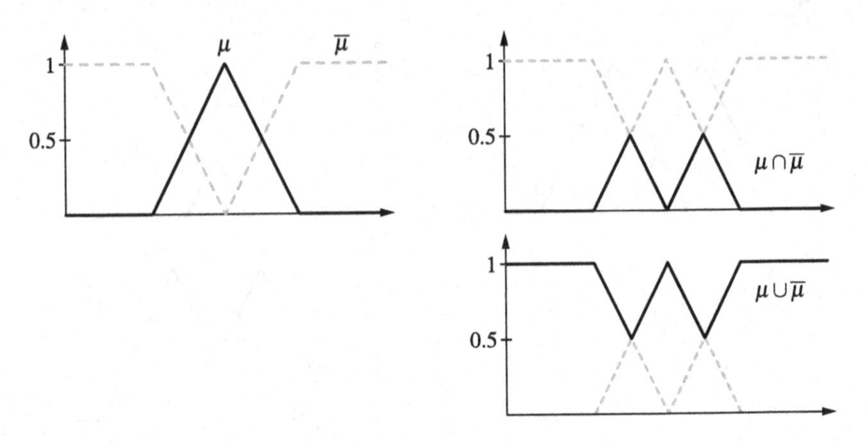

Fig. 15.17 Union and intersection of a fuzzy set with its complement

15.6.4 Linguistic Modifiers

Besides the complement as a unary operation on fuzzy sets, derived from the corresponding operation for ordinary sets, there are more fuzzy set specific unary operations, that have no counterpart in classical set theory. Normally, a fuzzy set represents a vague concept like "high velocity", "young" or "tall". From such concepts, we can derive other vague concepts applying **linguistic modifiers** ("linguistic hedges") like "very" or "more or less".

As an example, we consider the fuzzy set μ_{hv} of high velocities from Fig. 15.2 on page 298. How should the fuzzy set μ_{vhv} representing the concept of the "*very* high velocities" look like? A very high velocity is also a high velocity but not vice versa. Thus, the membership degree of a specific velocity v to the fuzzy set μ_{vhv} should not exceed its membership degree to the fuzzy set μ_{hv}. We can achieve this by understanding the linguistic modifier "very", similar to the negation, as a unary operator and assigning a suitable truth function to it, for instance $w_{\mathrm{very}}(\alpha) = \alpha^2$. In this way we obtain $\mu_{\mathrm{vhv}}(x) = (\mu_{\mathrm{hv}}(x))^2$. Now, a velocity which is to a degree of 1 a high velocity is also a very high velocity. A velocity which is not a high velocity (membership degree of 0) is not a very high velocity either. If the membership degree of a velocity to μ_{hv} is between 0 and 1, it is also a very high velocity but with a lower membership degree.

In the same way, we can assign a truth function to the modifier "more or less". This truth function should increase the degree of membership, for instance $w_{\text{more or less}}(\alpha) = \sqrt{\alpha}$.

Figure 15.18 shows the fuzzy set μ_{hv} of high velocities and the resulting fuzzy sets μ_{vhv} of very high velocities and μ_{mhv} of more or less high velocities.

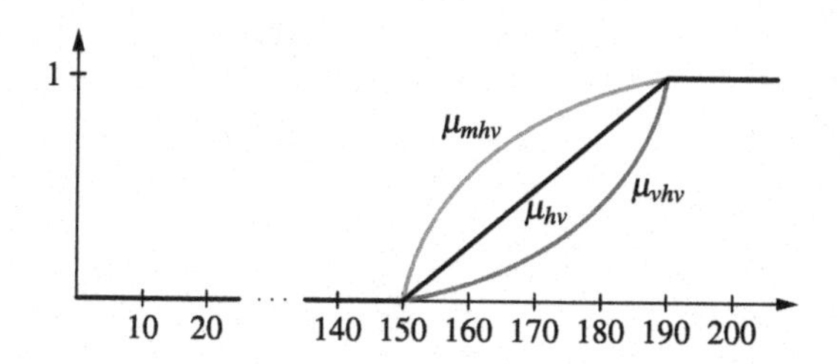

Fig. 15.18 The fuzzy sets μ_{hv}, μ_{vhv} and μ_{mhv} of high, very high and more or less high velocities

References

J.C. Bezdek. Fuzzy Models—What Are They, and Why? *IEEE Trans. on Fuzzy Systems* 1:1–5. IEEE Press, Piscataway, NJ, USA, 1993

D. Butnariu and E.P. Klement. *Triangular Norm-based Measures and Games with Fuzzy Coalitions*. Kluwer, Dordrecht, Netherlands, 1993

S. Gottwald (ed.) *A Treatise on Many-Valued Logic*. Research Studies Press, Baldock, UK, 2003

R. Kruse, J. Gebhardt, and F. Klawonn. *Foundations of Fuzzy Systems*. J. Wiley & Sons, Chichester, United Kingdom, 1994

A. Mayer, B. Mechler, A. Schlindwein and R. Wolke. *Fuzzy Logic*. Addison-Wesley, Bonn, Germany, 1993

L.A. Zadeh. Fuzzy Sets. *Information and Control* 8(3):338–353, 1965

H.-J. Zimmermann and P. Zysno. Latent Connectives in Human Decision Making and Expert Systems. *Fuzzy Sets and Systems* 4:37–51. Elsevier, Amsterdam, Netherlands, 1980

Chapter 16
The Extension Principle

In Sect. 15.6, we have discussed how set theoretic operations like intersection, union and complement can be generalized to fuzzy sets. This chapter is devoted to the issue of extending the concept of mappings or functions to fuzzy sets. These ideas allow us to define operations like addition, subtraction, multiplication, division or taking squares as well as set theoretic concepts like the composition of relations for fuzzy sets.

16.1 Mappings of Fuzzy Sets

As an example, let us consider the mapping $f : \mathbb{R} \to \mathbb{R}$ with $x \mapsto |x|$. The fuzzy set $\mu = \Lambda_{-1.5,-0.5,2.5}$ shown in Fig. 16.1 models the vague concept "about -0.5".

Which fuzzy set should represent "the absolute value of about -0.5" or, in other words, what is the image $f[\mu]$ of the fuzzy set μ? For a usual subset M of the universe of discourse X the image $f[M]$ under the mapping $f : X \to Y$ is defined as the subset of Y that contains all images of elements of M. Technically speaking, this means

$$f[M] = \{y \in Y \mid \exists x \in X : x \in M \wedge f(x) = y\},$$

or, in other words,

$$y \in f[M] \quad \Longleftrightarrow \quad (\exists x \in X)\big(x \in M \wedge f(x) = y\big). \tag{16.1}$$

For instance, for $M = [-1, 0.5] \subseteq \mathbb{R}$ and the mapping $f(x) = |x|$ we obtain the set $f[M] = [0, 1]$ as image of M under f.

Equation (16.1) allows us to define the image of a fuzzy set μ under a mapping f. As in Sect. 15.6 on the extension of the set-theoretic operations to fuzzy sets, we use the concepts of fuzzy logic again, which we introduced in Sect. 15.5. For fuzzy sets, Eq. (16.1) means

$$[\![y \in f[\mu]]\!] = [\![\exists x \in X : x \in \mu \wedge f(x) = y]\!].$$

R. Kruse et al., *Computational Intelligence*, Texts in Computer Science,
DOI 10.1007/978-1-4471-5013-8_16, © Springer-Verlag London 2013

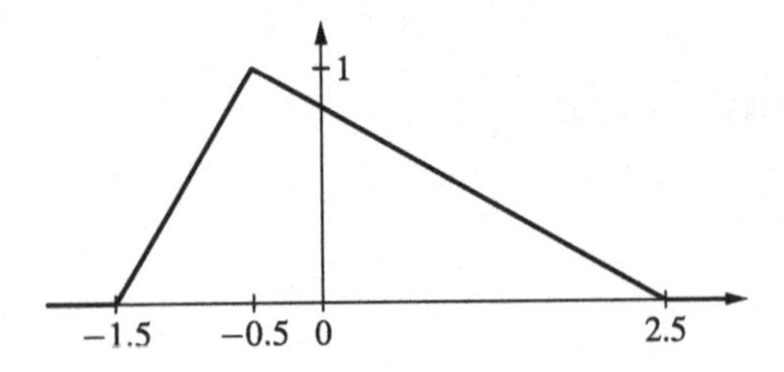

Fig. 16.1 The fuzzy set $\mu = \Lambda_{-1.5,-0.5,1.5}$ standing for "about −0.5"

As explained in Sect. 15.5, we evaluated the existential quantifier by the supremum and associate a t-norm t with the conjunction. Therefore, the fuzzy set

$$f[\mu](y) = \sup\{t(\mu(x), [\![f(x) = y]\!]) \mid x \in X\} \tag{16.2}$$

represents the image of μ under f. The choice of the t-norm t does not play any role in this case because the statement $f(x) = y$ is either true or false which means $[\![f(x) = y]\!] \in \{0, 1\}$, and therefore

$$t(\mu(x), [\![f(x) = y]\!]) = \begin{cases} \mu(x) & \text{if } f(x) = y, \\ 0 & \text{otherwise.} \end{cases}$$

Thus Eq. (16.2) is reduced to

$$f[\mu](y) = \sup\{\mu(x) \mid f(x) = y\}. \tag{16.3}$$

This definition says that the membership degree of an element $y \in Y$ to the image of the fuzzy set $\mu \in \mathcal{F}(X)$ under the mapping $f : X \to Y$ is the greatest possible membership degree of all x to μ that are mapped to y under f. This extension of a mapping to fuzzy sets is called **extension principle** (for a function with one argument).

From the example of the fuzzy set $\mu = \Lambda_{-1.5,-0.5,2.5}$ representing the vague concept "about −0.5", we obtain the fuzzy set shown in Fig. 16.2 as the image under the mapping $f(x) = |x|$. To illustrate the underlying principle, we determine the membership degree $f[\mu](y)$ for $y \in \{-0.5, 0, 0.5, 1\}$. Because of $f(x) = |x| \geq 0$ no value is mapped to $y = -0.5$ under f. So, we obtain $f[\mu](-0.5) = 0$. There is only one value that is mapped to $y = 0$, i.e. $x = 0$, hence we have $f[\mu](0) = \mu(0) = 5/6$. For $y = 0.5$, there exist two values ($x = -0.5$ and 0.5) mapped to y, which lead to

$$f[\mu](0.5) = \max\{\mu(-0.5), \mu(0.5)\} = \max\{1, 2/3\} = 1.$$

The values mapped to $y = 1$ are $x = -1$ and $x = 1$, respectively. Therefore, we obtain

$$\begin{aligned} f[\mu](1) &= \max\{\mu(-1), \mu(1)\} \\ &= \max\{0.5, 0.5\} = 0.5. \end{aligned}$$

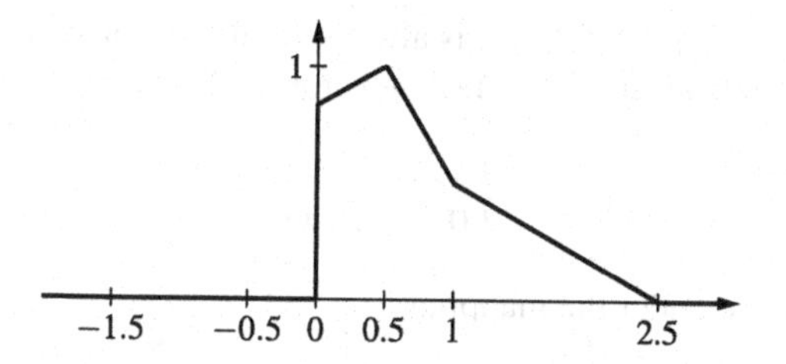

Fig. 16.2 The fuzzy set standing for the vague concept "the value of about −0.5"

Example 16.1 Let $X = X_1 \times \cdots \times X_n$, $i \in \{1, \dots, n\}$.

$$\pi_i : X_1 \times \cdots \times X_n \to X_i, \quad (x_1, \dots, x_n) \mapsto x_i$$

denotes the projection of the Cartesian product $X_1 \times \cdots \times X_n$ to the ith coordinate space X_i. According to the extension principle defined in Eq. (16.3), the projection of a fuzzy set $\mu \in \mathcal{F}(X)$ to the space X_i is given by

$$\pi_i[\mu](x) = \sup\{\mu(x_1, \dots, x_{i-1}, x, x_{i+1}, \dots, x_n) \mid$$
$$x_1 \in X_1, \dots, x_{i-1} \in X_{i-1}, x_{i+1} \in X_{i+1}, \dots, x_n \in X_n\}.$$

Figure 16.3 shows the projection of a fuzzy set which has nonzero membership degrees in two different regions.

16.2 Mapping of Level Sets

The membership degree of an element to the image of a fuzzy set can be computed on the basis of the membership degrees of all elements to the original one that are mapped to the considered element. Another way to characterize the image of a fuzzy set is to determine its level sets. Unfortunately, the level set of the image of a fuzzy set cannot be derived directly form the corresponding level set of the original fuzzy

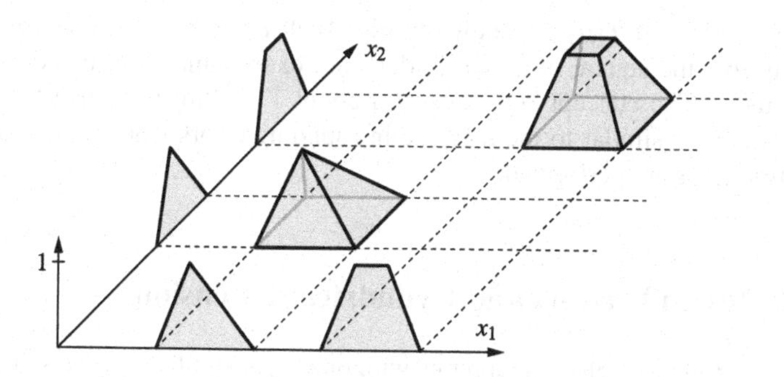

Fig. 16.3 Projection of a fuzzy set into the space X_2

set. The inclusion $[f[\mu]]_\alpha \supseteq f[[\mu]_\alpha]$ is always valid, but equality of these two sets is only satisfied in special cases. For instance, for the fuzzy set

$$\mu(x) = \begin{cases} x & \text{if } 0 \le x \le 1, \\ 0 & \text{otherwise} \end{cases}$$

we obtain as an image under the mapping

$$f(x) = I_{\{1\}}(x) = \begin{cases} 1 & \text{if } x = 1, \\ 0 & \text{otherwise} \end{cases}$$

the fuzzy set

$$f[\mu](y) = \begin{cases} 1 & \text{if } y \in \{0, 1\}, \\ 0 & \text{otherwise.} \end{cases}$$

Hence, we have $[f[\mu]]_1 = \{0, 1\}$ and $f[[\mu]_1] = \{1\}$ because of $[\mu]_1 = \{1\}$.

Provided that the universe of discourse $X = \mathbb{R}$ is the set of real numbers, the effect that the image of a level set is smaller than the corresponding level set of the image fuzzy set cannot happen, when the mapping f is continuous and for all $\alpha > 0$ the α-level sets of the original fuzzy set are compact. Therefore, in this case it is possible to characterize the image fuzzy set by the level sets.

Example 16.2 Let us consider the mapping $f : \mathbb{R} \to \mathbb{R}$, $x \mapsto x^2$. Obviously, the image of a fuzzy set $\mu \in \mathcal{F}(\mathbb{R})$ is given by

$$f[\mu](y) = \begin{cases} \max\{\mu(\sqrt{y}), \mu(-\sqrt{y})\} & \text{if } y \ge 0, \\ 0 & \text{otherwise.} \end{cases}$$

Let the fuzzy set $\mu = \Lambda_{0,1,2}$ represent the vague concept "about 1". The question, what "the square of about 1" is, can be answered by determining the level sets of the image fuzzy set $f[\mu]$ from the level sets of μ. Here, this is possible because the function f has compact level sets and the fuzzy set μ is continuous. Thus, we have $[\mu]_\alpha = [\alpha, 2 - \alpha]$ for all $0 < \alpha \le 1$ and we obtain

$$[f[\mu]]_\alpha = f[[\mu]_\alpha] = [\alpha^2, (2-\alpha)^2].$$

The fuzzy sets μ and $f[\mu]$ are shown in Fig. 16.4. Here, we can observe that the vague concept "the square of about 1" does not exactly match to the vague concept "about 1". The concept "the square of about 1" is "fuzzier" than "about 1". This effect is very similar to the increase in rounding errors when more and more computation steps are performed.

16.3 Cartesian Product and Cylindrical Extension

So far we have only extended mappings with one argument to fuzzy sets. To define operations like addition for fuzzy sets over real numbers, we need a concept how

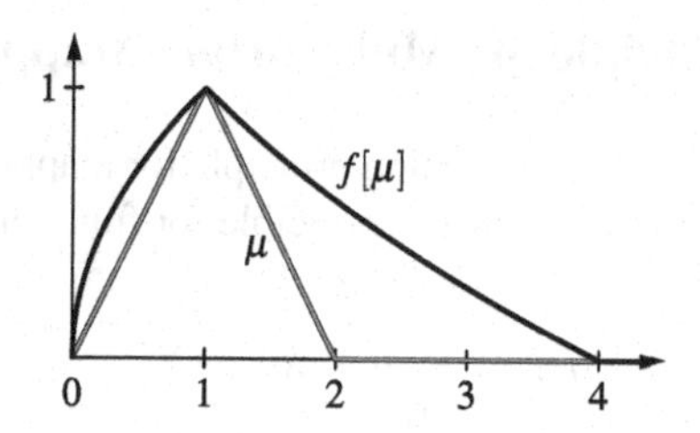

Fig. 16.4 The fuzzy sets μ and $f[\mu]$ that represent the vague concepts "about 1" and "square of about 1", respectively

to apply a mapping $f : X_1 \times \cdots \times X_n \to Y$ to a tuple $(\mu_1, \ldots, \mu_n) \in \mathcal{F}(X_1) \times \cdots \times \mathcal{F}(X_n)$ of fuzzy sets. Interpreting addition as a function in two arguments $f : \mathbb{R} \times \mathbb{R} \to \mathbb{R}$, $(x_1, x_2) \mapsto x_1 + x_2$, this concept will enable us to extend addition and other algebraic operations to fuzzy sets over real numbers.

In order to generalize the extension principle described in Eq. (16.3) to mappings with several arguments, we introduce the concept of the Cartesian product of fuzzy sets. Consider the fuzzy sets $\mu_i \in \mathcal{F}(X_i)$, $i = 1, \ldots, n$. The **Cartesian product** of the fuzzy sets $\mu_1, \ldots, \mu_n$ is the fuzzy set

$$\mu_1 \times \cdots \times \mu_n \in \mathcal{F}(X_1 \times \cdots \times X_n)$$

with

$$(\mu_1 \times \cdots \times \mu_n)(x_1, \ldots, x_n) = \min\{\mu_1(x_1), \ldots, \mu_n(x_n)\}.$$

This definition is motivated by the property

$$(x_1, \ldots, x_n) \in M_1 \times \cdots \times M_n \quad \iff \quad x_1 \in M_1 \wedge \cdots \wedge x_n \in M_n$$

of the Cartesian product of usual sets and corresponds to the formula

$$[\![(x_1, \ldots, x_n) \in \mu_1 \times \cdots \times \mu_n]\!] = [\![x_1 \in \mu_1 \wedge \cdots \wedge x_n \in \mu_n]\!],$$

where the minimum is chosen as truth function for the conjunction.

A special case of a Cartesian product is the **cylindrical extension** of a fuzzy set $\mu \in \mathcal{F}(X_i)$ to a product space $X_1 \times \cdots \times X_n$. The cylindrical extension is the Cartesian product of μ with the remaining universe of discourses X_j, $j \neq i$ or their characteristic functions:

$$\hat{\pi}_i(\mu) = I_{X_1} \times \cdots \times I_{X_{i-1}} \times \mu \times I_{X_{i+1}} \times \cdots \times I_{X_n},$$

$$\hat{\pi}_i(\mu)(x_1, \ldots, x_n) = \mu(x_i).$$

Obviously, projecting a cylindrical extension results in the original fuzzy set which means $\pi_i[\hat{\pi}_i(\mu)] = \mu$ provided that the sets $X_1, \ldots, X_n$ are nonempty. In general, $\pi_i[\mu_1 \times \cdots \times \mu_n] = \mu_i$ holds if the fuzzy sets μ_j, $j \neq i$ are **normal** which means $(\exists x_j \in X_j)(\mu_j(x_j)) = 1$.

16.4 Extension Principle for Multivariate Mappings

Using the Cartesian product, the extension principle for mappings with several arguments can be simplified to the extension principle for functions with one argument. Consider the mapping

$$f : X_1 \times \cdots \times X_n \to Y.$$

Then, the image of the tuple

$$(\mu_1, \ldots, \mu_n) \in \mathcal{F}(X_1) \times \cdots \times \mathcal{F}(X_n)$$

of fuzzy sets under f is the fuzzy set

$$f[\mu_1, \ldots, \mu_n] = f[\mu_1 \times \cdots \times \mu_n]$$

over the universe of discourse Y. That means

$$\begin{aligned} f[\mu_1, \ldots, \mu_n](y) \\ &= \sup_{(x_1,\ldots,x_n)\in X_1\times\cdots\times X_n} \{(\mu_1 \times \cdots \times \mu_n)(x_1, \ldots, x_n) f(x_1, \ldots, x_n) = y\} \\ &= \sup_{(x_1,\ldots,x_n)\in X_1\times\cdots\times X_n} \{\min\{\mu_1(x_1), \ldots, \mu_n(x_n)\} f(x_1, \ldots, x_n) = y\}. \end{aligned} \tag{16.4}$$

This formula represents the **extension principle** (Zadeh 1975a, 1975b, 1975c).

Example 16.3 Consider the mapping $f : \mathbb{R} \times \mathbb{R} \to \mathbb{R}$, $(x_1, x_2) \mapsto x_1 + x_2$ representing the addition. The fuzzy sets $\mu_1 = \Lambda_{0,1,2}$ and $\mu_2 = \Lambda_{1,2,3}$ model the vague concepts "about 1" and "about 2." Applying the extension principle, we obtain the fuzzy set $f[\mu_1, \mu_2] = \Lambda_{1,3,5}$ for the vague concept "about 1 + about 2" (cf. Fig. 16.5). We can observe the effect we already know from computing the square of "about 1" (see Example 16.2 and Fig. 16.4). The "fuzziness" of the resulting fuzzy set is greater than these of the original fuzzy sets to be added.

Analogously to the addition of fuzzy sets we can define subtraction, multiplication and division using the extension principle. These operations are continuous, therefore we can, like in Example 16.2, calculate the level sets of the resulting fuzzy sets directly from the level sets of the given fuzzy sets, provided that these are compact. When we have convex fuzzy sets, we carry out interval arithmetic on the corresponding levels. Interval arithmetic (Moore 1966, 1979) allows us to operate with intervals instead of real numbers.

Applying the extension principle we should be aware that we carry out two generalization steps at the same time: the extension of single elements to sets and the change from crisp to fuzzy sets. The extension principle cannot preserve all properties of the original mapping. This effect is not necessarily caused by the extension from crisp to fuzzy sets. Most of the problems are caused by the step from extending

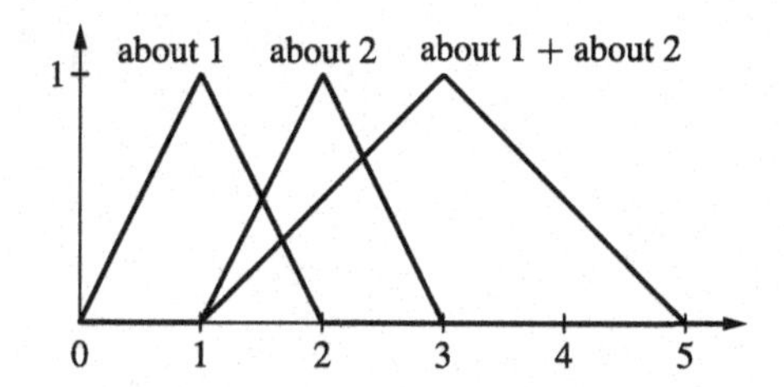

Fig. 16.5 The result of the extension principle for "about 1 + about 2"

a pointwise mapping to (crisp) sets. For example, in contrast to the standard addition, there is no inverse for the addition of fuzzy sets. There is no fuzzy set which added to the fuzzy set for "about 1 + about 2" from Fig. 16.5 which will yield back the fuzzy set "about 1." This phenomenon already occurs in interval arithmetic, so not the "fuzzification" of the addition is the problem, but the extension of the addition from numbers (points) to sets.

References

R.E. Moore. *Interval Analysis*. Prentice Hall, Englewood Cliffs, NJ, USA, 1966

R.E. Moore. *Methods and Applications of Interval Analysis*. Society for Industrial and Applied Mathematics, Philadelphia, PA, USA, 1979

L.A. Zadeh. The Concept of a Linguistic Variable and Its Application to Approximate Reasoning, Part I. *Information Sciences* 8:199–249, 1975a

L.A. Zadeh. The Concept of a Linguistic Variable and Its Application to Approximate Reasoning, Part II. *Information Sciences* 8:301–357, 1975b

L.A. Zadeh. The Concept of a Linguistic Variable and Its Application to Approximate Reasoning, Part III. *Information Sciences* 9:43–80, 1975c

Chapter 17
Fuzzy Relations

Relations can be used to model dependencies, correlations or connections between variables, quantities or attributes. Technically speaking, a (binary) relation over the universes of discourse X and Y is a subset R of the Cartesian product $X \times Y$ of X and Y. The pairs $(x, y) \in X \times Y$ belonging to the relation R are linked by a connection described by the relation R. Therefore, a common notation for $(x, y) \in R$ is also xRy.

We generalize the concept of relations to fuzzy relations. Fuzzy relations are useful for representing and understanding fuzzy controllers that describe a vague connection between input and output values. Furthermore, we can establish an interpretation of fuzzy sets and membership degrees on the basis of special fuzzy relations called similarity relations. This interpretation plays a crucial role in the context of fuzzy controllers. Similarity relations will be discussed in Chap. 18.

17.1 Crisp Relations

Before we introduce the definition of a fuzzy relation, we briefly review the fundamental concepts and mechanisms of crisp relations that are needed for understanding fuzzy relations.

Example 17.1 A house has six doors and each of them has a lock which can be unlocked by certain keys. Let the set of doors be $T = \{t_1, \dots, t_6\}$, the set of keys $S = \{s_1, \dots, s_5\}$. Key s_5 is the main key and fits to all doors. Key s_1 fits only to door t_1, s_2 to t_1 and t_2, s_3 to t_3 and t_4, s_4 to t_5. This situation can be formally described by the relation $R \subseteq S \times T$ ("fits to"). The pair $(s, t) \in S \times T$ is an element of R if and only if key s fits to door t that means

$$R = \big\{(s_1, t_1), (s_2, t_1), (s_2, t_2), (s_3, t_3), (s_3, t_4), (s_4, t_5), \\ (s_5, t_1), (s_5, t_2), (s_5, t_3), (s_5, t_4), (s_5, t_5), (s_5, t_6)\big\}.$$

Another way of describing the relation R is shown in Table 17.1. The entry 1 at position (s_i, t_j) indicates that $(s_i, t_j) \in R$ holds, 0 stands for $(s_i, t_j) \notin R$.

R. Kruse et al., *Computational Intelligence*, Texts in Computer Science,
DOI 10.1007/978-1-4471-5013-8_17, © Springer-Verlag London 2013

Table 17.1 The relation R: "key fits to door"

R	t_1	t_2	t_3	t_4	t_5	t_6
s_1	1	0	0	0	0	0
s_2	1	1	0	0	0	0
s_3	0	0	1	1	0	0
s_4	0	0	0	0	1	0
s_5	1	1	1	1	1	1

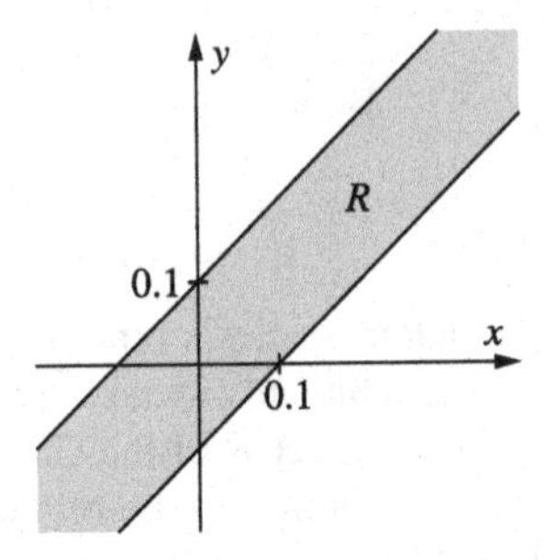

Fig. 17.1 The relation $y \mathrel{\hat{=}} x \pm 0.1$

Example 17.2 Let us consider a measuring instrument which can measure a quantity $y \in \mathbb{R}$ with a precision of ± 0.1. If x_0 is the measured value, we know the true value y_0 lies within the interval $[x_0 - 0.1, x_0 + 0.1]$. This can be described by the relation

$$R = \{(x, y) \in \mathbb{R} \times \mathbb{R} \mid |x - y| \leq 0.1\}.$$

A graphical representation of this relation is given in Fig. 17.1.

Mappings or their graphs can be considered as special cases of relations. If the function $f : X \to Y$ is a mapping of X to Y, the graph of f is the relation

$$\text{graph}(f) = \{(x, f(x)) \mid x \in X\}.$$

In order to be able to interpret a relation $R \subseteq X \times Y$ as a graph of a function we need that for each $x \in X$ there exists exactly one $y \in Y$ such that the pair (x, y) is contained in R.

17.2 Application of Relations and Deduction

So far we have used relations in a merely descriptive way. But similar to functions, relations can also be applied to elements or sets. If $R \subseteq X \times Y$ is a relation between the sets X and Y and $M \subseteq X$ is a subset of X, the image of M under R is the set

$$R[M] = \{y \in Y \mid \exists x \in X : (x, y) \in R \text{ and } x \in M\}. \tag{17.1}$$

$R[M]$ contains the elements from Y which are related to at least one element of the set M.

Table 17.2 Falk scheme for the calculation of $R[M]$

					1	0	0	0	0	0
					1	1	0	0	0	0
					0	0	1	1	0	0
					0	0	0	0	1	0
					1	1	1	1	1	1
1	1	1	1	0	1	1	1	1	1	0

If $f: X \to Y$ is a mapping, then applying the relation graph(f) to a one-element set $\{x\} \subseteq X$ we obtain the one-element set which contains the image of x under the function f:

$$\text{graph}(f)\big[\{x\}\big] = \big\{f(x)\big\}.$$

More generally, we have

$$\text{graph}(f)[M] = f[M] = \big\{y \in Y \mid \exists x \in X : x \in M \land f(x) = y\big\}$$

for arbitrary subsets $M \subseteq X$.

Example 17.3 Now we use the relation R from Example 17.1 in order to determine which doors can be unlocked if we have keys $s_1, \ldots, s_4$. All we have to do is to calculate all elements (doors) which are related (relation "fits to") to at least one of the keys $s_1, \ldots, s_4$. That means

$$R\big[\{s_1, \ldots, s_4\}\big] = \{t_1, \ldots, t_5\}$$

is the set of doors we want to know.

The set $R[\{s_1, \ldots, s_4\}]$ can be determined easily using the matrix in Table 17.1 in the following way. We encode the set $M = \{s_1, \ldots, s_4\}$ as a row vector with five components which contains the entry 1 at the ith place if $s_i \in M$ holds, and 0 in the case of $s_i \notin M$. Thus we obtain the vector $(1, 1, 1, 1, 0)$. Analogously to the **Falk scheme** for matrix multiplication of a vector by a matrix, we write the vector to the lower left of the matrix. Then, we transpose the vector and compare it with every column of the matrix. If we find at least one position during this comparison of the vector and a matrix column where the vector and the matrix have the entry 1,we write a one under this column, otherwise a zero. The resulting vector $(1, 1, 1, 1, 1, 0)$ below the matrix specifies the set $R[M]$we are looking for in an encoded form: It contains a 1 at place i if and only if $t_i \in R[M]$ holds. Table 17.2 illustrates this "Falk scheme" for relations.

Example 17.4 We follow up Example 17.2 and assume that we have the information that the measuring instrument indicated a value between 0.2 and 0.4. From this we can conclude that the true value is contained in the set $R[[0.2, 0.4]] = [0.1, 0.5]$ which is illustrated in Fig. 17.2.

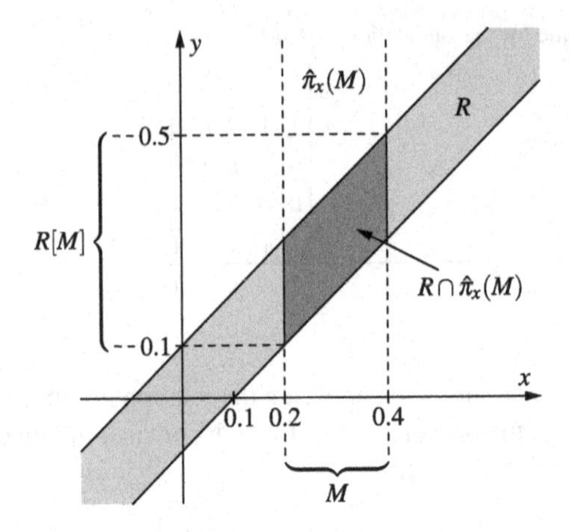

Fig. 17.2 How to determine the set $R[M]$ graphically

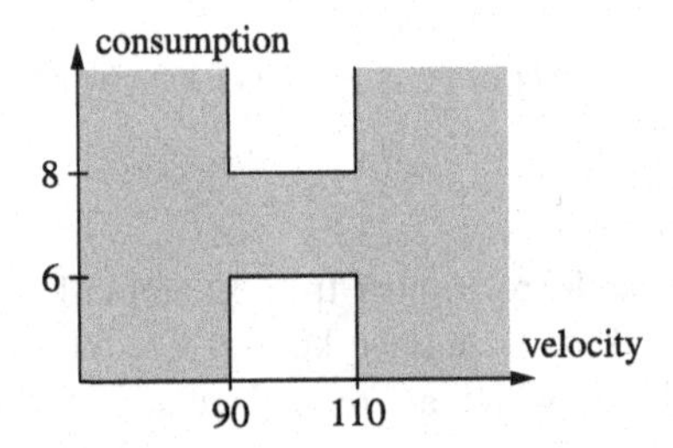

Fig. 17.3 Relation for the rule $v \in [90, 110] \to b \in [6, 8]$

In this figure, we can see that we obtain the set $R[M]$ as the projection of the intersection of the relation with the cylindrical extension of the set M which means

$$R[M] = \pi_y\big[R \cap \hat{\pi}_x(M)\big]. \tag{17.2}$$

Example 17.5 Logical deduction based on an implication of the form $x \in A \to y \in B$ can be modeled and computed by relations, too. All we have to do is to encode the rule $x \in A \to y \in B$ by the relation

$$R = \big\{(x, y) \in X \times Y \mid x \in A \to y \in B\big\} = (A \times B) \cup \bar{A} \times Y. \tag{17.3}$$

X and Y are the sets of possible values that x and y can attain. For the rule "If the velocity is between 90 km/h and 100 km/h, then the fuel consumption is between 6 and 8 liters" (as a logical formula: $v \in [90, 110] \to b \in [6, 8]$) we obtain the relation shown in Fig. 17.3.

If we know that the velocity has the value v, in the case $90 \le v \le 110$ we can conclude that for the consumption b we must have $6 \le b \le 8$. Otherwise and without further knowledge and any pieces of information than just the rule and the value for v, we cannot say anything about the value for the consumption which means that we obtain $b \in [0, \infty)$. We get the same result by applying the relation R to the one-element set $\{v\}$:

$$R[\{v\}] = \begin{cases} [6,8] & \text{if } v \in [90,110], \\ [0,\infty) & \text{otherwise.} \end{cases}$$

More generally, we have: When we know that the velocity attains a value in the set M, we can conclude in the case $M \subseteq [90, 110]$ that the consumption is between 6 and 8 liters, otherwise we only know $b \in [0, \infty)$. This coincides with the result we obtained by applying the relation R to the set M:

$$R[M] = \begin{cases} [6,8] & \text{if } M \subseteq [90,110], \\ \emptyset & \text{if } M = \emptyset, \\ [0,\infty) & \text{otherwise.} \end{cases}$$

17.3 Chains of Deductions

The example above shows how logical deduction can be represented in terms of a relation. Inferring new facts from rules and known facts usually means that we deal with chained deduction steps in the form of $\varphi_1 \to \varphi_2$, $\varphi_2 \to \varphi_3$ from which we can derive $\varphi_1 \to \varphi_3$. A similar principle can be formulated in the context of relations. Consider the relations $R_1 \subseteq X \times Y$ and $R_2 \subseteq Y \times Z$. An element x is indirectly related to an element $z \in Z$ if there exists an element $y \in Y$ such that x and y are in the relation R_1 and y and z are in the relation R_2. We can say that we go from x to z via y. In this way, the composition of the relations R_1 and R_2 can be defined as the relation

$$R_2 \circ R_1 = \{(x,z) \in X \times Z \mid \exists y \in Y : (x,y) \in R_1 \wedge (y,z) \in R_2\} \qquad (17.4)$$

between X and Z. Then, we have for all $M \subseteq X$

$$R_2[R_1[M]] = (R_2 \circ R_1)[M].$$

For the relations graph(f) and graph(g) induced by the mappings $f : X \to Y$ or $g : Y \to Z$, respectively, the composition of these relations is equal to the relation induced by the composition of the two mappings:

$$\text{graph}(g \circ f) = \text{graph}(g) \circ \text{graph}(f).$$

Example 17.6 We extend Example 17.1 of the keys and doors by considering a set $P = \{p_1, p_2, p_3\}$ of three people owning various keys. This is expressed by the relation

$$R' = \{(p_1,s_1),(p_1,s_2),(p_2,s_3),(p_2,s_4),(p_3,s_5)\} \subseteq P \times T.$$

$(p_i, s_j) \in R'$ means that person p_i owns the key s_j. The composition

$$R \circ R' = \{(p_1,t_1),(p_1,t_2),(p_2,t_3),(p_2,t_4),(p_2,t_5),\\ (p_3,t_1),(p_3,t_2),(p_3,t_3),(p_3,t_4),(p_3,t_5),(p_3,t_6)\}$$

of the relations R' and R contains the pair $(p,t) \in P \times T$ if and only if person p can unlock door t. For example, using the relation $R \circ R'$ we can determine which doors can be unlocked if the people p_1 and p_2 are present. The corresponding set of doors is

$$(R \circ R')[\{p_1,p_2\}] = \{t_1,\dots,t_5\} = R[R'[\{p_1,p_2\}]].$$

Example 17.7 In Example 17.2, we used the relation $R = \{(x,y) \in \mathbb{R} \times \mathbb{R} \mid |x-y| \le 0.1\}$ to model the fact that the measured value x represents the true value y with a precision of 0.1. When we can determine the quantity z from the quantity y with a precision of 0.2, we obtain the relation $R' = \{(y,z) \in \mathbb{R} \times \mathbb{R} \mid |x-y| \le 0.2\}$. The composition of R' and R results in the relation $R' \circ R = \{(x,z) \in \mathbb{R} \times \mathbb{R} \mid |x-z| \le 0.3\}$. If the measuring instrument indicates the value x_0, we can conclude that the value of the quantity z is in the set

$$(R' \circ R)[\{x_0\}] = [x_0 - 0.3, x_0 + 0.3].$$

Example 17.8 Example 17.5 demonstrated how an implication of the form $x \in A \to y \in B$ can be represented by a relation. When another rule $y \in C \to z \in D$ is known, in the case of $B \subseteq C$ we can derive the rule $x \in A \to z \in D$. Otherwise, knowing x does not provide any information about z in the context of these two rules. That means we obtain the rule $x \in X \to z \in Z$ in this case. Correspondingly, the composition of the relations R' and R representing the implications $x \in A \to y \in B$ and $y \in C \to z \in D$, respectively, results in the relation associated with the implication $x \in A \to z \in D$ and $x \in A \to z \in Z$, respectively:

$$R' \circ R = \begin{cases} (A \times D) \cup (\bar{A} \times Z) & \text{if } B \subseteq C, \\ (A \times Z) \cup (\bar{A} \times Z) = X \times Z & \text{otherwise.} \end{cases}$$

17.4 Simple Fuzzy Relations

After giving a general idea about the fundamental terminology and concepts of crisp relations, we can now introduce fuzzy relations.

Definition 17.1 A fuzzy set $\rho \in \mathcal{F}(X \times Y)$ is called (binary) *fuzzy relation* between the reference sets X and Y.

Table 17.3 The fuzzy relation ρ: "x is a financial fond with risk factor y"

ρ	l	m	h
s	0.0	0.3	1.0
f	0.6	0.9	0.1
e	0.8	0.5	0.2

In this sense, a fuzzy relation is a generalized crisp relation where two elements can be gradually related to each other. The greater the membership degree $\rho(x, y)$ the stronger is the relation between x and y.

Example 17.9 Let $X = \{s, f, e\}$ denote a set of financial fonds, devoted to shares (s), fixed-interest stocks (f) and real estates (e). The set $Y = \{l, m, h\}$ contains the elements low (l), medium (m) and high (h) risk. The fuzzy relation $\rho \in \mathcal{F}(X \times Y)$ in Table 17.3 shows for every pair $(x, y) \in X \times Y$ how much the fond x is considered having the risk factor y.

For example, the entry in column m and row e means that the fond dedicated to real estates is considered to have a medium risk with a degree of 0.5. Therefore, $\rho(e, m) = 0.5$.

Example 17.10 The measuring instrument from Example 17.2 had a precision of ± 0.1. However, it is not very realistic to assume that, given that the instrument shows the value x_0, all values from the interval $[x_0 - 0.1, x_0 + 0.1]$ are equally likely to represent the true value of the measured quantity. Instead of the crisp relation R from Example 17.2 for representing this fact, we can use a fuzzy relation, for instance

$$\rho : \mathbb{R} \times \mathbb{R} \to [0, 1], \quad (x, y) \mapsto 1 - \min\{10|x - y|, 1\},$$

yielding the membership degree of 1 for $x = y$. The membership degree to the relation decreases linearly with increasing distance $|x - y|$ until the difference between x and y exceeds the value 0.1.

In order to operate with fuzzy relations in a similar way as with usual relations, we have define what the image of a fuzzy set under a fuzzy relation is. This means we have to extend Eq. (17.1) to the framework of fuzzy sets and fuzzy relations.

Definition 17.2 For a fuzzy relation $\rho \in \mathcal{F}(X \times Y)$ and a fuzzy set $\mu \in \mathcal{F}(X)$ the image of μ under ρ is the fuzzy set

$$\rho[\mu](y) = \sup\{\min\{\rho(x, y), \mu(x)\} \mid x \in X\} \tag{17.5}$$

over the universe of discourse Y.

This definition can be justified in several ways. If ρ and μ are the characteristic functions of a usual relation R and the crisp set M, respectively, then $\rho[\mu]$ is the characteristic function of the image $R[M]$ of M under R. In this sense, the definition is a generalization of Eq. (17.1) for sets to fuzzy sets.

Equation (17.1) is equivalent to

$$y \in R[M] \iff \exists x \in X : (x, y) \in R \wedge x \in M.$$

We obtain Eq. (17.5) for fuzzy relations from this equivalence by assigning the minimum as truth function to the conjunction and evaluate the existential quantifier by the supremum. This means

$$\begin{aligned} \rho[\mu](y) &= [\![y \in \rho[\mu]]\!] \\ &= [\![\exists x \in X : (x, y) \in \rho \wedge x \in \mu]\!] \\ &= \sup\{\min\{\rho(x, y), \mu(x)\} \mid x \in X\}. \end{aligned}$$

Definition 17.2 can also be derived from the extension principle. In order to do this, we consider the partial mapping

$$f : X \times (X \times Y) \to Y, \quad (x, (x', y)) \mapsto \begin{cases} y & \text{if } x = x', \\ \text{undefined} & \text{otherwise.} \end{cases} \tag{17.6}$$

It is obvious that for a set $M \subseteq X$ and a relation $R \subseteq X \times Y$, we have

$$f[M, R] = f[M \times R] = R[M].$$

When we introduced the extension principle, we did not require that the mapping f, which has to be extended to fuzzy sets, must be defined everywhere. Therefore, the extension principle can also be applied to partial mappings. The extension principle for the mapping in Eq. (17.6), which can be used to compute the image of a set under a relation, leads to the formula specified in Definition 17.2, the formula for the image of a fuzzy set under a fuzzy relation.

Another justification of Definition 17.2 is based on the idea that was exploited in Example 17.4 and Fig. 17.2. There, we computed the image of a set under a relation as projection of the intersection of the cylindrical extension of the set with the relation (cf. Eq. (17.2)). Having this equation in mind, we replace the set M by a fuzzy set μ and the relation R by a fuzzy relation ρ and again obtain Eq. (17.5) if the intersection of fuzzy sets is computed using the minimum and the projection and the cylindrical extension for fuzzy sets are calculated as in Chap. 16.

Example 17.11 On the basis of the fuzzy relations from Example 17.9, we want to estimate the risk of a mixed fond which concentrates on shares but also invests a smaller part of its many into real estates. We represent this mixed fond over the universe of discourse $\{s, r, f\}$ as a fuzzy set μ with

$$\mu(s) = 0.8, \qquad \mu(f) = 0, \qquad \mu(r) = 0.2.$$

In order to determine the risk of this mixed fond, we compute the image of the fuzzy set μ under the fuzzy relation ρ from Table 17.3. We obtain

$$\rho[\mu](l) = 0.2, \qquad \rho[\mu](m) = 0.3, \qquad \rho[\mu](h) = 0.8.$$

Analogously to Example 17.3, the fuzzy set $\varrho[\mu]$ can be determined using a modified Falk scheme. The zeros and ones in Table 17.2 have to be replaced by the corresponding membership degrees. Below each column of the fuzzy relation, we obtain the membership degree of the corresponding value to the image fuzzy set $\varrho[\mu]$ in the following way. We first take the componentwise minimum of the vector representing the fuzzy set μ and the corresponding column of the matrix representing the fuzzy relation ϱ and then we compute the maximum of these minima.

In this sense, the calculation of the image of a fuzzy set μ under a fuzzy relation ϱ is similar to matrix multiplication of a matrix with a vector where the multiplication of the components is replaced by the minimum and the addition by the maximum.

Example 17.12 We have the information that the measuring instrument from Example 17.10 indicated a value of "about 0.3" which we represent by the fuzzy set $\mu = \Lambda_{0.2,0.3,0.4}$. For the true value y, we obtain the fuzzy set

$$\varrho[\mu](y) = 1 - \min\{5|y - 0.3|, 1\}$$

as image of the fuzzy set μ under the relation ϱ from Example 17.10.

Example 17.13 Example 17.5 illustrated how logic deduction on the basis of an implication of the form $x \in A \rightarrow y \in B$ can be represented using a relation. We generalize this method for the case that the sets A and B are replaced by the fuzzy sets μ or ν. Following Eq. (17.3) and using the formula $[\![(x, y) \in \varrho]\!] = [\![x \in \mu \rightarrow y \in \nu]\!]$ where we choose the Gödel implication as truth function for the implication, we define the fuzzy relation

$$\varrho(x, y) = \begin{cases} 1 & \text{if } \mu(x) \le \nu(y), \\ \nu(y) & \text{otherwise.} \end{cases}$$

The rule "If x is about 2, then y is about 3" leads to the fuzzy relation

$$\varrho(x, y) = \begin{cases} 1 & \text{if } \min\{|3 - y|, 1\} \le |2 - x|, \\ 1 - \min\{|3 - y|, 1\} & \text{otherwise,} \end{cases}$$

if we model "about 2" by the fuzzy set $\mu = \Lambda_{1,2,3}$ and "about 3" by the fuzzy set $\nu = \Lambda_{2,3,4}$. Knowing "x is about 2.5" represented by the fuzzy set $\mu' = \Lambda_{1.5,2.5,3.5}$ we obtain for y the fuzzy set

$$\varrho[\mu'](y) = \begin{cases} y - 1.5 & \text{if } 2.0 \le y \le 2.5, \\ 1 & \text{if } 2.5 \le y \le 3.5, \\ 4.5 - y & \text{if } 3.5 \le y \le 4.0, \\ 0.5 & \text{otherwise,} \end{cases}$$

shown in Fig. 17.4.

The membership degree of an element y_0 to this fuzzy set should be interpreted as how much one can believe that it is possible that the variable y attains the value y_0. This interpretation is a generalization of what we have obtained for the implication based on crisp sets in Example 17.5. In that case, only two sets were possible results of applying the deduction scheme: either the entire universe of discourse, if we could

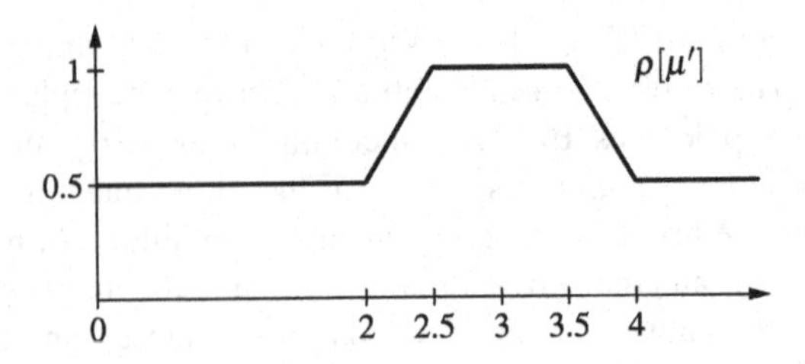

Fig. 17.4 The fuzzy set $\rho[\mu']$

not guarantee that the antecedent of the implication is satisfied, or the set determined specified in the consequent of the implication for the case that the antecedent is satisfied. The first case tells us that—only knowing the single rule—all values for y are still possible, since the rule is not applicable in this case. In the second case we know that the rule is applicable and only those values specified in the consequent of the rule are considered possible for Y. Extending this framework from crisp to fuzzy sets, the antecedent and the consequent of the implication can be partially satisfied. The consequence is that not only the whole universe of discourse and the fuzzy set in the consequent are possible results, but also fuzzy sets in between. In our example, all values y have a membership degree of at least 0.5 to the fuzzy set $\rho[\mu']$. The reason for this is that there exists a value, namely $x_0 = 2.0$ which has a membership degree of 0.5 to the fuzzy set μ' and a membership degree of 0 to μ. This means that the variable x can attain a value with a degree of 0.5, for which we cannot apply the rule, that is, y can attain any value. The membership degree 1 of the value $x_0 = 2.5$ to the fuzzy set μ' leads to the fact that all values of the interval $[2.5, 3.5]$ have a membership degree of 1 to $\rho[\mu']$. For $x_0 = 2.5$, we obtain $\mu(2.5) = 0.75$ which means that the antecedent of the implication is satisfied with the degree of 0.75. This implies that in order to satisfy the implication, a membership degree of at least 0.75 is required for the consequent. And the values in the interval $[2.5, 3.5]$ are those with a membership degree of at least 0.75 to the fuzzy set ν.

In a similar way, we can treat membership degrees between 0 and 1 to justify or compute the fuzzy set $\rho[\mu']$.

17.5 Composition of Fuzzy Relations

Now we are able to discuss the composition of fuzzy relations. The definition of an image of a fuzzy set under a fuzzy relation was motivated by Eq. (17.1) for crisp sets. Analogously, we define the composition of fuzzy relations based on Eq. (17.4) that describes composition in the case of crisp relations.

Definition 17.3 Let $\rho_1 \in \mathcal{F}(X \times Y)$ and $\rho_2 \in \mathcal{F}(Y \times Z)$ be fuzzy relations. The *composition* of the two fuzzy relations is the fuzzy relation

$$(\rho_2 \circ \rho_1)(x, z) = \sup\bigl\{\min\{\rho_1(x, y), \rho_2(y, z)\} \bigm| y \in Y\bigr\} \tag{17.7}$$

between the universes of discourse X and Z.

Table 17.4 The fuzzy relation ρ': "Given the risk y the profit/loss z is possible"

ρ'	hl	ll	lp	hp
l	0.0	0.4	1.0	0.0
m	0.3	1.0	1.0	0.4
h	1.0	1.0	1.0	1.0

This definition can be obtained from the equivalence

$$(x, z) \in R_2 \circ R_1 \quad \Longleftrightarrow \quad \exists y \in Y : (x, y) \in R_1 \wedge (y, z) \in R_2,$$

assigning the minimum as truth function to the conjunction and evaluating the existential quantifier by the supremum such that we obtain

$$\begin{aligned}(\rho_2 \circ \rho_1)(x, z) &= [\![(x, y) \in (\rho_2 \circ \rho_1)]\!] \\ &= [\![\exists y \in Y : (x, y) \in R_1 \wedge (y, z) \in R_2]\!] \\ &= \sup\{\min\{\rho_1(x, y), \rho_2(y, z)\} \mid y \in Y\}.\end{aligned}$$

Equation (17.7) can also be derived by applying the extension principle to the partial mapping

$$\begin{aligned}&f : (X \times Y) \times (Y \times Z) \to (X \times Y), \\ &((x, y), (y', z)) \mapsto \begin{cases} (x, z) & \text{if } y = y', \\ \text{undefined} & \text{otherwise,} \end{cases}\end{aligned}$$

on which the composition of crisp relations is based because we have

$$f[R_1, R_2] = f[R_1 \times R_2] = R_2 \circ R_1.$$

If ρ_1 and ρ_2 are the characteristic functions of the crisp relations R_1 or R_2, then $\rho_2 \circ \rho_1$ is the characteristic function of the relation $R_2 \circ R_1$. In this sense, Definition 17.3 generalizes the compositions of crisp relations to fuzzy relations.

For every fuzzy set $\mu \in \mathcal{F}(X)$, we have

$$(\rho_2 \circ \rho_1)[\mu] = \rho_2[\rho_1[\mu]].$$

Example 17.14 Let us come back to Example 17.11 analyzing the risk of financial fonds. Now, we extend the risk estimation of fonds by the set $Z = \{hl, ll, lp, hp\}$. The elements stand for "high loss", "low loss", "low profit", "high profit". The fuzzy relation $\rho' \in \mathcal{F}(Y \times Z)$ in Table 17.4 determines for each tuple $(y, z) \in Y \times Z$ the possibility to have a profit or loss of z under the risk y. The fuzzy relation resulting from the composition of the fuzzy relations ρ and ρ' is shown in Table 17.5.

In this case, where the universes of discourse are finite and the fuzzy relations can be represented as tables or matrices, the computation scheme for the composition of fuzzy relations is similar to matrix multiplication where we have to replace the

Table 17.5 The fuzzy relation $\rho' \circ \rho$: "With the yield object x the profit/loss z is possible"

ρ'	hl	ll	lp	hp
s	1.0	1.0	1.0	1.0
f	0.3	0.9	0.9	0.4
r	0.3	0.5	0.8	0.4

componentwise multiplication by the minimum and the addition by the maximum. For the mixed fond from Example 17.11 which was represented by the fuzzy set μ

$$\mu(s)=0.8, \qquad \mu(f)=0, \qquad \mu(r)=0.2,$$

we obtain

$$(\rho' \circ \rho)[\mu](hl)=(\rho' \circ \rho)[\mu](ll)=(\rho' \circ \rho)[\mu](lp)=(\rho' \circ \rho)[\mu](hp)=0.8$$

as fuzzy set describing the possible profit or loss.

Example 17.15 The precision of the measuring instrument from Example 17.10 was described by the fuzzy relation $\rho(x, y) = 1 - \min\{10|x - y|, 1\}$ which determines in how far the value y is the true value if x is the value indicated by the measuring instrument. We assume that we cannot exactly read the value from the (analog) instrument and therefore use the fuzzy relation $\rho'(a, x) = 1 - \min\{5|a - x|, 1\}$. This relation tells us in how far the value x is the value indicated by the measuring instrument when we read the value a. In order to estimate which value could be the true value y of the measured quantity, given we have read the value a, we have to compute the composition of the fuzzy relations ρ' and ρ

$$(\rho \circ \rho')(a, y) = 1 - \min\left\{\frac{10}{3}|a - y|, 1\right\}.$$

Assuming we have read $a = 0$, we obtain the fuzzy set

$$(\rho \circ \rho')[I_{\{0\}}] = \Lambda_{-0.3,0,0.3}$$

for the true value y.

Chapter 18
Similarity Relations

In this chapter, we discuss a special type of fuzzy relations, called similarity relations. They play an important role in interpreting fuzzy controllers and, more generally, can be used to characterize the inherent indistinguishability or vagueness of a fuzzy systems.

18.1 Similarity

Similarity relations are fuzzy relations that specify for pairs of elements or objects how indistinguishable or similar they are. From a similarity relation, we should expect that it is reflexive and symmetric which means that every element is similar to itself (with degree 1) and that x is as similar to y as y to x. In addition to these two minimum requirements, we also ask for the following weakened transitivity condition: If x is similar to y to a certain degree and y is similar to z to a certain degree, then x should also be similar to z to a certain (maybe lower) degree. Technically speaking, we define a similarity relation as follows:

Definition 18.1 A *similarity relation* $E : X \times X \to [0, 1]$ with respect to the t-norm t on the set X is a fuzzy relation over $X \times X$ which satisfies the conditions

$$
\begin{array}{lll}
\text{(E1)} & E(x,x) = 1, & \text{(reflexivity)} \\
\text{(E2)} & E(x,y) = E(y,x), & \text{(symmetry)} \\
\text{(E3)} & t(E(x,y), E(y,z)) \leq E(x,z) & \text{(transitivity)}
\end{array}
$$

for all $x, y, z \in X$.

The transitivity conditions for similarity relations can be understood in terms of fuzzy logic, as it was presented in Sect. 15.5, as follows: The truth value of the proposition

$$x \text{ and } y \text{ are similar AND } y \text{ and } z \text{ are similar}$$

R. Kruse et al., *Computational Intelligence*, Texts in Computer Science,
DOI 10.1007/978-1-4471-5013-8_18, © Springer-Verlag London 2013

should be at most as great as the truth value of the proposition

$$x \text{ and } z \text{ are similar}$$

where the t-norm t is used as the truth function for the conjunction AND.

In Example 17.10, we have already seen an example for a similarity relation, that is, the fuzzy relation

$$\varrho : \mathbb{R} \times \mathbb{R} \to [0, 1], \quad (x, y) \mapsto 1 - \min\{10|x - y|, 1\}$$

that indicates how indistinguishable two values are using a measuring instrument. We can prove easily that this fuzzy relation is a similarity relation with respect to the Łukasiewicz t-norm $t(\alpha, \beta) = \max\{\alpha + \beta - 1, 0\}$. More generally, an arbitrary pseudometric, that is, a distance measure $\delta : X \times X \to [0, \infty)$ satisfying the symmetry condition $\delta(x, y) = \delta(y, x)$ and the triangle inequality $\delta(x, y) + \delta(y, z) \geq \delta(x, z)$ induces a similarity relation with respect to the Łukasiewicz t-norm by

$$E^{(\delta)}(x, y) = 1 - \min\{\delta(x, y), 1\}$$

and vice versa, every similarity relation E with respect to the Łukasiewicz t-norm defines a pseudometric by

$$\delta^{(E)}(x, y) = 1 - E(x, y).$$

Furthermore, we have $E = E^{(\delta^{(E)})}$ and $\delta(x, y) = \delta^{(E^{(\delta)})}(x, y)$ if δ is bounded by one, that is, $\delta(x, y) \leq 1$ holds for all $x, y \in X$, such that similarity relations and pseudometrics (bounded by one) represent dual concepts.

Later on, other examples will provide a motivation to consider similarity relations with respect to other t-norms than the Łukasiewicz t-norm in order to characterize the vagueness or indistinguishability that is inherent in a fuzzy systems.

18.2 Fuzzy Sets and Extensional Hulls

If we assume that a similarity relation characterizes a certain indistinguishability, we should expect that elements that are (almost) indistinguishable should behave similar or have similar properties. For fuzzy systems, the (fuzzy) property of being an element of a (fuzzy) set is essential. Therefore, those fuzzy sets play an important role that are coherent with respect to a given similarity relation in the sense that similar elements have similar membership degrees. This property of a fuzzy set is called extensionality and is formally defined as follows:

Definition 18.2 Let $E : X \times X \to [0, 1]$ be a similarity relation with respect to the t-norm t on the set X. A fuzzy set $\mu \in \mathcal{F}(X)$ is called *extensional* with respect to E if

$$t(\mu(x), E(x, y)) \leq \mu(y)$$

holds for all $x, y \in X$.

In the view of fuzzy logic, the extensionality condition can be interpreted in the sense that the truth value of the proposition

x is an element of the fuzzy set μ AND
x and y are similar (indistinguishable)

should not be smaller than the truth value of the proposition

y is an element of the fuzzy set μ

where the t-norm t as truth function is assigned to the conjunction AND.

A fuzzy set can always be made extensional by adding all elements which are similar to at least one of its elements. If we formalize this idea, we obtain the following definition.

Definition 18.3 Let $E : X \times X \to [0, 1]$ be a similarity relation with respect to the t-norm t on the set X. The extensional hull $\hat{\mu}$ of the fuzzy set $\mu \in \mathcal{F}(X)$ (with respect to the similarity relation E) is given by

$$\hat{\mu}(y) = \sup\{t(E(x, y), \mu(x)) \mid x \in X\}.$$

If t is a continuous t-norm, then the extensional hull $\hat{\mu}$ of μ is the smallest extensional fuzzy set containing μ in the sense of $\mu \le \hat{\mu}$.

In principle, we obtain the extensional hull of a fuzzy set μ under the similarity relation E as the image of μ under the fuzzy relation E as in Definition 17.2. For the extensional hull, the minimum in Eq. (17.5) in Definition 17.2 is replaced by the t-norm t.

Example 18.1 Let us consider the similarity relation $E : \mathbb{R} \times \mathbb{R} \to [0, 1]$ defined by $E(x, y) = 1 - \min\{|x - y|, 1\}$ with respect to the Łukasiewicz t-norm which is induced by the usual metric $\delta(x, y) = |x - y|$ on the real numbers. A (crisp) set $M \subseteq \mathbb{R}$ can be viewed as a special type of fuzzy set when we consider its characteristic function I_M. In this way, we can also compute the extensional hulls of crisp sets.

The extensional hull of a point x_0, which means the one-element set $\{x_0\}$, with respect to the similarity relation E mentioned above is a fuzzy set with a triangular membership function $\Lambda_{x_0-1,x_0,x_0+1}$. The extensional hull of the interval $[a, b]$ is the trapezoidal function $\Pi_{a-1,a,b,b+1}$ (cf. Fig. 18.1).

This example establishes an interesting connection between fuzzy sets and similarity relations: Triangular and trapezoidal membership functions, that are very popular in most applications of fuzzy systems, can be interpreted as extensional hulls of points or intervals, that means as fuzzy points or fuzzy intervals in a vague environment which is characterized by a similarity relation induced by the standard metric on the real numbers.

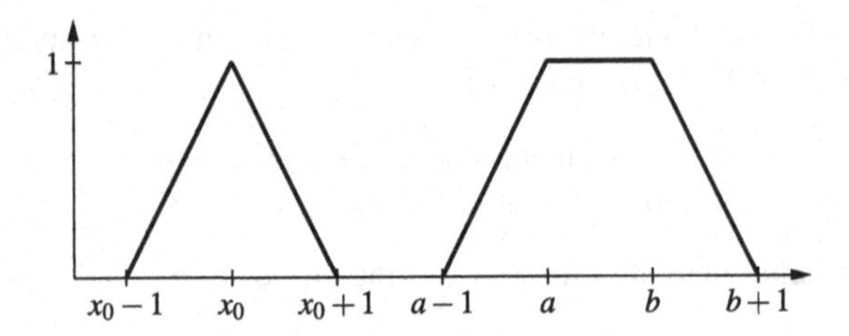

Fig. 18.1 Extensional hull of the point x_0 and the interval $[a, b]$

18.3 Scaling Concepts

The standard metric on the real numbers allows only very limited forms of triangular and trapezoidal functions as extensional hulls of points and intervals: the slopes where the membership degrees increase and decrease are always 1 and -1, respectively. However, it is reasonable to allow a scaling of the standard metric so that other forms of fuzzy sets as extensional hulls can result. This scaling can have two different meanings.

The degree of similarity of two measured values depends on the measuring unit. Two values measured in kilo-units can have a small distance and might be considered as almost indistinguishable or very similar, while the very same quantities measured in milli-units have a much greater distance and are distinguishable. Of course, the degree of similarity should not depend on the measurement unit. The height of two persons should have the same degree of similarity, no matter whether we measure the height in centimeters, meters or inches. In order to adapt a similarity relation to a measuring unit, the distance or the real axis has to be scaled by a constant factor $c > 0$, as in Example 17.10. In this way we obtain scaled metric $|c \cdot x - c \cdot y|$ which induces the similarity relation $E(x, y) = 1 - \min\{|c \cdot x - c \cdot y|, 1\}$.

An extension of the scaling concept is the use of varying scaling factors allowing a local scaling fitting to the problem.

Example 18.2 We want to describe the behavior of an air conditioner using fuzzy rules. It is neither necessary nor desirable to measure the room temperature, to which the air conditioner will react, as exactly as possible. However, the temperature values are of different importance for the air conditioner. Temperatures of 10 °C or 15 °C, for example, are much too cold, so the air conditioner should heat at full power, and the values 27 °C or 32 °C are much too hot and the air conditioner should cool at full power. Therefore, there is no need to distinguish between the values 10 °C and 15 °C or 27 °C and 32 °C for the control of the room temperature. Since we do not have to distinguish between 10 °C and 15 °C, we can choose a very small positive scaling factor—in the extreme case even the factor 0, then these temperatures are not distinguished at all. However, it would not be correct, if we chose a small scaling factor for the entire range of the temperature, because the air conditioner has to distinguish very well between the cold temperature 18.5 °C and the warm one 23.5 °C.

Table 18.1 Different sensitivities and scaling factors for controlling the room temperature

temperature (in °C)	scaling factor	interpretation
< 15	0.00	exact value insignificant (much too cold temperature)
15-19	0.25	too cold but near approximately OK, not very sensitive
19-23	1.50	very sensitive, near the optimum
23-27	0.25	too cold but near approximately OK, not very sensitive
> 27	0.00	exact value insignificant (much too hot temperature)

Table 18.2 Transformed distances based on the scaling factors and the induced similarity degrees

pair of variates	scaling factor	transf. distance	similarity degree
(x, y)	c	$\delta(x, y) =$ $\lvert c \cdot x - c \cdot y \rvert$	$E(x, y) =$ $1 - \min\{\delta(x, y), 1\}$
(13,14)	0.00	0.000	1.000
(14,14.5)	0.00	0.000	1.000
(17,17.5)	0.25	0.125	0.875
(20,20.5)	1.50	0.750	0.250
(21,22)	1.50	1.500	0.000
(24,24.5)	0.25	0.125	0.875
(28,28.5)	0.00	0.000	1.000

Instead of a global scaling factor we should choose individual scaling factors for different ranges of the temperature, so that we can make a fine distinction between temperatures which are near the optimum room temperature and a more rough one between temperatures which are much too cold or much too warm. Table 18.1 shows an example of a partition into five temperature ranges, each one having its individual scaling factor.

These scaling factors define a transformation of the range and the distances for the temperature, that can be used to define a similarity relation. Table 18.2 shows the transformed distances and the resulting similarity degrees for some pairs of values of the temperature. For each pair, the two values lie in a range where the scaling factor does not change. In order to understand how to determine the transformed distance and the resulting similarity degree for two temperatures that are not in a range with a constant scaling factor, we first analyze the effect of a single scaling factor.

Let us consider an interval $[a, b]$ where we measure the distance between two points based on the scaling factor c. Computing the scaled distance means that we apply stretching ($c > 1$) or shrinking ($0 \leq c < 1$) to the interval according to the factor c and calculate the distances between the points in the transformed (stretched or shrunk) interval. In order to take individual scaling factors for different ranges into account, we have to stretch or shrink each range, where the scaling factor is

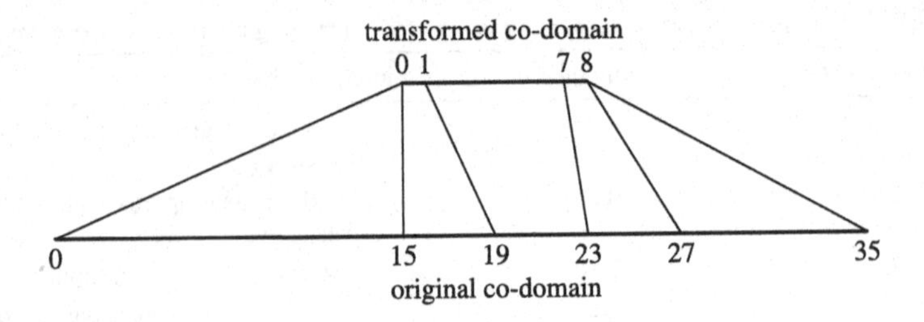

Fig. 18.2 Transformation of a co-domain with the help of scaling factors

constant, correspondingly. Gluing these transformed ranges together, we can now measure the distance between points that do not lie in a region with a constant scaling factor. The induced transformation is piecewise linear as shown in Fig. 18.2.

On the basis of the following three examples, we explain and illustrate the calculation of the transformed distance and the resulting similarity degrees. We determine the similarity degree between the values 18 and 19.2. The value 18 is in the interval [15, 19] with constant scaling factor 0.25. This interval of length 4 is now shrunk to an interval of length 1. Therefore, the distance of the value 18 to the right boundary of the interval, that is, 19, is also shrunk by the factor 0.25, such that the transformed distance between 18 and 19 is exactly 0.25. In order to calculate the transformed distance between 18 and 19.2, we have to add the transformed distance between 19 and 19.2. In this range the scaling factor is constantly 1.5, so the distance between 19 and 19.2 is stretched by the factor 1.5 and the resulting transformed distance is 0.3. Thus, the transformed distance between the values 18 and 19.2 is $0.25 + 0.3 = 0.55$ which leads to a similarity degree of $1 - \min\{0.55, 1\} = 0.45$.

In the second example, we consider the values 13 and 18. Because of the scaling factor 0, the transformed distance between 13 and 15 is also 0. In the range between 15 and 18, the scaling factor is 0.25 and therefore the transformed distance between 13 and 18 is 0.75. Since the transformed distance between 13 and 15 is zero, the overall transformed distance between 13 and 18 is 0.75 and the similarity degree between 13 and 18 is 0.25.

Finally, we determine the transformed distance and the similarity degree between the values 22.8 and 27.5. Here, we have to take three ranges with different scaling factors into account: the scaling factor between 22.8 and 23 is 1.5, between 23 and 27 it is 0.25 and between 27 and 27.5 it is constantly 0. Therefore, the transformed distances for the pairs $(22.8, 23)$, $(23, 27)$ and $(27, 27.5)$ are 0.3, 1 and 0, respectively. The sum of these distances and therefore the transformed distance between 22.8 and 27.5 is 1.3. The degree of similarity is $1 - \min\{1.3, 1\} = 0$.

The idea of using different scaling factors for different ranges can be extended by assigning a scaling factor to each single value which determines how to distinguish in the direct environment of this value. Instead of a piecewise constant scaling function as in Example 18.2 any (integrable) scaling functions $c : \mathbb{R} \to [0, \infty)$ can be used. The transformed distance between the values x and y under such a scaling function can be calculated using the following equation (Klawonn 1994):

$$\left| \int_x^y c(s)\, ds \right|. \tag{18.1}$$

18.4 Interpretation of Fuzzy Sets

We have seen that fuzzy sets can be interpreted as induced concepts based on similarity relation. Assuming that a similarity relation models a problem-specific indistinguishability, fuzzy sets are induced by crisp values or sets in a canonical way in the form of extensional hulls. In the following, we analyze the opposite view. Given a set of fuzzy sets, can we find a suitable similarity relation such that the fuzzy sets are nothing but extensional hulls of crisp values or sets? The results that we present here will be helpful in analyzing and understanding fuzzy controllers later on. In the context of fuzzy controllers a number of vague expression modeled by fuzzy sets are used for each considered variable. So for each domain X a set $\mathcal{A} \subseteq \mathcal{F}(X)$ of fuzzy sets is given. As we will see later, the inherent indistinguishability of these fuzzy sets can be characterized by similarity relations. The coarsest (greatest) similarity relation for which all fuzzy sets in $\mathcal{A}$ are extensional plays an important role. The following theorem (Klawonn and Castro 1995) describes how to compute this similarity relation.

Theorem 18.1 *Let t be a continuous t-norm and $\mathcal{A} \subseteq \mathcal{F}(X)$ a set of fuzzy sets. Then*

$$E_{\mathcal{A}}(x, y) = \inf\{\overset{\leftrightarrow}{t}\big(\mu(x), \mu(y)\big) \mid \mu \in \mathcal{A}\} \tag{18.2}$$

is the coarsest (greatest) similarity relation with respect to the t-norm t for which all fuzzy sets in $\mathcal{A}$ are extensional. $\overset{\leftrightarrow}{t}$ is the biimplication from Eq. (15.4) *associated with the t-norm t.*

Coarsest similarity relation means that for every similarity relation E for which all fuzzy sets in $\mathcal{A}$ are extensional, we have that $E_{\mathcal{A}}(x, y) \geq E(x, y)$ holds for all $x, y \in X$.

Equation (18.2) for the similarity relation $E_{\mathcal{A}}$ can be understood and interpreted within the framework of fuzzy logic. We interpret the fuzzy sets in $\mathcal{A}$ as representations of vague properties. Two elements x and y are similar with respect to a single property, if x has the property if and only if y has the property. Modeling this idea within fuzzy logic, this means that we have to interpret "x has the property associated with the fuzzy set μ" as the membership degree of x to μ. Then the similarity degree of x and y, taking only the property associated with the fuzzy set μ into account, is defined by $\overset{\leftrightarrow}{t}(\mu(x), \mu(y))$. When we can use all properties associated with the fuzzy sets in $\mathcal{A}$ to distinguish x and y, this means: x and y are similar if they are similar with respect to all properties in $\mathcal{A}$. Evaluating "for all" by the infimum, we obtain Eq. (18.2) for the similarity degree of two elements.

We have seen in Example 18.1 that typical fuzzy sets with triangular membership functions can be interpreted as extensional hulls of single points or values. This interpretation of fuzzy sets as vague values will be very useful in the context of fuzzy control. This is why we also study the issue, when a set $\mathcal{A} \subseteq \mathcal{F}(X)$ of fuzzy sets can be interpreted as extensional hulls of points.

Theorem 18.2 *Let t be a continuous t-norm and $\mathcal{A} \subseteq \mathcal{F}(X)$ a set of fuzzy sets. Let each $\mu \in \mathcal{A}$ have the property such that there exists an $x_\mu \in X$ with $\mu(x_\mu) = 1$. There exists a similarity relation E, such that for all $\mu \in \mathcal{A}$ the extensional hull of the point x_μ is equal to the fuzzy set μ if and only if*

$$\sup_{x \in X}\{t(\mu(x), \nu(x))\} \le \inf_{y \in X}\{\overset{\leftrightarrow}{t}(\mu(y), \nu(y))\} \tag{18.3}$$

holds for all $\mu, \nu \in \mathcal{A}$. In this case, $E = E_{\mathcal{A}}$ is the coarsest similarity relation for which all fuzzy sets in $\mathcal{A}$ are extensional hulls of points.

Condition (18.3) says that the degree of non-disjunction of any two fuzzy sets $\mu, \nu \in \mathcal{A}$ must not exceed their degree of equality. The corresponding formulas are obtained by interpreting the following conditions in terms of fuzzy logic:

- Two sets μ and ν are non-disjoint if and only if

$$\exists x : x \in \mu \wedge x \in \nu$$

  holds.
- Two sets μ and ν are equal if and only if

$$\forall y : y \in \mu \leftrightarrow y \in \nu$$

  holds.

Condition (18.3) in Theorem 18.2 is definitely satisfied when the fuzzy sets μ and ν are disjoint with respect to the t-norm t, that is, we have $t(\mu(x), \nu(x)) = 0$ for all $x \in X$. A proof for this theorem can be found in Kruse et al. (1994).

Most of the variables in the context of fuzzy control are numerical or continuous. Similarity relations on the real line can be defined in a simple and understandable way using scaling functions based on the concept of distance, as described in Eq. (18.1). When we require that the similarity relation in Theorem 18.2 should be definable in terms of a scaling function, the following result proved in Klawonn (1994) is important.

Theorem 18.3 *Let $\mathcal{A} \subseteq \mathcal{F}(\mathbb{R})$ be a non-empty, at most countable set of fuzzy sets such that for each $\mu \in \mathcal{A}$ the following conditions are satisfied:*

- *There exists an $x_\mu \in \mathbb{R}$ such that $\mu(x_\mu) = 1$.*
- *μ (as real-valued function) is non-decreasing in the interval $(-\infty, x_\mu]$.*
- *μ is non-increasing in the interval $[x_\mu, \infty)$.*
- *μ is continuous.*
- *μ is differentiable almost everywhere.*

There exists a scaling function $c : \mathbb{R} \to [0, \infty)$ such that for all $\mu \in \mathcal{A}$ the extensional hull of the point x_μ with respect to the similarity relation

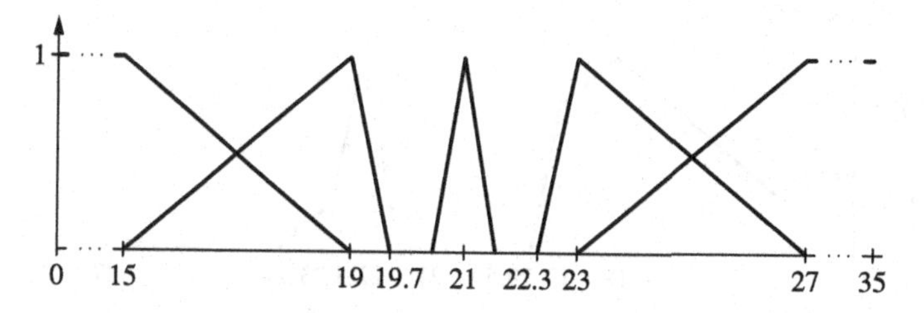

Fig. 18.3 The extensional hulls of the points 15, 19, 21, 23 and 27

$$E(x,y) = 1 - \min\left\{\left|\int_x^y c(s)\,ds\right|, 1\right\}$$

is equal to the fuzzy set μ *if and only if the condition*

$$\min\{\mu(x), \nu(x)\} > 0 \Rightarrow \left|\frac{d\mu(x)}{dx}\right| = \left|\frac{d\nu(x)}{dx}\right| \tag{18.4}$$

is satisfied for all $\mu, \nu \in \mathcal{A}$ *almost everywhere. In this case*

$$c : \mathbb{R} \to [0,\infty), \quad x \mapsto \begin{cases} \left|\frac{d\mu(x)}{dx}\right| & \text{if } \mu \in \mathcal{A} \text{ and } \mu(x) > 0, \\ 0 & \text{otherwise} \end{cases}$$

can be chosen as (almost everywhere well-defined) scaling function.

Example 18.3 In order to illustrate how extensional hulls of points with respect to a similarity relation induced by a piecewise constant scaling function might look like, we recall the scaling function

$$c : [0,35) \to [0,\infty), \quad s \mapsto \begin{cases} 0 & \text{if } 0 \le s < 15, \\ 0.25 & \text{if } 15 \le s < 19, \\ 1.5 & \text{if } 19 \le s < 23, \\ 0.25 & \text{if } 23 \le s < 27, \\ 0 & \text{if } 27 \le s < 35 \end{cases}$$

from Example 18.2. Figure 18.3 shows the extensional hulls of the points 15, 19, 21, 23 and 27 with respect to the similarity relation induced by the scaling function c.

These extensional hulls have triangular or trapezoidal membership functions. The reason for this is that the points are chosen in such a way that—moving away from the corresponding point—the membership degree drops to zero before the scaling function changes its value. When we choose a point close to a point where the scaling function jumps from one value to another we will not obtain a triangular or trapezoidal membership function for the extensional hull. In the case of a piecewise linear scaling function we can only guarantee that we obtain piecewise linear membership functions as extensional hulls as they are shown in Fig. 18.4.

It is quite common to use sets of fuzzy sets for fuzzy controllers as they are illustrated in Fig. 18.5. There, we choose values $x_1 < x_2 < \cdots < x_n$ and use triangular

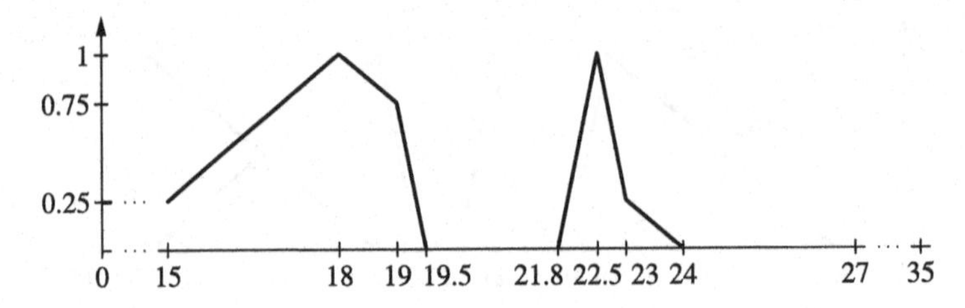

Fig. 18.4 The extensional hulls of the points 18.5 and 22.5

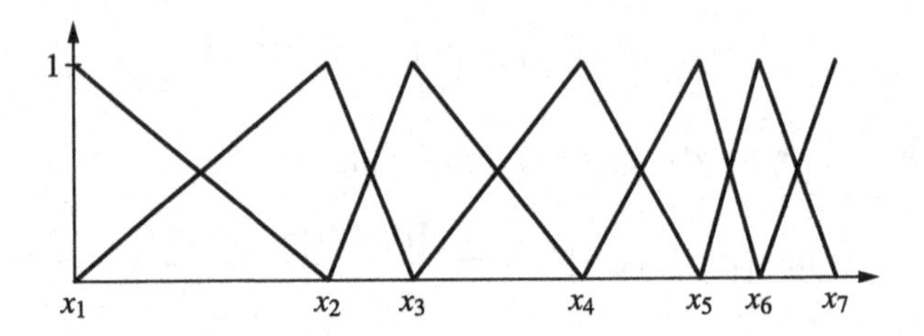

Fig. 18.5 Fuzzy sets for which we can define a scaling function

functions of the form $\Lambda_{x_{i-1},x_i,x_{i+1}}$ or the trapezoidal functions $\Pi_{-\infty,-\infty,x_1,x_2}$ and $\Pi_{x_{n-1},x_n,\infty,\infty}$ at the limits x_1 and x_n of the corresponding range, i.e.

$$\mathcal{A} = \{\Lambda_{x_{i-1},x_i,x_{i+1}} \mid 1 < i < n\} \cup \{\Pi_{-\infty,-\infty,x_1,x_2}, \Pi_{x_{n-1},x_n,\infty,\infty}\}.$$

In this case we can always define a scaling function c such that the fuzzy sets can be interpreted as the extensional hulls of the points $x_1, \ldots, x_n$, namely

$$c(x) = \frac{1}{x_{i+1} - x_i} \quad \text{if } x_i < x < x_{i+1}.$$

Now that we have introduced the basic concepts of similarity relations and their relation to fuzzy sets, we are able to discuss in more detail the conceptual background of fuzzy sets.

Using gradual membership degrees is the fundamental principle of fuzzy sets. Similarity relations are based on the fundamental concept of indistinguishability or similarity. The unit interval serves as the domain for gradual memberships as well as for similarity degrees. The values between 0 and 1 are interpreted in a mere intuitive way. There is no clear definition what exactly a membership or similarity degree of 0.8 or 0.9 means or what the quantitative difference is, apart from the fact that 0.9 is greater than 0.8.

Similarity relations with respect to the Łukasiewicz t-norm can be induced by pseudometrics. The concept of metric or distance is, at least in the case of the real line, elementary and does not need any further explanation or motivation.

In this sense, similarity relations, induced by the canonical metric—perhaps with some additional suitable scaling—can be understood as an elementary concept where the similarity degrees are interpreted dual to the concept of distances or metrics.

Fuzzy sets can be understood as concepts derived from similarity relations in the sense of extensional hulls of points or sets. In this interpretation, the membership degrees have a concrete meaning. The question is whether fuzzy sets should always be interpreted in this way. The answer is yes and no. Yes, because lacking an interpretation for membership degrees, the choice of the fuzzy sets and operations like t-norms becomes more or less arbitrary and a pure problem of parameter optimization. Yes, because at least in the context of fuzzy control, in most cases we have to deal with real numbers and the fuzzy sets usually fit quite well to the notion of an imprecise value, that is, an extensional hull of a point with respect to some suitable similarity relation. Also, the results on how fuzzy sets can be induced by similarity relations and how we can compute suitable similarity relations for given fuzzy sets speak in favor of an interpretation in terms of similarity relations.

18.5 Possibility Theory

Similarity relations are not the only and unique way to provide a more rigorous interpretation for fuzzy sets. Possibility theory offers an alternative interpretation for fuzzy sets that is also very useful especially in the case of discrete universes of discourse. It is out of the scope of this book to provide a rigorous introduction to possibility theory. But the ideas of a possibilistic interpretation of fuzzy sets can be illustrated by the following example.

Example 18.4 We consider an area where an automatic camera observes airplanes flying over this area. The recordings of a few days provide the following information: 20 airplanes of type A, 30 of type B and 50 of type C were observed. So next time, when we hear an airplane (before we can actually see it), we would think that with probability 20 %, 30 % and 50 % it is of type A, B and C, respectively.

Now we modify this purely probabilistic example slightly in order to explain the meaning of possibility distributions. In addition to the camera, we now have a radar unit and a microphone. The microphone has recorded 100 airplanes but we are not able to say anything about the type of airplane just by the sound. Because of poor visibility (fog or clouds) the camera could only identify 70 airplanes, 15 of type A, 20 of type B and 35 of type C. The radar unit was able to register 20 of the missing 30 airplanes but failed in the remaining 10 cases. From the radar unit, we know that 10 of these 20 airplanes were of type C because the radar unit can distinguish type C from type A and B since type C is much smaller than type A or B. The other 10 of the 20 airplanes registered by the radar unit are of type A or B but we cannot tell which of the two types since the radar image is to rough to distinguish between them.

This sample of 100 observed airplanes is very different from the 100 observations above where we were able to identify the type of airplane for each observation. Now

Table 18.3 Set-valued observations of airplane types

set	$\{A\}$	$\{B\}$	$\{C\}$	$\{A, B\}$	$\{A, B, C\}$
observed amount	15	20	45	10	10

a single observations can only be presented as a set of possible airplanes. How often each set was observed is shown in Table 18.3.

Without further knowledge or assumptions, we cannot estimate a probability for the single airplanes based on these observations without additional assumptions about the distribution of the airplane types of the observations in $\{A, B\}$ and $\{A, B, C\}$. In this case, (non-normalized) possibility distributions provide an alternative approach. Instead of a probability in the sense of a relative frequency we determine a **possibility degree**. The possibility degree for a certain type of an airplane is the ratio of the cases where the occurrence of the corresponding airplane is possible according to the observed set and the total number of all observations. Therefore, the corresponding possibility degrees are 35/100 for A, 40/100 for B and 55/100 for C, respectively. Then this "fuzzy set" over the universe of discourse $\{A, B, C\}$ is called a **possibility distribution**.

This example illustrates the difference between a possibilistic interpretation of fuzzy sets and one based on similarity relations. The possibilistic view is based on a form of uncertainty where the probabilistic concept of relative frequency is replaced by degrees of possibility. The underlying principle of similarity relations is not based on an uncertainty concept, but on an idea of indistinguishability or similarity, especially as a dual concept to the notion of distance. In fuzzy control the main issue is to model imprecision or vagueness in terms of "small distances" or similar values. Therefore, we do not further elaborate the details of possibility theory. Nonetheless, we refer the gentle reader to Dubois and Prade (1988, 2007) for a deeper look into this field.

References

D. Dubois and H. Prade. *Possibility Theory*. Plenum Press, New York, NY, USA, 1988

D. Dubois and H. Prade. Possibility theory. *Scholarpedia* 2(10):2074, 2007

F. Klawonn. Fuzzy Sets and Vague Environment. *Fuzzy Sets and Systems* 66:207–221. Elsevier, Amsterdam, Netherlands, 1994

F. Klawonn and J.L. Castro. Similarity in Fuzzy Reasoning. *Mathware and Soft Computing* 2:197–228. University of Granada, Granada, Spain, 1995

R. Kruse, J. Gebhardt, and F. Klawonn. *Foundations of Fuzzy Systems*. J. Wiley & Sons, Chichester, United Kingdom, 1994

Chapter 19
Fuzzy Control

The biggest success of fuzzy systems in the field of industrial and commercial applications has been achieved with fuzzy controllers. Fuzzy control is a way of defining a nonlinear table-based controller whereas its nonlinear transition function can be defined without specifying every single entry of the table individually. Fuzzy control does not result from classical control engineering approaches. In fact, its roots can be found in the area of rule-based systems. Fuzzy controllers simply comprise a set of vague rules that can be used for knowledge-based interpolation of a vaguely defined function (Moewes and Kruse 2012).

19.1 Mamdani Controllers

The first model of a fuzzy controller we introduce here was developed in 1975 by Mamdani (Mamdani and Assilian 1975) on the basis of the more general ideas of Zadeh published in Zadeh (1971, 1972, 1973).

The **Mamdani controller** is based on a finite set $\mathcal{R}$ of if-then-rules $R \in \mathcal{R}$ of the form

$$\begin{aligned} R: \quad & \text{If } x_1 \text{ is } \mu_R^{(1)} \text{ and } \dots \text{ and } x_n \text{ is } \mu_R^{(n)} \\ & \text{then } y \text{ is } \mu_R. \end{aligned} \tag{19.1}$$

$x_1, \dots, x_n$ are input variables of the controller and y is the output value. Usually, the fuzzy sets $\mu_R^{(i)}$ or μ_R stand for linguistic values, that is, for vague concepts like "about zero", "of average height" or "negative small" which, for their part, are represented by fuzzy sets. In order to simplify the notation, we do not distinguish between membership functions and linguistic values that they represent.

How to precisely interpret the rules is essential for understanding the Mamdani controller. Although the rules are formulated in terms of if-then statements, they should not be understood as logical implications but in the sense of a piecewise defined function. If the rule base $\mathcal{R}$ consists of the rules $R_1, \dots, R_r$, we should

R. Kruse et al., *Computational Intelligence*, Texts in Computer Science,
DOI 10.1007/978-1-4471-5013-8_19, © Springer-Verlag London 2013

understand it as a piecewise definition of a fuzzy function, that is

$$f(x_1, \dots, x_n) \approx \begin{cases} \mu_{R_1} & \text{if } x_1 \approx \mu_{R_1}^{(1)} \text{ and } \dots \text{ and } x_n \approx \mu_{R_1}^{(n)}, \\ \vdots & \\ \mu_{R_r} & \text{if } x_1 \approx \mu_{R_r}^{(1)} \text{ and } \dots \text{ and } x_n \approx \mu_{R_r}^{(n)}. \end{cases} \tag{19.2}$$

This is similar to the pointwise specification of a crisp function defined over a product space of finite sets in the form

$$f(x_1, \dots, x_n) \approx \begin{cases} y_1 & \text{if } x_1 = x_1^{(1)} \text{ and } \dots \text{ and } x_n = x_1^{(n)}, \\ \vdots & \\ y_r & \text{if } x_1 = x_r^{(1)} \text{ and } \dots \text{ and } x_n = x_r^{(n)}. \end{cases} \tag{19.3}$$

We obtain the graph of this function by

$$\text{graph}(f) = \bigcup_{i=1}^{r} \big(\hat{\pi}_1\big(\{x_i^{(1)}\}\big) \cap \dots \cap \hat{\pi}_n\big(\{x_i^{(n)}\}\big) \cap \hat{\pi}_Y\big(\{y_i\}\big)\big). \tag{19.4}$$

"Fuzzifying" this equation using the minimum for the intersection and the maximum (supremum) for the union, the fuzzy graph of the function described by the rule set $\mathcal{R}$ is the fuzzy set

$$\begin{aligned} &\mu_{\mathcal{R}} : X_1 \times \dots \times X_n \times Y \to [0, 1], \\ &(x_1, \dots, x_n, y) \mapsto \sup_{R \in \mathcal{R}} \big\{\min\big\{\mu_R^{(1)}(x_1), \dots, \mu_R^{(n)}(x_n), \mu_R(y)\big\}\big\} \end{aligned}$$

or

$$\begin{aligned} &\mu_{\mathcal{R}} : X_1 \times \dots \times X_n \times Y \to [0, 1], \\ &(x_1, \dots, x_n, y) \mapsto \max_{i \in \{1, \dots, r\}} \big\{\min\big\{\mu_{R_i}^{(1)}(x_1), \dots, \mu_{R_i}^{(n)}(x_n), \mu_{R_i}(y)\big\}\big\} \end{aligned}$$

in the case of a finite rule base $\mathcal{R} = \{R_1, \dots, R_r\}$.

If a concrete input vector $(a_1, \dots, a_n)$ for the input variables $x_1, \dots, x_n$ is given, then we obtain the fuzzy set

$$\mu_{\mathcal{R}, a_1, \dots, a_n}^{\text{output}} : Y \to [0, 1], \quad y \mapsto \mu_{\mathcal{R}}(a_1, \dots, a_n, y)$$

as the fuzzy "output value."

The fuzzy set $\mu_{\mathcal{R}}$ can be interpreted as fuzzy relation over the sets $X_1 \times \dots \times X_n$ and Y. Therefore, the fuzzy set $\mu_{\mathcal{R}, a_1, \dots, a_n}^{\text{output}}$ corresponds to the image of the one-element set $\{(a_1, \dots, a_n)\}$ or its characteristic function under the fuzzy relation $\mu_{\mathcal{R}}$. So in principle, instead of a sharp input vector we could also use a fuzzy set as input. For this reason, it is very common to call the procedure of feeding concrete input values to a fuzzy controller as **fuzzification**, that is, the input vector $(a_1, \dots, a_n)$ is

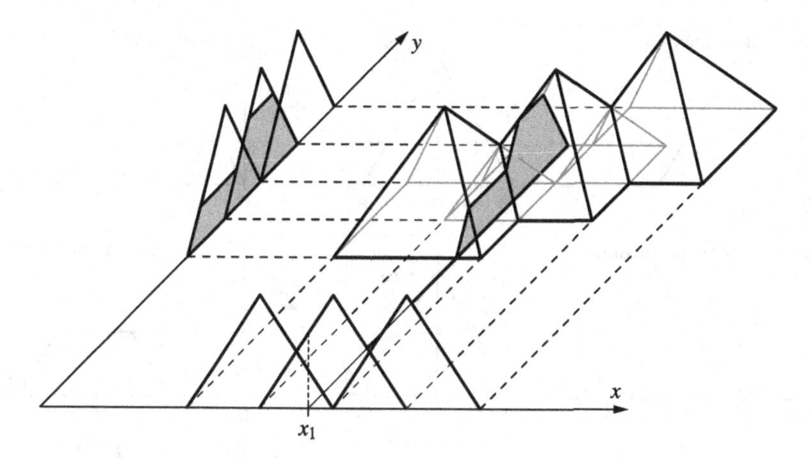

Fig. 19.1 The projection of the input value x_1 onto the output axis y

transformed into a fuzzy set which is nothing else than a representation of a one-element set in terms of its characteristic function.

Fuzzification can also be interpreted in another way. In Chap. 17, we have seen that we can obtain the image of a fuzzy set under a fuzzy relation computing the cylindrical extension of the fuzzy set, intersecting the cylindrical extension with the fuzzy relation and projecting the result to the output space. In this sense, we can understand the cylindrical extension of the measured tuple or the corresponding characteristic function as a the fuzzification procedure which is necessary for intersecting it with the fuzzy relation.

Figure 19.1 shows this procedure. In order to make a graphical representation possible, we consider only one input and output variable. Three rules are shown in the figure:

$$\text{If } x \text{ is } A_i, \text{ then } y \text{ is } B_i, \quad (i = 1, 2, 3).$$

The fuzzy relation $\mu_{\mathcal{R}}$ is represented by the three pyramids. If the input value x is given, the cylindrical extension of $\{x\}$ defines a plane cutting through the pyramids. The projection of this cutting plane onto the y-axis (pointing from the front to the back) yields us the fuzzy set $\mu_{\mathcal{R},x}^{\text{output}}$ which describes the desired output value (in fuzzy terms).

We can illustrate the computation of the output value by the following scheme. Figure 19.2 shows two rules of a Mamdani controller with two input variables and one variable. At first, we consider only one of the two rules, say the rule R. The degree to which the antecedent is satisfied for the given input values is determined by taking the minimum of the membership degrees of the corresponding fuzzy sets. Then, the fuzzy set in the consequent of the rule is "cut off" at this level. This means that the membership degree to the actual output fuzzy set is the minimum of the membership degrees to the fuzzy set in the antecedent of the rule and the firing strength of the rule.

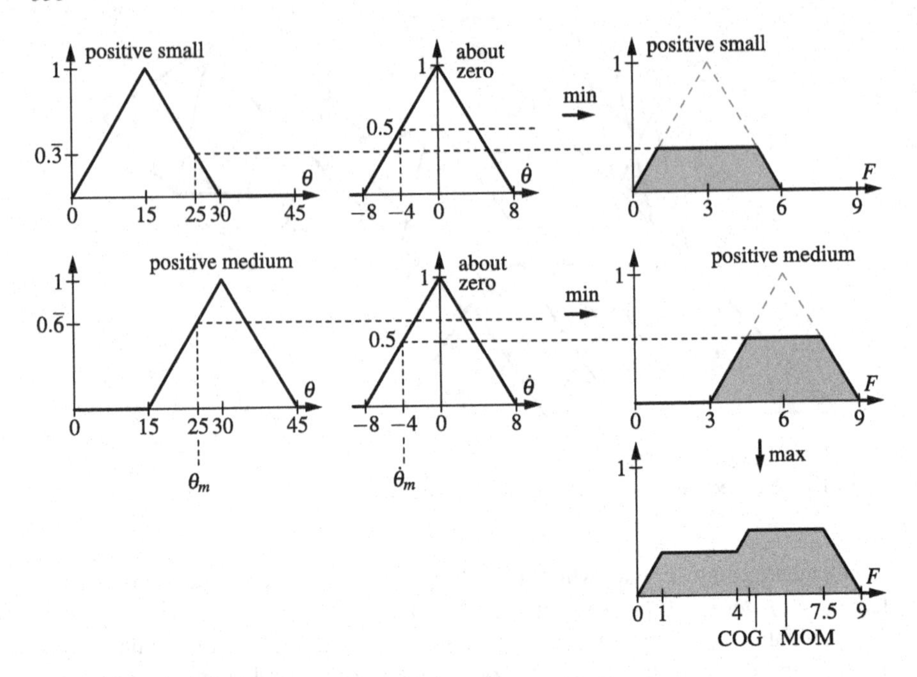

Fig. 19.2 Illustration of the Mamdani controller computation scheme

If the firing strength of rule is 1, then we exactly obtain the fuzzy set in the consequent as result, that is, $\mu_R = \mu^{\text{output}}_{R,a_1,\ldots,a_n}$. When the rule is not applicable, meaning that the firing strength is 0, we obtain $\mu^{\text{output}}_{R,a_1,\ldots,a_n} = 0$, that is, based on this rule nothing can be said about the output value.

All other rules are treated in the same way—Fig. 19.2 shows just two rules—so, for every rule R we obtain a fuzzy set $\mu^{\text{output}}_{R,a_1,\ldots,a_n}$. In real applications, usually only a few rules have a nonzero firing strength at the same time and thus contribute to the final output fuzzy set. In the second step, these output fuzzy sets derived from single rules have to be joint to a single fuzzy set which characterizes the described output value.

In order to explain in which way this aggregation is carried out, we reconsider the interpretation of the rule base of a fuzzy controller in the sense of a piecewise definition of a function (cf. Eq. (19.2)). For a crisp piecewisely defined function, the r different cases have to be disjoint or, if they overlap they must provide the same output because otherwise the function value is not well defined. We assume that each single case describes a "function value" in the form of a set for every input value: If a considered input value satisfies the condition of the corresponding case, then we obtain a one-element set with the specified function value. Otherwise, we obtain the empty set for this case. With this interpretation, the actual function value or the one-element set which contains this function value is given by the union of the sets resulting from the single cases.

For this reason, the (disjoint) union of the fuzzy sets $\mu^{\text{output}}_{R,a_1,\ldots,a_n}$ resulting from the single rules should be computed for the overall fuzzy output. In order to compute the union of fuzzy sets, we use a t-conorm, for example, the maximum. Thus, we obtain

$$\mu^{\text{output}}_{\mathcal{R},a_1,\ldots,a_n} = \max_{R\in\mathcal{R}}\left\{\mu^{\text{output}}_{R,a_1,\ldots,a_n}\right\} \tag{19.5}$$

as the final output fuzzy set given both the rule base $\mathcal{R}$ and the input vector $(a_1,\ldots,a_n)$. In this way, the two rules shown in Fig. 19.2 lead to the output fuzzy set which is also shown there.

In order to obtain a single crisp output value, the output fuzzy set has to be *defuzzified*. A large variety of heuristic defuzzification strategies has been proposed in the literature. However, without specifying more precisely how the fuzzy sets and rules should be interpreted, defuzzification remains a matter of pure heuristics. Therefore, we restrict our considerations to one defuzzification strategy and will discuss the issue of defuzzification in more detail after the introduction of conjunctive rule systems.

In order to understand the fundamental idea of defuzzification applied in the case of Mamdani controllers, we consider the output fuzzy set determined in Fig. 19.2 once again. The fuzzy sets in the consequents of the two rules are interpreted as vague or imprecise values. In the same sense, the resulting output fuzzy set is a vague or imprecise description of the desired output value. Intuitively, the output fuzzy set in Fig. 19.2 can be understood in the sense that we should favor a larger output value, but should also consider a smaller output value with a lower preference. This interpretation is justified by the fact that the first rule, voting for a larger output value, fires with a higher degree than the second rule that points to a lower output value. Therefore, we should choose an output value that is large, but not too large, that is, a compromise of the proposed outputs of the two rules, however, putting a higher emphasis on the first rule.

A defuzzification strategy which satisfies this criterion is the center of gravity (COG) method which is also called center of area (COA). The output value of this method is the center of gravity (or, to be precise, its projection onto the output axis) of the area under the output fuzzy set, that is,

$$\text{COA}\left(\mu^{\text{output}}_{\mathcal{R},a_1,\ldots,a_n}\right) = \frac{\int_Y \mu^{\text{output}}_{\mathcal{R},a_1,\ldots,a_n} \cdot y \, dy}{\int_Y \mu^{\text{output}}_{\mathcal{R},a_1,\ldots,a_n} \, dy}. \tag{19.6}$$

This method requires the implicit condition that the functions $\mu^{\text{output}}_{\mathcal{R},a_1,\ldots,a_n} \cdot y$ and $\mu^{\text{output}}_{\mathcal{R},a_1,\ldots,a_n}$ are integrable which is always satisfied, provided that the membership functions appearing in the consequents of the rules are chosen in a reasonable way, for example, when they are continuous.

19.1.1 Remarks on Fuzzy Controller Design

When choosing the fuzzy sets for the input variables, one should make sure that the domain of each input variable is completely covered, this means, for every possible value there exists at least one fuzzy set to which it has a nonzero membership degree. Otherwise, the fuzzy controller cannot determine an output value for this input value.

The fuzzy sets should represent vague or imprecise values or ranges. Therefore, convex fuzzy sets are preferable. Triangular and trapezoidal membership functions are especially suitable, because they have a simple parametric representation and determining the membership degrees can be achieved with low computational effort. In ranges where the controller has to react very sensitively to small changes of an input value we should choose very narrow fuzzy sets in order to distinguish between the values well enough. We should, however, take into account that the number of possible rules grows very fast with the number of fuzzy sets. If we have k_i fuzzy sets for the ith input variable, then a complete rule base that assigns to all possible combinations of input fuzzy sets an output fuzzy set, contains $k_1 \cdot \dots \cdot k_n$ rules when we have n input variables. If we have four input values with only five fuzzy sets for each of them, then we already get 625 rules.

Concerning the choice of the fuzzy sets for the output variable similar constraints as for the input variables should be considered. They should be convex and in the ranges, where a very exact output value is important, narrow fuzzy sets should be used. Additionally, the choice of fuzzy sets for the output value strongly depends on the defuzzification strategy. It should be noted that the defuzzification of asymmetrical triangular membership functions of the form $\Lambda_{x_0-a,x_0,x_0+b}$ with $a \neq b$ does not always correspond to what we might expect. If only one single rule fires with the degree 1 and all others with degree 0, we obtain the corresponding fuzzy set in the consequent of this rule as the output fuzzy set of the controller before defuzzification. If this is an asymmetrical triangular membership function $\Lambda_{x_0-a,x_0,x_0+b}$, we have $\mathrm{COA}(\Lambda_{x_0-a,x_0,x_0+b}) \neq x_0$, that is, not the point where the triangular membership function reaches the membership degree 1.

Another problem of the center of gravity method is that it can never return a boundary value of the interval of the output variable. This means, the minimum and maximum value of the output domain are not accessible for the fuzzy controller. A possible solution of this problem is to extend the fuzzy sets beyond the interval limits for the output variable. However, in this case we have to ensure that the controller will never yield an output value outside the range of permitted output values.

For the design of the rule base, completeness is an important issue. We have to make sure that for any possible input vector there is at least one rule that fires. This does not mean that for every combination of fuzzy sets of input values we have to formulate a rule with these fuzzy sets in the antecedent. On the one hand, a sufficient overlapping of the fuzzy sets guarantees that there will still be rules firing, even if we have not specified a rule for all possible combinations of input fuzzy sets. On the other hand, there might be combinations of input values that correspond to a state which the system can or must never reach. For these cases, it is not necessary to specify rules. We should also avoid contradicting rules. Rules with identical antecedents and different consequents should be avoided.

The Mamdani controller we have introduced here is also called max-min controller because of Eq. (19.5) for computing the output fuzzy set $\mu_{\mathcal{R},a_1,\ldots,a_n}^{\text{output}}$. Maximum and minimum were used to calculate the union and the intersection, respectively, in Eq. (19.4).

Of course, also other t-norms and t-conorms can be considered instead of minimum or maximum. In applications the product t-norm is often preferred and the bounded sum $s(\alpha, \beta) = \min\{\alpha + \beta, 1\}$ is sometimes used as the corresponding t-conorm. The disadvantage of minimum and maximum is the idempotency property. The output fuzzy set $\mu_{R,a_1,\ldots,a_n}^{\text{output}}$ of a rule R depends only on the input variable for which the minimum membership degree to the corresponding fuzzy set in the antecedent is obtained. A change of another input variable will not affect the output of this rule, unless the change is large enough to let the membership degree of this input value to its corresponding fuzzy set drop below the membership degree of the other input values.

If the fuzzy sets $\mu_{R,a_1,\ldots,a_n}^{\text{output}}$ of different rules support a certain output value to some degree, the aggregation based on the maximum will only take the largest of these membership degrees into account. It might be desirable that such an output value obtains a higher support than another one that is supported to the same degree, but only by a single rule. In this case, a t-conorm like the bounded sum should be preferred to the maximum.

In principle, we could also compute the firing strength of a rule and the influence of this degree to the fuzzy set in the consequent of this rule based on different t-norms. Some approaches even choose an individual t-norm for each rule.

Sometimes, even t-conorms are used to compute the firing strength of a rule. In this case, the corresponding rule must read as

$$\begin{aligned} R: \quad & \text{If } x_1 \text{ is } \mu_R^{(1)} \text{ or } \ldots \text{ or } x_n \text{ is } \mu_R^{(n)} \\ & \text{then } y \text{ is } \mu_R. \end{aligned}$$

In the sense of our interpretation of the rules as a piecewise definition of a fuzzy function, this rule can be replaced by the following n rules.

$$\begin{aligned} R_i: \quad & \text{If } x_i \text{ is } \mu_R^{(i)} \\ & \text{then } y \text{ is } \mu_R. \end{aligned}$$

In some commercial programs, weighted rules are allowed. The resulting output fuzzy set of a rule is then multiplied by the assigned weight. Such weights increase the number of the free parameters of a fuzzy controller. But the same effect can be achieved directly by an adequate choice of the fuzzy sets in the antecedent or the consequent of the rule without any weights. In most cases, weights make it more difficult to interpret a fuzzy controller.

The fundamental idea of the Mamdani controller as a piecewise definition of a fuzzy function requires implicitly that the antecedents of the rules represent disjoint

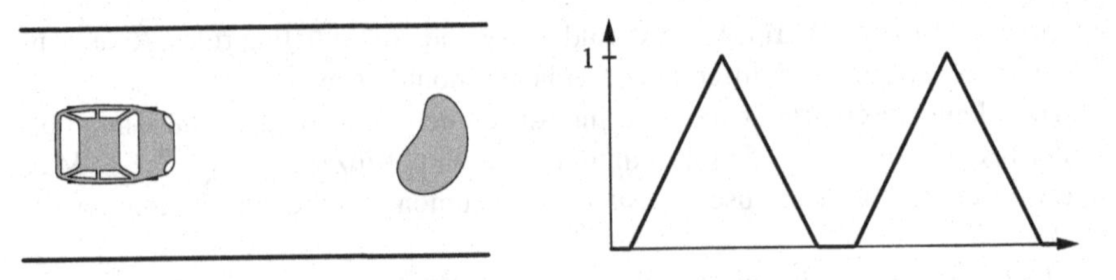

Fig. 19.3 A car driving towards an object (*left*) and possible output fuzzy set of a steering controller consisting of two disjoint fuzzy (*right*)

fuzzy situations or cases. At this point we do not to formalize this concept of disjointness for fuzzy sets exactly. However, when this assumption is ignored, this can lead to an undesired behavior of the fuzzy controller. For instance, refining the control actions cannot be achieved by merely adding rules without changing the already existing fuzzy sets. As an extreme example we consider the rule

$$\text{If } x \text{ is } I_X, \text{ then } y \text{ is } I_Y,$$

where the fuzzy sets I_X and I_Y are the characteristic functions of the input and output domain, respectively, yielding membership degree 1 for any value. The rule can be read as

If x is *anything*, then y is *anything*.

With the rule, the output fuzzy set will always be constantly 1, the characteristic function of the whole output domain. It does not depend on or change with other rules that we add to the rule base. We will discuss this problem again, when we introduce conjunctive rule systems.

Another problem of the disjoint fuzzy cases is illustrated in Fig. 19.3 which shows an output fuzzy set where the defuzzification is difficult.

Should we interpolate between the two fuzzy values represented by the triangles like it would be done by the center of gravity method? That would mean that a defuzzification yields a value whose membership degree to the output fuzzy set is 0 which surely does not fit to our intuition. Or do the two triangles represent two alternative output values from which we have to choose one? For example, the illustrated fuzzy set could be the output fuzzy set of a controller which is supposed to drive a car around an obstacle. Then, the fuzzy set says that we have to evade the obstacle either to the left or to the right, but not keep on driving straight ahead. The latter solution would be proposed by the center of gravity method, leading to the effect that the car will bump into the obstacle. The interpretation of two alternative outputs is contradictory to the underlying philosophy of the Mamdani controller as a piecewise definition of a fuzzy function, because, in this case, the function is not well-defined, since two fuzzy values are assigned at the same time to only one input.

19.1.2 Defuzzification Methods

In recent years, numerous defuzzification methods were proposed which were developed more or less intuitively on the basis of the fact that a fuzzy set, but no further information is given. But what is missing is a systematic approach which is based on a more rigorous interpretation of the fuzzy set which has to be defuzzified.

A general defuzzification strategy has to fulfill two tasks at the same time. It has to turn a fuzzy set into a set and it has to choose one (fuzzy) value from a set of (fuzzy) values. It is not obvious in which order these two steps should be carried out. For example, the fuzzy set from Fig. 19.3 could be defuzzified by first choosing one of the two fuzzy values it represents, that is, by choosing one of the two triangles. In the second step, the corresponding triangle, representing only a single fuzzy value must be turned into a suitable crisp value. We could also exchange the order of these two steps. Then we would first turn the fuzzy set into a crisp set that in this case would contain two elements. Afterwards, we have to pick one of the values from this crisp set. These considerations are not taken into account in the axiomatic approach for defuzzification in Runkler and Glesner (1993) nor by most defuzzification methods which implicitly assume that the fuzzy set which is to be defuzzified represents just a single vague value and not a set of vague values.

The underlying semantics or interpretation of the fuzzy controller or fuzzy system is also essential for the choice of the defuzzification strategy. In the following section, we will explain in further detail that the Mamdani controller is based on an interpolation philosophy. Other approaches do not fit into this interpolation scheme, as we will see in the section on conjunctive rule systems.

At this point, we explain some other defuzzification strategies and their properties in order to better understand the issue of defuzzification. The *mean of maxima* (MOM) method is a very simple defuzzification strategy which chooses as output value the mean values of the values with maximum membership degree to the output fuzzy set. This method is rarely applied in practice because for symmetrical fuzzy sets it leads to a discontinuous controller behavior. Given the input values, the output value depends only on the output fuzzy set which belongs to the rule with the highest firing strength—provided that there are not two or more rule which by chance fire to the same degree and which have different fuzzy sets in their consequents. When we use symmetrical fuzzy sets in the consequents of the rules in the context of the mean of maxima method, the output value will always be on of the centers of the fuzzy sets, except in the rare case when two or more rules fire with the same maximum degree. Therefore, the rule with the highest firing strength will determine the constant output until another rule takes over. Then the controller output will jump directly or with one intermediate step to the output of the rule that has now the maximum firing strength.

Both the center of area method as well es MOM will result in the possibly non-desired mean value in the defuzzification problem shown in Fig. 19.3. There, they always choose the right-most value of the values with maximum membership degree to the output fuzzy set. (Alternatively one can always take the left-most one.)

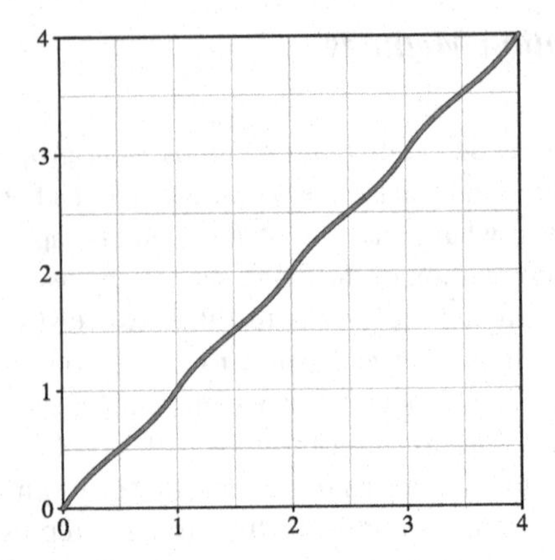

Fig. 19.4 Interpolation of a straight line using the center of gravity method

According to Kahlert and Frank (1994), the authors hold a patent on this method. But similar to MOM, it can also lead to discontinuous changes of the output value.

The center of gravity method requires relatively high computational costs and does not have the interpolation properties we would expect. To illustrate this, we consider a Mamdani controller with the following rule base:

If x is 'about 0',	then y is 'about 0'
If x is 'about 1',	then y is 'about 1'
If x is 'about 2',	then y is 'about 2'
If x is 'about 3',	then y is 'about 3'
If x is 'about 4',	then y is 'about 4'

The terms "about 0", ..., "about 4" are represented by fuzzy sets in the form of symmetrical triangular membership functions of width two, that is, $\Lambda_{-1,0,1}$, $\Lambda_{0,1,2}$, $\Lambda_{1,2,3}$, $\Lambda_{2,3,4}$ and $\Lambda_{3,4,5}$, respectively. It seems that these rules describe the function $y = x$, that is, a straight line. Applying the center of gravity method the resulting function matches the straight line only at the values $0, 0.5, 1, 1.5, \ldots, 3.5$ and 4. At all of the other points, the control function differs a little bit from the simple straight line as shown in Fig. 19.4.

These and other undesirable effects, which, for example, can appear using asymmetrical membership functions in the consequent, can be avoided by using rules with a crisp value in the consequent. For describing the input values, we still use fuzzy sets, but the outputs are specified not in terms of fuzzy sets, but as a single value for each rule. In this case, the defuzzification is also very simple. We take the weighted mean of the rule outputs, that is, each rule assigns its firing strength as weight to its crisp output value in the consequent. In this way, we obtain

$$y = \frac{\sum_R \mu^{\text{output}}_{R,a_1,\ldots,a_n} \cdot y_R}{\sum_R \mu^{\text{output}}_{R,a_1,\ldots,a_n}} \tag{19.7}$$

as the output of the fuzzy controller. The rules have the form

$$R: \quad \text{If } x_1 \text{ is } \mu_R^{(1)} \text{ and} \ldots \text{and } x_n \text{ is } \mu_R^{(n)}, \text{ then } y \text{ is } y_R$$

with crisp output values y_R. $a_1, \ldots, a_n$ are the measured input values of the input variables $x_1, \ldots, x_n$ and $\mu_{R,a_1,\ldots,a_n}^{\text{output}}$ denotes—as it used to do so—the firing strength of the rule R with these input values.

The question how to defuzzify output fuzzy sets is actively discussed even today: Recently, kernel functions have been introduced to generalize existing defuzzification methods (Runkler 2012), for example, in order to improve smoothness properties.

19.2 Takagi–Sugeno–Kang Controllers

Takagi–Sugeno or Takagi–Sugeno–Kang controllers (TS or TSK models) (Sugeno 1985; Takagi and Sugeno 1985) use rules of the form

$$R: \quad \text{If } x_1 \text{ is } \mu_R^{(1)} \text{ and} \ldots \text{and } x_n \text{ is } \mu_R^{(n)}, \text{ then } y = f_R(x_1, \ldots, x_n). \tag{19.8}$$

In the same manner as in the case of the Mamdani controller (cf. Eq. (19.1)), the input values in the rules are described by fuzzy sets. However, using a TSK model, the consequent of a single rule consists no longer of a fuzzy set, but determines a function with the input variables as arguments. The basic idea is that the corresponding function is a good local control function for the fuzzy region that is described by the antecedent of the rule. For instance, if we use linear functions, the desired input/output behavior of the controller is described locally (in fuzzy regions) by linear models. At the boundaries between single fuzzy regions, we have to interpolate in a suitable way between the corresponding local models. This is done by

$$y = \frac{\sum_R \mu_{R,a_1,\ldots,a_n} \cdot f_R(x_1, \ldots, x_n)}{\sum_R \mu_{R,a_1,\ldots,a_n}} \tag{19.9}$$

where $a_1, \ldots, a_n$ are the measured input values of the input variables $x_1, \ldots, x_n$, and $\mu_{R,a_1,\ldots,a_n}$ denotes the firing strength of rule R which results given these input values.

A special case of the TSK models is the variant of the Mamdani controller where the fuzzy sets in the consequent of the rules are replaced by constant values and the output value is calculated by Eq. (19.7). In this case, the functions f_R are constant.

In order to maintain the interpretability of a TSK controller in terms of local models f_R for fuzzy regions, a strong overlap of these regions should be avoided, since otherwise the interpolation formula given in Eq. (19.9) can completely blur the single models and mix them together into one complex model, that might have a good control behavior, but looses interpretability completely. As an example we consider the following rules:

$$\begin{array}{ll}\text{If } x \text{ is 'very small'}, & \text{then } y = x\\ \text{If } x \text{ is 'small'}, & \text{then } y = 1\\ \text{If } x \text{ is 'large'}, & \text{then } y = x - 2\\ \text{If } x \text{ is 'very large'}, & \text{then } y = 3\end{array}$$

First, the terms 'very small', 'small', 'large' and 'very large' are modeled by the four non-overlapping fuzzy sets in Fig. 19.5. In this case, the four functions or local models $y = x$, $y = 1$, $y = x - 2$ and $y = 3$ defined in the rules are reproduced exactly as shown in Fig. 19.5. If we choose only slightly overlapping fuzzy sets, the TSK model yields the control function in Fig. 19.6. Finally, Fig. 19.7 shows the result of the TSK model which uses fuzzy sets that overlap even more.

We can see that the TSK model can lead to slight overshoots as shown in Fig. 19.6, even if the fuzzy sets overlap just slightly. If we choose an overlap of the fuzzy sets typical for Mamdani controllers, then the single local model are not visible anymore as Fig. 19.7 shows.

A suitable way to avoid this undesirable effect is to use trapezoidal membership functions instead of triangular ones, when working with TSK models. When we choose trapezoidal membership functions in such a way that an overlap occurs only at the edges of the trapezoidal functions, the corresponding local models are reproduced exactly in the regions where membership degree is 1.

19.3 Logic-Based Controllers

In this section, we discuss the consequences resulting from an interpretation of the rules of a fuzzy controller in the sense of logical implications. In Example 17.13 on page 337, we have already seen how logical inference can be modeled on the basis of fuzzy relations. This concept is now applied to fuzzy control. In order to simplify the notation, we first restrict our considerations to fuzzy controllers with only one input and one output variable. The rules have the form

$$\text{If } x \text{ is } \mu, \text{ then } y \text{ is } \nu.$$

With one single rule of this form and a given input value x, we obtain an output fuzzy set according the computation scheme from Example 17.13. When the input value x has a membership degree of 1 to the fuzzy set μ, the resulting output fuzzy matches exactly the fuzzy set μ, just as in the case of the Mamdani controller. However, in contrast to the Mamdani controller, the output fuzzy set becomes larger, the less the antecedent of the rule is satisfied, that is, the lower the value $\mu(x)$ is. In the extreme case $\mu(x) = 0$, we obtain as output the fuzzy set which is constantly 1. For a Mamdani controller, we would obtain the fuzzy set which is constantly 0. Therefore, using a logic-based controller we should interpret the output fuzzy set as set of the yet possible values. If the antecedent does not apply at all ($\mu(x) = 0$), the rule does not provide any information about the output value and all output values are possible. If the rule applies to 100 % ($\mu(x) = 1$), only the values from the (fuzzy) set μ are possible. Therefore, a single rule provides a (fuzzy) constraint on the set

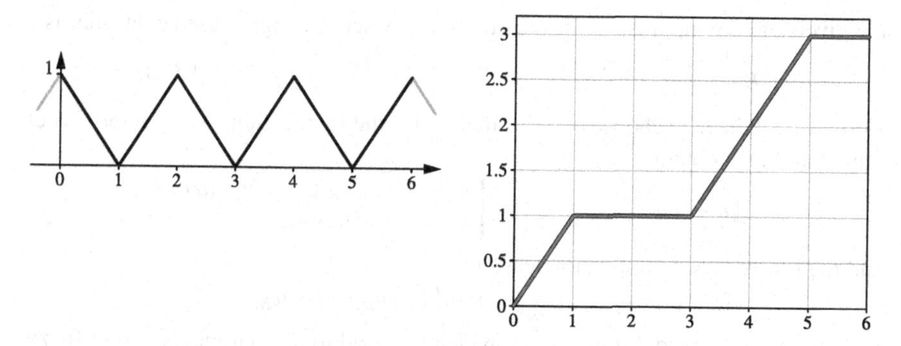

Fig. 19.5 Four non-overlapping fuzzy sets: exact reproduction of the local models

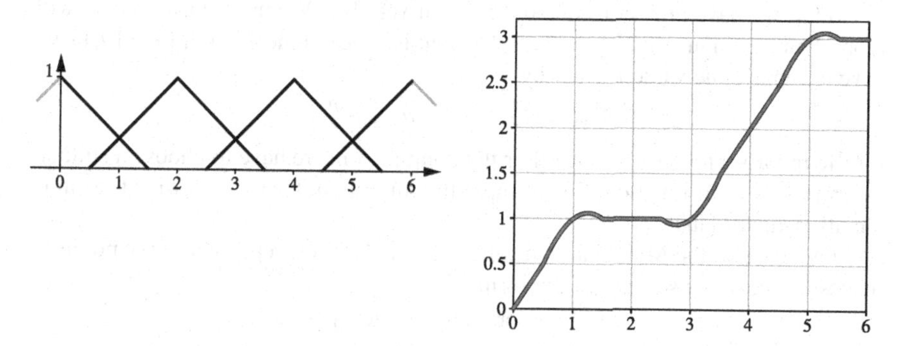

Fig. 19.6 Four slightly overlapping fuzzy sets: slight mixture of the local models

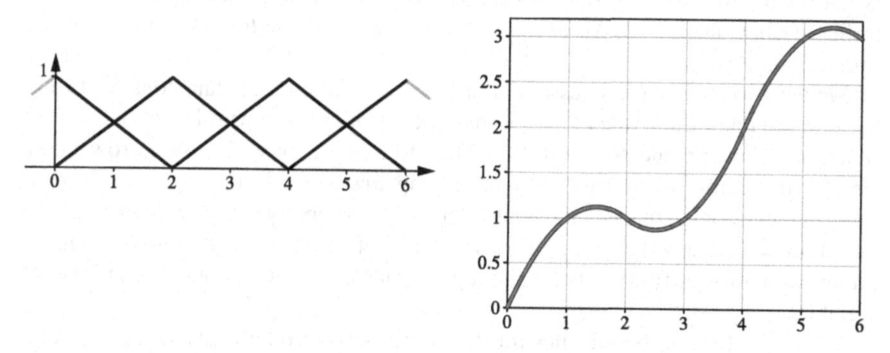

Fig. 19.7 Four strongly overlapping fuzzy sets: almost complete mix of the local models

of possible values. Since all rules are considered to be correct (true), all fuzzy constraints specified by the rules have to be satisfied, that is, the fuzzy sets resulting from the single rules have to be intersected with each other. If r rules of the form

$$R_i: \quad \text{If } x \text{ is } \mu_{R_i}, \text{ then } y \text{ is } \nu_{R_i} \qquad (i = 1, \ldots, r)$$

are given and the input is $x = a$, the output fuzzy set of a logic-based controller is

$$\mu_{\mathcal{R},a}^{\text{out,logic}} : Y \to [0,1], \quad y \mapsto \min_{i \in \{1,\ldots,r\}} \{ [\![a \in \mu_{R_i} \to y \in \nu_{R_i}]\!] \}.$$

Here, we still have to choose a truth function for the implication $\to$. With the Gödel implication, we obtain

$$[\![a \in \mu_{R_i} \to y \in \nu_{R_i}]\!] = \begin{cases} \nu_{R_i}(y) & \text{if } \nu_{R_i}(y) < \mu_{R_i}(a), \\ 1 & \text{otherwise,} \end{cases}$$

and the Łukasiewicz implication leads to

$$[\![a \in \mu_{R_i} \to y \in \nu_{R_i}]\!] = \min\{1 - \nu_{R_i}(y) + \mu_{R_i}(a), 1\}.$$

In contrast to the Gödel implication, which can lead to discontinuous output fuzzy sets, the output fuzzy sets of the Łukasiewicz implication are always continuous, provided that the involved fuzzy sets are continuous (as real-valued functions).

So far, we have only considered one input variable. When we have to deal with more than one input variable in the rules, that is, rules in the form of Eq. (19.1), we have to replace the value $\mu_{R_i}(a)$ by

$$[\![a_1 \in \mu_{R_i}^{(1)} \wedge \cdots \wedge a_n \in \mu_{R_i}^{(n)}]\!]$$

for the input vector $(a_1, \ldots, a_n)$. For the conjunction, we have to choose a suitable t-norm as its truth function, for example the minimum, the Łukasiewicz t-norm or the algebraic product.

In the case of the Mamdani controller, where the rules represent fuzzy points, it makes no sense to use rules of the form

If x_1 is μ_1 or x_2 is μ_2, then y is ν.

But if we use a logic-based controller, we can determine an arbitrary logic expression with predicates (fuzzy sets) over the input variables as antecedent, so that it is reasonable to have also rules with disjunction or negation for logic-based controllers (Klawonn 1992). We only have to specify suitable truth functions for the logical connectives.

We want to emphasize an essential difference between Mamdani controllers and logic-based ones. Each rule of a logic-based controller is a constraint for the control function (Klawonn and Novák 1996). Therefore, the choice of very narrow fuzzy sets for the output and (strongly) overlapping fuzzy sets in the input can lead to contradictory constraints and the controller yields the empty fuzzy set (constantly 0) as output. While specifying the fuzzy sets this fact should be taken into account by preferring narrower fuzzy sets for the input variables and wider ones for the output variables.

Increasing the number of rules for the Mamdani controller leads, in general, to a more fuzzy output, because the output fuzzy set is the union of the output fuzzy sets resulting from the single rules. In the extreme case, the trivial but empty rule

If x is *anything*, then y is *anything*,

where *anything* is modeled by the fuzzy set which is constantly 1, causes that the output fuzzy set to be constantly 1. This is independent of other rules occurring in the rule base of the Mamdani controller. In a logic-based controller this true but useless rule has no effect and does not destroy the control function completely.

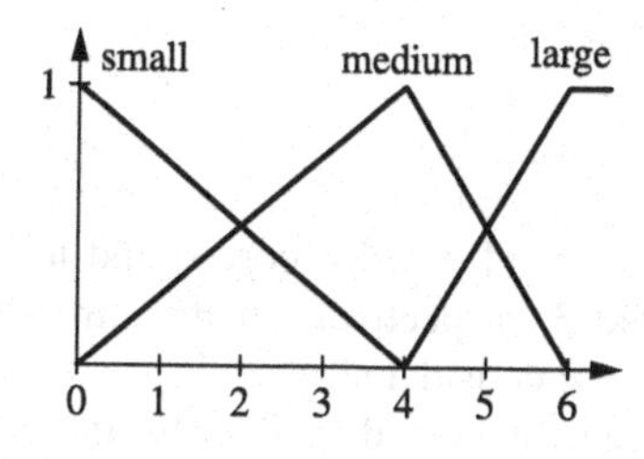

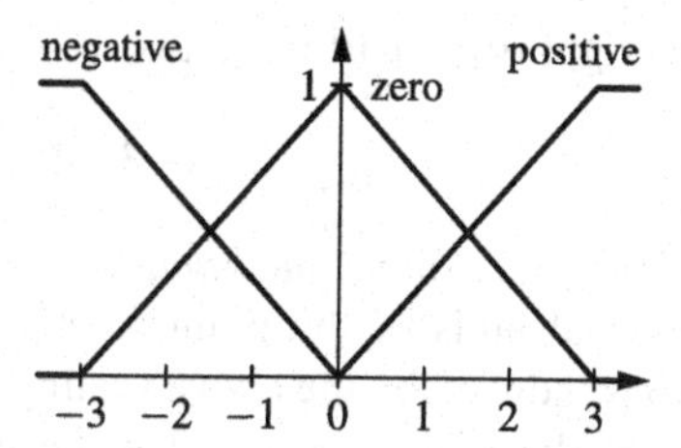

Fig. 19.8 Two fuzzy partitions

19.4 Mamdani Controller and Similarity Relations

When we introduced the Mamdani controllers, we have already seen that the fuzzy rules there represent fuzzy points on the graph of the control or transfer function which is described by the controller. Based on the concept of similarity relations as discussed in Chap. 18, fuzzy sets, such as the ones appearing in the Mamdani controller, can be interpreted as fuzzy points. Here we discuss this interpretation of the Mamdani controller and its consequences in detail.

19.4.1 Interpretation of a Controller

At first, we consider a given Mamdani controller. Let us assume that the fuzzy sets defined for the input and output domains satisfy the conditions of Theorem 18.2 or even better of Theorem 18.3. In this case, we can derive similarity relations such that the fuzzy sets can be interpreted as extensional hulls of single points.

Example 19.1 For a Mamdani controller with two input variables x and y and one output variable z, we use the left fuzzy partition from Fig. 19.8 for the input variables and the right one for the output variable. The rule base consists of four rules.

R_1: If x is *small*	and y is *small*	then z is *positive*
R_2: If x is *medium*	and y is *small*	then z is *zero*
R_3: If x is *medium*	and y is *big*	then z is *zero*
R_4: If x is *big*	and y is *big*	then z is *negative*

These fuzzy partitions satisfy the conditions of Theorem 18.3 such that suitable scaling functions can be found. For the left fuzzy partition in Fig. 19.8, the scaling function is

$$c_1 : [0, 6] \to [0, \infty), \qquad x \mapsto \begin{cases} 0.25 & \text{if } 0 \le x < 4, \\ 0.5 & \text{if } 4 \le x \le 6, \end{cases}$$

for the right fuzzy partition, it is

$$c_2 : [-3, 3] \to [0, \infty), \quad x \mapsto \frac{1}{3}.$$

The fuzzy sets *small, medium, big, negative, zero* and *positive* correspond to the extensional hulls of the points $0, 4, 6, -3, 0$ and 3, respectively, if the similarity relations induced by the above scaling functions are considered.

Then, the four rules say that the graph of the function described by the controller should pass through the corresponding points $(0, 0, 3)$, $(4, 0, 0)$, $(4, 6, 0)$ and $(6, 6, -3)$.

The interpretation on the basis of similarity relations in the example above specifies four points on the graph of the control function as well as additional information encoded in the similarity relations. The construction of the whole control function is therefore an interpolation task. We want to find a function which passes through the given points and maps similar values to similar values in the sense of the similarity relations.

If we, for example, want to find a suitable output value for the input $(1, 1)$, we can see that $(0, 0)$ is the most similar input pair for which we know an output value, i.e. 3 according to rule R_1. The similarity degree of 1 to 0 is nothing but the membership degree of the value 1 to the extensional hull of 0, that is, to the fuzzy set *small*, which is 0.75. The input $(1, 1)$ also similar to the input pair $(4, 0)$ in rule R_2, however, much less similar than to $(0, 0)$. The similarity degree of 1 to 4 is 0.25, the similarity of 1 to 0 is again 0.75. Thus, the output value for $(1, 1)$ should be quite similar to the output value 3 for the input $(0, 0)$ (rule R_1) but also a little bit similar to the output value 0 of the input $(4, 0)$ (rule R_2).

So far we have left the question open, how the two similarity degrees, which we obtain for the two components x and y of the input pair, should be aggregated. A t-norm is a suitable choice in this case, for example, the minimum. As an example, let us determine how well the output value 2 fits to the input $(1, 1)$. In order to answer this question, we have to compute the similarity degrees to the points defined by the four rules. Each of these similarity degrees is simply the minimum of the membership degrees of the three components 1, 1 and 2 to the three fuzzy sets in the corresponding rule.

In this way, for the point induced by rule R_1 we obtain a similarity degree of $2/3 = \min\{3/4, 3/4, 2/3\}$. For R_2 the result is $0.25 = \min\{1/4, 3/4, 2/3\}$. For the two rules R_3 and R_4 the similarity degree is 0 because already the considered input values do not fit to these rules at all. The similarity degree according to the four given points or rules corresponds to the best possible value, that is 2/3. Repeating the calculation for any value z in the input/output tuple $(1, 1, z)$, we obtain a function

$$\mu : [-3, 3] \to [0, 1]$$

for the given input $(1, 1)$. This function can be interpreted as a fuzzy set over the output range. When we compare this computation scheme with the calculations carried for the Mamdani controller, we obtain exactly the output fuzzy set (cf. Eq. (19.5)) of the corresponding Mamdani controller.

19.4.2 Construction of a Controller

Instead of determining the scaling functions or similarity relations and the corresponding interpolation points indirectly from a Mamdani controller we can also specify them directly and then derive the corresponding Mamdani controller. The advantage of this procedure is on the one hand that we can no longer specify arbitrary fuzzy sets, but only fuzzy sets which are consistent in a certain way. And on the other hand the interpretation of the scaling functions and especially of the interpolation points that have to be specified is very simple. The scaling functions can be interpreted in the sense of Example 18.2 on page 344. In ranges where the control is very sensitive to a change of the value, we should distinguish very exactly between the single values. So we should choose a larger scaling factor. For ranges where the exact values are less important for the control, a small scaling factor is sufficient. This leads to very narrow fuzzy sets in ranges where very precise control actions have to be taken, while larger fuzzy sets are used in regions where a rough control is sufficient. In this way we can explain, why fuzzy sets near the operating point of a controller are usually chosen to be very narrow in contrast to the other ranges. The operating point requires very exact control actions in most cases. In contrast to this, if the process is far from the operating point, first of all rough and strong actions have to be carried out to force the process closer to its operating point.

Using scaling function it is also obvious which additional hidden assumptions are used for the design of a Mamdani controller. The fuzzy partitions are defined for the single domains and are then used in the rules. In the sense of scaling functions this means that the scaling functions for the different domains are assumed to be independent from each other. The similarity of the values in a domain does not depend on the values in other domains. In order to illustrate this issue, we consider a simple PD controller which uses as input variable the error—the difference from the reference variable—and the change of the error. For a very small error value, it is obviously very important for the controller whether the change of the error is slightly positive or slightly negative. Therefore, we would choose a larger scaling factor near zero in the domain of the change of error, resulting in narrow fuzzy sets. On the other hand, if the error value is large, it is not very important whether the change of error tends to be slightly positive or slightly negative. First of all, we have to reduce the error. This fact speaks in favor for a small scaling factor near zero in the domain of the change of error, that is, we should use wider fuzzy sets. In order to solve this problem, there are three possibilities:

1. We specify a similarity relation in the product space of the domains error and change of error which models the dependence described above. But this seems to be quite difficult because the similarity relation in the product space cannot be specified using scaling functions.
2. We choose a high scaling factor near zero in the domain of the change of error. However, in this case, it might be necessary to specify many almost identical rules for the case of a large error value. The rules differ only in the fuzzy sets for the change of the error, like

If error is *big* and change is *positive small*, then y is *negative*.
If error is *big* and change is *zero*, then y is *negative*.
If error is *big* and change is *negative small*, then y is *negative*.

3. We define rules that do not use all input values, for instance

If error is *big*, then y is *negative*.

The interpretation of the Mamdani controller in the sense of similarity relations also explains why it is very convenient that adjacent sets of a fuzzy partition meet each other at the level of 0.5. A fuzzy set is a vague value which will be used for the interpolation points. If a certain value has been specified, this value provides also some information about similar values, where similar should be interpreted in the sense of the corresponding similarity relation. So, as long as the similarity degree does not drop to zero, there is still some information available. Therefore, once the similarity degree is zero, new information is needed, that is, a new interpolation point must be introduced. Following this principle, the fuzzy sets will exactly overlap at the level of 0.5. Of course, the interpolation points could be chosen closer, provided sufficiently detailed knowledge about the process is available. This would lead to strongly overlapping fuzzy sets. However, this does not make sense, when we desire a representation of the expert's knowledge on the process that is as compact as possible. Therefore, new interpolation points are only introduced, when they are needed.

Even if a Mamdani controller does not satisfy the conditions of one of the Theorems 18.2 or 18.3, calculating the similarity relations from Theorem 18.1 will still provide important pieces of information. The corresponding similarity relations always exist and make the fuzzy sets extensional, even though they might not be interpreted as extensional hulls of points. The following results are taken from Klawonn and Castro (1995), Klawonn and Kruse (2004).

1. The output of a Mamdani controller does not change if we use instead of a crisp input value its extensional hull as input.
2. The output fuzzy set of a Mamdani controller is always extensional.

This means that we cannot overcome the indistinguishability or vagueness that is inherently coded in the fuzzy partitions.

19.5 Hybrid Systems to Tune Fuzzy Controllers

A very interesting field of application for fuzzy control is to combine them with computational intelligence method in order to complete the advantages of a fuzzy controller (i.e., linguistic interpretability) by the ones of a neural networks (i.e., ability to learn) or evolutionary algorithms (i.e., ability to evolve). Such combinations exist in many different varieties and are referred to as hybrid fuzzy systems. Their goal is to fine tune fuzzy control rules by optimizing certain objective functions. In the following, we present a couple of the most known approaches of hybrid fuzzy systems. First, in Sect. 19.5.1, we talk about neuro-fuzzy control that exploits the learning techniques of artificial neural networks to obtain fuzzy rules from data. We remark that other machine learning techniques (e.g., support vector machines or a rough set approach) have been used as well to accomplish this task (Moewes and Kruse 2011, 2013). Then, in Sect. 19.5.2 we explain how evolutionary algorithms can be used to evolve chromosomes where each of it represents either one fuzzy rule or an entire fuzzy rule base. For further details about this and other topics of fuzzy control, we refer the interested reader to Michels et al. (2006).

19.5.1 Neuro-Fuzzy Control

Promising approaches for optimization of existing fuzzy controllers and to learn a fuzzy system from scratch are techniques that combine fuzzy systems and learning methods of artificial neural networks. These methods are called neuro-fuzzy systems. Nowadays, there are many specialized models. Beside the models for system control, especially systems for classification and more general models for function approximation have been developed. For a detailed introduction to this broad field see, for example, Nauck et al. (1997), Nauck and Nürnberger (2012). In this section, we only introduce systematics and discuss some approaches.

Fuzzy systems provides the ability of interpreting a fuzzy controller and to introduce *a priori* knowledge whereas and neural networks contribute its capability of learning and therefore the ability of automatic optimization or automatic generation of the whole controller. Due to the possibility to introduce *a priori* knowledge by fuzzy rules into the system, we expect to strongly reduce both the optimization time and the amount of training data required to train the system in comparison with pure neural network based controllers. If we have only some training data available, the prior knowledge might be even necessary in order to be able to create a controller. Furthermore, with neuro-fuzzy systems we are—in principle—able to learn a controller and then to interpret its control strategy by analyzing the learned fuzzy rules and fuzzy sets and, if necessary, revise them.

Essentially, neuro-fuzzy controllers can be divided in cooperative and hybrid models. In cooperative models the neural network and the fuzzy controller operate separately. The neural network generates (offline) or optimizes (online, i.e., during control) some parameters (Kosko 1992; Nomura et al. 1992). Hybrid models try to unite the structures of neural networks and fuzzy controllers. Thus, a hybrid fuzzy controller can be interpreted as neural network and can even be implemented with the help of a neural network. Hybrid models have the advantage of an integrated structure which does not require any communication between the two different models. Therefore, the system is principally able to learn online as well as offline and thus these approaches have become much more accepted than the cooperative models (Halgamuge and Glesner 1994; Jang 1993; Nauck and Kruse 1993).

The idea of hybrid methods is to map fuzzy sets and fuzzy rules to a neural network structure. This principle is explained in the following. For that, we reconsider

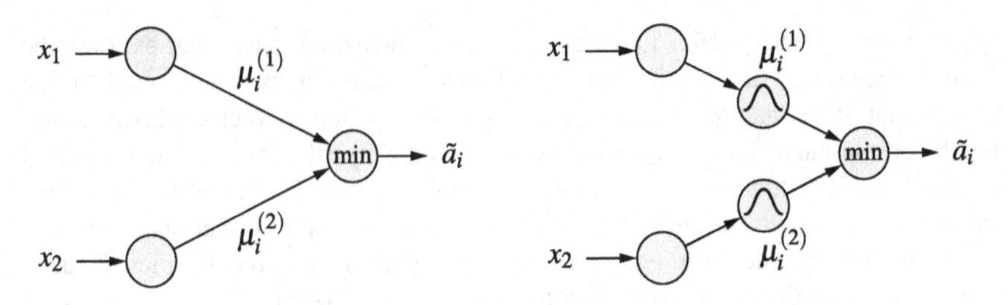

Fig. 19.9 Example of a neural network for calculating the activation of the antecedent of a fuzzy rule: Modeling of the fuzzy sets as weights (*left*) and as activation functions of a neuron (*right*)

the fuzzy rules R_i of a Mamdani controller defined in Eq. (19.1), that is,

$$R_i: \quad \text{If } x_1 \text{ is } \mu_i^{(1)} \text{ and} \dots \text{and } x_n \text{ is } \mu_i^{(n)}$$
$$\text{then } y \text{ is } \mu_i,$$

or the fuzzy rules R'_i of a TSK controller (cf. Eq. (19.8)), i.e.

$$R'_i: \quad \text{If } x_1 \text{ is } \mu_i^{(1)} \text{ and} \dots \text{and } x_n \text{ is } \mu_i^{(n)}, \text{ then } y = f_i(x_1, \dots, x_n).$$

The activation $\tilde{a}_i$ of these rules can be calculated by a t-norm. With given input values $x_1, \dots, x_n$, we obtain for $\tilde{a}_i$ using the minimum t-norm

$$\tilde{a}_i(x_1, \dots, x_n) = \min\{\mu_i^{(1)}(x_1), \dots, \mu_i^{(n)} x_n\}.$$

One way to represent such a rule with a neural network is to replace each real-valued connection weight w_{ji} from an input neuron u_j to an inner neuron u_i by a fuzzy set $\mu_i^{(j)}$. Therefore, an inner neuron represents a rule and the connections from the input units represent the fuzzy sets of the antecedents of the rules. In order to calculate the rule activation of the inner neurons, we just have to modify their network input functions. If we, for example, choose the minimum as t-norm, then we define

$$\text{net}_i = \min\{\mu_i^{(1)}(x_1), \dots, \mu_i^{(n)} x_n\}$$

as network input function.

If we finally replace the activation function of the neuron by the identity, then the activation of the neuron corresponds to the rule activation $\tilde{a}_i$. Therefore, the neuron can be used directly to compute the rule activity of any fuzzy rule. A graphical representation of such a structure for a rule with two inputs is shown in Fig. 19.9 (on the left).

We obtain another representation if the fuzzy sets of the antecedent are modeled as separate neurons. For that, the network input function is the identity and the activation function of the neuron is the membership function of the fuzzy set. Thus the neuron calculates for each input the degree of membership to the fuzzy set represented by the activation function. For this representation, we need two layers of neurons in order to model the antecedent of a fuzzy rule (see Fig. 19.9 (on the right)). The advantage of this representation is that the fuzzy sets can be directly used in several rules in order to ensure the interpretability of the entire rule base. In this representation, the network weights $w_i j$ in the connections from the fuzzy sets to the rule neuron are initialized with 1 and are regarded as constant. The weights of the input values to the fuzzy sets can be used for scaling the input values.

If the evaluation of the entire rule base shall to be modeled as well, then we have to decide whether a Mamdani or a TSK controller is needed. For the TSK controller, various realizations are possible. But, in principle, for each rule we get one more unit for evaluating the output function f_i—which will then be implemented as network input function—and it will be connected to all of the input units $(x_1, \ldots, x_n)$. In an output neuron, the outputs of these units will be combined with the rule activations $\tilde{a}_i$ which are calculated by the rule neurons. This output neuron will finally calculate the output of the TSK controller with the help of the network input function

$$\text{out} = \frac{\sum_{i=1}^{r} \tilde{a}_i \cdot f_i(x_i, \ldots, x_n)}{\sum_{i=1}^{r} \tilde{a}_i}.$$

The connection weights between the neurons are again constantly 1 and the identity is used as activation function.

For the Mamdani controller, the concrete implementation depends on the chosen t-conorm and the defuzzification method. In any case, a common output neuron combines the activations of the rule neurons and calculates a crisp output value with the help of a modified network input function based on the corresponding fuzzy sets in the consequents of the rules.

The transfer of a fuzzy rule base into a network structure can be summarized by the following steps:

1. For every input variable x_i, we create a neuron of the same denotation in the input layer.
2. For every fuzzy set $\mu_i^{(j)}$, we create a neuron of the same denotation and connect it to the corresponding x_i.
3. For every output variable y_i, we create a neuron of the same denotation.
4. For every fuzzy rule R_i, we create an inner (rule) neuron of the same denotation and we specify a t-norm for calculating the rule activation.
5. Every rule neuron R_i is connected according to its fuzzy rule to the corresponding neurons that represent the fuzzy sets of the antecedent.
6. Mamdani controller: Every rule neuron is connected to the output neuron according to the consequent of its fuzzy rule. As connection weight, we have to choose the consequent of the corresponding fuzzy set. Furthermore, a t-conorm and the defuzzification method have to be integrated adequately into the output neurons.

   TSK controller: For every rule unit, one more neuron is created to calculate the output function. These neurons are connected to the corresponding output neuron. Furthermore, all of the input units are connected to the neurons for the calculation of the output function and all of the rule neurons are connected to the output neuron.

Having described the mapping of a rule base as a network, learning algorithms of artificial neural networks can be applied to this structure. However, the learning methods usually have to be modified due to several reasons. First of all, the network input and activation functions changed. Second, not the real-valued network weights but the parameter of the fuzzy sets have to be learned. In the following sections, we discuss two hybrid neuro-fuzzy systems in more detail. Furthermore, we explain the principles and problems of neuro-fuzzy architectures, especially in terms of applications in system control.

19.5.1.1 Models with Supervised Learning Methods

Neuro-fuzzy models with supervised learning methods try to optimize the fuzzy sets and—for a TSK model—the parameters of the output function of a given rule base with the help of known input-output tuples. Therefore, applying supervised models is convenient if we already have a description of the system to be control but need the control behavior—and thus the rule base—to be more exact. If measured data of the system to be modeled exist (tuple of state, output and control variables), they can be used to retrain the system.

Neuro fuzzy models with supervised learning methods are also convenient if an existing controller is to be replaced by a fuzzy controller, that is, that measured data of the control behavior of the real controller are available. Here, an existing rule base is also presupposed. The learning methods can then be used in order to optimize the approximation of the original controller. If no initial fuzzy rule base is available describing the system that should be approximated and if an approximate rule base also cannot be created manually, we might use fuzzy clustering methods (see Chap. 20 on page 389) or evolutionary algorithms (to be discussed) in order to obtain an initial rule base if measured data for training is available.

In the following, we discuss a typical example for a neuro-fuzzy system with supervised learning, that is, the ANFIS model. Beside this one, there are several other approaches which are based on similar principles. For an overview of other models see, for example, Nauck et al. (1997).

The ANFIS Model

In Jang (1993), the neuro-fuzzy system *ANFIS* (Adaptive-Network-based Fuzzy Inference System) was developed which by now has been integrated in many controllers and simulation tools. The ANFIS model is based on a hybrid structure, that is, it can be interpreted as neural network and as fuzzy system. The model uses the fuzzy rules of a TSK controller. Figure 19.10 shows an example for a model with the three fuzzy rules

$$R_1: \quad \text{If } x_1 \text{ is } A_1 \text{ and } x_2 \text{ is } B_1 \text{ then } y = f_1(x_1, x_2)$$
$$R_2: \quad \text{If } x_1 \text{ is } A_1 \text{ and } x_2 \text{ is } B_2 \text{ then } y = f_2(x_1, x_2)$$
$$R_3: \quad \text{If } x_1 \text{ is } A_2 \text{ and } x_2 \text{ is } B_2 \text{ then } y = f_3(x_1, x_2)$$

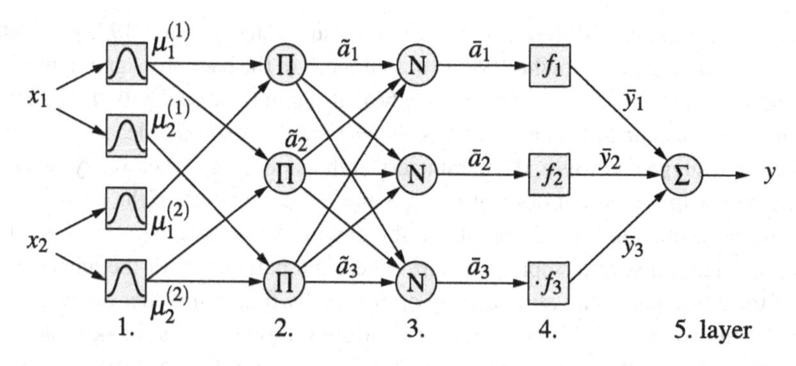

Fig. 19.10 Structure of an ANFIS network with three rules

where A_1, A_2, B_1 and B_2 are linguistic expressions which are assigned to the corresponding fuzzy sets $\mu_j^{(i)}$ in the antecedents. The functions f_i in the consequent of the ANFIS model are defined by linear combination of the input variables, that is, in the example above by

$$f_i = p_i x_1 + q_i x_2 + r_i.$$

Here, we use the product t-norm for the evaluation of the antecedent, that is, the neurons in the layer 2 calculate activation $\tilde{a}_i$ of rule R_i by

$$\tilde{a}_i = \prod_j \mu_i^{(j)}(x_j).$$

In the ANFIS model, the evaluation of the consequent and the calculation of the output value is split up to layers 3 to 5. Layer 3 calculates the contribution $\bar{a}_i$ each rule made to creating the total output by taking into account the activations $\tilde{a}_i$. Therefore, the neurons of layer 3 calculate

$$\bar{a}_i = a_i = \text{net}_i = \frac{\tilde{a}_i}{\sum_j \tilde{a}_j}.$$

Then, the neurons of layer 4 calculate the weighed control outputs based on the input variables x_k and the relative control activations $\bar{a}_i$ of the previous layer:

$$\bar{y}_i = a_i = \text{net}_i = \bar{a}_i f_i(x_1, \dots, x_n).$$

Finally, the output neuron u_{out} of layer 5 calculates the total output of the network and the fuzzy system, respectively:

$$y = a_{\text{out}} = \text{net}_{\text{out}} = \sum_i \bar{y}_i = \frac{\sum_i \tilde{a}_i f_i(x_1, \dots, x_n)}{\sum_i \tilde{a}_i}.$$

For learning, the ANFIS model needs a fixed learning exercise. Therefore for training, it is necessary that a sufficient amount of input/output tuples exists. Based on these training data, model parameters are determined, that is, the parameters of the fuzzy sets and the parameters of the output function f_i.

As learning method, different approaches are suggested in Jang (1993). Besides the pure gradient descent method which is analogous to the backpropagation method for neural networks (see Sect. 5.5 on page 66), also combinations with methods for solving overdetermined linear equation systems (i.e., the method of least square (estimate)) are suggested. Here, the parameters of the antecedents (i.e., fuzzy sets) are determined with a gradient descent method and the parameters of the consequents (i.e., linear combination of the input variables) with the least squares method. The learning occurs in several separated steps where the parameters of the antecedents and of the consequents are alternatingly optimized by fixing the other ones.

In the first step, all input vectors are propagated through the network to layer 3. For every input vector, the control activations are stored. Based on these values, for the parameters of the functions f_i in the consequent, a overdetermined equation system is created.

Let r_{ij} be the parameters of the output function f_i, $x_i(k)$ the input values, $y(k)$ the output value of the kth training pair and $\bar{a}_i(k)$ the relative control activation. Then, we obtain

$$y(k) = \sum_i \bar{a}_i(k) y_i(k) = \sum_i \bar{a}_i(k) \left(\sum_{j=1}^{n} r_{ij} x_j(k) + r_{i0} \right), \quad \forall i, k.$$

Therefore, with $\hat{x}_i(k) := [1, x_1(k), \dots, x_n(k)]^T$ we obtain the overdetermined linear equation system

$$\mathbf{y} = \bar{\mathbf{a}} \mathbf{R} \mathbf{X}$$

for a sufficient number m of training data ($m > (n+1) \cdot r$, where r is the number of rules and n the number of input variables).

Therefore, the parameters of the linear equation system, which is built in this way,—the parameters of the output function f_i in the matrix $\mathbf{R}$—can be determined with the method of least squares after the propagation of all of the training data. Finally, the error is determined in the output units based on the new calculated output functions and with the help of a gradient descent method the parameters of the fuzzy sets are optimized. The combination of the two methods leads to an improved convergence because the method of least squares already has an optimal solution (i.e., the one with the least error squares) for the parameters of the output function with regard to the initial fuzzy sets.

Unfortunately, the ANFIS model has no restrictions for the optimization of the fuzzy sets in the antecedents, that is, it is not ensured that the input range will still be covered completely with fuzzy sets after the optimization. Thus, definition gaps can appear. This has necessarily to be checked after optimizing. Fuzzy sets can also can change, independently form each other and can also exchange their order and thus their importance. We have to pay attention to this, especially if an initial rule base was set manually and the controller has to be interpreted afterwards.

19.5.1.2 Models with Reinforcement Learning

The fundamental idea of the models of reinforcement learning (Barto et al. 1983) is to determine a controller with knowing as little as possible about the system. The aim is that the learning process can be managed with a minimum of information about the aim of the control. In the extreme case, the learning method gets merely the information whether the system is still stable or the controller had failed.

The main problem of these approaches is that to rework the control action such that it can be used for learning and optimizing the controller. In general, we cannot assume that the last control action has the greatest influence on the current system state. This problem is also called *credit assignment problem* (Barto et al. 1983), that is, the problem of assigning to a control action the (long-term) effect which it has on the system.

By now, many models have been suggested in the field of reinforcement learning. All of these are essentially based on the principle of dividing the learning problem into two systems: A criticizing system (critic) and a system which stores a description of the control strategy and applies this to the system (actor). The critic evaluates the current state considering the previous states and control action, evaluates as well the output of the actor based on these information and, if necessary, adapts its control strategy.

The methods suggested by now for using the principles of reinforcement learning are mostly based on the combination with neural nets (Kaelbling et al. 1996). Very promising are the methods which use dynamic programming in order to determine an optimal control strategy. For a more detailed discussion about this topic, see, for example, Sutton and Barto (1998).

In the field of neuro-fuzzy systems there are many approaches. But by now none of these have achieved the quality of the systems based on neural networks. Examples of such approaches are GARIC (Berenji and Khedkar 1992), FYNESSE (Riedmiller et al. 1999) and the NEFCON model (Nürnberger et al. 1999) which is shortly presented below.

The NEFCON Model

The principal aim of the NEFCON model (neuro-fuzzy controller) is to detect online a suitable and interpretable rule base with a the least amount of training cycles. Furthermore, it should be possible to bring previous knowledge into the training process in the easiest way in order to speed up the learning process. This is the difference to the most reinforcement learning approaches which try to generate the most optimal controller and therefore lose much time in the long learning phases. Furthermore, in the NEFCON model has also heuristic approaches for learning a rule base. In this point is different to the most of the other neuro fuzzy systems which, in general, can only be used for optimizing a rule base.

The NEFCON model is a hybrid model of a neuro-fuzzy controller. We assume the definition of a Mamdani controller and obtain the network structure if we—analogously to the description in Sect. 19.5.1—interpret the fuzzy sets as weights

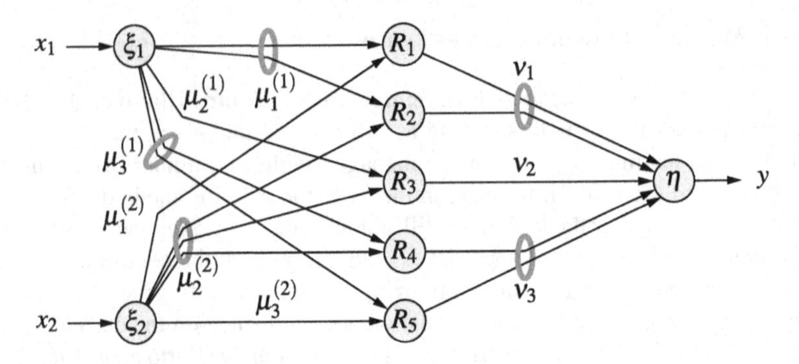

Fig. 19.11 A NEFCON system with two input variables and 5 rules

R_1: if x_1 is $A_1^{(1)}$ and x_2 is $A_1^{(2)}$ then y is B_1
R_2: if x_1 is $A_1^{(1)}$ and x_2 is $A_2^{(2)}$ then y is B_1
R_3: if x_1 is $A_2^{(1)}$ and x_2 is $A_2^{(2)}$ then y is B_2
R_4: if x_1 is $A_3^{(1)}$ and x_2 is $A_2^{(2)}$ then y is B_3
R_5: if x_1 is $A_3^{(1)}$ and x_2 is $A_3^{(2)}$ then y is B_3

Table 19.1 The rule base of the NEFCON system shown in Fig. 19.11

and the measuring and regulating variables as well as the rules as operating units. Then, the net has the structure of a multilayer perceptron and can be interpreted as three-layer fuzzy perceptron. The fuzzy perceptron arises from a perceptron by modeling the weights, the network inputs and the activation of the output unit as fuzzy sets. An example for a fuzzy controller with 5 rule units, two measuring variables and one regulating variable is shown in Fig. 19.11.

The inner units $R_1, \ldots, R_5$ represent the rules, the units x_1, x_2 and y the measuring and regulating variables and $\mu_r^{(i)}$ and ν_r the fuzzy sets for the antecedents and the consequents, respectively. The connection with common weights denote equal fuzzy sets. A changing of these weights makes it necessary that all connections with this weight have to be adapted in order to ensure that the same fuzzy sets keep represented by the same weights. Thus, the rule base defined by the network structure can also be formulated in the form of the fuzzy rules listed in Table 19.1.

The learning process of the NEFCON model can be divided into two main phases. The first phase is designed to learn an initial rule base, if no prior knowledge about the system is available. Furthermore, it can be used to complete a manually defined rule base. The second phase optimizes the rules by shifting or modifying the fuzzy sets of the rules. Both phases use a fuzzy error e, which describes the quality of the current system state, to learn or to optimize the rule base. The fuzzy error plays the role of the critic element in reinforcement learning models. In addition, the sign of the optimal output value must be known. Thus, the extended fuzzy error E is defined as

$$E(x_1, \ldots, x_n) = \text{sgn}(y_{\text{opt}}) \cdot e(x_1, \ldots, x_n),$$

with the input $(x_1, \ldots, x_n)$. At the end of this section we briefly discuss some methods to describe the system error.

Learning a Rule Base

If for the system which has to be controlled an adequate rule base does not exist or cannot be determined manually, it has to be created by an appropriate rule learning method. Methods to learn an initial rule base can be divided into three classes: Methods starting with an empty rule base, methods starting with a "full" rule base (combination of every fuzzy set in the antecedents with every consequent) and methods starting with a random rule base. In the following, we briefly present algorithms of the first two classes. The algorithms do not require a given fixed learning problem but they try to determine a suitable rule base based on the extended system error E (see also Nürnberger et al. 1999). However, both methods require an appropriate fuzzy partitioning of the input and output variables.

If for the system which has to be controlled does not exist an adequate rule base, or it cannot be determined, it has to be created by an appropriate rule learning method. Following, we present some algorithms for learning a rule base, they do not need a given fixed learning exercises but based on the system error they try to determine a suitable rule base (see also Nürnberger et al. 1999). Both of the methods need suitable fuzzy partitioning of the measuring and regulating variable ranges (see also the discussion in Sect. 19.4.2).

An Elimination Method for Learning a Rule Base

The elimination method starts with a complete overdetermined rule base, that is, the rule base consists of all rules which can be defined by the combination of all fuzzy sets in the initial fuzzy partitions of the input and output variables.

The algorithm can be divided into two phases which are executed during a fixed period of time or a fixed number of iteration steps. During the first phase, rules with an output sign different from that of the optimal output value are removed. During the second phase, a rule base is constructed for each control action by selecting randomly one rule from every group of rules with identical antecedents. The error of each rule (the output error of the whole network weighted by the activation of the individual rule) is accumulated. At the end of the second phase from each group of rule nodes with identical antecedents the rule with the least error value remains in the rule base. All other rule nodes are deleted. In addition, rules used very rarely are removed from the rule base.

A disadvantage of the elimination method is that it starts with a very large rule base and, thus, for systems with many state variables or fine grained partitions, i.e. many fuzzy sets are used to model the input and output values, it requires much memory and is computationally expensive. Therefore it is advisable to use the incremental rule learning procedure for larger rule bases.

Incremental Learning the Rule Base

This learning methods starts with an empty rule base. However, an initial fuzzy partitioning of the input and output intervals must be given. The algorithm can be divided into two phases. During the first phase, the rules' antecedents are determined by classifying the input values, that is, finding that membership function for each variable that yields the highest membership value for the respective input value. Then the algorithm tries to "guess" the output value by deriving it from the current fuzzy error with a heuristics. It is assumed that input patterns with similar error values require similar output (control) values. The so found rule is inserted into the rule base. In the second phase we try to optimize the rule consequent by applying the learned rule base to the system and, based on the detected error values, exchanging the fuzzy sets in the consequents, if necessary.

The used heuristics maps the extended error E merely linearly to the interval of the control variable. Thus, we assume a direct dependency between the error and the control variable. This is might be critical, especially for systems which would require an integrator for control, that is, which require a control value unequal 0 in order to achieve the control goal or to keep it. This heuristic also assumes that the error in not determined only from the control deviation but that the error also take the following system states into account.

Through the incremental learning method it is easily possible to introduce previous knowledge into the rule base. Missing rules are added during learning. However, because of the already discussed problems, both of the presented heuristics cannot provide a rule base which is appropriate for all systems.

The rule bases learned with the methods presented above should be checked manually for its consistence—at least if the optimization method, which we discuss in the following paragraph, is unable to provide a satisfying solution. In any case, the possibility to introduce prior knowledge should be considered, that is, known rules should be inserted into the rule base before learning and the rule learning methods should not change the manually defined rules afterwards.

Optimization of the Rule Base

The NEFCON learning algorithm for optimizing a rule base is based on the idea of the backpropagation algorithm for the multilayer perceptron. The error is, beginning with the output unit, propagated backwards trough the network and used locally for adapting the fuzzy sets.

The optimization of the rule base is done by changing the fuzzy sets in the antecedents and consequents. Depending on the contribution to the control action and the resulting error the fuzzy sets of a rule are 'rewarded' or 'punished'. Thus, the principle of the reinforcement learning is applied. A 'reward' or 'punishment' can be done by displacing or reducing/enlarging the support of the fuzzy set. These adaptations are made iteratively, that is, during the learning process the controller is used for controlling the system, and after every control action an evaluation of the new state is made as well as a incremental adaption of the controller.

The principal problem of this approach is that the system error has to be determined very carefully in order to avoid the problems of evaluating a control action which we discussed at the beginning of this section. In many cases, this can be very difficult or even impossible. Nevertheless, the discussed methods can be applied to simple systems, and they can very helpful for designing fuzzy controllers for more complex systems. We have to pay attention to the fact that then such a controller should be checked very carefully for its stability.

19.5.2 Evolutionary Fuzzy Control

If a fuzzy controller should be generated automatically by an evolutionary algorithms, at first, we have to define the objective function which has to be optimized by the evolutionary algorithm. If measure data of a controller are given—for example, measuring data obtained from observing a human operator—the fuzzy controller should approximate the underlying control function as good as possible. As the error function, which has to be minimized, we could use the mean squared error or the absolute error as well as the maximum deviation of the control function from the measured output. When the measured data come from different operators, an approximation can lead to a very bad behavior of the controller. If the single operators use effective, but different control strategies and the controller is forced to approximate the data as good as possible, it will interpolate between the different strategies at each point. In the worst case, this mixed or interpolated control function might not work at all. If, for example, a car shall drive around an obstacle, and in the data for half of the cases an operator (driver) chose to avoid the obstacle to the right side and for the other half to the left side, interpolation will result in a strategy that keeps on driving straight ahead. If possible, the observed data should always be checked for consistency.

When we have a simulation model of the system or process to be controlled, we can define various criteria that a good controller should satisfy, for example, the time or the energy the controller needs to bring the process from different initial states to the desired state, some kind of evaluation of overshoots etc. If the evolutionary algorithm uses a simulation model with such an objective function, it is often better to slowly tighten the conditions of the objective function. In a random initial population, it is very probable that no individual (controller) at all will be able to bring the process very close to a desired state. Therefore, at first the objective function might only measure the time how long a controller is able to keep the process in a larger vicinity of the desired state (Hopf and Klawonn 1994). With an increasing number of generations, the objective function is chosen more strict until it reflects the actual criterion.

The parameters of a fuzzy controller which can be learned with an evolutionary algorithm can be divided into three groups, described in more detail in the following.

19.5.2.1 The Rule Base

Let us first assume that the fuzzy sets are specified in advance or are optimized at simultaneously by another method. For instance, if the controller has two input variables for which n_1, respectively n_2 fuzzy sets are defined, for each of the possible $n_1 \cdot n_2$ combinations an output can be defined. For a Mamdani controller with n_o fuzzy sets for the output variables, we could use a chromosome with $n_1 \cdot n_2$ parameters (genes), where each of these genes can take one of n_o values. But the coding of the rule table as linear vector with $n_1 \cdot n_2$ components, which is required for the genetic algorithm, is not optimal for the crossover. Crossover should enable the genetic algorithm to combine to solutions (chromosomes) that have optimized different parameters (genes) to one good solution.

For the optimization of a rule base of a fuzzy controller the conditions needed to benefit from crossover are satisfied, that is, that the parameters are independent to a certain degree. Adjacent rules in a rule base operate on overlapping regions, so that they contribute to the output of the controller at the same time, meaning that they interact and are dependent. Non-adjacent rules do not interact and never fire simultaneously. In this sense they are independent. If there are two fuzzy controllers in a population, which found, each one in a different part of the rule table, well-performing entries for the output of the rules, the combination of the two parts of the table will result in a better controller. However, a (fuzzy) region does not correspond to a row or a column in a rule table, but to a rectangular region in the table. A traditional genetic algorithm would only exchange linear parts in the crossover process. In the case of optimizing the rule base of a controller with two input variables it makes sense, to deviate from the linear chromosome structure and to use a planar structure for the chromosome. Crossover should then exchange smaller rectangles within the rule table (Kinzel et al. 1994). Here we discussed the case of only two input variables. This idea can be generalized to more input variables in a straight forward manner. For k input variables, we would use a k-dimensional hyperbox as the chromosome structure.

In order to guarantee small changes in the mutation process, an output fuzzy set should not be replaced by an arbitrary other fuzzy set, but by an adjacent one.

For a TSK model for the rule base, we have to determine output functions instead of output fuzzy sets. Usually, these functions are given in parametrized form, for example,

$$f(x, y; a_R, b_R, c_R) = a_R + b_R x + c_R y$$

for input variables x and y as well as three parameters a_R, b_R and c_R that have to be determined for each rule R. For a rule table with—like above—$n_1 \cdot n_2$ entries, we had to determine, in total, $3 \cdot n_1 \cdot n_2$ real parameters for the rule table. In this case, we should apply an evolution strategy (see Sect. 13.2 on page 239) since the parameters are not discrete but continuous.

If we do not want to fill the rule table completely and want to limit the number of rules, we could assign to each rule an additional binary gene (parameter) which tells us whether the rule of the controller is used or not. For a TSK model, we have a proper evolutionary algorithm, because we have to deal with continuous and discrete parameters at the same time. The number of active rules can be fixed in advance, where we have to ensure that this number is not changed by mutation or crossover. Mutation could always activate one rule and deactivate another one at the same time, so that the number of active rules is not changed by mutation. For crossover a repairing algorithm is needed. If we have too many active rules after crossover, then, for example, rules could be deactivated randomly, until the desired number of active rule is reached again.

A better strategy is not to fix the number of active rules in advance. Since fuzzy controllers with a smaller number of rules are to be preferred due to better interpretability, we could introduce an additional term in the objective function penalizing higher numbers of rules. This additional term should have a suitable weight. If the weight is too large, the emphasis is put more or less completely on keeping the number of rules small, ignoring the performance of the controller. A weight chosen too small will not contribute enough to keep the rule base small.

19.5.2.2 The Fuzzy Sets

Usually, the fuzzy sets are represented in a parametrized form like triangular, trapezoidal or Gaussian membership functions. The corresponding real parameters are suitable for an optimization based on evolution strategies. However, giving complete freedom to the evolution to optimize these parameters, seldom to meaningful results. The optimized fuzzy controller might perform perfectly but the fuzzy sets overlap completely, so that it is hardly possible to assign meaningful linguistic terms to them and to formulate interpretable rules. In this case the fuzzy controller corresponds to a black box, as a neural network, without any chance to explain or interpret its control strategy.

It is recommended to choose the parameter set in such a way that the interpretability of the fuzzy controller is always guaranteed. One possible way would be the restriction to triangular functions which are chosen in such a way that the left and the right neighbor of a fuzzy set get the value 1 at the point where the membership degree of the fuzzy set in the middle drops to 0. In this case, the evolution strategy would have for each input or output variable as many real parameters as fuzzy sets are used for the corresponding domain. The respective real parameters indicate where the corresponding triangular function assume the value 1.

Even with this parametrization undesired effects can occur, for example, if the fuzzy set “approximately zero”, because of mutation, overtakes the fuzzy set “positive small”. A simple change in the coding of the parameters can avoid this effect: the value of the parameter k determines no longer the point where the triangular membership function is 1, but how far it is shifted to the left relative to its left neighbor. The disadvantage of this coding is that a change (mutation) of the first value leads to a new position for fuzzy sets and therefore a quite great change of the total behavior of the controller. If the fuzzy sets are parametrized independently, a mutation only has local effect. Therefore, we should stick to the independent parametrization, but prohibit mutations which lead to overtaking fuzzy sets. In this way, mutations causes small changes and the interpretability of the fuzzy controller is preserved.

19.5.2.3 Additional Parameters

With evolutionary algorithms we can—if we want to—adjust also other parameters of a fuzzy controller. For example, we can use a parametrized t-norm for the aggregation of the rule antecedents and for each rule adjust the parameter of the t-norm individually. The same approach can also be used for a parametrized defuzzification strategy. Such parameters tend to cause problems in the interpretability of a fuzzy controller and will not be further pursued here.

So far, we have not answered the question, whether the rule base and the fuzzy sets should be optimized at the same time or separately. As long as the rule base can change drastically, it does not make sense to fine-tune the fuzzy set. The rule base functions as the skeleton of a fuzzy controller. The concrete choice of the fuzzy sets is responsible for fine-tuning. In order to keep the number of parameters to be optimized by the evolutionary algorithm small, it is often better to learn at first the rule base on the basis of standard fuzzy partitions and then optimize the fuzzy sets with a fixed rule base.

19.5.2.4 A Genetic Algorithm for Learning a TSK Controller

In order to illustrate the principle of parameter coding, in the following section we present a genetic algorithm for learning a TSK controller which was suggested in Lee and Takagi (1993). The algorithm tries to optimize all of the parameters of the controller, that is, the rule base, the form of the fuzzy sets and the parameters of the consequents at the same time.

In order to learn the rules

$$R_r: \quad \text{If } x_1 \text{ is } \mu_R^{(1)} \text{ and } \ldots \text{ and } x_n \text{ is } \mu_R^{(n)} \text{ then } y = f_r(x_1, \ldots, x_n),$$

of a Takagi–Sugeno controller with

$$f_r(x_1, \ldots, x_n) = p_0^r + x_1 \cdot p_1^r + \cdots + x_n \cdot p_n^r,$$

we have to encode the fuzzy sets of the input values and the parameters $p_0, \ldots, p_n$ of each rule.

In this approach, a triangular fuzzy set is described by three binary coded parameters (**membership function chromosome (MFC)**):

leftbase	center	rightbase
10010011	10011000	11101001

The parameters *leftbase*, *rightbase* and *center* are no absolute quantities but denote the distances to the reference point. *leftbase* and *rightbase* refer to the *center* of a fuzzy set and the *center* refers to the distance between the center and the left neighbor of the fuzzy set. If these parameters are positive, passing and abnormal fuzzy sets can be avoided.

The parameters $p_0, \ldots, p_n$ of a rule are encoded directly by binary numbers and result in the **rule-consequent parameters chromosome (RPC)**:

p_0	...	p_n
10010011	...	11101001

Based on these parameter encodings, the complete rule base of a TSK controller is encoded in the form of a bit string:

variable 1	...	variable n	parameters of the consequents
$\text{MFC}_{1\ldots m_1}$	...	$\text{MFC}_{1\ldots m_n}$	$\text{RPC}_{1\ldots(m_1\cdot\ldots\cdot m_n)}$

Besides the parameter optimization, the algorithm tries to minimize the amount of fuzzy sets assigned to a variable and thus also tries to minimize the amount of rules in a rule base. For that, we assume a maximum amount of fuzzy sets. We eliminate fuzzy sets which are no longer in the permitted domain of a variable. Furthermore, among controllers of the same performance the selection process prefers the ones with less rules.

In Lee and Takagi (1993), this approach was tested with a inverted pendulum (i.e., pole balancing problem). For a rule base with five fuzzy sets for each of the two input variables and 8-bit binary numbers, we obtain a chromosome of length $2 \cdot (5 \cdot 3 \cdot 8) + (5 \cdot 5) \cdot (3 \cdot 8) = 840$. For the evaluation of the learning process, the controllers were tested with eight different starting conditions and the time which the controller needed to bring the pendulum to the vertical position was measured. For this, we have to distinguish three cases:

1. If the controller brings the pendulum to the vertical position within a certain period of time, it obtains the more points the faster it was done.
2. If the controller is not able to bring the pendulum to the vertical position within this period of time, it obtains a predetermined amount of points which is lower than in the first case.
3. If the pendulum falls over within the period of simulation, the controller obtains the more points the longer the pendulum did not fall but less than in the two previous cases.

The authors of Lee and Takagi (1993) report that for learning a ‘usable’ controller, more than 1000 generations were needed. This quite big amount of generations results from the great chromosome length. Furthermore, for encoding, neighbor relations are hardly used. Thus, it can happen that the antecedent of a rule is determined by the fuzzy set which is coded at the beginning of the chromosome but the corresponding consequent is at the end of the chromosome. Therefore, the probability that the crossover destroys a good rule is quite big.

The interesting point of this approach is the ability of minimizing the required amount of rules. Here, the main idea is not only to optimize the controller’s behavior but to determine the important rules for the controller.

References

A.G. Barto, R.S. Sutton, and C.W. Anderson. Neuronlike Adaptive Elements that Can Solve Difficult Learning Control Problems. *IEEE Transactions on Systems, Man and Cybernetics*, 13(5):834–846. IEEE Press, Piscataway, NJ, USA, 1983

H.R. Berenji and P. Khedkar. Learning and Tuning Fuzzy Logic Controllers Through Reinforcements. *IEEE Transactions on Neural Networks*, 3(5):724–740. IEEE Press, Piscataway, NJ, USA, 1992

S.K. Halgamuge and M. Glesner. Neural Networks in Designing Fuzzy Systems for Real World Applications. *Fuzzy Sets and Systems*, 65(1):1–12. Elsevier, Amsterdam, Netherlands, 1994

J. Hopf and F. Klawonn. Learning the Rule Base of a Fuzzy Controller by a Genetic Algorithm. In: R. Kruse, J. Gebhardt and R. Palm (eds.). *Fuzzy Systems in Computer Science*, 63–74. Vieweg, Braunschweig, Germany, 1994

J.-S.R. Jang. ANFIS: Adaptive-Network-Based Fuzzy Inference System. *IEEE Transactions on Systems, Man and Cybernetics*, 23(3):665–685. IEEE Press, Piscataway, NJ, USA, 1993

L.P. Kaelbling, M.H. Littman, and A.W. Moore. Reinforcement Learning: A Survey. *Journal of Artificial Intelligence Research*, 4:237–285. AI Access Foundation and Morgan Kaufman Publishers, El Segundo/San Francisco, CA, USA, 1996

J. Kahlert and H. Frank. *Fuzzy-Logik und Fuzzy-Control*, 2nd edition (in German). Vieweg, Braunschweig, Germany, 1994

J. Kinzel, F. Klawonn, and R. Kruse. Modifications of Genetic Algorithms for Designing and Optimizing Fuzzy Controllers. In: *Proc. IEEE Conf. on Evolutionary Computation (ICEC'94, Orlando, FL)*, 28–33. IEEE Press, Piscataway, NJ, USA, 1994

F. Klawonn. On a Lukasiewicz Logic Based Controller. In: *Proc. Int. Seminar on Fuzzy Control Through Neural Interpretations of Fuzzy Sets (MEPP'92)*, 53–56. Åbo Akademi, Turku, Finland, 1992

F. Klawonn and J.L. Castro. Similarity in Fuzzy Reasoning. *Mathware and Soft Computing*, 2:197–228. University of Granada, Granada, Spain, 1995

F. Klawonn and R. Kruse. The Inherent Indistinguishability in Fuzzy Systems. In: W. Lenski (ed.) Logic Versus Approximation: Essays Dedicated to Michael M. Richter on the Occasion of His 65th Birthday, 6–17. Springer-Verlag, Berlin, Germany, 2004

F. Klawonn and V. Novák. The Relation Between Inference and Interpolation in the Framework of Fuzzy Systems. *Fuzzy Sets and Systems*, 81:331–354. Elsevier, Amsterdam, Netherlands, 1996

B. Kosko (ed.). *Neural Networks for Signal Processing*. Prentice Hall, Englewood Cliffs, NJ, USA, 1992

M. Lee and H. Takagi. Integrating Design Stages of Fuzzy Systems Using Genetic Algorithms. In: *Proc. IEEE Int. Conf. on Fuzzy Systems (San Francisco, CA)*, 612–617. IEEE Press, Piscataway, NJ, USA, 1993

E.H. Mamdani and S. Assilian. An Experiment in Linguistic Synthesis with a Fuzzy Logic Controller. *International Journal of Man-Machine Studies* 7:1–13. Academic Press, Waltham, MA, USA, 1975

K. Michels, F. Klawonn, R. Kruse, and A. Nürnberger. *Fuzzy Control: Fundamentals, Stability and Design of Fuzzy Controllers*. Studies in Fuzziness and Soft Computing, vol. 200. Springer-Verlag, Berlin/Heidelberg, Germany, 2006

C. Moewes and R. Kruse. On the Usefulness of Fuzzy SVMs and the Extraction of Fuzzy Rules from SVMs. In: S. Galichet, J. Montero, and G. Mauris (eds.) *Proc. 7th Conf. of Europ. Soc. for Fuzzy Logic and Technology (EUSFLAT-2011) and LFA-2011*, 943–948. Advances in Intelligent Systems Research, vol. 17. Atlantis Press, Amsterdam/Paris, Netherlands/France, 2011

C. Moewes and R. Kruse. Fuzzy Control for Knowledge-Based Interpolation. In: E. Trillas, P.P. Bonissone, L. Magdalena, and J. Kacprzyk (eds.) *Combining Experimentation and Theory: A Hommage to Abe Mamdani*, 91–101. Springer-Verlag, Berlin/Heidelberg, Germany, 2012

C. Moewes and R. Kruse. Evolutionary Fuzzy Rules for Ordinal Binary Classification with Monotonicity Constraints. In: R.R. Yager, A.M. Abbasov, M.Z. Reformat, and S.N. Shahbazova (eds.)

Soft Computing: State of the Art Theory and Novel Applications, 105–112. Studies in Fuzziness and Soft Computing, vol. 291. Springer-Verlag, Berlin/Heidelberg, Germany, 2013

D. Nauck and R. Kruse. A Fuzzy Neural Network Learning Fuzzy Control Rules and Membership Functions by Fuzzy Error Backpropagation. In: *Proc. IEEE Int. Conf. on Neural Networks (ICNN'93, 1993, San Francisco, CA)*, 1022–1027. IEEE Press, Piscataway, NJ, USA, 1993

D.D. Nauck and A. Nürnberger. Neuro-Fuzzy Systems: A Short Historical Review. In: C. Moewes and A. Nürnberger (eds.) *Computational Intelligence in Intelligent Data Analysis*, 91–109. Studies in Computational Intelligence, vol. 445. Springer-Verlag, Berlin/Heidelberg, Germany, 2012

D.D. Nauck, F. Klawonn, and R. Kruse. *Foundations of Neuro-Fuzzy Systems*. Wiley, Chichester, 1997

H. Nomura, I. Hayashi, and N. Wakami. A Learning Method of Fuzzy Inference Rules by Descent Method. In: *Proc. IEEE Int. Conf. on Fuzzy Systems 1992*, 203–210. San Diego, CA, USA, 1992

A. Nürnberger, D.D. Nauck, and R. Kruse. Neuro-Fuzzy Control Based on the NEFCON-Model: Recent Developments. *Soft Computing*, 2(4):168–182. Springer-Verlag, Berlin/Heidelberg, Germany, 1999

M. Riedmiller, M. Spott, and J. Weisbrod. FYNESSE: A Hybrid Architecture for Selflearning Control. In: I. Cloete and J. Zurada (eds.) *Knowledge-Based Neurocomputing*, 291–323. MIT Press, Cambridge, MA, USA, 1999

T.A. Runkler. Kernel Based Defuzzification. In: C. Moewes and A. Nürnberger (eds.) *Computational Intelligence in Intelligent Data Analysis*, 61–72. Springer-Verlag, Berlin/Heidelberg, Germany, 2012

T.A. Runkler and M. Glesner. A Set of Axioms for Defuzzification Strategies—Towards a Theory of Rational Defuzzification Operators. In: *Proc. 2nd IEEE Int. Conf. on Fuzzy Systems (FUZZ-IEEE'93, San Francisco, CA)*, 1161–1166. IEEE Press, Piscataway, NJ, USA, 1993

M. Sugeno. An Introductory Survey of Fuzzy Control. *Information Sciences*, 36:59–83. Elsevier, New York, NY, USA, 1985

R.S. Sutton and A.G. Barto. *Reinforcement Learning: An Introduction*. MIT Press, Cambridge, MA, USA, 1998

T. Takagi and M. Sugeno. Fuzzy Identification of Systems and Its Applications to Modeling and Control. *IEEE Transactions on Systems, Man and Cybernetics* 15:116–132. IEEE Press, Piscataway, NJ, USA, 1985

L.A. Zadeh. Towards a Theory of Fuzzy Systems. In: R.E. Kalman and N. de Claris (eds.) *Aspects of Networks and System Theory*, 469–490. Rinehart and Winston, New York, USA, 1971

L.A. Zadeh. A Rationale for Fuzzy Control. *Journal of Dynamic Systems, Measurement, and Control*, 94(1):3–4. American Society of Mechanical Engineers (ASME), New York, NY, USA, 1972

L.A. Zadeh. Outline of a New Approach to the Analysis of Complex Systems and Decision Processes. *IEEE Transactions on Systems, Man and Cybernetics*, 3:28–44. IEEE Press, Piscataway, NJ, USA, 1973

Chapter 20
Fuzzy Clustering

After a brief overview of fuzzy methods in data analysis, this chapter focuses on fuzzy cluster analysis as the oldest fuzzy approach to data analysis. Fuzzy clustering comprises a family of prototype-based clustering methods that can be formulated as the problem of minimizing an objective function. These methods can be seen as "fuzzifications" of, for example, the classical c-means algorithm, which strives to minimize the sum of the (squared) distances between the data points and the cluster centers to which they are assigned. However, in order to "fuzzify" such a crisp clustering approach, it is not enough to merely allow values from the unit interval for the variables encoding the assignments of the data points to the clusters: the minimum is still obtained for a crisp data point assignment. As a consequence, additional means have to be employed in the objective function in order to obtain actual degrees of membership. This chapter surveys the most common fuzzification means and examines and compares their properties.

20.1 Fuzzy Methods in Data Analysis

So far, we considered fuzzy methods for modeling purposes, for which it is beneficial to incorporate vague concepts. As a consequence, the created (fuzzy) models are designed by domain experts and thus result from a purely knowledge-driven approach. However, fuzzy models may also be derived (automatically) from data if a sufficient amount of suitable data is available (data-driven approach).

Purely knowledge-driven and purely data-driven approaches for constructing models can be seen as the extreme ends of a range of strategies. It is possible, however, to combine them on any level: to provide a knowledge-based model and adjust only a few parameters—for instance, the exact location of triangular fuzzy sets—or to use domain knowledge only to choose a very general model class from which a specific model with concrete parameters is chosen based on the data. Neural networks are a typical example for the latter strategy where only the type or structure of the neural network is fixed and then the neural network must be trained with data.

R. Kruse et al., *Computational Intelligence*, Texts in Computer Science,
DOI 10.1007/978-1-4471-5013-8_20, © Springer-Verlag London 2013

With a data-driven approach it is no wonder that fuzzy techniques are also used in the context of data analysis, data mining and machine learning. Note that here the term **fuzzy data analysis** may be interpreted in two fundamentally different ways:

- **Analysis of Fuzzy Data**
  Measured data are often imprecise or under the influence of noise. Answers in questionnaires usually involve terms like "strongly agree", "agree", "neither agree or disagree", "disagree", "strongly disagree", which are essentially fuzzy terms. One way to handle this inherent fuzziness in the data is to model the data by fuzzy sets and then to analyze the fuzzy data. There are two main interpretations of fuzzy sets in the context of statistical data analysis (Dubois 2012): in the **epistemic view** fuzzy sets are used to represent incomplete knowledge about an underlying object or a precise quantity (Kwakernaak 1978; Kruse 1987). In the **ontic view** fuzzy sets are considered as real complex humped entities (Puri and Ralescu 1986; Blanco-Fernández et al. 2012). Note that due to the different semantics of the data also the statistical methods have to be different. We do not consider this topic in more detail here. An interested reader can find an abundance of information on this topic in the textbooks Kruse and Meyer (1987), Bandemer and Näther (1992), Viertl (2011) and the proceedings of the conference series Soft Methods in Probability and Statistics (Kruse et al. 2012).
- **Fuzzy Techniques for the Analysis of (Crisp) Data**
  Although data might be noisy or imprecise, they remain crisp in this approach and fuzzy techniques are applied to analyze the data. In this case, the objective is often to obtain fuzzy models that describe the crisp data. As an example for the application of fuzzy methods in data analysis, we focus in this chapter on fuzzy cluster analysis as the oldest fuzzy approach to data analysis.

Both research areas are still discussed lively: two recent conferences on these topics were organized by three of the authors (Borgelt et al. 2013; Kruse et al. 2012).

In order to analyze (crisp) data, a large variety of fuzzy techniques have been developed. For example, there is vast collection of methods to tune the parameters or to learn the rules of a fuzzy system based on data. But there are also applications in the context of data mining and machine learning. It is beyond the scope of this book to discuss these approaches in detail. An interested reader can find excellent overviews in Hüllermeier (2005, 2011).

20.2 Clustering

The general objective of *clustering* or *cluster analysis* (Everitt 1981; Jain and Dubes 1988; Kaufman and Rousseeuw 1990; Höppner et al. 1999) is to group given objects in such a way that objects from the same cluster are as similar as possible, while objects from different clusters are as dissimilar as possible. In order to formalize the notion of similarity, so that it becomes mathematically treatable, it is usually expressed as a *distance measure* between points (or vectors) representing the objects in a metric space, usually $\mathbb{R}^m$. Two objects are then seen as the more similar, the smaller the distance between the data points that represent them.

A common approach to describe the clusters is to use *prototypes* that capture the location and possibly also the shape and size of the clusters in the data space. With such an approach, the general objective of clustering can be reformulated as the task to find a set of cluster prototypes together with an assignment of the data points to them, so that the data points are as close as possible to their assigned prototypes. By formalizing this approach, and using for the prototypes only points in the data space that represent the *cluster centers*, one obtains immediately the objective function of classical c-means clustering (Ball and Hall 1967; Hartigan and Wong 1979; Lloyd 1982): simply sum the (squared) distances of the data points to the center of the cluster to which they are assigned. The c-means algorithm then strives to minimize this objective function by iteratively updating the assignment of the data points to the clusters and recomputing the cluster centers.

Unfortunately, c-means clustering always partitions the data, that is, each data point is assigned to one cluster and one cluster only. This is often inappropriate, as it can lead to somewhat arbitrary cluster boundaries and certainly does not treat points properly that lie between two (or more) clusters without belonging to any of them unambiguously. Solutions to this problem consist in either using a probabilistic approach, like applying the expectation maximization (EM) algorithm to a mixture of Gaussians (see, for example, Dempster et al. 1977; Everitt and Hand 1981; Bilmes 1997), or to employ one of the different "fuzzifications" of the classical crisp scheme (see, for instance, Ruspini 1969; Dunn 1973; Bezdek 1981; Bezdek et al. 1999; Höppner et al. 1999; Borgelt 2005).

In this chapter we focus on the latter approach, that is, on how the objective function of classical c-means clustering can be modified in order to obtain graded cluster memberships (so-called *fuzzy clustering*). We survey different methods that have been suggested in the literature and examine and compare their properties.

20.3 Presuppositions and Notation

We are given a data set $\mathbf{X} = \{\mathbf{x}_1, \ldots, \mathbf{x}_n\}$ with n data points, each of which is an m-dimensional real-valued vector, that is, $\forall j; 1 \leq j \leq n: \mathbf{x}_j = (x_{j1}, \ldots, x_{jm}) \in \mathbb{R}^m$. These data points are to be grouped into c clusters, each of which is described by a prototype $\mathbf{c}_i$, $i = 1, \ldots, c$. The set of all prototypes is denoted by $\mathbf{C} = \{\mathbf{c}_1, \ldots, \mathbf{c}_c\}$. We confine ourselves here to cluster prototypes that consist merely of a cluster center, that is, $\forall i; 1 \leq i \leq c$: $\mathbf{c}_i = (c_{i1}, \ldots, c_{im}) \in \mathbb{R}^m$. The assignment of the data points to the cluster centers is encoded as a $c \times n$ matrix $\mathbf{U} = (u_{ij})_{1 \leq i \leq c; 1 \leq j \leq n}$, which is often called the *partition matrix*. In the crisp case, a matrix element $u_{ij} \in \{0, 1\}$ states whether data point $\mathbf{x}_j$ belongs to cluster $\mathbf{c}_i$ or not. In the fuzzy case, $u_{ij} \in [0, 1]$ states the degree to which $\mathbf{x}_j$ belongs to $\mathbf{c}_i$ (degree of membership).

Furthermore, we confine ourselves to the (squared) Euclidean distance as the measure for the distance between a data point $\mathbf{x}_j$ and a cluster center $\mathbf{c}_i$, that is,

$$d_{ij}^2 = d^2(\mathbf{c}_i, \mathbf{x}_j) = (\mathbf{x}_j - \mathbf{c}_i)^\top (\mathbf{x}_j - \mathbf{c}_i) = \sum_{k=1}^{m} (x_{jk} - c_{ik})^2.$$

A common alternative is the (squared) Mahalanobis distance with a cluster specific covariance matrix Σ_i (Gustafson and Kessel 1979; Gath and Geva 1989), that is, $d_{ij}^2 = (\mathbf{x}_j - \mathbf{c}_i)^\top \Sigma_i^{-1} (\mathbf{x}_j - \mathbf{c}_i)$. However, this choice adds at least a shape parameter and in some approaches also a size parameter to the cluster prototypes (see, for example, Bezdek et al. 1999; Höppner et al. 1999; Borgelt 2005). Nevertheless, extending the approaches to this distance measure is usually fairly straightforward. An extension to the L_1-distance (Jajuga 2003), that is, to $d_{ij} = \sum_{k=1}^{m} |x_{jk} - c_{ik}|$, or to other Minkowski metrics, although certainly useful in specific cases, is less simple to achieve and clearly beyond the scope of this chapter.

20.4 Classical c-Means Clustering

As already stated, classical c-means clustering strives to find, for a given data set $\mathbf{X}$, a set $\mathbf{C}$ of cluster centers and a partition matrix $\mathbf{U}$, such that the objective function

$$J(\mathbf{X}, \mathbf{C}, \mathbf{U}) = \sum_{i=1}^{c} \sum_{j=1}^{n} u_{ij}\, d_{ij}^2$$

is minimized under the constraints $\forall i; 1 \le i \le c : \forall j; 1 \le j \le n : u_{ij} \in \{0, 1\}$ and $\forall j; 1 \le j \le n : \sum_{i=1}^{c} u_{ij} = 1$. These constraints ensure that each data point is assigned to one cluster and to one cluster only (crisp partition of the data set).

Since the minimum cannot be found directly using analytical means, an *alternating optimization* scheme is employed. At the beginning the cluster centers are initialized randomly, for example, by selecting c data points arbitrarily or by sampling c points from some distribution on the data space. Then the two steps of *partition matrix update* (data point assignment) and *cluster center update* are iterated until convergence, that is, until the cluster centers do not change anymore.

In the partition matrix update, each data point $\mathbf{x}_j$ is assigned to the cluster $\mathbf{c}_i$, the center of which is closest to it. That is, the partition matrix is updated according to

$$u_{ij} = \begin{cases} 1, & \text{if } i = \operatorname{argmin}_{i=1}^{c} d_{ij}^2, \\ 0, & \text{otherwise.} \end{cases}$$

In the cluster center update, each cluster center is recomputed as the mean of the data points that were assigned to it (hence the name c-means clustering), that is,

$$\mathbf{c}_i = \frac{\sum_{j=1}^{n} u_{ij}\, \mathbf{x}_j^2}{\sum_{j=1}^{n} u_{ij}}.$$

This update process is guaranteed to converge and usually does so after fairly few steps. However, it is fairly sensitive to the initial conditions (i.e., the initial cluster

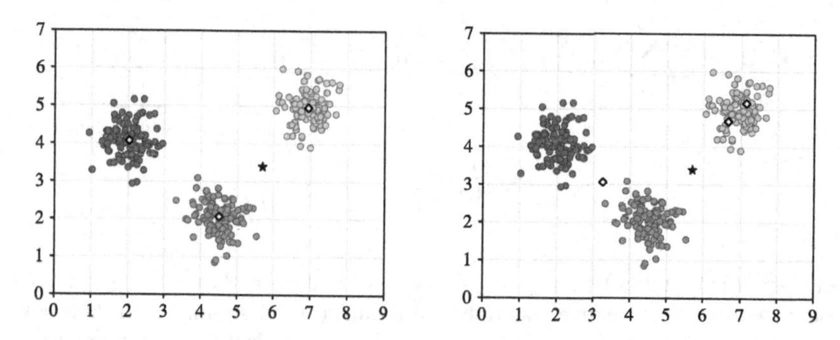

Fig. 20.1 A data set with three clusters and an additional point (marked with a *star*). Result of a successful c-means clustering (*left*) and a local optimum (*right*); *diamonds* mark cluster centers

centers), due to which it can yield undesired results, which are caused by local minima of the objective function. In order to handle this drawback, it is usually recommended to execute the clustering algorithm multiple times and take the best result, that is, the result that yields the smallest value of the objective function.

As an illustration of the problem of local minima, consider the simple two-dimensional data set shown in Fig. 20.1. Visual inspection clearly tells us that there are three clusters (ignore the additional data point marked with a star for now) and hence we expect a clustering result as it is shown on the left: the diamonds mark the cluster centers. Although this result is obtained for a suitable initialization, another initialization yields the result shown on the right. Since two centers are located in the same cluster (the one on the top right), while the other two clusters are captured by a single center that is located between them, this result is clearly undesirable. It corresponds to a local optimum of the objective function, though, and thus the update process may converge to it (depending on the initialization).

However, even if c-means clustering correctly identifies the three clusters (as in Fig. 20.1 on the left), it is not quite clear to which cluster the additional point (marked with a star) should be assigned, because it lies in the middle between two clusters. Although it is clearly counter-intuitive to assign this point uniquely to one cluster, c-means clustering has to do so, because it always yields a (crisp) partition of the data set. Clearly, it would be more appropriate to be able to assign this point with degrees of membership to more than one cluster, for example, with a degree of about 0.5 to both of the clusters the centers of which are closest to it.

To obtain such *degrees of membership*, it may seem, at first sight, to be sufficient to simply extend the allowed range of values of the u_{ij} from the set $\{0, 1\}$ to the real interval $[0, 1]$, but to make no changes to the objective function itself. However, this is not the case: the optimum of the objective function is obtained for a crisp assignment, regardless of whether we enforce a crisp assignment or not.

This can easily be demonstrated as follows: let $k_j = \operatorname{argmin}_{i=1}^{c} d_{ij}^2$, that is, let k_j be the index of the cluster center closest to the data point $\mathbf{x}_j$. Then it is

$$J(\mathbf{X}, \mathbf{C}, \mathbf{U}) = \sum_{i=1}^{c} \sum_{j=1}^{n} u_{ij} d_{ij}^2 \geq \sum_{i=1}^{c} \sum_{j=1}^{n} u_{ij} d_{k_j j}^2 = \sum_{j=1}^{n} d_{k_j j}^2 \underbrace{\sum_{i=1}^{c} u_{ij}}_{=1 \text{ (due to the constraints)}}$$

$$= \sum_{j=1}^{n} \left(1 \cdot d_{k_j j}^2 + \sum_{\substack{i=1 \\ i \neq k_j}}^{c} 0 \cdot d_{ij}^2 \right).$$

Therefore, it is best to set $\forall j; 1 \leq j \leq n$: $u_{k_j j} = 1$ and $u_{ij} = 0$ for $1 \leq i \leq c, i \neq k_j$. In other words: the objective function is minimized by assigning each data point crisply to the closest cluster, even though we allowed for degrees of membership.

20.5 Fuzzification by Membership Transformation

Since we cannot obtain degrees of membership by merely expanding the range of values of the u_{ij}, we have to modify the objective function if we desire graded assignments. The most common approach is to apply a transformation to the membership degrees, that is, to use an objective function of the form

$$J(\mathbf{X}, \mathbf{C}, \mathbf{U}) = \sum_{i=1}^{c} \sum_{j=1}^{n} h(u_{ij}) d_{ij}^2,$$

where h is a convex function on the real interval [0, 1]. This general form was first studied by Klawonn and Höppner (2003), where the convexity of h was derived as follows: for simplicity, we confine ourselves to two clusters $\mathbf{c}_1$ and $\mathbf{c}_2$ and consider the terms of the objective function that refer to a single data point $\mathbf{x}_j$. That is, we consider $J(\mathbf{x}_j, \mathbf{c}_1, \mathbf{c}_2, u_{1j}, u_{2j}) = h(u_{1j}) d_{1j}^2 + h(u_{2j}) d_{2j}^2$ and study how it behaves for different values u_{1j} and u_{2j}. Note that a crisp assignment should not be ruled out categorically, namely if the distances d_{1j} and d_{2j} differ significantly. Hence, we assume that d_{1j} and d_{2j} differ only slightly, so that a graded assignment is desired.

$J(\mathbf{x}_j, \mathbf{c}_1, \mathbf{c}_2, u_{1j}, u_{2j})$ is minimized by choosing u_{1j} and u_{2j} appropriately. Exploiting $\sum_{i=1}^{c} u_{ij} = 1$ yields $J(\mathbf{x}_j, \mathbf{c}_1, \mathbf{c}_2, u_{1j}) = h(u_{1j}) d_{1j}^2 + h(1 - u_{1j}) d_{2j}^2$. A necessary condition for a minimum is $\frac{\partial}{\partial u_{1j}} J(\mathbf{x}_j, \mathbf{c}_1, \mathbf{c}_2, u_{1j}) = h'(u_{1j}) d_{1j}^2 - h'(1 - u_{1j}) d_{2j}^2 = 0$, where $'$ denotes taking the derivative w.r.t. the argument of the function. This leads to $h'(u_{1j}) d_{1j}^2 = h'(1 - u_{1j}) d_{2j}^2$, which yields another argument that a graded assignment cannot be optimal without any function h: if h is the identity, we have $h'(u_{1j}) = h'(1 - u_{1j}) = 1$ and thus the equation cannot hold if the distances differ.

For the further analysis let us assume, without loss of generality, that $d_{1j} < d_{2j}$, which implies $h'(u_{1j}) > h'(1 - u_{1j})$. In addition, we know that $u_{1j} > u_{2j} = 1 - u_{1j}$, because the degree of membership should be higher for the cluster that is closer. In other words, the function h must be the steeper, the greater its argument. Therefore it must be a convex function on the unit interval (Klawonn and Höppner 2003).

Since we confine ourselves to the Euclidean distance (see Sect. 20.3), we can already derive the update rule for the cluster centers, namely by exploiting that a necessary condition for a minimum of the objective function J is that the partial derivatives w.r.t. the cluster centers vanish. Therefore, we have $\forall k; 1 \leq k \leq c$:

$$\begin{aligned}\nabla_{\mathbf{c}_k} J(\mathbf{X}, \mathbf{C}, \mathbf{U}) &= \nabla_{\mathbf{c}_k} \sum_{i=1}^{c} \sum_{j=1}^{n} h(u_{ij})(\mathbf{x}_j - \mathbf{c}_i)^\top (\mathbf{x}_j - \mathbf{c}_i) \\ &= -2 \sum_{j=1}^{n} h(u_{ij})(\mathbf{x}_j - \mathbf{c}_i) \stackrel{!}{=} 0.\end{aligned}$$

Independent of the function h, it follows immediately

$$\mathbf{c}_i = \frac{\sum_{j=1}^{n} h(u_{ij})\mathbf{x}_j}{\sum_{j=1}^{n} h(u_{ij})}.$$

This update rule already shows one of the core drawbacks of a fuzzification by membership transformation, namely that the transformation function enters the update of the cluster centers. It would be more intuitive to use the membership degrees directly as the weights for the mean computation, which would also ensure that all data points enter with the same total unit weight (since $\sum_{i=1}^{c} u_{ij} = 1$ by definition). However, the weights are rather the transformed membership degrees $h(u_{ij})$, which gives unequal weight to the data points as they need not sum to 1.

It may be argued, though, that this effect can actually be desirable: due to the convexity of the function h the total weight $\sum_{i=1}^{c} h(u_{ij})$ of data points $\mathbf{x}_j$ with a less ambiguous assignment is higher than that of more ambiguously assigned data points. (The maximum 1 is obtained for a crisp assignment.) Hence in this scheme the locations of the cluster centers depend more strongly on the data points that are "typical" for the clusters. Such an effect is very much in the spirit of, for instance, robust regression techniques, in which data points receive a lower weight if they do not fit well to the regression function. This connection to robust statistical methods was explored in more detail, for example, in Davé and Krishnapuram (1997).

In order to derive the update rule for the partition matrix (and thus for the membership degrees u_{ij}) we need to know the exact form of the function h. The most common choice is $h(u_{ij}) = u_{ij}^2$, which leads to the standard objective function of fuzzy clustering (Dunn 1973). The more general form $h(u_{ij}) = u_{ij}^w$ was introduced by Bezdek (1981). The exponent w, $w > 1$, is called the *fuzzifier*, since it controls the "fuzziness" of the data point assignments: the higher w, the softer the boundaries between the clusters. This leads to the commonly used objective function (Bezdek 1981; Bezdek et al. 1999; Höppner et al. 1999; Borgelt 2005)

$$J(\mathbf{X}, \mathbf{U}, \mathbf{C}) = \sum_{i=1}^{c} \sum_{j=1}^{n} u_{ij}^w d_{ij}^2.$$

The update rule for the membership degrees is now derived by incorporating the constraints $\forall j; 1 \leq j \leq n : \sum_{i=1}^{c} u_{ij} = 1$ with Lagrange multipliers into the objective function. This yields the Lagrange function

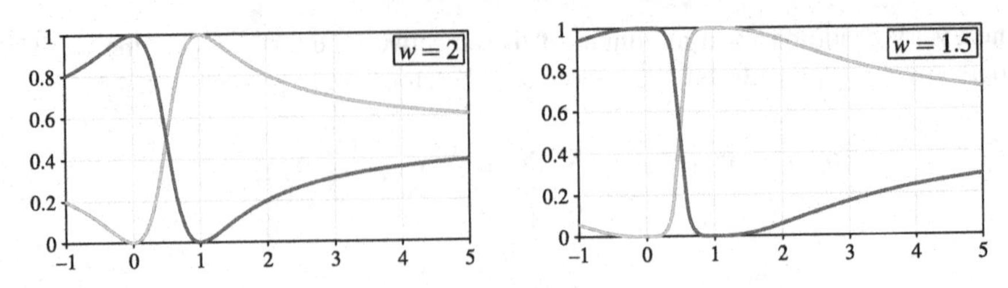

Fig. 20.2 Membership degrees for two clusters with fuzzifiers $w = 2$ (*left*) and $w = 1.5$ (*right*)

$$L(\mathbf{X}, \mathbf{U}, \mathbf{C}, \Lambda) = \underbrace{\sum_{i=1}^{c}\sum_{j=1}^{n} u_{ij}^{w}\, d_{ij}^{2}}_{=J(\mathbf{X},\mathbf{U},\mathbf{C})} + \sum_{j=1}^{n} \lambda_j \left(1 - \sum_{i=1}^{c} u_{ij}\right),$$

where $\Lambda = (\lambda_1, \ldots, \lambda_n)$ are the Lagrange multipliers, one per constraint.

Since a necessary condition for a minimum of the Lagrange function is that the partial derivatives w.r.t. the membership degrees vanish, we obtain

$$\frac{\partial}{\partial u_{kl}} L(\mathbf{X}, \mathbf{U}, \mathbf{C}, \Lambda) = w\, u_{kl}^{w-1}\, d_{kl}^{2} - \lambda_l \stackrel{!}{=} 0 \quad \text{and thus} \quad u_{kl} = \left(\frac{\lambda_l}{w\, d_{kl}^{2}}\right)^{\frac{1}{w-1}}.$$

Summing these equations over the clusters (in order to be able to exploit the corresponding constraints on the membership degrees, which are recovered from the fact that it is a necessary condition for a minimum that the partial derivatives of the Lagrange function w.r.t. the Lagrange multipliers vanish), we get

$$1 = \sum_{i=1}^{c} u_{ij} = \sum_{i=1}^{c} \left(\frac{\lambda_j}{w\, d_{ij}^{2}}\right)^{\frac{1}{w-1}} \quad \text{and thus} \quad \lambda_j = \left(\sum_{i=1}^{c} \left(w\, d_{ij}^{2}\right)^{\frac{1}{1-w}}\right)^{1-w}.$$

Therefore, we finally have for the membership degrees $\forall i; 1 \le i \le c$: $\forall j; 1 \le j \le n$:

$$u_{ij} = \frac{d_{ij}^{\frac{2}{1-w}}}{\sum_{k=1}^{c} d_{kj}^{\frac{2}{1-w}}} \quad \text{and thus for } w = 2\text{:} \quad u_{ij} = \frac{d_{ij}^{-2}}{\sum_{k=1}^{c} d_{kj}^{-2}}.$$

This rule is fairly intuitive, as it updates the membership degrees according to the relative inverse squared distances of the data points to the cluster centers.

The effect of the fuzzifier w is illustrated in Fig. 20.2, which shows the membership degrees of points on the x-axis for two clusters, which are located at $x = 0$ and $x = 1$, for the fuzzifiers $w = 2$ (left) and $w = 1.5$ (right). Clearly, the larger fuzzifier $w = 2$ yields a smoother transition between the two clusters as the degree of membership rises (or falls) less steeply than for the fuzzifier $w = 1.5$.

Unfortunately, the above rule has the disadvantage that it necessarily yields a graded assignment. Regardless of how far a data point is from a cluster center, it always receives a non-vanishing degree of membership to the corresponding cluster.

Even worse: the farther a data point is from all clusters, the more the membership degrees become equal. This effect can also be seen in Fig. 20.2: the farther we get to the right in the diagrams, the more the degrees of membership to the two clusters approach each other. The undesirable results that can be caused by this property in the presence of clusters with fairly uneven numbers of members have been demonstrated clearly by Klawonn and Höppner (2003).

In addition, it was revealed in Klawonn and Höppner (2003) that the reason lies essentially in the fact that $h'(u_{ij}) = \frac{\mathrm{d}}{\mathrm{d}u_{ij}} u_{ij}^w = w\, u_{ij}^{w-1}$ vanishes at $u_{ij} = 0$. This suggests the idea to use a transformation function that does not have this property and thus allows, at least for sufficiently large distance relationships, a crisp assignment of data points to cluster centers. In Klawonn and Höppner (2003) the function $h(u_{ij}) = \alpha u_{ij}^2 + (1-\alpha)u_{ij}$, $\alpha \in (0, 1]$, or, with a more easily interpretable parametrization, $h(u_{ij}) = \frac{1-\beta}{1+\beta} u_{ij}^2 + \frac{2\beta}{1+\beta} u_{ij}$, $\beta \in [0, 1)$, was suggested as such a transformation. It relies on the standard function $h(u_{ij}) = u_{ij}^2$ and mixes it with the identity to avoid a vanishing derivative at zero. The parameter β is, for two clusters, the ratio of the smaller to the larger squared distance, at and below which we get a crisp assignment (Klawonn and Höppner 2003). It therefore takes the place of the fuzzifier w: the smaller β, the softer the boundaries between the clusters.

The update rule for the membership degrees is derived in essentially the same way as for $h(u_{ij}) = u_{ij}^w$, although one has to pay attention to the fact that crisp assignments are now possible and thus some membership degrees may vanish. The detailed derivation, which we omit here, can be found in Klawonn and Höppner (2003) or in Borgelt (2005). It yields

$$u_{ij} = \frac{u'_{ij}}{\sum_{k=1}^{c} u'_{kj}} \quad \text{with } u'_{ij} = \max\left\{0,\ d_{ij}^{-2} - \frac{\beta}{1+\beta(c_j-1)} \sum_{k=1}^{c_j} d_{\varsigma(k)j}^{-2}\right\},$$

where $\varsigma : \{1, \ldots, c\} \to \{1, \ldots, c\}$ is a mapping function for the cluster indices such that $\forall i; 1 \le i < c : d_{\varsigma(i)j} \le d_{\varsigma(i+1)j}$ (that is, ς sorts the distances ascendingly) and

$$c_j = \max\left\{k \;\middle|\; d_{\varsigma(k)j}^{-2} > \frac{\beta}{1+\beta(k-1)} \sum_{i=1}^{k} d_{\varsigma(i)j}^{-2}\right\}$$

is the number of clusters to which the data point $\mathbf{x}_j$ has a non-vanishing membership. This update rule is fairly interpretable, as it still assigns membership degrees essentially according to the relative inverse squared distances to the clusters, but subtracts an offset from them, which makes crisp assignments possible.

20.6 Fuzzification by Membership Regularization

We have seen that transforming the membership degrees in the objective function has the disadvantage that the transformation function appears in the update rule for the cluster centers. In order to avoid this drawback, one may try to achieve a fuzzification by leaving the membership degrees in their weighting of the (squared) distances untouched. Graded memberships are rather achieved by adding a regularization term to the objective function, which pushes the minimum away from a crisp assignment. Most commonly, the objective function then takes the form

$$J(\mathbf{X}, \mathbf{C}, \mathbf{U}) = \sum_{i=1}^{c} \sum_{j=1}^{n} u_{ij} d_{ij}^2 + \gamma \sum_{i=1}^{c} \sum_{j=1}^{n} f(u_{ij}),$$

where f is a convex function on the real interval $[0, 1]$. The parameter γ takes the place of the fuzzifier w: the higher γ, the softer the boundaries between the clusters.

To analyze this objective function, we use the same basic means as in the preceding section: we confine ourselves to two clusters $\mathbf{c}_1$ and $\mathbf{c}_2$ and consider the terms of the objective function that refer to a single data point $\mathbf{x}_j$, that is, we consider $J(\mathbf{x}_j, \mathbf{c}_1, \mathbf{c}_2, u_{1j}, u_{2j}) = u_{1j} d_{1j}^2 + u_{2j} d_{2j}^2 + \gamma f(u_{1j}) + \gamma f(u_{2j})$. Since $u_{2j} = 1 - u_{1j}$, it is $J(\mathbf{x}_j, \mathbf{c}_1, \mathbf{c}_2, u_{1j}) = u_{1j} d_{1j}^2 + (1 - u_{1j}) d_{2j}^2 + \gamma f(u_{1j}) + \gamma f(1 - u_{1j})$. A necessary condition for a minimum is $\frac{\partial}{\partial u_{1j}} J(\mathbf{x}_j, \mathbf{c}_1, \mathbf{c}_2, u_{1j}) = d_{1j}^2 - d_{2j}^2 + \gamma f'(u_{1j}) - \gamma f'(1 - u_{1j}) = 0$, where $'$ denotes taking the derivative w.r.t. the argument of the function. This leads to the simple condition $d_{1j}^2 + \gamma f'(u_{1j}) = d_{2j}^2 + \gamma f'(1 - u_{1j})$.

We now assume again, without loss of generality, that $d_{1j} < d_{2j}$, which implies $f'(u_{1j}) > f'(1 - u_{1j})$. In addition, we know $u_{1j} > u_{2j} = 1 - u_{1j}$, because the degree of membership should be higher for the cluster that is closer. In other words, the function f must be the steeper, the greater its argument. Hence, it must be a convex function on the unit interval in order to allow for graded memberships.

More concretely, we obtain $(d_{2j}^2 - d_{1j}^2)/\gamma = f'(u_{1j}) - f'(1 - u_{1j})$ as a condition for a minimum. Since f is a convex function on the unit interval, the maximum value of the right hand side is $f'(1) - f'(0)$. If $f'(1) - f'(0) < \infty$, we have the possibility of crisp assignments, because in this case there exist values for d_{1j}^2, d_{2j}^2 and γ such that the minimum of the function $J(\mathbf{x}_j, \mathbf{c}_1, \mathbf{c}_2, u_{1j})$ w.r.t. u_{ij} either does not exist or lies outside the unit interval. In such a situation, the best choice is the crisp assignment $u_{1j} = 1$ and $u_{2j} = 0$ (still assuming that $d_{1j} < d_{2j}$).

To obtain the update rule for the cluster centers, we can simply transfer the result from the preceding section, since the regularization term does not refer to the cluster centers. Therefore, we have the simple rule (because here $h(u_{ij}) = u_{ij}$)

$$\mathbf{c}_i = \frac{\sum_{j=1}^{n} u_{ij} \mathbf{x}_j}{\sum_{j=1}^{n} u_{ij}}.$$

This demonstrates the advantage of a membership regularization approach, because the membership degrees are directly the weights with which the data points enter the mean computation that yields the new cluster center.

In order to derive the update rule for the membership degrees, we have to respect the constraints $\forall j; 1 \le j \le n : \sum_{i=1}^{c} u_{ij} = 1$. This is achieved in the usual way (cf. the preceding section) by incorporating them with Lagrange multipliers into the objective function. The resulting Lagrange function is

$$L(\mathbf{X}, \mathbf{U}, \mathbf{C}, \Lambda) = \underbrace{\sum_{i=1}^{c}\sum_{j=1}^{n} u_{ij}d_{ij}^2 + \gamma \sum_{i=1}^{c}\sum_{j=1}^{n} f(u_{ij})}_{=J(\mathbf{X},\mathbf{C},\mathbf{U})} + \sum_{j=1}^{n} \lambda_j \left(1 - \sum_{i=1}^{c} u_{ij}\right),$$

where $\Lambda = (\lambda_1, \ldots, \lambda_n)$ are the Lagrange multipliers, one per constraint.

Since a necessary condition for a minimum of the Lagrange function is that the partial derivatives w.r.t. the membership degrees vanish, we obtain

$$\frac{\partial}{\partial u_{kl}} L(\mathbf{X}, \mathbf{U}, \mathbf{C}) = d_{kl}^2 + \gamma f'(u_{kl}) - \lambda_l \stackrel{!}{=} 0 \quad \text{and thus} \quad u_{kl} = f'^{-1}\left(\frac{\lambda_l - d_{kl}^2}{\gamma}\right),$$

where $'$ denotes taking the derivative w.r.t. the argument of the function and f'^{-1} denotes the inverse of the derivative of the function f. In analogy to Sect. 20.5, the constraints on the membership degrees are now exploited to obtain $1 = \sum_{k=1}^{c} u_{kj} = \sum_{k=1}^{c} f'^{-1}((\lambda_j - d_{kj}^2)/\gamma)$. This equation has to be solved for λ_j and the result has to be used to substitute λ_l in the expression for the u_{kl} derived above. However, in order to do so, we need to know the exact form of the regularization function f.

The regularization functions f that have been suggested in the literature (concrete examples are studied below) can be seen as derived from a maximum entropy approach. That is, the term of the objective function that forces the u_{ij} to minimize the weighted sum of squared distances is complemented by a term that forces them to maximize the entropies of the distributions over the clusters, the u_{ij} describe for each data point. Thus the u_{ij} are pushed away from a crisp assignment, which has minimum entropy, and towards a uniform assignment, which has maximum entropy. Generally, such an approach starts from the objective function

$$J(\mathbf{X}, \mathbf{C}, \mathbf{U}) = \sum_{i=1}^{c}\sum_{j=1}^{n} u_{ij}d_{ij}^2 - \gamma \sum_{j=1}^{n} H(\mathbf{u}_j),$$

where $\mathbf{u}_j = (u_{1j}, \ldots, u_{cj})$ comprises the degrees of membership the data point $\mathbf{x}_j$ has to the different clusters. H computes their entropy, as $\mathbf{u}_j$ is, at least formally, a probability distribution, since it satisfies $\forall i; 1 \leq i \leq c : u_{ij} \in [0, 1]$ and $\sum_{i=1}^{c} u_{ij} = 1$.

In order to develop the maximum entropy approach in more detail, we consider the generalized entropy proposed by Daróczy (1970). Let $\mathbf{p} = (p_1, \ldots, p_r)$ be a probability distribution over r values. Then *Daróczy entropy* is defined as

$$H_\beta(\mathbf{p}) = \frac{2^{\beta-1}}{2^{\beta-1} - 1} \sum_{i=1}^{r} p_i \left(1 - p_i^{\beta-1}\right) = \frac{2^{\beta-1}}{2^{\beta-1} - 1} \left(1 - \sum_{i=1}^{r} p_i^\beta\right).$$

From this general formula, the well-known *Shannon entropy* (Shannon 1948) can be derived as the limit for $\beta \to 1$, that is, as

$$H_1(\mathbf{p}) = \lim_{\beta \to 1} H_\beta(\mathbf{p}) = -\sum_{i=1}^{r} p_i \log_2 p_i.$$

Employing it in the entropy-regularized objective function leads to

$$J(\mathbf{X}, \mathbf{C}, \mathbf{U}) = \sum_{i=1}^{c} \sum_{j=1}^{n} u_{ij} d_{ij}^2 + \gamma \sum_{i=1}^{c} \sum_{j=1}^{n} u_{ij} \ln u_{ij},$$

where the factor $1/\ln 2$ (which stems from the relation $\log_2 u_{ij} = \ln u_{ij}/\ln 2$) is incorporated into the factor γ, as the natural logarithm allows for easier mathematical treatment. That is, we have $f(u_{ij}) = u_{ij} \ln u_{ij}$ (Karayiannis 1994; Li and Mukaidono 1995; Miyamoto and Mukaidono 1997; Boujemaa 2000) and therefore obtain $f'(u_{ij}) = 1 + \ln u_{ij}$ and $f'^{-1}(y) = e^{y-1}$. Using the latter in the formulas obtained above for deriving the update rule for the membership degrees yields

$$u_{ij} = \frac{e^{-d_{ij}^2/\gamma}}{\sum_{k=1}^{c} e^{-d_{kj}^2/\gamma}}.$$

As was pointed out in Mori et al. (2003), Honda and Ichihashi (2005), this update rule relates the approach very closely to the expectation maximization (EM) algorithm for Gaussian mixtures (Dempster et al. 1977; Everitt and Hand 1981; Bilmes 1997), since by setting $\gamma = 2\sigma^2$, we obtain exactly the formula for the expectation step. As a consequence, this update rule can be interpreted as computing the probability that a data point $\mathbf{x}_j$ was sampled from a Gaussian distribution centered at $\mathbf{c}_i$ and having the variance σ^2. In addition, since the update rule for the cluster centers coincides with the maximization step, this form of fuzzy clustering is actually indistinguishable from the expectation maximization algorithm for a mixture of Gaussians.

It should be noted that $f'(u_{ij}) = 1 + \ln u_{ij}$ implies $f'(1) - f'(0) = \infty$ and thus Shannon entropy regularization always yields graded assignments. However, this drawback is less harmful here, because $e^{-d_{ij}^2/\gamma}$ is much "steeper" than d_{ij}^{-2} and thus is less prone to produce undesired results (cf. the discussion in Döring et al. 2005).

Another commonly used special case of Daróczy entropy is so-called *quadratic entropy*, which results if we set the parameter $\beta = 2$, that is,

$$H_2(\mathbf{p}) = 2\sum_{i=1}^{r} p_i(1 - p_i) = 2 - 2\sum_{i=1}^{r} p_i^2.$$

Employing it in the entropy-regularized objective function leads to

$$J(\mathbf{X}, \mathbf{C}, \mathbf{U}) = \sum_{i=1}^{c} \sum_{j=1}^{n} u_{ij} d_{ij}^2 + \gamma \sum_{i=1}^{c} \sum_{j=1}^{n} u_{ij}^2,$$

as the constant term 2 has no influence on the location of the minimum and thus can be discarded, and the factor 2 can be incorporated into the factor γ. That is, we have

$f(u_{ij}) = u_{ij}^2$ (Miyamoto and Umayahara 1998) and therefore obtain $f'(u_{ij}) = 2u_{ij}$ and finally $f'^{-1}(y) = \frac{y}{2}$ for the needed inverse function of the derivative.

In order to derive the update rule for the memberships, one has to pay attention to the fact that $f'(1) - f'(0) = 2$. Therefore, crisp assignments are possible and some membership degrees may vanish. However, the detailed derivation can easily be found by following, for example, the same lines as for the analogous approach in the preceding section, which also allowed for vanishing membership degrees.

The resulting membership degree update rule is $\forall i : 1 \le i \le c : \forall j : 1 \le j \le n$:

$$u_{ij} = \max\left\{0, \frac{1}{c_j}\left(1 + \sum_{k=1}^{c_j} \frac{d_{\varsigma(k)j}^2}{2\gamma}\right) - \frac{d_{ij}}{2\gamma}\right\},$$

where $\varsigma : \{1, \dots, c\} \to \{1, \dots, c\}$ is a mapping function for the cluster indices such that $\forall i; 1 \le i < c : d_{\varsigma(i)j} \le d_{\varsigma(i+1)j}$ (that is, ς sorts the distances ascendingly) and

$$c_j = \max\left\{k \,\middle|\, \sum_{i=1}^{k} d_{\varsigma(i)j}^2 > k\, d_{kj} - 2\gamma\right\}$$

is the number of clusters to which the data point $\mathbf{x}_j$ has a non-vanishing membership. In this update rule 2γ can be interpreted as a reference distance relative to which all distances are judged. For two clusters, 2γ is the difference between the distances of a data point to the cluster centers, at and above which a crisp assignment is used. Clearly, this is equivalent to saying that the distances, if measured in 2γ units, must differ by less than 1 in order to obtain a graded assignment.

A disadvantage of this update rule is that it refers to the difference of the distances rather than their ratio, which seems more intuitive. As a consequence, a data point that has distance x to one cluster and distance y to the other is assigned in exactly the same way as a data point that has distance $x + z$ to the first cluster and distance $y + z$ to the second, regardless of the value of z (provided $z \ge -\min\{x, y\}$).

Alternatives to the discussed approaches modified the Shannon entropy term, using, for instance, $f(u_{ij}) = u_{ij} \ln u_{ij} + (1 - u_{ij}) \ln(1 - u_{ij})$ (Yasuda et al. 2001), or replaced it with Kullback–Leibler information divergence (Kullback and Leibler 1951) to the (estimated) cluster probability distribution (Ichihashi et al. 2001), that is, $f(u_{ij}) = u_{ij} \ln \frac{u_{ij}}{p_i}$ with $p_i = \frac{1}{n}\sum_{j=1}^{n} u_{ij}$.

It has also been tried to use $f(u_{ij}) = u_{ij}^w$ (Yang 1993; Özdemir and Akarun 2002), but combined with $h(u_{ij}) = u_{ij}^w$ (to avoid technical complications), so that the objective function is effectively

$$J(\mathbf{X}, \mathbf{C}, \mathbf{U}) = \sum_{i=1}^{c} \sum_{j=1}^{n} u_{ij}^w \left(d_{ij}^2 + \gamma\right).$$

Hence this is actually a hybrid approach that combines membership transformation and regularization. Another hybrid approach, proposed in Wei and Fahn (2002), combines $h(u_{ij}) = u_{ij}^w$ and Shannon entropy regularization $f(u_{ij}) = u_{ij} \ln u_{ij}$.

Finally, a generalized objective function was presented in Bezdek and Hathaway (2003) and analyzed in more detail in Yu and Yang (2007).

It should be noted, though, that the approach of Frigui and Krishnapuram (1997), which is covered by the generalized objective function of Bezdek and Hathaway (2003) and based on

$$J(\mathbf{X}, \mathbf{C}, \mathbf{U}) = \sum_{i=1}^{c} \sum_{j=1}^{n} u_{ij}^{w} d_{ij}^{2} - \gamma \sum_{i=1}^{c} p_i^2 \quad \text{with } p_i = \frac{1}{n} \sum_{j=1}^{n} u_{ij},$$

is *not* a membership regularization scheme, as it yields crisp assignments unless $w > 1$. In this approach the entropy term (which is added rather than subtracted) serves the purpose to choose the number of clusters automatically.

A closely related approach is *possibilistic clustering* (Krishnapuram and Keller 1993, 1996), which eliminates the membership constraints $\forall j; 1 \leq j \leq n : \sum_{i=1}^{c} u_{ij} = 1$ and is based on the objective function

$$J(\mathbf{X}, \mathbf{C}, \mathbf{U}) = \sum_{i=1}^{c} \sum_{j=1}^{n} u_{ij}^{w} d_{ij}^{2} + \sum_{i=1}^{c} \eta_i \sum_{j=1}^{n} (1 - u_{ij})^{w}.$$

Here the η_i are suitable positive numbers (one per cluster $\mathbf{c}_i$, $1 \leq i \leq c$) that determine the distance at which the membership degree of a point to a cluster is 0.5. They are usually initialized, based on the result of a preceding run of standard fuzzy clustering, as the average fuzzy intra-cluster distance $\eta_i = \sum_{j=1}^{n} u_{ij}^{w} d_{ij}^{2} / \sum_{j=1}^{n} u_{ij}^{w}$ and may or may not be updated in each iteration (Krishnapuram and Keller 1993).

Although this approach is useful in certain applications, it should be noted that the objective function of possibilistic clustering is truly optimized only if all clusters are identical (Timm et al. 2004), because the missing constraints decouple the clusters. Thus, it actually *requires* that the optimization process gets stuck in a local optimum in order to yield useful results, which is a somewhat strange property.

20.7 Comparison

Since classical c-means clustering does not yield graded data point assignments, even if one allows the membership variables to take values in the unit interval, the objective function has to be modified if graded assignments are desired. There are two fundamental approaches to this: transforming the membership degrees or adding a membership regularization term. In both cases variants can be derived that allow partially crisp assignments, that is, allow for vanishing membership degrees, as well as variants that enforce graded assignments regardless of the data. All of these variants have advantages and disadvantages: membership transformation suffers generally from the fact that the transformation function enters the cluster center update, but uses a fairly intuitive relative inverse squared distance scheme for the membership updates. Quadratic entropy regularization allows for vanishing membership degrees, but refers to distance differences rather than more intuitive distance ratios. Shannon entropy regularization leads to a procedure that is equivalent to the expectation maximization (EM) algorithm for a mixture of Gaussian and thus is not a specifically "fuzzy" approach anymore. However, judging from the discussion in Döring et al. (2005) due to which the forced graded assignment is unproblematic, its practical advantages make it, in our opinion, the most recommendable approach.

References

G.H. Ball and D.J. Hall. A Clustering Technique for Summarizing Multivariate Data. *Behavioral Science* 12(2):153–155. J. Wiley & Sons, Chichester, United Kingdom, 1967

H. Bandemer and W. Näther. *Fuzzy Data Analysis*. Kluwer, Dordrecht, Netherlands, 1992

J.C. Bezdek. *Pattern Recognition with Fuzzy Objective Function Algorithms*. Plenum Press, New York, NY, USA, 1981

J.C. Bezdek and R.J. Hathaway. Visual Cluster Validity (VCV) Displays for Prototype Generator Clustering Methods. *Proc. 12th IEEE Int. Conf. on Fuzzy Systems (FUZZ-IEEE 2003, Saint Louis, MO)*, 2:875–880. IEEE Press, Piscataway, NJ, USA, 2003

J.C. Bezdek and N. Pal. *Fuzzy Models for Pattern Recognition*. IEEE Press, New York, NY, USA, 1992

J.C. Bezdek, J.M. Keller, R. Krishnapuram, and N. Pal. *Fuzzy Models and Algorithms for Pattern Recognition and Image Processing*. Kluwer, Dordrecht, Netherlands, 1999

J. Bilmes. A Gentle Tutorial on the EM Algorithm and Its Application to Parameter Estimation for Gaussian Mixture and Hidden Markov Models. *Tech. Report ICSI-TR-97-021*. University of Berkeley, CA, USA, 1997

A. Blanco-Fernández, M.R. Casals, A. Colubi, R. Coppi, N. Corral, S. Rosa de Sáa, P. D'Urso, M.B. Ferraro, M. García-Bárzana, M.A. Gil, P. Giordani, G. González-Rodríguez, M.T. López, M.A. Lubiano, M. Montenegro, T. Nakama, A.B. Ramos-Guajardo, B. Sinova, and W. Trutschnig. Arithmetic and Distance-Based Approach to the Statistical Analysis of Imprecisely Valued Data. In: Borgelt et al. (2013), 1–18

C. Borgelt. *Prototype-based Classification and Clustering*. Habilitationsschrift, Otto-von-Guericke-University of Magdeburg, Germany, 2005

C. Borgelt, M.A. Gil, J.M.C. Sousa, and M. Verleysen (eds.). *Towards Advanced Data Analysis by Combining Soft Computing and Statistics*. Studies in Fuzziness and Soft Computing, vol. 285. Springer-Verlag, Berlin/Heidelberg, Germany, 2013

N. Boujemaa. Generalized Competitive Clustering for Image Segmentation. *Proc. 19th Int. Meeting North American Fuzzy Information Processing Society (NAFIPS 2000, Atlanta, GA)*, 133–137. IEEE Press, Piscataway, NJ, USA, 2000

Z. Daróczy. Generalized Information Functions. *Information and Control* 16(1):36–51. Academic Press, San Diego, CA, USA, 1970

R.N. Davé and R. Krishnapuram. Robust Clustering Methods: A Unified View. *IEEE Transactions on Fuzzy Systems* 5(1997):270–293. IEEE Press, Piscataway, NJ, USA, 1997

A.P. Dempster, N. Laird and D. Rubin. Maximum Likelihood from Incomplete Data via the EM Algorithm. *Journal of the Royal Statistical Society. Series B* 39:1–38. Blackwell, Oxford, United Kingdom, 1977

C. Döring, C. Borgelt, and R. Kruse. Effects of Irrelevant Attributes in Fuzzy Clustering. *Proc. 14th IEEE Int. Conf. on Fuzzy Systems (FUZZ-IEEE'05, Reno, NV)*, 862–866. IEEE Press, Piscataway, NJ, USA, 2005

D. Dubois. Statistical Reasoning with set-Valued Information: Ontic vs. Epistemic Views. In: Borgelt et al. (2013), 119–136

J.C. Dunn. A Fuzzy Relative of the ISODATA Process and Its Use in Detecting Compact Well-Separated Clusters. *Journal of Cybernetics* 3(3):32–57, 1973. American Society for Cybernetics, Washington, DC, USA. Reprinted in Bezdek and Pal (1992), 82–101

B.S. Everitt. *Cluster Analysis*. Heinemann, London, United Kingdom, 1981

B.S. Everitt and D.J.Hand. *Finite Mixture Distributions*. Chapman & Hall, London, United Kingdom, 1981

H. Frigui and R. Krishnapuram. Clustering by Competitive Agglomeration. *Pattern Recognition* 30(7):1109–1119. Pergamon Press, Oxford, United Kingdom, 1997

I. Gath and A.B. Geva. Unsupervised Optimal Fuzzy Clustering. *IEEE Transactions on Pattern Analysis and Machine Intelligence* 11:773–781, 1989. IEEE Press, Piscataway, NJ, USA. Reprinted in Bezdek and Pal (1992), 211–218

E.E. Gustafson and W.C. Kessel. Fuzzy Clustering with a Fuzzy Covariance Matrix. *Proc. of the IEEE Conf. on Decision and Control (CDC 1979, San Diego, CA)*, 761–766. IEEE Press, Piscataway, NJ, USA, 1979. Reprinted in Bezdek and Pal (1992), 117–122

J.A. Hartigan and M.A. Wong. A k-Means Clustering Algorithm. *Applied Statistics* 28:100–108. Blackwell, Oxford, United Kingdom, 1979

K. Honda and H. Ichihashi. Regularized Linear Fuzzy Clustering and Probabilistic PCA Mixture Models. *IEEE Transactions on Fuzzy Systems* 13(4):508–516. IEEE Press, Piscataway, NJ, USA, 2005

F. Höppner, F. Klawonn, R. Kruse, and T. Runkler. *Fuzzy Cluster Analysis*. J. Wiley & Sons, Chichester, United Kingdom, 1999

E. Hüllermeier. Fuzzy-Methods in Machine Learning and Data Mining: Status and Prospects. *Fuzzy Sets and Systems* 156(3):387–407. Elsevier, Amsterdam, Netherlands, 2005

E. Hüllermeier. Fuzzy Sets in Machine Learning and Data Mining. *Applied Soft Computing* 11(2):1493–1505. Elsevier, Amsterdam, Netherlands, 2011

H. Ichihashi, K. Miyagishi and K. Honda. Fuzzy c-Means Clustering with Regularization by K-L Information. *Proc. 10th IEEE Int. Conf. on Fuzzy Systems (FUZZ-IEEE 2001, Melbourne, Australia)*, 924–927. IEEE Press, Piscataway, NJ, USA, 2001

A.K. Jain and R.C. Dubes. *Algorithms for Clustering Data*. Prentice Hall, Englewood Cliffs, NJ, USA, 1988

K. Jajuga. L_1-norm Based Fuzzy Clustering. *Fuzzy Sets and Systems* 39(1):43–50. Elsevier, Amsterdam, Netherlands, 2003

N.B. Karayiannis. MECA: Maximum Entropy Clustering Algorithm. *Proc. 3rd IEEE Int. Conf. on Fuzzy Systems (FUZZ-IEEE 1994, Orlando, FL)*, I:630–635. IEEE Press, Piscataway, NJ, USA, 1994

L. Kaufman and P. Rousseeuw. *Finding Groups in Data: An Introduction to Cluster Analysis*. J. Wiley & Sons, New York, NY, USA, 1990

F. Klawonn and F. Höppner. What is Fuzzy about Fuzzy Clustering? Understanding and Improving the Concept of the Fuzzifier. *Proc. 5th Int. Symposium on Intelligent Data Analysis (IDA 2003, Berlin, Germany)*, 254–264. Springer-Verlag, Berlin, Germany, 2003

R. Krishnapuram and J.M. Keller. A Possibilistic Approach to Clustering. *IEEE Transactions on Fuzzy Systems* 1(2):98–110. IEEE Press, Piscataway, NJ, USA, 1993

R. Krishnapuram and J.M. Keller. The Possibilistic c-Means Algorithm: Insights and Recommendations. *IEEE Transactions on Fuzzy Systems* 4(3):385–393. IEEE Press, Piscataway, NJ, USA, 1996

R. Kruse. On the Variance of Random Sets. *Journal of Mathematical Analysis and Applications* 122:469–473. Elsevier, Amsterdam, Netherlands, 1987

R. Kruse and K.D. Meyer. *Statistics with Vague Data*. D. Reidel Publishing Company, Dordrecht, Netherlands, 1987

R. Kruse, M.R. Berthold, C. Moewes, M.A. Gil, P. Grzegorzewski, and O. Hryniewicz (eds). *Synergies of Soft Computing and Statistics for Intelligent Data Analysis*. Advances in Intelligent Systems and Computing, vol. 190. Springer-Verlag, Heidelberg/Berlin, Germany, 2012

S. Kullback and R.A. Leibler. On Information and Sufficiency. *Annals of Mathematical Statistics* 22:79–86. Institute of Mathematical Statistics, Hayward, CA, USA, 1951

H. Kwakernaak. Fuzzy Random Variables—I. Definitions and Theorems. *Information Sciences* 15:1–29. Elsevier, Amsterdam, Netherlands, 1978

H. Kwakernaak. Fuzzy Random Variables—II. Algorithms and Examples for the Discrete Case. *Information Sciences* 17:252–278. Elsevier, Amsterdam, Netherlands, 1979

R.P. Li and M. Mukaidono. A Maximum Entropy Approach to Fuzzy Clustering. *Proc. 4th IEEE Int. Conf. on Fuzzy Systems (FUZZ-IEEE 1994, Yokohama, Japan)*, 2227–2232. IEEE Press, Piscataway, NJ, USA, 1995

S. Lloyd. Least Squares Quantization in PCM. *IEEE Transactions on Information Theory* 28:129–137. IEEE Press, Piscataway, NJ, USA, 1982

S. Miyamoto and M. Mukaidono. Fuzzy c-Means as a Regularization and Maximum Entropy Approach. *Proc. 7th Int. Fuzzy Systems Association World Congress (IFSA'97, Prague, Czech Republic)*, II:86–92, 1997

S. Miyamoto and K. Umayahara. Fuzzy Clustering by Quadratic Regularization. *Proc. IEEE Int. Conf. on Fuzzy Systems/IEEE World Congress on Computational Intelligence (WCCI 1998, Anchorage, AK)*, 2:1394–1399. IEEE Press, Piscataway, NJ, USA, 1998

Y. Mori, K. Honda, A. Kanda, and H. Ichihashi. A Unified View of Probabilistic PCA and Regularized Linear Fuzzy Clustering. *Proc. Int. Joint Conf. on Neural Networks (IJCNN 2003, Portland, OR)*, I:541–546. IEEE Press, Piscataway, NJ, USA, 2003

D. Özdemir and L. Akarun. A Fuzzy Algorithm for Color Quantization of Images. *Pattern Recognition* 35:1785–1791. Pergamon Press, Oxford, United Kingdom, 2002

M. Puri and D. Ralescu. Fuzzy Random Variables. *Journal of Mathematical Analysis and Applications* 114:409–422. Elsevier, Amsterdam, Netherlands, 1986

E.H. Ruspini. A New Approach to Clustering. *Information and Control* 15(1):22–32, 1969. Academic Press, San Diego, CA, USA. Reprinted in Bezdek and Pal (1992), 63–70

C.E. Shannon. The Mathematical Theory of Communication. *The Bell System Technical Journal* 27:379–423. Bell Laboratories, Murray Hill, NJ, USA, 1948

H. Timm, C. Borgelt, C. Döring, and R. Kruse. An Extension to Possibilistic Fuzzy Cluster Analysis. *Fuzzy Sets and Systems* 147:3–16. Elsevier Science, Amsterdam, Netherlands, 2004

R. Viertl. *Statistical Methods for Fuzzy Data*. John Wiley & Sons, Chichester, UK, 2011

C. Wei and C. Fahn. The Multisynapse Neural Network and Its Application to Fuzzy Clustering. *IEEE Transactions on Neural Networks* 13(3):600–618. IEEE Press, Piscataway, NJ, USA, 2002

M.S. Yang. On a Class of Fuzzy Classification Maximum Likelihood Procedures. *Fuzzy Sets and Systems* 57:365–375. Elsevier, Amsterdam, Netherlands, 1993

M. Yasuda, T. Furuhashi, M. Matsuzaki and S. Okuma. Fuzzy Clustering Using Deterministic Annealing Method and Its Statistical Mechanical Characteristics. *Proc. 10th IEEE Int. Conf. on Fuzzy Systems (FUZZ-IEEE 2001, Melbourne, Australia)*, 2:797–800. IEEE Press, Piscataway, NJ, USA, 2001

J. Yu and M.S. Yang. A Generalized Fuzzy Clustering Regularization Model with Optimality Tests and Model Complexity Analysis. *IEEE Transactions on Fuzzy Systems* 15(5):904–915. IEEE Press, Piscataway, NJ, USA, 2007

Part IV
Bayes Networks

Chapter 21
Introduction to Bayes Networks

Relational database systems are amongst the most wide-spread data management systems in today's businesses. A database typically consists of several tables that contain data about business objects such as customer data, sales orders or product information. Each table row represents a description of a single object with each table column representing an attribute of that object. Relations between these objects are also modeled via tables. Please note that we use the notions *table* and *relation* interchangeably. A major part of database theory is concerned with the task to represent data with as little redundancy as possible.

21.1 A Fictitious Example

Let us illustrate the concepts in the setting of a fictitious car manufacturer. For each part of a car, a table with different suppliers is maintained. For the sake of simplicity, let us assume there are only three different parts: engine, transmission and brakes.

Tables 21.1, 21.2 and 21.3 show example values for these parts. The first column denotes the primary key, that is, the identifying table entry. Let us assume further that all attributes' value combinations are possible and hence there are 36 different unique car models. The above-mentioned decomposition for redundancy reduction is not shown here. For example, in real life there would be another table containing the address and contact data for each supplier which in turn are referenced in Tables 21.1–21.3 by their name.

Apart from just querying the database, we would like to go further than this and draw *inferences* like "Supplier X can only deliver transmission t_4 at the moment. Which consequences have to be inferred for the procurement of brakes and engines?". Answering such questions involves historical information (e.g., installation rates of single car parts from the past) and expert knowledge (e.g., the allowed technical combinations of parts) that are modeled and exploited via probability theory. Such a probability distribution is exhaustively (together with all marginal distributions) displayed in Fig. 21.1. Table 21.4 sketches how they are stored inside the

R. Kruse et al., *Computational Intelligence*, Texts in Computer Science,
DOI 10.1007/978-1-4471-5013-8_21, © Springer-Verlag London 2013

Table 21.1 Table Engines

EID	Power	Type	...
e_1	100kW	Diesel	...
e_2	150kW	Otto	...
e_3	200kW	Otto	...

Table 21.2 Table Brakes

BID	Material	Manufacturer	...
b_1	Steal	Firm 1	...
b_2	Steal	Firm 2	...
b_3	Ceramic	Firm 2	...

Table 21.3 Table Transmissions

TID	Gears	Automatic	...
t_1	4	n	...
t_2	5	n	...
t_3	5	y	...
t_4	6	y	...

Table 21.4 Sketch of the ternary relation with the relative frequencies of all 36 possible car combinations. The full relation can be found in Fig. 21.1

TID	EID	BID	$P(\cdot)$
t_1	e_1	b_1	0.084
t_1	e_1	b_2	0.056
⋮	⋮	⋮	⋮
t_4	e_3	b_2	0.072
t_4	e_3	b_3	0.080

database. The above question how the distribution of brakes and engines changes if only transmission t_4 is available is shown in Fig. 21.2. Obviously, all combinations with $T \neq t_4$ become impossible. Since the result needs to be another probability distribution, all remaining entries of the "slice" $T = t_4$ were normalized to sum up to one. This was done by dividing all entries by the marginal installation rate $P(T = t_4) = 0.280$. Now, let $\text{dom}(M)$ be the domain of attribute M, that is, the set of all possible values M can assume. To determine for example, the new installation rates for brake b_1, the following implicit steps are required:

$$P(B = b_1 \mid T = t_4) = \frac{\sum_{m \in \text{dom}(M)} P(M = m,\, B = b_1,\, T = t_4)}{\sum_{m \in \text{dom}(M)} \sum_{b \in \text{dom}(B)} P(M = m,\, B = b\,,\, T = t_4)}$$

$$= \frac{8 + 17 + 9}{80 + 17 + 3 + 72 + 68 + 6 + 8 + 17 + 9} \approx 0.122.$$

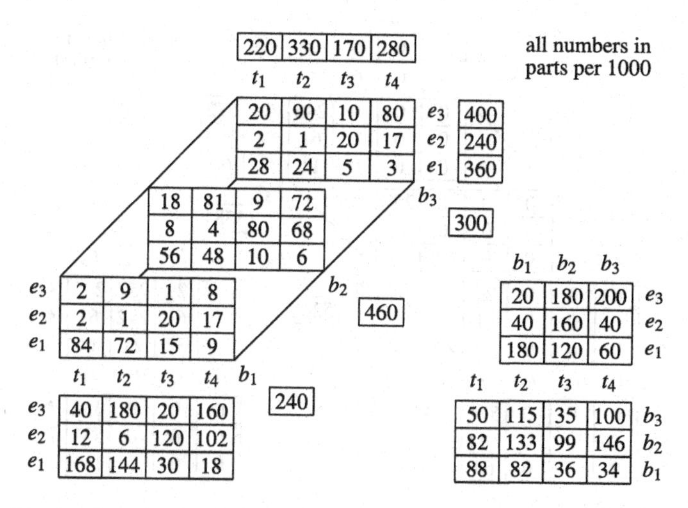

Fig. 21.1 Three-dimensional probability distribution over the attributes Transmission, Engine and Brakes. Additionally, the marginal distributions (sums over rows, columns or both) are shown

These summations clearly show what effort it takes to compute the conditional distributions directly. Dealing with a three-dimensional database, such a procedure is still feasible. In real-world applications, however, one has to deal with hundreds of attributes having considerably larger domains. Assuming a more realistic number of attributes of 200 and only three values per attribute, the state space has a cardinality of $3^{200} \approx 2.6 \cdot 10^{95}$ and thus contains more unique combinations than there are elementary particles in the universe, that is, an estimated number of 10^{87} elementary particles. Even a summation over subsets of the distribution is impossible. Another challenge is to efficiently store the distribution since a enumeration of all possible combinations is infeasible and most of them are likely never used at all. This can be seen as follows: assume that all vehicles on the planet were built by our manufacturer and no two vehicles are identical (w.r.t. their attribute values). Even then the estimated number of just below a billion of those vehicles is negligible compared to the number 3^{200} of theoretically possible ones.

The main idea to efficiently store and use a high-dimensional probability distribution p (containing the knowledge about a certain area of application) is to decompose it into a set $\{p_1, \ldots, p_s\}$ of lower-dimensional and possibly overlapping distributions. If this is achieved one can infer from it the same conclusions as from the original distribution p. As one can easily verify, the following criterion is valid inside the example distribution $P(T, E, B)$ from Fig. 21.1:

$$P(T=t, E=e, B=b) = \frac{P(T=t, E=e) \cdot P(E=e, B=b)}{P(E=e)}. \tag{21.1}$$

Obviously, it is sufficient to store the two-dimensional distributions over attributes B and E, and T and E in order to correctly reconstruct the original three-dimensional distribution since the one-dimensional distribution over E can be computed via (affordable) summation over one of the two two-dimensional distributions. The at-

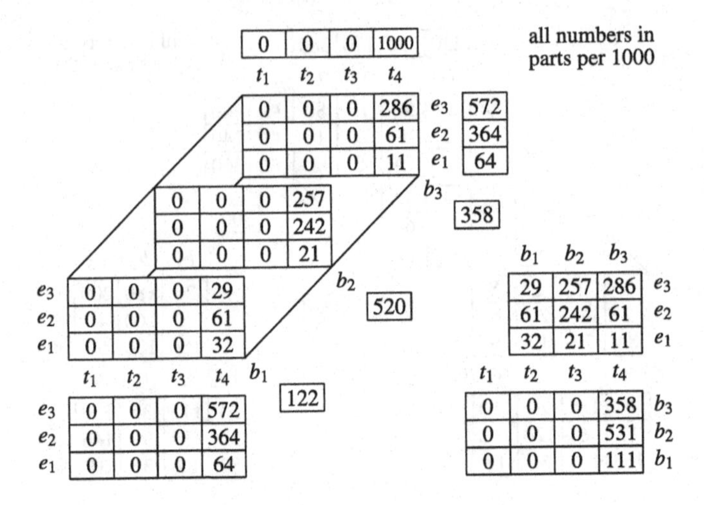

Fig. 21.2 Conditional distribution(s) given the condition $T = t_4$

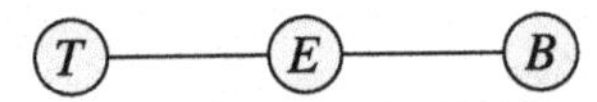

Fig. 21.3 Graphical model of the relation shown in Fig. 21.1

tribute E apparently plays an important role w.r.t. the decomposability since it is contained in both two-dimensional distributions. We will later refer to the criterion in Eq. (21.1) as *conditional independence* (of the attributes T and B given attribute E). Such conditional independences will later be intuitively represented by (directed or undirected) graphs. The objective is to be able to infer all valid stochastic statements from the underlying probability distribution by only using graph-theoretic criteria. The undirected graph for our small example is depicted in Fig. 21.3. The attributes for which we need a joint distribution are connected by an edge. Another advantage of a graphical representation will be the ability to use it to determine the paths and steps needed to update the remaining attributes in the presence of evidence (like known attribute values such as the value t_4 of attribute T in our example) without the need to reconstruct the original distribution.

The graph in Fig. 21.3 describes a path over which information about attribute T can be transferred via attribute E to infer a changed probability distribution of attribute B. Thus, only those two lower-dimensional distributions are necessary that are represented by the edges:

$$\begin{aligned}
P(B = b_1 \mid T = t_4) &= \frac{1}{P(t_4)} \cdot \sum_{m \in \mathrm{dom}(M)} \frac{P(t_4, m) \cdot P(m, b_1)}{P(m)} \\
&= \frac{1000}{280} \cdot \left(\frac{18 \cdot 180}{360 \cdot 1000} + \frac{102 \cdot 40}{240 \cdot 1000} + \frac{160 \cdot 20}{400 \cdot 1000} \right) \\
&= \frac{34}{280} \approx 0.122.
\end{aligned}$$

One can easily understand that in a scenario with a reasonable (large) number of attributes, only exploiting the previously sketched concepts allows for a feasible evidence propagation. Concluding the above ideas, the following list of questions will be answered in the next chapters:

1. How can expert knowledge about complex domains be efficiently represented? We will look into knowledge representations based on directed and undirected Bayes and Markov network.
2. How can inferences be drawn inside these representations? The graphical representations will provide the paths over which the evidence needs to be propagated inside the respective Bayes or Markov networks.
3. How can such graphical representations be automatically learned from data? We will discuss learning principles for Bayes and Markov networks and look deeper into an example for Bayes networks.

Chapter 22
Elements of Probability and Graph Theory

This chapter introduces required theoretical concepts for the definition of Bayes and Markov networks. After important elements of probability theory—especially (conditional) independences—are discussed, we present relevant graph-theoretic notions with emphasis on so-called separation criteria. These criteria will later allow us to capture probabilistic independences with an undirected or directed graph.

22.1 Probability Theory

The classical notion of probability and its interpretation in terms of relative frequencies are deeply embedded in our intuition. Modern mathematics embraced an axiomatic methodology which abstracts from specific meanings of objects. Rather, it assumes all objects to be given with no further property than their own identity (i.e., the objects are mutually distinguishable) and studies the relations amongst these objects resulting from postulated axioms.

Probability theory is formed in that axiomatic way by employing the so-called Kolmogorov axioms (Kolmogorov 1933). An event in this nomenclature is simply a set of elementary events that are distinguishable, i.e., that have an identity. A probability is then a number assigned to an event satisfying certain criteria that are defined by the above-mentioned axioms. Let us first define the fundamental notions of an *event algebra* and a *σ-algebra*.

Definition 22.1 (Event algebra) Let Ω be an event space (i.e., a universal set of elementary events). A system of subsets $\mathcal{S}$ over Ω is called *event algebra* if and only if the following conditions hold:

- The certain event Ω and the impossible event $\emptyset$ are in $\mathcal{S}$.
- For every $A \in \mathcal{S}$ the complement $\overline{A} = \Omega \backslash A$ is also contained in $\mathcal{S}$.
- If A and B are in $\mathcal{S}$, then $A \cup B$ and $A \cap B$ are also in $\mathcal{S}$.

The following condition may also hold:

R. Kruse et al., *Computational Intelligence*, Texts in Computer Science,
DOI 10.1007/978-1-4471-5013-8_22, © Springer-Verlag London 2013

- If for all $i \in \mathbb{N}$ the event A_i is in $\mathcal{S}$, then the events $\bigcup_{i=1}^{\infty} A_i$ and $\bigcap_{i=1}^{\infty} A_i$ are also in $\mathcal{S}$.

In that case $\mathcal{S}$ is called a *σ-algebra.*

The semantic of $A \cup B$ is the event that occurs if A or B occurs. The intersection $A \cap B$ occurs if and only if A and B occur. The complement $\overline{A}$ occurs if and only if A does not occur. Two events A and B are *disjoint* if and only if they cannot occur simultaneously, i.e., if their intersection is the impossible event: $A \cap B = \emptyset$.

To assign a probability to an event, the so-called Kolmogorov axioms are used:

Definition 22.2 (Kolmogorov axioms) Let $\mathcal{S}$ be an event algebra over a finite event space Ω.

- The *probability* $P(A)$ of an event $A \in \mathcal{S}$ is a uniquely defined non-negative number of value at most one, i.e., if $0 \leq P(A) \leq 1$ holds.
- The certain event Ω has probability one: $P(\Omega) = 1$.
- *Addition axiom*: If the events A and B are disjoint ($A \cap B = \emptyset$), then $P(A \cup B) = P(A) + P(B)$ holds.

In event spaces Ω containing infinitely many elementary events, $\mathcal{S}$ has to be a σ-algebra and the addition axiom has to be replaced by:

- *Extended addition axiom*: If $A_1, A_2, \ldots$ are countably infinitely many pairwise disjoint events, then

$$P\left(\bigcup_{i=1}^{\infty} A_i\right) = \sum_{i=1}^{\infty} A_i.$$

These three axioms already implies the following (incomplete) list of properties:

- $\forall A \in \mathcal{S} : P(\overline{A}) = 1 - P(A)$.
- $P(\emptyset) = 0$.
- For pairwise disjoint events, $A_1, \ldots, A_n$ holds:

$$P\left(\bigcup_{i=1}^{n} A_i\right) = \sum_{i=1}^{n} P(A_i).$$

- For any (not necessarily disjoint) events, A and B holds:

$$P(A \cup B) = P(A) + P(B) - P(A \cap B).$$

The Kolmogorov axioms are consistent since there exist systems that satisfy all axioms. Kolmogorov's axioms allow for an embedding of probability theory into measure theory and to interpret the probability as an non-negative normalized additive set function, i.e. as a measure.

Since the definitions of event algebra and Kolmogorov axioms are not unique but represent a class of set systems and functions, respectively, one has to specify precisely for every application the underlying objects. This is done with the notion of a *probability space*.

Definition 22.3 (Probability space) Let Ω be an event space, $\mathcal{S}$ a σ-algebra over Ω and P a probability on $\mathcal{S}$. The triple $(\Omega, \mathcal{S}, P)$ is called a *probability space.*

Up to now, we only computed the probabilities of events without discussing the change of this probability when new information (in form of, again, events) becomes known. That is, we now ask for the probability of an event given the knowledge that one or more other events have (or have not) occurred.

Definition 22.4 (Conditional probability) Let A and B be events with $P(B) > 0$. Then

$$P(A \mid B) = \frac{P(A \cap B)}{P(B)}$$

is called the *conditional probability* of A given (the condition) B.

The following theorem directly follows.

Theorem 22.1 (Product theorem/Multiplication theorem) *For any two events A and B holds*:

$$P(A \cap B) = P(A \mid B) \cdot P(B) = P(B \mid A) \cdot P(A).$$

For a set U of events together with a total ordering $\prec$ on this set, we can generalize the product theorem by induction over the events:

$$P\left(\bigcap_{A \in U} A\right) = \prod_{A \in U} P\left(A \,\middle|\, \bigcap_{B \prec A} B\right).$$

If there is no $B \in U$ with $B \prec A$, then the intersection in the condition of the right-hand side is not empty but it is not computed at all, leading to an implicit Ω:

$$P\left(A \,\middle|\, \bigcap_{B \prec A} B\right) = P\left(A \,\middle|\, \Omega \cap \bigcap_{B \prec A} B\right) = P(A \mid \Omega) = P(A).$$

For $U = \{A, B\}$ with $B \prec A$ the above Theorem 22.1 follows. Further, multiple events can make up the condition.

Theorem 22.2 *Let U, V and W be non-empty sets of events with $U = V \cup W$ and $V \cap W = \emptyset$. Then the following statement holds*:

$$P\left(\bigcap_{A \in U} A\right) = P\left(\bigcap_{A \in V} A \,\middle|\, \bigcap_{A \in W} A\right) \cdot P\left(\bigcap_{A \in W} A\right).$$

A conditional probability satisfies all Kolmogorov axioms. With this we get the following theorem.

Theorem 22.3 *For any fixed event B with $P(B) > 0$ the function P_B defined as*

$$P_B(A) = P(A \mid B)$$

constitutes a probability function that satisfies the condition $P_B(\overline{B}) = 0$.

Definition 22.5 (Event partition) Let U be a set of events. The events in U form an *event partition* if all events are pairwise disjoint (that is, if $\forall A, B \in U : A \neq B \Leftrightarrow A \cap B = \emptyset$ holds) and if $\bigcup_{A \in U} = \Omega$ holds (that is, they cover the entire event space).

Theorem 22.4 (Total probability) *Let U be a set of events that form an event partition. Then the probability of any event B can be written as*

$$P(B) = \sum_{A \in U} P(B \mid A) P(A).$$

If one replaces the $P(B)$ in the right-hand side of the equation in Definition 22.4 by an event partition we arrive at the Bayes theorem.

Theorem 22.5 (Bayes theorem) *Let U be a set of events that form an event partition. Further let B be an event with $P(B) > 0$. Then the following equality holds:*

$$\forall A \in U : P(A \mid B) = \frac{P(B \mid A) P(A)}{P(B)} = \frac{P(B \mid A) P(A)}{\sum_{A' \in U} P(B \mid A') P(A')}.$$

This equation is also known as the equation on the probability of hypotheses because it is possible to compute the probability of hypotheses (e.g. diseases), given the knowledge about the probabilities with which the respective hypotheses (here: A) lead to the events $B \in U$ (e.g. symptoms).

22.1.1 Random Variables and Random Vectors

Until now, we combined elements of the event space to make up events without discussing a specific way how to determine the elements of an event A. Further, we still lack the ability to specify properties of the elementary events. Let's assume the event space Ω be the set of all students at the University of Magdeburg. We are interested in, say, the attributes Gender, Year and Course. To assign values to these attributes we use functions defined on Ω having a reasonable domain (e.g., {male, female} or {CS, Math, Econ, ...}). The preimage of such a function is a set of elementary events (e.g., the set of computer science students or all female students). If these preimages constitute proper events (w.r.t. an underlying event algebra) we will refer to these functions as *random variables*.

Definition 22.6 ((Discrete) Random variable) A function X defined on an event space Ω with domain $\text{dom}(X)$ is called a *random variable* if the preimage of any subset of its domain has a probability. A subset $W \subseteq \text{dom}(X)$ has w.r.t. X the following preimage:

$$X^{-1}(W) = \{\omega \in \Omega \mid X(\omega) \in W\} \overset{\text{abbr.}}{=} \{X \in W\}.$$

Note that despite the name random *variable* and the traditional uppercase letter we are dealing with a function here. In the remainder, we refer to the domain of any function X (including random variables, of course) by $\text{dom}(X)$. The concept of a random variable can be generalized to sets of random variables.

Definition 22.7 (Random vector) Let $X_1, \dots, X_n$ be random variables over the same event space Ω and the same event algebra $\mathcal{S}$. Then the vector $\mathbf{X} = (X_1, \dots, X_n)$ is called a *random vector*.

We will use the following conjunctive interpretation for the computation of the probability of the value of a random vector. Simultaneously, we will introduce some shorthand notations:

$$\begin{aligned} \forall \mathbf{x} \in \mathop{\times}_{i=1}^{n} \text{dom}(X_i): \quad & P(\mathbf{X} = \mathbf{x}) \\ & \equiv P(\mathbf{x}) \\ & \equiv P(x_1, \dots, x_n) \\ & \equiv P(X_1 = x_1, \dots, X_n = x_n) \\ & \equiv P\left(\bigwedge_{i=1}^{n} X_i = x_i\right) \\ & := P\left(\bigcap_{i=1}^{n} \{X_i = x_i\}\right) \\ & = P\big(X_1^{-1}(x_1) \cap \dots \cap X_n^{-1}(x_n)\big). \end{aligned}$$

Given a set of random variables, we can consider the probabilities of all (domain) value combinations a structured representation of the underlying probability space and will refer to them as a probability distribution. We first define distributions for a single random variable.

Definition 22.8 ((Probability) distribution) A random variable X with a finite or countably infinite domain $\text{dom}(X)$ is called discrete. The entirety p_X of all pairs

$$\big(x_i, P(X = x_i)\big) \quad \text{with } x_i \in \text{dom}(X)$$

is called the *(probability) distribution* of the discrete random variable X. We use the notation

$$p_X(x_i) = P(X = x_i) \quad \text{for all } x_i \in \text{dom}(X).$$

The generalization of this notion to sets of random variables is straightforward and not given here.

Until now, we used vectors to represent higher dimensions (of random variables). The (domain) value combinations thus were vectors, that is, elements of the Cartesian product of the domains of the respective random variables. In order to simplify notations later on, we need to do away with the implicit order that underlies a Cartesian product. We choose to use tuples (instead of a vector) as functions on the *set* of random variables. With this rationale, the order of the random variables becomes irrelevant.

Definition 22.9 (Tuple) Let $V = \{A_1, \ldots, A_n\}$ be a finite set of random variables with the respective domains $\operatorname{dom}(A_i)$, $i = 1, \ldots, n$. An *instantiation* of the random variables in V or a *tuple* over V is a mapping

$$\mathbf{t}_V : V \to \bigcup_{A \in V} \operatorname{dom}(A),$$

that satisfies the following condition:

$$\forall A \in V: \quad \mathbf{t}_V(A) \in \operatorname{dom}(A).$$

The bold notation illustrates that the tuple assigns values to multiple random variables. A tuple that assigns a value to only a single random variable will be denoted in its scalar form: t. The index V is dropped if the set V is clear from context. A tuple over the set $\{A, B, C\}$ of random variables which assigns to A the value a_1, to B the value b_2 and to C the value c_2, is denoted as

$$\mathbf{t} = (A \mapsto a_1, B \mapsto b_2, C \mapsto c_2),$$

or shorter (if one can infer the attribute from its value):

$$\mathbf{t} = (a_1, b_2, c_2).$$

For two tuples to be equal, they must be declared on the same sets of random variables and map to identical values:

$$\mathbf{t}_V = \mathbf{t}'_U \quad \Leftrightarrow \quad V = U \wedge \forall A \in V : t(A) = t'(A).$$

The domain of a tuple is restricted using a projection that is defined as follows:

Definition 22.10 (Projection (of a tuple)) Let $\mathbf{t}_X$ be a tuple over a set X of random variables and $Y \subseteq X$. Then $\operatorname{proj}_Y^X(\mathbf{t}_X)$ be the *projection* of tuple t_X to Y. That is, the mapping $\operatorname{proj}_Y^X(\mathbf{t}_X)$ assigns values only to elements of Y.

We, again, drop the index X if it is clear from context.

Up to now, we always considered all random variables for computation of probabilities. If only fewer random variables are required, we marginalize (sum) over all value combinations of the variables to be eliminated.

Definition 22.11 (Marginalization, Marginal distribution) Let $V = \{X_1, \ldots, X_n\}$ be a set of random variables over the same probability space and p_V a probability distribution over V. For any subset $M \subset V$ the *marginalization over* M is defined as the distribution $p_{V \setminus M}$ that results when summing over all values of all random variables in M, i.e., if the following holds:

$$\forall x_1 \in \operatorname{dom}(X_1) : \cdots \forall x_n \in \operatorname{dom}(X_n) :$$

$$p_{V \setminus M}\Bigl(\bigwedge_{X_i \in V \setminus M} X_i = x_i\Bigr) = \sum_{\substack{\forall X_j \in M: \\ \forall x_j \in \operatorname{dom}(X_j)}} p_V\Bigl(\bigwedge_{X_j \in M} X_j = x_j, \bigwedge_{X_i \in V \setminus M} X_i = x_i\Bigr).$$

For $V \backslash M = \{X\}$ p_X is called the *marginal distribution* of X.

Using this definition, we can easily verify the values of the marginal distributions in Table 22.1:

$$P(\mathsf{G}=\mathsf{m}) = 0.5 \qquad P(\mathsf{Sm}=\mathsf{sm}) = 0.3$$
$$P(\mathsf{G}=\mathsf{f}) = 0.5 \qquad P(\mathsf{Sm}=\overline{\mathsf{sm}}) = 0.7$$

$$P(\mathsf{Sm}=\mathsf{sm}, \mathsf{Pr}=\mathsf{pr}) = 0.01 \qquad P(\mathsf{Sm}=\mathsf{r}, \mathsf{Pr}=\overline{\mathsf{pr}}) = 0.29$$
$$P(\mathsf{Sm}=\overline{\mathsf{sm}}, \mathsf{Pr}=\mathsf{pr}) = 0.04 \qquad P(\mathsf{Sm}=\overline{\mathsf{sm}}, \mathsf{Pr}=\overline{\mathsf{pr}}) = 0.66$$

In the last paragraphs, we used the notion random variable and attribute interchangeably. Additional synonyms that we will use in the remainder, are random variable, property and dimension. Further, we will from now on only use probability statements w.r.t. random variables; no directly specified events (as subsets of Ω) will be used anymore. Therefore, we need to emphasize the difference between the probability statements: $P(A)$ together with a specific event $A \subseteq \Omega$ stands for a specific probability, i.e., $P(A) \in [0, 1]$. If the object A is a random variable (as it will be the case from now on) the proposition $P(A)$ represents an all-quantified statement over all values of the domain of A. For two random variables A and B, the equation

$$P(A \mid B) = \frac{P(A, B)}{P(B)}$$

is a shorthand notation for the following verbose statement:

$$\forall a \in \operatorname{dom}(A) : \forall b \in \operatorname{dom}(B) : P(A = a \mid B = b) = \frac{P(A = a, B = b)}{P(B = b)}.$$

22.1.2 Independences

In Chap. 21, we motivated the decomposition of a high-dimensional distribution into several lower-dimensional distributions. The exploited property for that process is the (conditional) independence between attributes. Let us start with the so-called marginal (i.e., unconditional) independence. As the name suggests, we need a criterion which tells that for two attributes A and B it is irrelevant for the probabilities of A to know anything about B. Formally, the marginal probability distribution of A shall not be any different from the conditional distribution of A given B:

Definition 22.12 (Independence of random variables) The random variable A is (stochastically) independent of random variable B with $0 < P(B) < 1$ if and only if

$$P(A \mid B) = P(A)$$

or, equivalently, if

$$P(A, B) = P(A) \cdot P(B).$$

Table 22.1 Example distribution with conditional independence

p_{orig}	G = m		G = f	
	Sm = sm	Sm = $\overline{\text{sm}}$	Sm = sm	Sm = $\overline{\text{sm}}$
Pr = pr	0	0	0.01	0.04
Pr = $\overline{\text{pr}}$	0.2	0.3	0.09	0.36

The last proposition is obtained by applying Definition 22.4 which was declared on events to $P(A \mid B)$ and resolve for $P(A, B)$ where we use the implicitly all-quantified version over all values of the attributes. Note that the (stochastic) independence is symmetric, i.e., if A is (stochastically) independent of B, then B is (stochastically) independent of A. The notion of (stochastic) independence is easily generalized to more than two events:

Definition 22.13 (Full (stochastic) independence) Let U be a set of random variables. The random variables in U are fully (stochastically) independent, if the following holds:

$$\forall V \subseteq U: \quad P\left(\bigcap_{A \in V} A\right) = \prod_{A \in V} P(A).$$

As none of the probability statements had a condition, we refer to the above concept as *unconditional* or *marginal* independence.

Table 22.1 shows a three-dimensional example distribution over the Boolean attributes Gender, Pregnant and Smoker. The distribution after marginalizing over the attribute Gender is depicted in Table 22.2(a). Marginalizing further to the one-dimensional distributions for Smoker and Pregnant, we can constructing the two-dimensional joint distribution via multiplication as depicted in Table 22.2(b). Even though the values are close, both two-dimensional distributions over Smoker and Pregnant (the original and the reconstructed one) are different. Hence, attributes Smoker and Pregnant are *not* independent.

However, if we consider the columns for the values of attribute Gender in Table 22.1 separately and renormalize the probabilities to one, we arrive at the conditional distributions as shown in Table 22.2(c) and Table 22.2(d), respectively.

Testing these distributions for independence reveals that in both cases the attributes Pregnant and Smoker are independent. We observe an independence under the condition that the value of the third attribute Gender is known. Hence, we refer to this concept as *conditional independence of* Pregnant *and* Smoker *given* (*the condition*) Gender.

The mathematical formulation results by inserting into the probability functions of Definition 22.12 one or more additional conditions:

$$P(A \mid B, C) = P(A \mid C) \quad \Leftrightarrow \quad P(A, B \mid C) = P(A \mid C)\, P(B \mid C).$$

Note that the independence has to hold for all conditions, that is, for all attribute values of C, in order to conclude a conditional independence given C:

Table 22.2 Distributions to illustrate conditional independence of "Pregnant is conditionally independent of Smoker given Gender"

p_1	Sm = sm	Sm = $\overline{\text{sm}}$	
Pr = pr	0.01	0.04	0.05
Pr = $\overline{\text{pr}}$	0.29	0.66	0.95
	0.30	0.70	

(a) Distribution $p_1 = P(\text{Sm}, \text{Pr})$

p_2	Sm = sm	Sm = $\overline{\text{sm}}$	
Pr = pr	0.015	0.035	0.05
Pr = $\overline{\text{pr}}$	0.285	0.665	0.95
	0.300	0.700	

(b) Distribution $p_2 = P(\text{Sm}) \cdot P(\text{Pr})$

p_3	Sm = sm	Sm = $\overline{\text{sm}}$	
Pr = pr	0	0	0
Pr = $\overline{\text{pr}}$	0.4	0.6	1.0
	0.4	0.6	

(c) Distribution $p_3 = P(\text{Sm}, \text{Pr} \mid \text{G} = \text{m})$

p_4	Sm = sm	Sm = $\overline{\text{sm}}$	
Pr = pr	0.02	0.08	0.1
Pr = $\overline{\text{pr}}$	0.18	0.72	0.9
	0.20	0.80	

(d) Distribution $p_4 = P(\text{Sm}, \text{Pr} \mid \text{G} = \text{f})$

$$\forall a \in \text{dom}(A) : \forall b \in \text{dom}(B) : \forall c \in \text{dom}(C) :$$
$$P(A = a, B = b \mid C = c) = P(A = a \mid C = c)\, P(B = b \mid C = c).$$

In the remainder, we will use the generalized notion of conditional independence which is extended to sets of attributes.

Definition 22.14 (Conditional independence of random variables) Let $X = \{A_1, \dots, A_k\}$, $Y = \{B_1, \dots, B_l\}$ and $Z = \{C_1, \dots, C_m\}$ be three pairwise disjoint sets of random variables. X and Y are conditionally independent given Z (w.r.t. a given distribution p)—written as $X \perp\!\!\!\perp_p Y \mid Z$—if and only if the following holds:

$$\begin{aligned}
&\forall a_1 \in \text{dom}(A_1) : \cdots \forall a_k \in \text{dom}(A_k) : \\
&\quad \forall b_1 \in \text{dom}(B_1) : \cdots \forall b_l \in \text{dom}(B_l) : \\
&\quad\quad \forall c_1 \in \text{dom}(C_1) : \cdots \forall c_m \in \text{dom}(C_m) : \\
&\quad\quad\quad P(A_1 = a_1, \dots, A_k = a_k \mid B_1 = b_1, \dots, B_l = b_l, C_1 = c_1, \dots, C_m = c_m) \\
&\quad\quad\quad = P(A_1 = a_1, \dots, A_k = a_k \mid C_1 = c_1, \dots, C_m = c_m).
\end{aligned}$$

Or in shorthand notation:

$$P(A_1, \dots, A_k \mid B_1, \dots, B_l, C_1, \dots, C_m) = P(A_1, \dots, A_k \mid C_1, \dots, C_m).$$

22.2 Graph Theory

In order to represent Bayes and Markov networks, we will need *directed, acyclic graphs* (short: *DAGs*) and undirected graphs. This section introduces the necessary graph-theoretic notions.

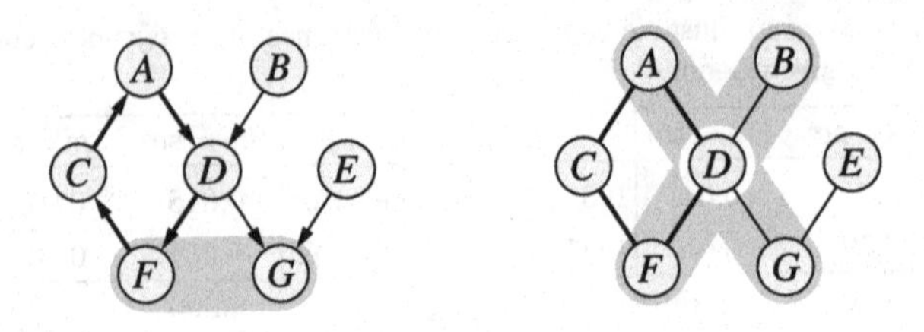

Fig. 22.1 The adjacency sets of node D w.r.t. the directed and undirected case are shaded in *gray*. The closed path A–D–F–C represents a cycle and circle, respectively

22.2.1 Background

Definition 22.15 ((Simple) graph) A *simple graph*—in the remainder just referred to as *graph*—is a tuple $G = (V, E)$ where $V = \{A_1, \dots, A_n\}$ is a finite set of n *vertices* or *nodes* and $E \subseteq (V \times V) \setminus \{(A, A) \mid A \in V\}$ is a set of *edges*.

Such a graph is called *simple* because no multiple edges or loops (edges from a node to itself) are allowed.

Definition 22.16 (Directed edge) Let $G = (V, E)$ be a simple graph.
An edge $e = (A, B) \in E$ is called a *directed edge* if

$$(A, B) \in E \Rightarrow (B, A) \notin E.$$

Such an edge points from A to B which is denoted as $A \to B$. The node A is called *parent node* of B while B is called the *child node* of A.

Definition 22.17 (Undirected edge) Let $G = (V, E)$ be a graph. Two pairs (A, B) and (B, A) from E comprise a single *undirected edge* between nodes A and B if

$$(A, B) \in E \Rightarrow (B, A) \in E.$$

We denote such an edge as $A - B$ or $B - A$.

Definition 22.18 (Adjacency set) Let $G = (V, E)$ be a graph. The set of nodes that are reachable from a given node A is called *adjacency set* of A:

$$\text{adj}(A) = \{B \in V \mid (A, B) \in E\}.$$

Definition 22.19 (Path) Let $G = (V, E)$ be a graph. A series ρ of r pairwise different nodes

$$\rho = \langle A_{i_1}, \dots, A_{i_r} \rangle$$

is called a *path* from A_i to A_j if

- $A_{i_1} = A_i$
- $A_{i_r} = A_j$
- $(A_{i_k}, A_{i_{k+1}}) \in E$ or $(A_{i_{k+1}}, A_{i_k}) \in E$, $1 \le k < r$

The symmetric formulation of the last item allows paths that run against the edge direction (which will be important later on). A path consisting only of undirected edges is called *undirected path* and is denoted as

$$\rho = A_{i_1} - \cdots - A_{i_r}$$

while paths consisting only of directed edges that run exclusively in edge direction are called *directed paths* and denoted as

$$\rho = A_{i_1} \to \cdots \to A_{i_r}.$$

Paths with only directed edges that also might run in opposite edge directed are called *mixed paths* and are denoted according to the contained edges. The left graph of Fig. 22.1 contains for example, the mixed path $F \leftarrow D \rightarrow G \leftarrow E$. If two nodes A and B are connected via a directed path ρ in a graph G, we denote it as $A \overset{\rho}{\underset{G}{\rightsquigarrow}} B$. In case of an undirected path, we write $A \overset{\rho}{\underset{G}{\leftrightsquigarrow}} B$.

A graph with only undirected edges is called *undirected graph*. A graph with only directed edges is called *directed graph*. Note that the path Definition 22.19 allows edges from the last node to the first one. In that case, we call them cycles or circles for directed or undirected graphs, respectively.

Definition 22.20 (Cycle) Let $G = (V, E)$ be a directed graph. A path

$$\rho = X_1 \to \cdots \to X_k$$

with $X_k \to X_1 \in E$ is called a *cycle*.

Definition 22.21 Circle Let $G = (V, E)$ be an undirected graph. A path

$$\rho = X_1 - \cdots - X_k$$

with $X_k - X_1 \in E$ is called *circle*.

In Fig. 22.1, the path *A–D–F–C* is a cycle in the directed graph and a circle in the undirected graph.

Definition 22.22 (Tree) An undirected graph in which any pair of nodes is connected by exactly one path is called a *tree*.

Definition 22.23 (Minimum spanning tree) Let $G = (V, E)$ be an undirected graph and w a function assigning to each edge in E a weight:

$$w : E \to \mathbb{R}.$$

A graph $G' = (V, E')$ is called *minimum spanning tree* if

- G' is a tree
- $E' \subseteq E$
- $\sum_{e \in E'} w(e) = \min$

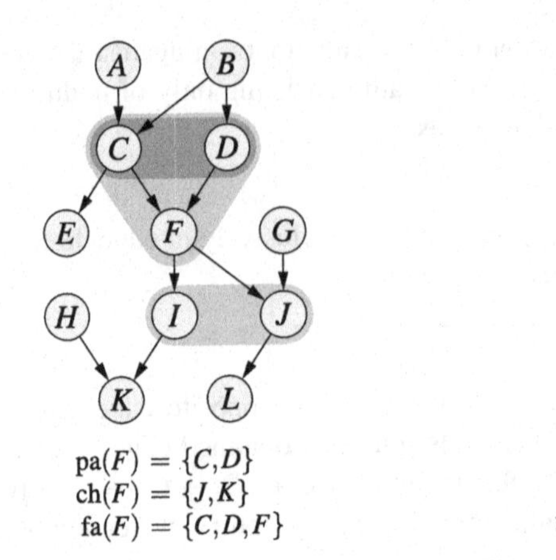

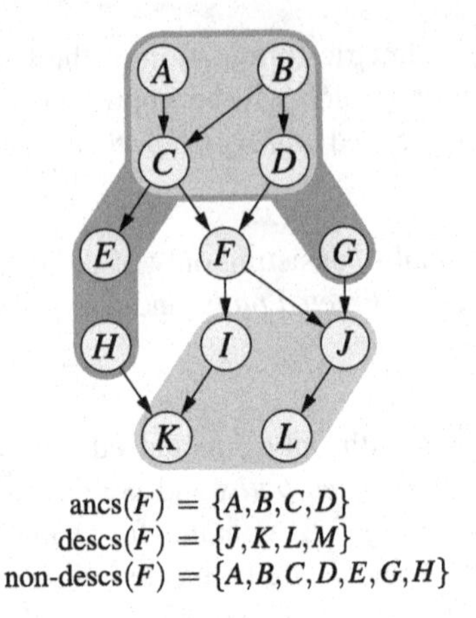

Fig. 22.2 Node relations in directed graphs

That is, there is no other tree over all nodes V with a smaller edge weight sum. There may be, however, multiple trees with equal minimal edge weight sums.

Along these lines *maximal spanning trees* can be defined: The sum of the edge weights has then to be maximal. Two widely known algorithms for constructing minimal or maximal spanning trees are:

- KRUSKAL algorithm (Kruskal 1956)
- PRIM algorithm (Prim 1957)

We now introduce notions for directed graphs.

Definition 22.24 (Parent nodes, child nodes, family) Let $G = (V, E)$ be a directed graph and $A \in V$ a node.

$$\mathrm{pa}(A) = \{B \in V \mid B \to A \in E\}.$$

Analogous, the set of *child nodes* of A is defined as:

$$\mathrm{ch}(A) = \{B \in V \mid A \to B \in E\}.$$

The *family* of a node A consists of the node A itself together with its parent nodes:

$$\mathrm{fa}(A) = \{A\} \cup \mathrm{pa}(A).$$

Definition 22.25 Directed acyclic graph (DAG) A directed graph $G = (V, E)$ is called *acyclic* if for each path $X_1 \to \cdots \to X_k$ in G holds:

$$X_k \to X_1 \notin E.$$

Definition 22.26 (Ancestors, Descendants, Non-descendants) Let $G = (V, E)$ be a directed acyclic graph and $A \in V$ a node. In contrast to Definition 22.24, we require

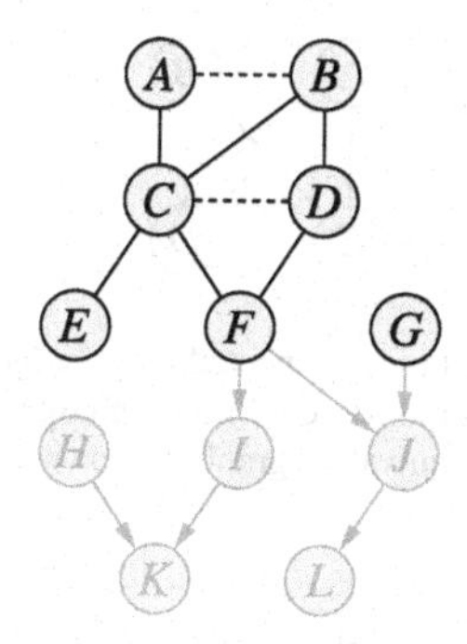

Fig. 22.3 Moral graph of the minimal ancestral graph induced from Fig. 22.2 by the node set $\{E, F, G\}$. Edges inserted during moralization are *drawn dashed*

Table 22.3 Properties of the example graph in Fig. 22.2

	Ancestors	Descendants	Non-descendants
A	$\emptyset$	$\{C, E, F, J, K, L, M\}$	$\{B, D, G, H\}$
B	$\emptyset$	$\{C, D, E, F, J, K, L, M\}$	$\{A, G, H\}$
C	$\{A, B\}$	$\{E, F, J, K, L, M\}$	$\{A, B, D, G, H\}$
D	$\{B\}$	$\{F, J, K, L, M\}$	$\{A, B, C, E, G, H\}$
E	$\{A, B, C\}$	$\emptyset$	$\{A, B, C, D, F, G, H, J, K, L, M\}$
F	$\{A, B, C, D\}$	$\{J, K, L, M\}$	$\{A, B, C, D, E, G, H\}$
G	$\emptyset$	$\{K, M\}$	$\{A, B, C, D, E, F, H, J, L\}$
H	$\emptyset$	$\{L\}$	$\{A, B, C, D, E, F, G, J, K, M\}$
J	$\{A, B, C, D, F\}$	$\{L\}$	$\{A, B, C, D, E, F, G, H, K, M\}$
K	$\{A, B, C, D, F, G\}$	$\{M\}$	$\{A, B, C, D, E, F, G, H, J, L\}$
L	$\{A, B, C, D, F, H, J\}$	$\emptyset$	$\{A, B, C, D, E, F, G, H, J, K, M\}$
M	$\{A, B, C, D, F, G, K\}$	$\emptyset$	$\{A, B, C, D, E, F, G, H, J, K, L\}$

acyclicity. Otherwise, nodes could be ancestors or descendants of themselves. The set of *ancestors* of A is defined as:

$$\mathrm{ancs}(A) = \left\{B \in V \mid \exists \rho : B \overset{\rho}{\underset{G}{\leadsto}} A\right\}.$$

The set of *descendants* of A is defined as:

$$\mathrm{descs}(A) = \left\{B \in V \mid \exists \rho : A \overset{\rho}{\underset{G}{\leadsto}} B\right\}.$$

The set of *non-descendants* of node A is defined as:

$$\text{non-descs}(A) = V \setminus \{A\} \setminus \mathrm{descs}(A).$$

Example 22.1 The left graph in Fig. 22.2 shows a directed acyclic graph (DAG) together with the parents, children and the family of node F. The right graph shows the DAG together with the ancestors, descendants and non-descendants of node F. Table 22.3 contains the ancestors, descendants and non-descendants of all nodes of the DAG in Fig. 22.2.

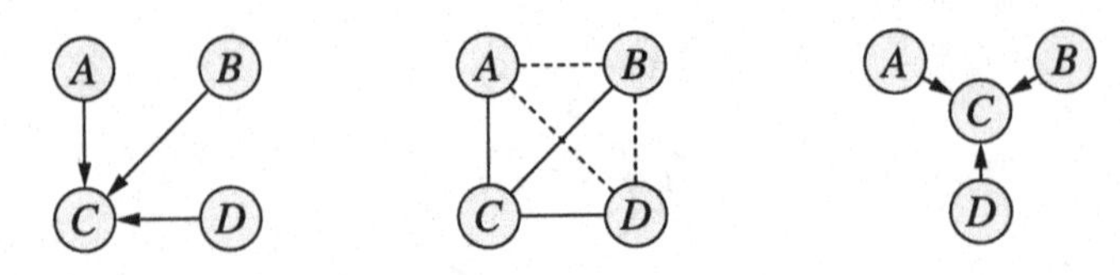

Fig. 22.4 All pairs of parents have to be connected by an edge during moralization. It might be helpful to change the graph layout to observe all missing edges more easily. The edge A–D may be overlooked in the left graph. An equivalent graph is shown on the right which clearly shows all needed edges for moralization

Definition 22.27 ((Induced) subgraph) Let $G = (V, E)$ be an undirected graph and $W \subseteq V$ a set of nodes. Then $G_W = (W, E_W)$ with

$$E_W = \{(u, v) \in E \mid u, v \in W\}$$

is called *subgraph of G induced by W*.

Definition 22.28 (Minimal ancestral graph) Let $G = (V, E)$ be a directed acyclic graph and $M \subseteq V$ a set of nodes. The smallest subgraph of G that contains all ancestors of all nodes of M is called *minimal ancestral graph*. That is, it is the subgraph of G induced by the following set:

$$M \cup \bigcup_{A \in M} \operatorname{ancs}(A).$$

Definition 22.29 (Moral graph, Moralization) Let $G = (V, E)$ be a directed acyclic graph. Its moral graph G' is an undirected graph with the same node set that is obtained by first adding (arbitrarily directed) edges between unconnected parent nodes of all families and then replacing all edges by undirected ones. This transformation is known as *moralization*.

This notion goes back to Lauritzen and Spiegelhalter (1988). The intuition being that "unmarried" parents of a common child node are "married". This rather conservative naming is unfortunate as child nodes may have more than two parents that by definition have to be married to each other. The notion, however, caught on.

Example 22.2 (Moralized minimal ancestral graph) We first consider the minimal ancestral graph induced by the node set $\{E, F, G\}$ in Fig. 22.2. It is drawn black in Fig. 22.3. The dropped descendants of the nodes of the inducing set are drawn in gray. This graph shall now be moralized: All unconnected parents have to be connected after which all edge directions are dropped. This applies to the parents $\{C, D\}$ of node F and parents $\{A, B\}$ of node C. Note that in case of more than two parents, all possible parent pairs have to be connected! Fig. 22.4 illustrates this. The edge A–D must not be overlooked here.

Definition 22.30 (Complete graph) An undirected graph $G = (V, E)$ is called *complete* if and only if each pair of (different) nodes from V is connected by an edge.

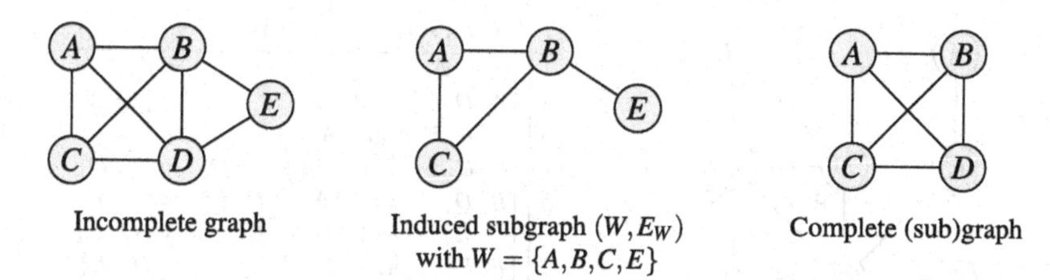

Fig. 22.5 Induction of two subgraphs from the left graph

Definition 22.31 (Complete set, Clique) Let $G = (V, E)$ be an undirected graph. A set $W \subseteq V$ is called *complete* if and only if it induces a complete subgraph. Additionally, W is called a *clique* if and only if it is maximal, i.e., if it is impossible to add a node to W without violating the completeness.

We will later apply the notion of a clique to subgraphs. We then refer to the node set of the respective subgraph.

Example 22.3 (Cliques) The three graphs in Fig. 22.5 contain the following cliques:

left: $\{A, B, C, D\}$ and $\{B, D, E\}$
middle: $\{A, B, C\}$ and $\{B, E\}$
right: $\{A, B, C, D\}$

In a tree (e.g., the graph in Fig. 22.2 without edge (B, C) and without edge directions), each edge represents a clique.

Definition 22.32 (Ordering) Let $G = (V, E)$ be an (arbitrary) graph and α a bijective function with

$$\alpha : V \to \{1, \dots, |V|\}.$$

Then α is called an *ordering*.

Definition 22.33 (Topological ordering) Let α be an ordering on a directed acyclic graph $G = (V, E)$. α is called a *topological ordering* if

$$\forall A \in V : \forall B \in \text{descs}(A) : \alpha(A) < \alpha(B).$$

A directed acyclic graph may have multiple topological orderings.

Definition 22.34 (Perfect ordering) Let $G = (V, E)$ be an undirected graph with n nodes and a total ordering $\alpha = \langle v_1, \dots, v_n \rangle$ on V. α is called *perfect* if the sets

$$\text{adj}(v_i) \cap \{v_1, \dots, v_{i-1}\}, \quad i = 1, \dots, n$$

are complete.

An undirected graph can have multiple perfect orderings or none at all.

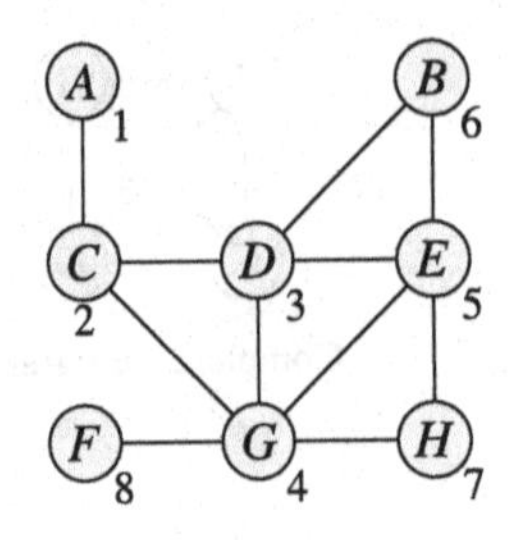

i	$\text{adj}(v_i) \cap \{v_1, \dots, v_{i-1}\}$	
1	$\{C\} \cap \emptyset$	$= \emptyset$
2	$\{A, D, F\} \cap \{A\}$	$= \{A\}$
3	$\{C, B, E, F\} \cap \{A, C\}$	$= \{C\}$
4	$\{G, C, D, E, H\} \cap \{A, C, D\}$	$= \{C, D\}$
5	$\{B, D, F, H\} \cap \{A, C, D, F\}$	$= \{D, F\}$
6	$\{D, E\} \cap \{A, C, D, F, E\}$	$= \{D, E\}$
7	$\{F, E\} \cap \{A, C, D, F, E, B\}$	$= \{F, E\}$
8	$\{F\} \cap \{A, C, D, F, E, B, H\}$	$= \{F\}$

Fig. 22.6 α is a perfect ordering

Figure 22.6 depicts an undirected graph and a node ordering α that is perfect w.r.t. the graph. The table shows the perfectness criterion for each node. The intersections on the right side show that for each step the criterion from Definition 22.34 is satisfied: the two-element sets correspond to edges that are contained in the graph. All single-element sets are nodes and thus trivially complete.

Definition 22.35 (Chord of a circle) A chord of a circle is an edge between two nodes of the circle which is not contained in the circle itself.

The circle B–D–F–H–E of the graph in Fig. 22.6 has two chords: D–E and F–E. Obviously, only circles of length greater than tree can have chords.

Chords subdivide larger circles into smaller ones. It will be beneficial in the later sections not to have chord-less circles with more than three nodes.

Definition 22.36 (Triangulated graph) An undirected graph is called *triangulated* if and only if every simple circle (i.e., a path with its nodes occurring at most once (except for the start/end node, of course)) with more than three nodes has a chord.

Note that a triangulated graph not necessarily needs to consists solely of triangles. For example, the graph in the center of Fig. 22.5 is triangulated: the edge C–E need not be inserted. Contrary, not every graph that consists of triangles is also triangulated: The top part of Fig. 22.7 depicts an obvious triangulation. The same insertion, however, must also be applied in the bottom part. The circle A–B–E–C has no chord in both scenarios.

Definition 22.37 (Maximum cardinality search (MCS)) Let $G = (V, E)$ be an undirected graph. A node ordering α by maximum cardinality search is constructed as follows:

1. Choose an arbitrary start node from V and assign it number 1.
2. Assign the next higher order number to the node that is adjacent to the largest number of already numbered nodes.

Obviously, MCS is not unique. Figure 22.8 shows an example. Node A gets assigned number 1, node C then has to follow with number 2 as it is the only neighbor. For number 3, we have the choice of node D and F because both are adjacent to one already numbered node (C). Let's choose D as the third node which leads im-

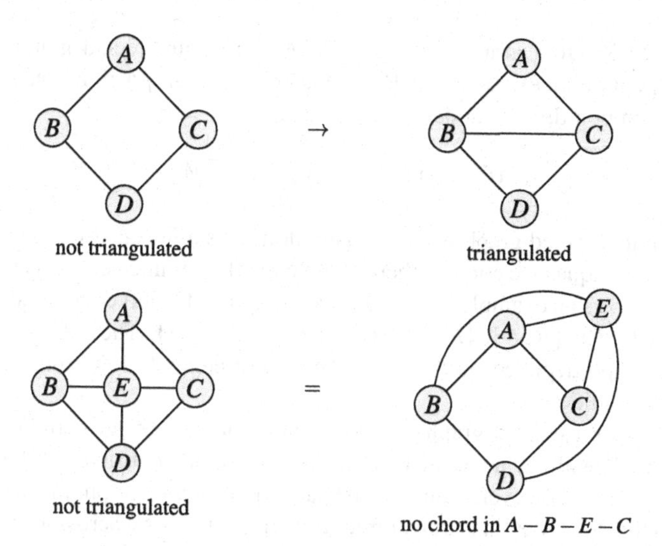

Fig. 22.7 *Top*: triangulation is achieved by insertion of edge B–C (or A–D). *Bottom*: graph is not triangulated as can be easier seen in the alternative drawing

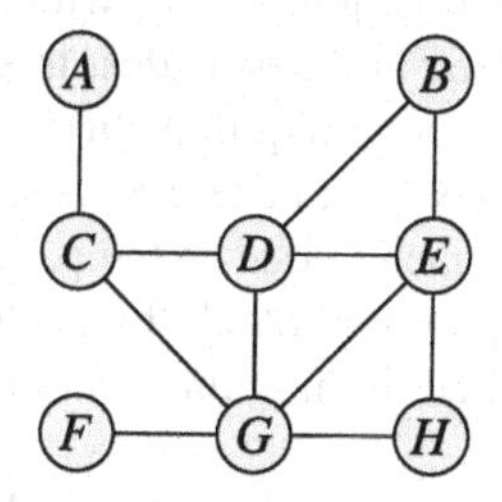

Fig. 22.8 Example of a maximum cardinality search starting with node A. Number 3 can be assigned to nodes D or F, for number 6 we have the choice between nodes H and B

mediately to node F with number 4 as it is the only one that is now connected to two already numbered nodes (C and D). The same arguments leads to number 5 of node E. For number 6, we again have two choices: node B or H. Setting $\alpha(B) = 6$ leads to $\alpha(H) = 7$ and finally to $\alpha(G) = 8$.

22.2.2 Join Graphs

The (undirected) graph concept discussed so far provided an algebraic structure to describe the relations amongst single objects (nodes). It is a relation in the algebraic sense, because the edge set is a subset of $V \times V$. The relations coincide with the edges. We now want to extend this concept to relations between *sets of nodes* and introduce so-called join graphs (or cluster graphs). Since we only define notions that we need later on, we refrain from an exhaustive introduction of the matter. For more details the reader is referred to Castillo et al. (1997). Instead allowing arbitrary subsets of V to be the new nodes, we restrict them to be cliques of a graph.

Definition 22.38 (Join graph) Let $G = (V, E)$ be an undirected graph and $\mathcal{C} = \{C_1, \dots, C_p\}$ its cliques. $G' = (\mathcal{C}, E')$ is called a *join graph* if E' only contains edges between non-disjoint nodes, i.e., if holds:

$$(C_i, C_j) \in E' \Rightarrow C_i \cap C_j \neq \emptyset.$$

Given an undirected graph, we can easily derive its associated join graph: After identifying all cliques we connect them by an edge if their intersection is nonempty. Figure 22.9 shows an example. The undirected graph on the left contains six cliques: $\{A, C\}$, $\{C, D, F\}$, $\{B, D\}$, $\{B, E\}$, $\{G, F\}$ and $\{E, F, H\}$. These form the nodes of the join graph depicted on the right. The seven edges follow from the mutual intersection conditions.

We will later use join graphs to propagate evidence about certain attributes (which coincide with the nodes in V, obviously) to all other attributes. The underlying algorithm will, however, require a special form of a join graph. First, it must be guaranteed that information is transferred from clique to clique across a unique path. And second, any change of an attribute contained in a clique must be transferable to any other clique containing that attribute. The first requirement can be achieved by using so-called join trees, that is, join graphs with tree structure. Since in a tree there is exactly one path between any two nodes, this will be the unique path for the evidence propagation. The second requirement can be formulated as follows: If two cliques share some attributes (i.e., have a non-empty intersection), then these attributes must be contained in each clique of the path connecting the two cliques. In that way, no "gaps" on the path will block the evidence propagation. This latter property is known as the so-called running intersection property.

Definition 22.39 (Running intersection property (RIP)) Let $G = (V, E)$ be an undirected graph with r cliques. An ordering of these cliques has *running intersection property* if for each $j > 1$ there exists an $i < j$ for which the condition

$$C_j \cap (C_1 \cup \dots \cup C_{j-1}) \subseteq C_i$$

holds.

Before illustrating the running intersection property, we need to introduce the notion of a join tree.

Definition 22.40 (Join tree) A join graph with tree structure whose cliques satisfy the running intersection property, is called a *join tree*.

Let us discuss the RIP with an example. Consider the graph in Fig. 22.10. Ordering the cliques ascendingly w.r.t. their index will yield an ordering satisfying RIP which can be validated in Table 22.4. This allows us to use the RIP to construct the tree: For each clique, we immediately know that there is at least one preceding clique with at least one common attribute. Therefore, each clique (except the first

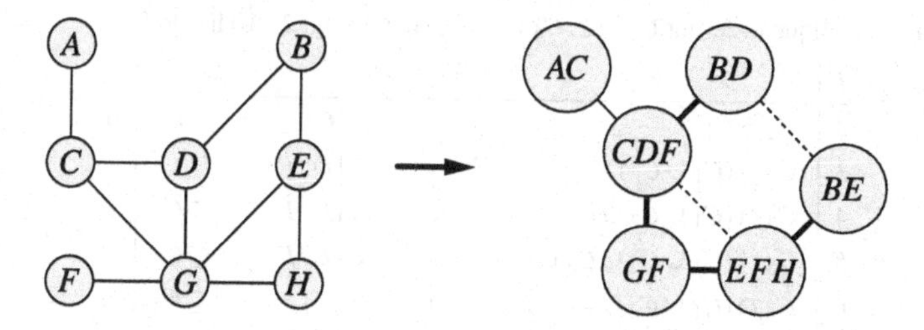

Fig. 22.9 Corresponding join graph of an undirected graph for which there is no join tree. If we delete e.g. the two dashed edges to form a tree, the RIP will not be satisfied: The marked path from BD to BE does not contain the attribute B. It is not possible here to achieve the RIP by deleting any other edges

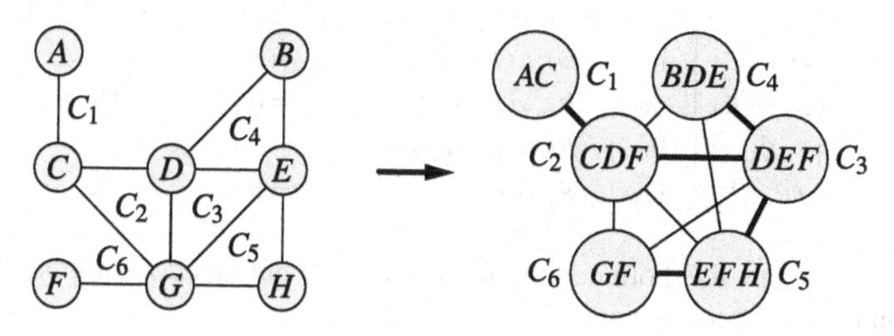

Fig. 22.10 Corresponding join graph of an undirected, triangulated graph

one, of course) has a neighbor candidate. We do not need to worry about circles as we only consider preceding cliques (w.r.t. the ordering). Hence, we can use the following scheme to form a join tree: Starting with the last clique of the ordering (that satisfies the RIP), we connect it to that preceding clique that shares the largest number of attributes. Ties are resolved arbitrarily. Figure 22.10 shows a possible join tree which was created as follows:

1. Starting with the last clique C_6 of the ordering, we look for preceding cliques satisfying the RIP (see also Table 22.4). There are three candidates: C_5, C_3 and C_2. All three have an equally large (single-element) intersection $\{F\}$ with C_6. We (arbitrarily) choose a clique to be the neighbor in the join tree: C_5.
2. C_5 forms the largest intersection $\{E, F\}$ with C_3 and hence is connected to it.
3. C_4 also forms the largest intersection $\{D, E\}$ with C_3 and is connected to it.
4. Clique C_3 forms a non-empty intersection with only C_2 and is therefore connected to that clique.
5. C_2 is connected trivially to C_1 as it is the last remaining clique with non-empty intersection.

The created join tree is shown on the right of Fig. 22.10 by the bold edges.

To create a tree structure from a given join graph is rather simple as only some edges need to be deleted. The question arises whether the RIP can always be achieved. This is not the case as it can be seen from Fig. 22.9. Two edges need to be deleted to arrive at a tree structure. However, none of the two resulting trees satisfies the RIP. If we delete for example, the edges CDF–EFH and BD–BE,

Table 22.4 The clique ordering $C_1, \ldots, C_6$ of the graph in Fig. 22.10 has RIP

j				i
2	$C_2 \cap C_1$	$= \{C\}$	$\subseteq C_1$	1
3	$C_3 \cap (C_1 \cup C_2)$	$= \{D, F\}$	$\subseteq C_2$	2
4	$C_4 \cap (C_1 \cup C_2 \cup C_3)$	$= \{D, E\}$	$\subseteq C_3$	3
5	$C_5 \cap (C_1 \cup C_2 \cup C_3 \cup C_4)$	$= \{E, F\}$	$\subseteq C_3$	3
6	$C_6 \cap (C_1 \cup C_2 \cup C_3 \cup C_4 \cup C_5)$	$= \{F\}$	$\subseteq C_5$	5

Table 22.5 Generating a clique ordering with RIP from a perfect node ordering α of Fig. 22.6

Clique	Rank		
$\{A, C\}$	$\max\{\alpha(A), \alpha(C)\}$	$= 2$	$\to C_1$
$\{C, D, F\}$	$\max\{\alpha(C), \alpha(D), \alpha(F)\}$	$= 4$	$\to C_2$
$\{D, E, F\}$	$\max\{\alpha(D), \alpha(E), \alpha(F)\}$	$= 5$	$\to C_3$
$\{B, D, E\}$	$\max\{\alpha(B), \alpha(D), \alpha(E)\}$	$= 6$	$\to C_4$
$\{F, E, H\}$	$\max\{\alpha(F), \alpha(E), \alpha(H)\}$	$= 7$	$\to C_5$
$\{F, G\}$	$\max\{\alpha(F), \alpha(G)\}$	$= 8$	$\to C_6$

the cliques BD and BE both contain the attribute B but this does not hold for the cliques on their connecting path.

The question whether there is a structural property of a join graph (or its underlying undirected graph) that guarantees the existence of a corresponding join tree was positively answered in Jensen (1988): An undirected graph G has a join tree if and only if G is triangulated. Unfortunately, the definition of the RIP is not constructive, i.e., it gives us a criterion to test a given clique ordering for RIP but it does not produce an algorithm to find such a clique ordering. We exploit the following relation to solve the problem: If an undirected graph has a perfect ordering and if we order its cliques ascendingly w.r.t. the largest perfect number of its contained attributes, then the resulting clique ordering will satisfy the RIP. The clique ordering from Table 22.4 was generated from the perfect ordering in Fig. 22.6. Table 22.5 illustrates the assignment.

We just reduced the problem of finding an appropriate clique ordering to the problem of finding a perfect node ordering, which might seem little beneficial at first sight since there exist more nodes than cliques. The objective thus is to find a perfect ordering on the nodes of an undirected graph. For that task, the following relationship is helpful: The nodes of an undirected graph have at least one perfect ordering if and only if the graph is triangulated.

The last building block in this chain is the construction of such a perfect ordering. We exploit the following statement: A node ordering induced by a maximum cardinality search on a triangulated graph is perfect. Let us summarize this chain of reasoning:

1. Undirected, triangulated graph $G = (V, E)$
2. Node ordering induced by MCS on V is perfect
3. Clique ordering w.r.t. largest perfect node number has RIP
4. Construct join tree using RIP

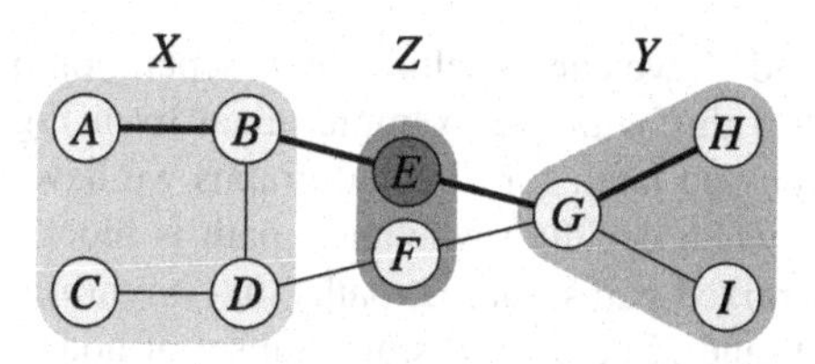

Fig. 22.11 The node sets $\{A, B, C, D\}$ and $\{G, H, J\}$ are u-separated by $\{E, F\}$. The emphasized path A–B–E–G–H is blocked by node E

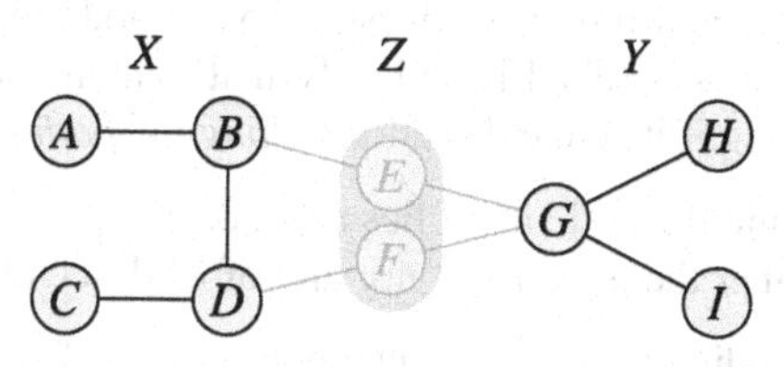

Fig. 22.12 After deleting the nodes of set Z from the graph, there is no path left from X to Y: both sets are u-separated by Z

22.2.3 Separations

One objective of graphical models is to capture as many (conditional) independences of a high-dimensional probability distribution as possible in an undirected or directed graph in order to infer probabilistic statements solely by exploiting graph-theoretic criteria. The fact that there exist certain (conditional) independences amongst a set of attributes shall be represented by means of separation criteria of the corresponding nodes inside a graph. We will discuss such separation criteria for both directed and undirected graphs.

Let us begin with an undirected graph $G = (V, E)$ and three mutually disjoint node subsets X, Y and Z. The set Z shall u-separate X and Y (the u stands for *u*ndirected) if and only if each path from a node in X to a node in Y is blocked. A path is blocked if and only if it contains a blocking node. And, finally, a node is blocking if and only if it is contained in Z. This somewhat artificial definition of u-separation will help us later to define the corresponding concept of d-separation for directed graphs.

Definition 22.41 (u-separation) Let $G = (V, E)$ be an undirected graph and X, Y and Z three disjoint subsets of V. The sets X and Y are *u-separated* by Z in G, written $X \perp\!\!\!\perp_G Y \mid Z$, if and only if each path from a node in X to a node in Y contains at least one node of Z. A path that contains at least one node of Z is called *blocked*, otherwise it is *active*.

Let us consider an example in Fig. 22.11. The set $Z = \{E, F\}$ separates the node sets $\{A, B, C, D\}$ and $\{G, H, J\}$ since all paths from the one set to the other are running through Z as is shown exemplary with path A–B–E–G–H.

An alternative but equivalent way to test for u-separation is as follows: the nodes (and all adjacent edges) of set Z are deleted from the graph. If no path is left that connects the sets X and Y, we can conclude the u-separation of them by the set Z. Figure 22.12 illustrates this with the example taken from Fig. 22.11. Because of semantic reasons discussed later, for directed graphs we have to take the edge directions into account when deciding whether a path is blocked or not. From now on, we will consider mixed paths, that is, paths that might run against edge directions. Again, let X, Y and Z be three disjoint subset of nodes of a directed graph. We use the same reasoning as above for u-separation, only the blocking criteria are modified for the directed case. Again, the sets X and Y are d-separated by Z (the d stands for *d*irected) if and only if each path from a node in X to a node in Y is blocked (by Z). A path is blocked if and only if it contains at least one blocking node. A node is blocking if its edge directions *along the path* are

- serial or diverging and the node itself lies in Z, or
- converging and neither the node itself nor any of its descendants lies in Z.

The four possible edge directions at a node are grouped as follows:

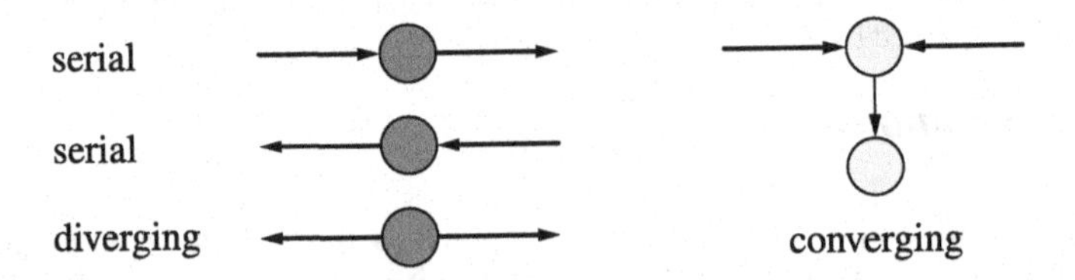

The type of a node (i.e., serial, converging or diverging) depends on the path in which it is contained. In Fig. 22.13, the node E is serial w.r.t. the path $C \rightarrow E \rightarrow G$ while it is converging w.r.t. the path $C \rightarrow E \leftarrow D$. The same node is finally diverging in the path $F \leftarrow E \rightarrow G$.

Definition 22.42 (d-separation) Let $G = (V, E)$ be a directed graph and X, Y and Z three disjoint subsets of V. X and Y are *d-separated* by Z in G, written $X \perp\!\!\!\perp_G Y \mid Z$, if and only if there is no path from a node in X to a node in Y along which the following criteria are satisfied:

1. Every node with converging edges (along the path) is Z or has a descendant in Z.
2. Every other node is not in Z.

A path satisfying both above criteria is called *active*, else *blocked* (by Z).

Note the two equivalent but mutually negated definitions of d-separation. The first description defines the blockade of a path and requires for a d-separation every path to be blocked. The second description defines d-separation as the absence of any active path. We mentioned both version here as both are found in literature.

Let us consider some examples. In the following figures, the sets X and Y will be shaded light and medium gray, while Z will be drawn in dark gray.

Example 22.4 In Fig. 22.13, both X and Y contain just one node and Z is empty. Since we deal with a tree, only the path $A \rightarrow C \rightarrow E \leftarrow D$ needs to be checked.

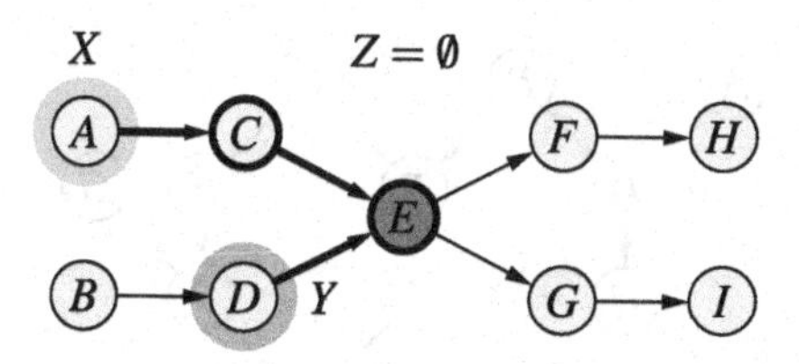

Fig. 22.13 Node E blocks the only path from A to D: $A \perp\!\!\!\perp_G D \mid \emptyset$

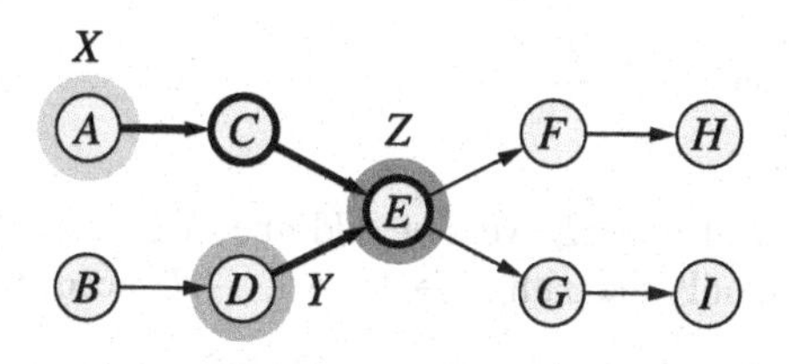

Fig. 22.14 The path from A to E is active because of $E \in Z$: $A \not\perp\!\!\!\perp_G D \mid E$

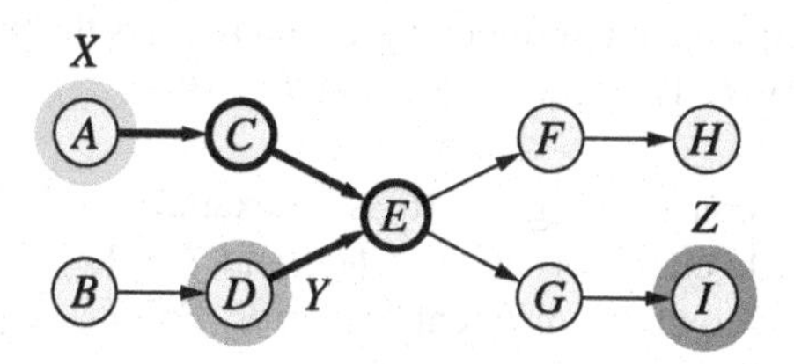

Fig. 22.15 J activates the path from A to D: $A \not\perp\!\!\!\perp_G D \mid J$

- C is a serial node in the path and not in Z. It is therefore non-blocking.
- E is a converging node in the path and not in Z. Also, none of its descendants F, H, G and J is in Z. Thus, E is blocking.

The existence of one blocking node is sufficient for the blockade of the entire path. Hence, we can conclude $A \perp\!\!\!\perp_G D \mid \emptyset$.

Example 22.5 In contrast to the Example 22.4, the set Z now contains the node E. Figure 22.14 depicts this situation. Again, only the path $A \to C \to E \leftarrow D$ needs to be checked.

- C is a serial node in the path and not in Z. Therefore, it is non-blocking.
- E is a converging node in the path and contained in Z. Thus, it is non-blocking as well.

All nodes of the only path are non-blocking which renders the path active. Hence, A and D are not d-separated and therefore $A \not\perp\!\!\!\perp_G D \mid E$.

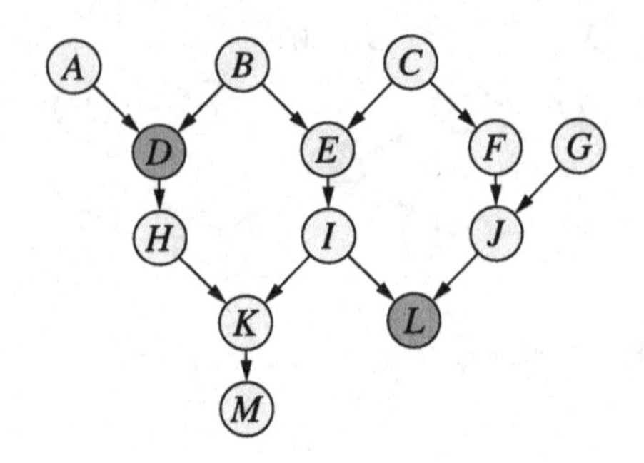

Fig. 22.16 Graph for Example 22.7

Example 22.6 Instead of node E we now add one of its descendants to the set Z, see Fig. 22.15. Again, only the path $A \rightarrow C \rightarrow E \leftarrow D$ needs to be checked.

- C is a serial node in the path and not in Z. Therefore, it is non-blocking.
- E is a converging node in the path but not in Z. However, one of its descendants, namely J, is in Z. Hence, E is non-blocking.

All nodes of the only path are non-blocking which renders the path active. A and D are not d-separated by J. Therefore $A \not\perp\!\!\!\perp_G D \mid J$.

Example 22.7 Let us consider Fig. 22.16 as an example with more than one path between the (still single element) sets $X = \{D\}$ and $Y = \{L\}$. We will successively add nodes to the set Z to check different scenarios. Following paths exist between the nodes D and L:

1. $D \rightarrow H \rightarrow K \leftarrow I \rightarrow L$
2. $D \leftarrow B \rightarrow E \rightarrow I \rightarrow L$
3. $D \leftarrow B \rightarrow E \leftarrow C \rightarrow F \rightarrow J \rightarrow L$

- $Z = \emptyset$

  Path 1: blocked $K, M \notin Z$
  Path 2: active $B, E, I \notin Z$ $\Rightarrow$ $D \not\perp\!\!\!\perp_G L \mid \emptyset$
  Path 3: blocked $E, I \notin Z$

- $Z = \{E\}$

  Path 1: blocked $K, M \notin Z$
  Path 2: blocked $E \in Z$ $\Rightarrow$ $D \not\perp\!\!\!\perp_G L \mid E$
  Path 3: active $E \in Z$

- $Z = \{E, J\}$

  Path 1: blocked $K, M \notin Z$
  Path 2: blocked $E \in Z$ $\Rightarrow$ $D \perp\!\!\!\perp_G L \mid E, J$
  Path 3: blocked $J \in Z$

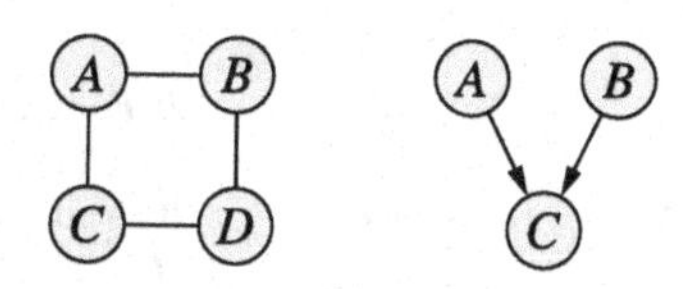

Fig. 22.17 Undirected graph without equivalent directed graph and directed graph without equivalent undirected graph

Note that u-separation (in contrast to d-separation) is monotonic. If a set Z u-separates two node sets X and Y, then X and Y are also u-separated by all supersets of Z. Formally we have:

$$X \perp\!\!\!\perp Y \mid Z \quad \Rightarrow \quad X \perp\!\!\!\perp Y \mid Z \cup W, \tag{22.1}$$

where W is a fourth disjoint node set. This is immediately clear as paths can only be blocked by nodes in Z. Adding more nodes to Z may block further paths but it can never activate a blocked path. This is completely different to d-separation where adding nodes to Z may indeed activate a path and thus render previously d-separated sets X and Y connected. In Example 22.7 the set $Z = \{E, J\}$ d-separates the nodes D and L. If we added the node K to Z, it would activate the path $D \rightarrow H \rightarrow K \leftarrow I \rightarrow L$ and thus prevents the d-separation of D and L:

$$D \perp\!\!\!\perp L \mid E, J \quad \text{but} \quad D \not\!\perp\!\!\!\perp L \mid E, J, K.$$

Also, u-separation and d-separation have different expressiveness. Consider the undirected graph in Fig. 22.17. Via u-separation we can easily conclude the following separations:

$$A \perp\!\!\!\perp C \mid B, D \quad \text{and} \quad B \perp\!\!\!\perp D \mid A, C.$$

However, there exists no directed graph encoding exactly the same separations. We need to require an exact match here to prevent any additional separations. Otherwise, an edge-less graph would encode all possible separations and thus also both above-mentioned ones. If we directed the edges starting at A via B and D to C, we would actually get the separation $A \perp\!\!\!\perp C \mid B, D$, but this would imply $B \not\!\perp\!\!\!\perp D \mid A, C$ because C would be a converging node which activates the path $D \rightarrow C \leftarrow B$.

Likewise, there is not necessarily for every directed graph an equivalent undirected one (w.r.t. the encoded separations) as illustrated on the right of Fig. 22.17. It encodes the d-separation

$$A \perp\!\!\!\perp B \mid \emptyset.$$

Again, we cannot represent this separation in an undirected graph without introducing new separations (which are not valid in the directed graph).

Finally, we recall the running intersection property (RIP) which defines two other important node sets:

C_i	R_i	S_i
C_1	$\{A,C\}$	$\emptyset$
C_2	$\{D,F\}$	$\{C\}$
C_3	$\{E\}$	$\{D,F\}$
C_4	$\{B\}$	$\{D,E\}$
C_5	$\{H\}$	$\{E,F\}$
C_6	$\{G\}$	$\{F\}$

⟶

$D,F \perp\!\!\!\perp_G A \mid C$
$E \perp\!\!\!\perp_G A,C \mid D,F$
$B \perp\!\!\!\perp_G A,C,F \mid D,E$
$H \perp\!\!\!\perp_G A,B,C,D \mid E,F$
$G \perp\!\!\!\perp_G A,B,C,D,E,H \mid F$

Fig. 22.18 Residual and separator sets of the graph in Fig. 22.10 together with the encoded u-separations

Definition 22.43 (Residual set, Separator set) Let $C_1, \ldots, C_n$ be a clique ordering satisfying the RIP. The sets

$$S_i = C_i \cap (C_1 \cup \cdots \cup C_{i-1}), \quad i = 1, \ldots, n, \quad S_1 = \emptyset$$

are called *separator sets* and the sets

$$R_i = C_i \setminus S_i, \quad i = 1, \ldots, n$$

are called *residual sets.*

With these definitions together with the RIP we can easily conclude the following separation criteria inside join graphs:

$$R_i \perp\!\!\!\perp (R_1 \cup \cdots \cup R_{i-1}) \backslash S_i \mid S_i, \quad i = 2, \ldots, n. \tag{22.2}$$

Example 22.8 Let us clarify these concepts with the graphs in Fig. 22.10. Figure 22.18 lists all residual and separator sets of all cliques. The u-separations that are derivable via Eq. (22.2) are given as well.

References

E. Castillo, J.M. Gutierrez, and A.S. Hadi. *Expert Systems and Probabilistic Network Models.* Springer-Verlag, New York, NY, USA, 1997

F.V. Jensen. *Junction Trees and Decomposable Hypergraphs.* Research Report, JUDEX Data Systems, Aalborg, Denmark, 1988

A.N. Kolmogorov. *Grundbegriffe der WahrScheinLichKeitsRechnung.* Springer-Verlag, Heidelberg, 1933. English edition: *Foundations of the Theory of Probability.* Chelsea, New York, NY, USA, 1956

J.B. Kruskal. On the Shortest Spanning Subtree of a Graph and the Traveling Salesman Problem. *Proceedings of the American Mathematical Society* 7(1):48–50. American Mathematical Society, Providence, RI, USA, 1956

S.L. Lauritzen and D.J. Spiegelhalter. Local Computations with Probabilities on Graphical Structures and Their Application to Expert Systems. *Journal of the Royal Statistical Society, Series B* 2(50):157–224. Blackwell, Oxford, United Kingdom, 1988

R.C. Prim. Shortest Connection Networks and Some Generalizations. *The Bell System Technical Journal* 36:1389–1401. Bell Laboratories, Murray Hill, NJ, USA, 1957

Chapter 23
Decompositions

The objective of this chapter is to connect the concepts of conditional independence with the separation in graphs. Both can be represented by a ternary relation $(\cdot \perp\!\!\!\perp \cdot \mid \cdot)$ on either the set of attributes or nodes and it seems to be promising to investigate how to represent the probabilistic properties of a distribution by the means of a graph. The idea then is to use only graph-theoretic criteria (separations) to draw inferences about (conditional) independences because it is them what enables us to decompose a high-dimensional distribution and propagate evidence.

We first introduce an axiomatic approach to the concept of conditional independence (or separation) that goes back to Dawid (1979) and Pearl and Paz (1987). With these axioms, we can syntactically derive new separations or independences from a given set without checking for graph-theoretic nor probabilistic conditions.

Definition 23.1 (Semi-graphoid and Graphoid Axioms) Let V be a set of (mathematical) objects and $(\cdot \perp\!\!\!\perp \cdot \mid \cdot)$ a ternary relation of subsets of V. Furthermore, let W, X, Y and Z be four disjoint subsets of V. The four statements

(a) symmetry: $(X \perp\!\!\!\perp Y \mid Z) \Rightarrow (Y \perp\!\!\!\perp X \mid Z)$
(b) decomposition: $(W \cup X \perp\!\!\!\perp Y \mid Z) \Rightarrow (W \perp\!\!\!\perp Y \mid Z) \wedge (X \perp\!\!\!\perp Y \mid Z)$
(c) weak union: $(W \cup X \perp\!\!\!\perp Y \mid Z) \Rightarrow (X \perp\!\!\!\perp Y \mid Z \cup W)$
(d) contraction: $(X \perp\!\!\!\perp Y \mid Z \cup W) \wedge (W \perp\!\!\!\perp Y \mid Z) \Rightarrow (W \cup X \perp\!\!\!\perp Y \mid Z)$

are called the *semi-graphoid axioms*. A ternary relation $(\cdot \perp\!\!\!\perp \cdot \mid \cdot)$ that satisfies the semi-graphoid axioms for all W, X, Y and Z is called a *semi-graphoid*. The above four statements together with

(e) intersection: $(W \perp\!\!\!\perp Y \mid Z \cup X) \wedge (X \perp\!\!\!\perp Y \mid Z \cup W) \Rightarrow (W \cup X \perp\!\!\!\perp Y \mid Z)$

are called the *graphoid axioms*. A ternary relation $(\cdot \perp\!\!\!\perp \cdot \mid \cdot)$ that satisfies the graphoid axioms for all W, X, Y and Z is called a *graphoid*.

The axioms (b) to (e) are illustrated in Fig. 23.1.

If we talk about sets $\mathcal{I}$ of independence or separation statements, we refer to the same structures in an algebraic way—only the origin is different.

R. Kruse et al., *Computational Intelligence*, Texts in Computer Science,
DOI 10.1007/978-1-4471-5013-8_23, © Springer-Verlag London 2013

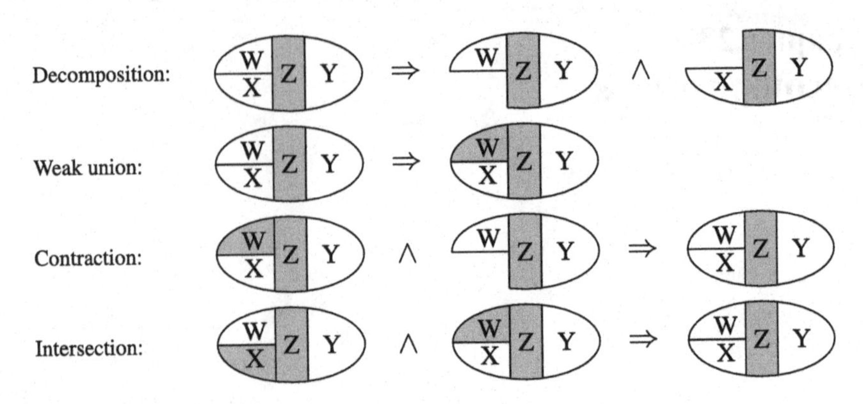

Fig. 23.1 Illustration of the graphoid axioms and separation in graphs

Definition 23.2 Let $V = \{A_1, \ldots, A_n\}$ be a set of attributes. Furthermore, let p be a distribution over the attributes V. Then $\mathcal{I}_p \subseteq 2^V \times 2^V \times 2^V$ be the set of all (conditional) independence statements that are valid in p.

Definition 23.3 Let $V = \{A_1, \ldots, A_n\}$ be as set of nodes. Furthermore, let G be an undirected graph with node set V. Then $\mathcal{I}_G$ be the set of all separation statements that can be read from G via u-separation.

Definition 23.4 Let $V = \{A_1, \ldots, A_n\}$ be a set of nodes. Furthermore, let G be a directed acyclic graph with node set V. Then $\mathcal{I}_{\mathrm{G}}$ be the set of all separation statements that can be read from G via d-separation.

Given a set of axioms $\mathcal{A}$, we can syntactically derive from a set $\mathcal{I}$ of conditional independence statements additional statements. If a statement I can be derived from $\mathcal{I}$ by application of $\mathcal{A}$ we write

$$\mathcal{I} \vdash_{\mathcal{A}} I.$$

If a conditional independence statement follows semantically (w.r.t. an underlying theory $\mathcal{T}$) from a set $\mathcal{I}$ of statements we denote this with

$$\mathcal{I} \models_{\mathcal{T}} I.$$

In our case, the theories $\mathcal{T}$ refer to the probability theory, d- and u-separation. The semantic deducibility of a conditional independence I from a set of independence statements $\mathcal{I}$ then means that I is valid in all distributions in which all statements of $\mathcal{I}$ are valid:

$$\mathcal{I} \models_P I \quad \Leftrightarrow \quad \forall p : \mathcal{I} \subseteq \mathcal{I}_p \Rightarrow I \in \mathcal{I}_p.$$

The semantic deducibility for graphs is defined in an analogous way.

For every system of axioms it needs to be answered whether the deductions are *sound* and *complete*. That is, whether each deducible statement is correct w.r.t. the

Table 23.1 Summary of soundness and completeness of the (semi-)graphoid axioms w.r.t. four different independence and separation criteria

		Soundness	Completeness
conditional independence	general distributions	semi-graphoid axioms	no*
	strictly positive dists.	graphoid axioms	
u-separation		graphoid axioms	no†
d-separation		graphoid axioms	no‡

*Conjectured in Pearl (1988), refuted in Studeny (1989, 1990)

†If weak union is replaced by two other axioms (Pearl and Paz 1987), the resulting axiom set is sound and complete

‡Counterexample in Borgelt et al. (2009, p. 102)

theory $\mathcal{T}$ and whether all statements that can be inferred from $\mathcal{I}$ via theory $\mathcal{T}$ can also be deduced by applying the axioms.

$$\begin{aligned} \mathcal{I} \vdash_{\mathcal{A}} I \quad &\Rightarrow \quad \mathcal{I} \models_{\mathcal{T}} I \quad \text{(Soundness)} \\ \mathcal{I} \models_{\mathcal{T}} I \quad &\Rightarrow \quad \mathcal{I} \vdash_{\mathcal{A}} I \quad \text{(Completeness)} \end{aligned}$$

After these theoretical considerations, we turn to specific answers w.r.t. conditional independence and both separation criteria.

Theorem 23.1 *Conditional stochastic independence satisfies the semi-graphoid axioms. For strictly positive probability distributions it satisfies the graphoid axioms.*

Theorem 23.2 *Both u-separation and d-separation satisfy the graphoid axioms.*

Both theorems obviously contain soundness statements: the semi-graphoid axioms are sound for arbitrary probability distributions whereas the graphoid axioms are sound for strictly positive probability distributions, d- and u-separation. The axioms are, however, not complete in general Studeny (1989, 1990). Table 23.1 summarizes these facts.

We observed that both u- and d-separation satisfy the same axioms as the notion of conditional independence (of a strictly positive probability distribution). The idea suggests itself to use a (directed or undirected) graph to encode the conditional independence statements that hold in a given distribution such that deducible separations directly correspond to valid conditional independence statements.

Unfortunately, there is in general no isomorphism of both notion (that is, between the conditional independence and one of the separations). One reason for that is that u-separation satisfies stronger axioms than the graphoid axioms. As an example, consider the discrepancy between the (semi-)graphoid axiom of weak union and the monotonicity of the u-separation: On the one hand, the Eq. (22.1) on page 439 resembles the weak union axiom with the difference that the node set W does not have to be u-separated beforehand. The Eq. (22.1) is also called strong union axiom (Pearl and Paz 1987). On the other hand, the conditional independence in general only satisfies the semi-graphoid axioms (see Theorem 23.1). That is, applying the intersection axiom can already lead to incorrect results as the following example illustrates:

Example 23.1 (Cond. indep. does not satisfy intersection axiom in general) We consider the following (not strictly positive) three-dimensional probability distribution over the binary attributes A, B and C with $P(A = a_1, B = b_1, C = c_1) = P(A = a_2, B = b_2, C = c_2) = 0.5$ (all other value combinations have probability zero). In this distribution, the following conditional independences hold:

$$A \perp\!\!\!\perp B \mid C, \qquad A \perp\!\!\!\perp C \mid B \quad \text{and} \quad B \perp\!\!\!\perp C \mid A.$$

This can be easily seen as follows: The left table shows the distribution p_{ABC} while the other two show the conditional distributions given the attribute C. Both distributions also show the marginal distributions $p_{A|C}$ and $p_{B|C}$.

p_{ABC}	c_1		c_2	
	a_1	a_2	a_1	a_2
b_1	$^1/_2$	0	0	0
b_2	0	0	0	$^1/_2$

$p_{AB\|c_1}$	a_1	a_2	
b_1	1	0	1
b_2	0	0	0
	1	0	

$p_{AB\|c_2}$	a_1	a_2	
b_1	0	0	0
b_2	0	1	1
	0	1	

Obviously, the following relationship holds which corresponds to the above conditional independences:

$$\begin{aligned} &\forall a \in \mathrm{dom}(A) : \forall b \in \mathrm{dom}(B) : \forall c \in \mathrm{dom}(C) : \\ &\quad P(A = a, B = b \mid C = c) = P(A = a \mid C = c) \cdot P(B = b \mid C = c). \end{aligned}$$

Applying the intersection axiom (with $Z = \emptyset$) to the above independences results in:

W	X	Y	Inference
$\{A\}$	$\{C\}$	$\{B\}$	$AC \perp\!\!\!\perp B \mid \emptyset$
$\{A\}$	$\{B\}$	$\{C\}$	$AB \perp\!\!\!\perp C \mid \emptyset$
$\{B\}$	$\{C\}$	$\{A\}$	$BC \perp\!\!\!\perp A \mid \emptyset$

p_{AB}	a_1	a_2	
b_1	$^1/_2$	0	$^1/_2$
b_2	0	$^1/_2$	$^1/_2$
	$^1/_2$	$^1/_2$	

p'_{AB}	a_1	a_2	
b_1	$^1/_4$	$^1/_4$	$^1/_2$
b_2	$^1/_4$	$^1/_4$	$^1/_2$
	$^1/_2$	$^1/_2$	

Applying the decomposition axiom on the previous independences, we finally arrive at the following statements:

$$A \perp\!\!\!\perp B \mid \emptyset, \qquad B \perp\!\!\!\perp C \mid \emptyset \quad \text{and} \quad C \perp\!\!\!\perp A \mid \emptyset.$$

None of these three (here: marginal) independences holds true in p_{ABC} as can be easily verified: The distribution p_{AB} cannot be reconstructed from the two marginal distributions p_A and p_B as the comparison between p_{AB} and p'_{AB} shows. The same holds also for the other attribute combinations.

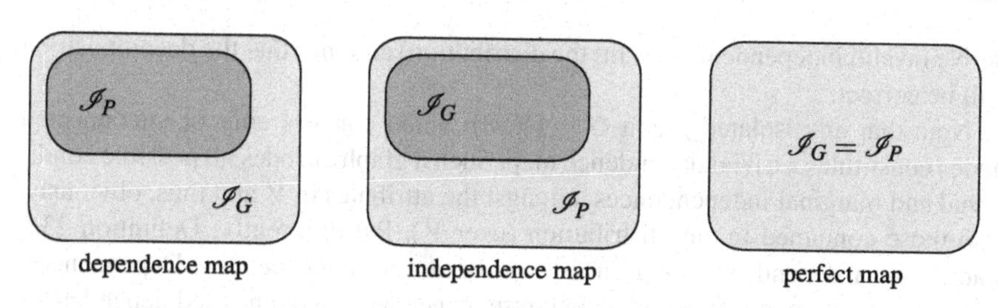

Fig. 23.2 Relationships between deducibility of conditional independences in different map types

23.1 Dependence Graphs and Independence Graphs

Since we cannot rely on isomorphism between the separation and independence concepts, we will have to confine ourselves to weaker statements. For that, we first consider the following definitions.

Definition 23.5 (Dependence, Independence, and Perfect Map) Let $(\cdot \perp\!\!\!\perp_p \cdot \mid \cdot)$ be a ternary relation representing the conditional independences of a given distribution p over the attribute set V. An undirected (directed) graph $G = (V, E)$ is called *conditional dependence graph* or *dependence map* w.r.t. p if and only if for all disjoint subsets $X, Y, Z \subset V$ the implication

$$X \perp\!\!\!\perp_p Y \mid Z \quad \Rightarrow \quad X \perp\!\!\!\perp_G Y \mid Z$$

holds. That is, if G describes via u-separation (d-separation) all conditional independences of p and thus describes only sound dependences.

An undirected (directed) graph $G = (V, E)$ is called *conditional independence graph* or *independence map* w.r.t. p if and only if for all disjoint subsets $X, Y, Z \subset V$ the implication

$$X \perp\!\!\!\perp_G Y \mid Z \quad \Rightarrow \quad X \perp\!\!\!\perp_p Y \mid Z$$

holds. That is, if G describes via u-separation (d-separation) only those conditional independences that are valid in p. G is called a *perfect map* of the conditional (in)dependences in p if and only if it is both a dependence map and an independence map.

A dependence graph represent all conditional independences that are valid in the distribution p but may contain additional ones that are not valid in p. An independence graph, however, encodes only those conditional independences that are valid in p but maybe not all of them. Figure 23.2 illustrates these relationships.

We saw in Chap. 21 that we can decompose a distribution by exploiting conditional independences. Consequently, me must make sure that no independence can be read from the graph that is not valid inside the distribution. Since we cannot hope for perfect maps in general, we will confine ourselves to independence maps. We may not be able to derive all valid independences from them (and, hence, the decomposition might not be as efficient as it could be) but we can be sure that we do not derive invalid independences (w.r.t. the distribution) and thus that the decomposition will be correct.

Note that any isolated graph $G = (V, \emptyset)$ which consists only of unconnected nodes constitutes a trivial dependence map. Such a graph encodes all possible conditional and marginal independences amongst the attributes in V and thus, obviously, also those contained in any distribution (over V). Put differently: Definition 23.5 stated that a dependence map encodes only correct dependences. Dependences, however, require connected nodes. In consequence, the set of encoded dependences in an isolated graph is empty. And this empty set is (trivially) a subset of any set of dependences of any distribution. Contrary, each complete graph (that is, a clique V) constitutes a trivial independence map. Since independences are encoded via missing edges, the set $\mathcal{I}_G$ is obviously empty in a complete graph G. Thus it is (trivial) subset of any set $\mathcal{I}_p$.

A trivial independence map is—albeit correct—of little use. Therefore, we are interested in graphs that encode as many independences of a given distribution as possible, that is, where $|\mathcal{I}_G|$ is large, but without violating the inclusion $\mathcal{I}_G \subseteq \mathcal{I}_p$. Although we maximize the number of conditional independences that are contained in $\mathcal{I}_G$, they are referred to as *minimal* independence maps in the literature because the number of edges is minimized. We will follow this convention here and define:

Definition 23.6 (Minimal Independence Map) Let $G = (V, E)$ be a (directed or undirected) independence map of a probability distribution p_V. G is called a *minimal independence map* or *minimal independence graph* if it is not possible to delete an edge from E without introducing a conditional independence that is invalid in p_V.

An analogous definition for dependence maps is possible as well. We omit it here, as we do not use this notion in the remainder. As we now have defined the type of graphs that are of interest to us, we can now define the notion of decomposition or decomposability (which was used rather colloquially up to now). Again, we distinguish between the two graph types.

Definition 23.7 (Decomposable w.r.t. an Undirected Graph) A probability distribution p_V over an attribute set $V = \{A_1, \dots, A_n\}$ is called decomposable or factorizable w.r.t. to an undirected graph $G = (V, E)$ if and only if it can be written as a product of non-negative functions that are defined on the cliques of G. Precisely: Let $\mathcal{C}$ be a family (that is, a set) of subsets of V such that the subgraphs induced by the sets $C \in \mathcal{C}$ coincide with the cliques of G. Let further be $\mathcal{E}_C$ the set of events that can be described by assigning values to the attributes in C. Then p_V is called decomposable or factorizable w.r.t. G if there exist functions $\phi_C : \mathcal{E}_C \to \mathbb{R}_0^+$, $C \in \mathcal{C}$, such that the following holds:

$$\forall a_1 \in \text{dom}(A_1) : \cdots \forall a_n \in \text{dom}(A_n) :$$

$$p_V\Big(\bigwedge_{A_i \in V} A_i = a_i\Big) = \prod_{C \in \mathcal{C}} \phi_C\Big(\bigwedge_{A_i \in C} A_i = a_i\Big).$$

Definition 23.8 (Decomposable w.r.t. a Directed Acyclic Graph) A probability distribution p_V over an attribute set $V = \{A_1, \ldots, A_n\}$ is called decomposable or factorizable w.r.t. a directed acyclic graph $G = (V, E)$ if and only if it can be written as the product of the conditional probabilities of the attributes given their parent attributes in G. That is, if the following holds:

$$\forall a_1 \in \operatorname{dom}(A_1) : \cdots \forall a_n \in \operatorname{dom}(A_n) :$$
$$p_V\left(\bigwedge_{A_i \in V} A_i = a_i\right) = \prod_{A_i \in V} P\left(A_i = a_i \,\middle|\, \bigwedge_{A_j \in \mathrm{pa}_G(A_i)} A_j = a_j\right).$$

Let us consider two examples to illustrate the decomposition of a distribution w.r.t. an undirected and a directed acyclic graph.

Example 23.2 (Decomposition w.r.t. an undirected graph) The undirected graph in Fig. 23.3 contains the following cliques:

$$C_1 = \{B, C, E, G\}, C_2 = \{A, B, C\}, C_3 = \{C, F, G\}, C_4 = \{B, D\}, C_5 = \{G, F, H\}.$$

The decomposition induced by these cliques reads:

$$\begin{aligned}
&\forall a \in \operatorname{dom}(A) : \cdots \forall h \in \operatorname{dom}(H) : \\
&p_V(A = a, \ldots, H = h) = \phi_{C_1}(B = b, C = c, E = e, G = g) \\
&\qquad \cdot \phi_{C_2}(A = a, B = b, C = c) \\
&\qquad \cdot \phi_{C_3}(C = c, F = f, G = g) \\
&\qquad \cdot \phi_{C_4}(B = b, D = d) \\
&\qquad \cdot \phi_{C_5}(G = g, F = f, H = h)
\end{aligned}$$

Example 23.3 (Decomposition w.r.t. a directed acyclic graph) The directed acyclic graph in Fig. 23.3 induces the following decomposition:

$$\begin{aligned}
&\forall a \in \operatorname{dom}(A) : \cdots \forall h \in \operatorname{dom}(H) : \\
&p_V(A = a, \ldots, H = h) = P(H = h \mid G = g, F = f) \cdot P(G = g \mid B = b, E = e) \\
&\qquad \cdot P(F = f \mid C = c) \cdot P(E = e \mid B = b, C = c) \\
&\qquad \cdot P(D = d \mid B = b) \cdot P(C = c \mid A = a) \\
&\qquad \cdot P(B = b \mid A = a) \cdot P(A = a)
\end{aligned}$$

As both the notions decomposability w.r.t. a graph and independence map are now defined, we still need the connection between them. Our objective is to infer correct (sound) conditional independences of an underlying distribution from the separation criteria read from an independence map. This connection is established by the following two theorems.

Theorem 23.3 *Let p_V be a strictly positive probability distribution over a set V of (discrete) attributes. p_V is factorizable w.r.t. an undirected graph $G = (V, E)$, if and only if G is a conditional independence map of p_V.*

Theorem 23.4 *Let p_V be probability distribution over a set V of (discrete) attribute. p_V is factorizable w.r.t. a directed acyclic graph $G = (V, E)$ if and only if G is a conditional independence map of p_V.*

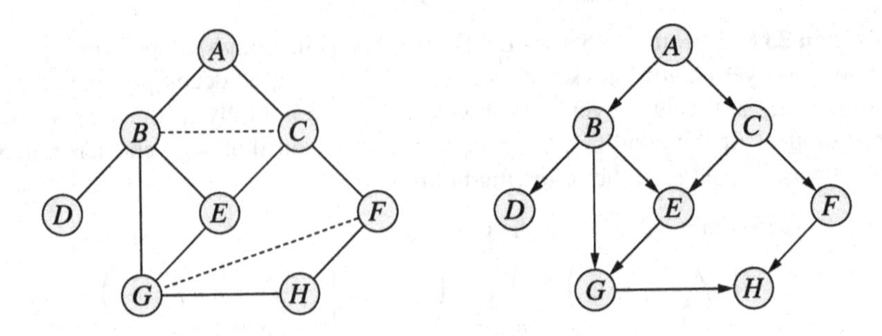

Fig. 23.3 Two graphs to illustrate the decomposition of a distribution

Both theorems can be summarized as follows (taking care of the restriction w.r.t. strict positivity of p in Theorem 23.3):

$$G \text{ factorizes } p \quad \Leftrightarrow \quad G \text{ is an independence map of } p.$$

Let us clarify which conclusion we can draw from this equivalence. Assume, we observe that a distribution p_V can be decomposed w.r.t. a given graph G. It is immediately clear that G must be an independence map of p_V (implication from left to right). This in turn allows us to read every separation in G as a valid conditional independence in p_V (implication from right to left).

Until now, most definitions and theorems assumed a graph and a distribution to be given. Usually, this is not the case. Therefore, we will now answer the two questions how to construct a minimal independence map (directed and undirected) given a distribution and how to construct a distribution for which a given (directed and undirected) graph is an independence map. Hence, we will now introduce four algorithms that establish the constructive connection between distributions and the decomposition w.r.t. a graph.

Algorithm 23.1 (Decomposition w.r.t. an Undirected Triangulated Graph)

Input: undirected triangulated graph $G = (V, E)$
Output: decomposition of a distribution p_V with independence map G

1. Determine all cliques of G.
2. Determine a clique ordering $C_1, \dots, C_r$ with the running intersection property (see Definition 22.39).
3. Determine the sets $S_j = C_j \cap (C_1 \cup \dots \cup C_{j-1})$ and $R_j = C_j \setminus S_j$.
4. Return

$$\forall a_1 \in \text{dom}(A_1) : \cdots \forall a_n \in \text{dom}(A_n) :$$

$$p_V\Bigl(\bigwedge_{A_i \in V} A_i = a_i\Bigr) = \prod_{C_j \in \mathcal{C}} P\Bigl(\bigwedge_{A_i \in R_j} A_i = a_i \Bigm| \bigwedge_{A_i \in S_j} A_i = a_i\Bigr)$$

Example 23.4 (Illustration of Algorithm 23.1) Consider again the undirected triangulated graph in Fig. 23.3. The following clique ordering has running intersection property (see Definition 22.43 on page 440) which allows us to state the following residual and separator sets:

i	C_i	R_i	S_i
1	$\{B, C, E, G\}$	$\{B, C, E, G\}$	$\emptyset$
2	$\{A, B, C\}$	$\{A\}$	$\{B, C\}$
3	$\{C, F, G\}$	$\{F\}$	$\{C, G\}$
4	$\{B, D\}$	$\{D\}$	$\{B\}$
5	$\{F, G, H\}$	$\{H\}$	$\{F, G\}$

This leads to the following decomposition as described by Algorithm 23.1 (for the sake of clarity, we omit the all-quantifiers over the respective attributes domains):

$$\begin{aligned}
&p_V(A, B, C, D, E, F, G, H) \\
&\quad = P(R_1 \mid S_1) \cdot P(R_2 \mid S_2) \cdot P(R_3 \mid S_3) \cdot P(R_4 \mid S_4) \cdot P(R_5 \mid S_5) \\
&\quad = P(B, C, E, G) \cdot P(A \mid B, C) \cdot P(F \mid C, G) \cdot P(D \mid B) \cdot P(H \mid F, G) \\
&\quad = \frac{P(B, C, E, G)}{1} \cdot \frac{P(A, B, C)}{P(B, C)} \cdot \frac{P(F, C, G)}{P(C, G)} \cdot \frac{P(D, B)}{P(B)} \cdot \frac{P(H, F, G)}{P(F, G)} \\
&\quad = \frac{P(C_1)}{1} \cdot \frac{P(C_2)}{P(S_2)} \cdot \frac{P(C_3)}{P(S_3)} \cdot \frac{P(C_4)}{P(S_4)} \cdot \frac{P(C_5)}{P(S_5)}
\end{aligned}$$

The last line in this equation shows an alternative representation without explicitly stating conditional probabilities (by just replacing them by their definition). For a general clique, set $\mathcal{C}$ satisfying RIP we get:

$$\forall a_1 \in \operatorname{dom}(A_1) : \cdots \forall a_n \in \operatorname{dom}(A_n) :$$

$$p_V\left(\bigwedge_{A_i \in V} A_i = a_i\right) = \frac{\prod_{j=1}^{r} P(\bigwedge_{A_i \in C_j} A_i = a_i)}{\prod_{j=2}^{r} P(\bigwedge_{A_i \in S_j} A_i = a_i)}.$$

The start index 2 in the denominator excludes the probability of the formally empty separator set S_1. We will use the above decomposition formula in join trees. The separator sets are the intersections of neighboring cliques. Since there only $n - 1$ edges in a tree with n nodes, this explains the "absence" of one separator set.

Algorithm 23.2 (Decomposition w.r.t. a Directed Acyclic Graph)

Input: directed acyclic graph $G = (V, E)$
Output: decomposition of a distribution p_V with independence map G

1. Determine the parent sets $\mathrm{pa}_G(A_i)$.
2. Return

$$\forall a_1 \in \operatorname{dom}(A_1) : \cdots \forall a_n \in \operatorname{dom}(A_n) :$$

$$p_V\left(\bigwedge_{A_i \in V} A_i = a_i\right) = \prod_{A_i \in V} P\left(A_i = a_i \,\middle|\, \bigwedge_{A_j \in \mathrm{pa}_G(A_i)} A_j = a_j\right).$$

This algorithm basically coincides with Definition 23.8. In contrast to a decomposition by an undirected graph where appropriate clique potentials have to be found, they are immediately clear here. Example 23.3 serves as an illustration.

Algorithm 23.3 (Minimal Undirected Independence Map)

Input: strictly positive distribution p_V over a set $V = \{A_1, \ldots, A_n\}$ of attributes
Output: minimal undirected independence map $G = (V, E)$ of p_V.

1. Start with $G = (V, E)$ as fully connected graph: $E = V \times V$.
2. For each edge $(A, B) \in E$ compute:

$$p_{V\setminus\{A\}}\Bigg(\bigwedge_{A_i\in V\setminus\{A\}} A_i = a_i\Bigg) = \sum_{a\in\mathrm{dom}(A)} p_V\Bigg(\bigwedge_{A_i\in V} A_i = a_i\Bigg)$$

$$p_{V\setminus\{B\}}\Bigg(\bigwedge_{A_i\in V\setminus\{B\}} A_i = a_i\Bigg) = \sum_{b\in\mathrm{dom}(B)} p_V\Bigg(\bigwedge_{A_i\in V} A_i = a_i\Bigg)$$

$$p_{V\setminus\{A,B\}}\Bigg(\bigwedge_{A_i\in V\setminus\{A,B\}} A_i = a_i\Bigg) = \sum_{a\in\mathrm{dom}(A)} \sum_{b\in\mathrm{dom}(B)} p_V\Bigg(\bigwedge_{A_i\in V} A_i = a_i\Bigg)$$

If $p_V\Big(\bigwedge_{A_i\in V} A_i = a_i\Big) \cdot p_{V\setminus\{A,B\}}\Big(\bigwedge_{A_i\in V\setminus\{A,B\}} A_i = a_i\Big)$

$$= p_{V\setminus\{A\}}\Bigg(\bigwedge_{A_i\in V\setminus\{A\}} A_i = a_i\Bigg) \cdot p_{V\setminus\{B\}}\Bigg(\bigwedge_{A_i\in V\setminus\{B\}} A_i = a_i\Bigg),$$

then delete edge (A, B) from E (and also delete (B, A)).
3. Return G.

The independence test in step 2 exploits the following fact of undirected graphs: the absence of an edge (A, B) in an undirected graph $G = (V, E)$ implies that those two nodes A and B are u-separated by all other nodes:

$$\forall A, B \in V, A \neq B: \quad (A, B) \notin E \Rightarrow A \perp\!\!\!\perp_G B \mid V \setminus \{A, B\}.$$

Since the graph G shall be an independence map w.r.t. a given distribution p_V we must ensure that deducible u-separations also hold true in form of conditional independence statements in p_V. The separation $A \perp\!\!\!\perp_G B \mid V\setminus\{A, B\}$ in G has to have the following corresponding probabilistic counterpart in p_V:

$$P\big(A, B \mid V \setminus \{A, B\}\big) = P\big(A \mid V \setminus \{A, B\}\big) \cdot P\big(B \mid V \setminus \{A, B\}\big).$$

Applying equivalence transformations (multiply twice with $P(V \setminus \{A, B\})$) we arrive at the test criterion of step 2:

$$\begin{aligned} P\big(A, B \mid V \setminus \{A, B\}\big) &= P\big(A \mid V \setminus \{A, B\}\big) \cdot P\big(B \mid V \setminus \{A, B\}\big) \\ P(V) &= P\big(A \mid V \setminus \{A, B\}\big) \cdot P\big(V \setminus \{A\}\big) \\ P(V) \cdot P\big(V\setminus\{A, B\}\big) &= P\big(V \setminus \{B\}\big) \cdot P\big(V \setminus \{A\}\big) \end{aligned}$$

The resulting independence map is obviously minimal: no further edge can be deleted (because we tested them all) without encoding an invalid independence.

Algorithm 23.4 (Minimal Directed Independence Map)

Input: distribution p_V over a set $V = \{A_1, \dots, A_n\}$ of attributes
Output: minimal directed independence map $G = (V, E)$ of p_V.

1. Determine an arbitrary attribute ordering $A_1, \dots, A_n$.
2. Find for each A_i a minimal predecessor set Π_i, that renders A_i conditionally independent of $\{A_1, \dots, A_{i-1}\} \setminus \Pi_i$.
3. Start with $G = (V, \emptyset)$ and insert for each A_i an edge from each node in Π_i to A_i.
4. Return G.

Example 23.5 Consider the three-dimension distribution of Table 22.1 on page 422 which is repeated as Table 23.2. Let us assume the following node order:

$$\mathsf{G} \prec \mathsf{Pr} \prec \mathsf{Sm}.$$

We know from Sect. 22.1.2 (and the distributions in Table 22.2) that only the following conditional independence holds in the distribution p_{GPrSm} (and, of course, its symmetric counterpart):

$$\mathsf{Pr} \perp\!\!\!\perp_{p_{\mathsf{GPrSm}}} \mathsf{Sm} \mid \mathsf{G}.$$

Let us now start to determine the sets Π_i. We always start with all predecessors as the candidate set: $\Pi_i = \{A_1, \dots, A_{i-1}\}$. We then delete as many attributes from Π_i until the conditional independence would be violated. The remaining attributes will become the parents of A_i. It may happen, that we cannot exclude any attribute from the initial set Π_i. We then add all predecessors $A_1, \dots, A_{i-1}$ of A_i as its parents.

1. G has no predecessor. Hence it is (and remains) $\Pi_{\mathsf{G}} = \emptyset$.
2. For Pr we start with $\Pi_{\mathsf{Pr}} = \{\mathsf{G}\}$. The first (and only) reduction of Π_{Pr} is the one leading to the empty set. Consequently, we test whether $\mathsf{Pr} \perp\!\!\!\perp_{p_{\mathsf{GPrSm}}} \mathsf{G} \mid \emptyset$. This is not the case as can be easily verified. There are no other options to set Π_{S}, hence the initial set remains: $\Pi_{\mathsf{Pr}} = \{\mathsf{G}\}$.
3. We start with $\Pi_{\mathsf{Sm}} = \{G, Pr\}$ and test the following three options:

   $$\mathsf{R} \perp\!\!\!\perp_{p_{\mathsf{GPrSm}}} \mathsf{Pr} \mid \mathsf{G}, \qquad \mathsf{R} \perp\!\!\!\perp_{p_{\mathsf{GPrSm}}} \mathsf{G} \mid \mathsf{Pr} \quad \text{and} \quad \mathsf{R} \perp\!\!\!\perp_{p_{\mathsf{GPrSm}}} \mathsf{G}, \mathsf{Pr} \mid \emptyset$$

   The only valid independence is $\mathsf{Sm} \perp\!\!\!\perp_{p_{\mathsf{GPrSm}}} \mathsf{Pr} \mid \mathsf{G}$ which leads to $\Pi_{\mathsf{Sm}} = \{G\}$.

We arrive at the graph shown in Fig. 23.4, which is not only a minimal independence map but also a perfect map in this case.

Finally, we can define the core structures covered by the term *graphical models*.

Definition 23.9 (Markov Network) A *Markov network* is an undirected conditional independence graph $G = (V, E)$ of a probability distribution p_V together with a family of non-negative functions ϕ_M of the factorization induced by the graph.

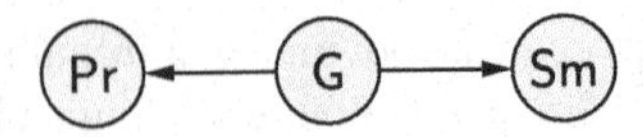

Fig. 23.4 Resulting graph for Example 23.5

Table 23.2 Three-dimensional example distribution

p_{orig}	G = m		G = w	
	Sm = sm	Sm = $\overline{\text{sm}}$	Sm = sm	Sm = $\overline{\text{sm}}$
Pr = pr	0	0	0.01	0.04
Pr = $\overline{\text{pr}}$	0.2	0.3	0.09	0.36

The undirected graph in Fig. 23.3 together with the decomposition from Example 23.2 (and appropriately selected clique potentials) is a Markov network.

Definition 23.10 (Bayes Network) A *Bayes network* is a directed conditional independence graph of a probability distribution p_V together with a family of conditional probabilities of the factorization induced by the graph.

The directed acyclic graph in Fig. 23.3 together with the decomposition from Example 23.3 (and appropriately selected conditional distributions) is a Bayes network.

23.2 A Real-World Application

We now have enough mechanisms to answer the first of the three main questions raised on page 413:

How can expert knowledge about complex domains be represented efficiently?

The expert knowledge is formally represented by a distribution p_V over the set of relevant attributes V (customer data, product data, order information, etc.) Conditional independences are used to decompose this distribution into lower-dimensional distributions by means of a Markov or Bayes network. Figure 23.5 shows a Markov Network of a real-world application at the Volkswagen AG (Gebhardt and Kruse 2005). The different attributes of a car are described by 204 attributes which are anonymized by numbers in the figure. The Markov network in the figure consists of 123 at most 18-dimensional cliques (51 attributes of the original 204 are independent, that is, they are not connected to the join tree and omitted here for brevity, otherwise the figure would show 174 cliques). Let us make clear the efficiency gain of such a representation by comparing the theoretical necessary storage for the original 204-dimensional distribution p_{orig} with the distribution p_{net} encoded by the Markov network. To simplify the estimation, we assume that the domains of all attributes contain only five values which comes close to the average number of

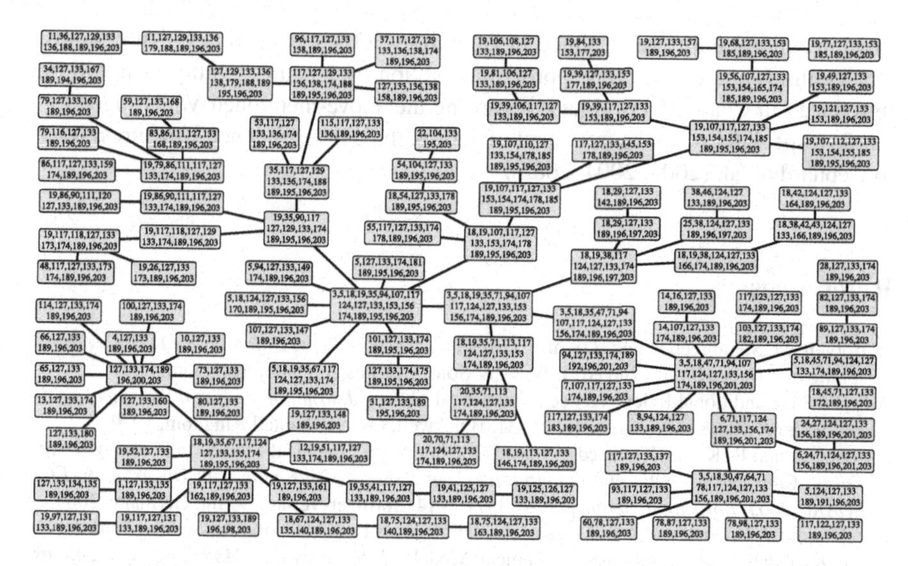

Fig. 23.5 Markov Network of a real-world application with 204 attributes (51 of which are independent and therefore not shown). The distribution can be approximated by 123 cliques (or 174 cliques if the independent attributes are taken into account) with at most 18 dimensions

domain sizes in typical industrial projects. Our argumentation w.r.t. the efficiency gain does not change if we assume a different domain size. With this assumption, we arrive at the following number of parameters for each representation (i.e., single probabilities or clique potential entries):

Distribution	Number of parameters
p_{orig}	$1 \cdot 5^{204} \approx 4 \cdot 10^{142}$
p_{net}	$\sum_{i=1}^{174} 5^{\|C_i\|} \approx 1 \cdot 10^{13}$

The number of parameters for p_{orig} exceeds any imaginable measure. The size of p_{net}, however, can be handled if certain additional observations are taken into account. The high-dimensional cliques, for example, are usually quite sparse which allows for better representations than a full joint distribution.

For reasons of fairness, we have to admit that this sparseness criterion also applies to the distribution p_{orig}, which would be extremely sparse in reality. Even if every human being on earth would own a unique Volkswagen, then p_{orig} would only contain around 7 billion nonzero probabilities which would be almost negligible in contrast to the 10^{142} possible combinations. Hence, we would most likely use a data structure that accounts for the sparsity which would lead to some kind of list which in turn would grow with the number of cars. The Markov network representation, however, can cater for any car combination without growing as long as the representation is compatible with the conditional (in)dependences encoded by the network structure. It is immediately clear that any new data (in form of adjusted car configuration frequencies) have to satisfy the (conditional) independences because it is these independences that allow for the efficient representation. In practice, the exact decomposability (as claimed in Definitions 23.7 and 23.8) never fully holds. One confines oneself with an approximate notion of decomposability as it will be introduced in Chap. 25 For details regarding the above-mentioned Volkswagen example (especially w.r.t. the treatment of approximate independence structures) refer to Gebhardt et al. (2003, 2004, 2006).

References

C. Borgelt, M. Steinbrecher, and R. Kruse. *Graphical Models—Representations for Learning, Reasoning and Data Mining*, 2nd ed. J. Wiley & Sons, Chichester, United Kingdom, 2009

A.P. Dawid. Conditional Independence in Statistical Theory. *Journal of the Royal Statistical Society, Series B (Methodological)* 41(1):1–31. Blackwell, Oxford, United Kingdom, 1979

J. Gebhardt and R. Kruse. Knowledge-Based Operations for Graphical Models in Planning. *Proc. Europ. Conf. on Symbolic and Quantitative Approaches to Reasoning with Uncertainty (ECSQARU 2005, Barcelona, Spain)*, LNAI 3571:3–14. Springer-Verlag, Berlin, Germany, 2005

J. Gebhardt, H. Detmer, and A.L. Madsen. Predicting Parts Demand in the Automotive Industry—An Application of Probabilistic Graphical Models. *Proc. Bayesian Modelling Applications Workshop at Int. Joint Conf. on Uncertainty in Artificial Intelligence (UAI 2003, Acapulco, Mexico)*, 2003

J. Gebhardt, C. Borgelt, R. Kruse, and H. Detmer. Knowledge Revision in Markov Networks. *Mathware and Soft Computing* 11(2–3):93–107. University of Granada, Granada, Spain, 2004

J. Gebhardt, A. Klose, H. Detmer, F. Rügheimer, and R. Kruse. Graphical Models for Industrial Planning on Complex Domains. In: D. Della Riccia, D. Dubois, R. Kruse, and H.-J. Lenz (eds.) *Decision Theory and Multi-Agent Planning*, CISM Courses and Lectures 482:131–143. Springer-Verlag, Berlin, Germany, 2006

J. Pearl. *Probabilistic Reasoning in Intelligent Systems: Networks of Plausible Inference*. Morgan Kaufmann, San Mateo, CA, USA, 1988

J. Pearl and A. Paz. Graphoids: A Graph Based Logic for Reasoning About Relevance Relations. In: B.D. Boulay, D. Hogg, and L. Steels (eds.) *Advances in Artificial Intelligence 2*, 357–363. North Holland, Amsterdam, Netherlands, 1987

M. Studeny. Multiinformation and the Problem of Characterization of Conditional Independence Relations. *Problems of Control and Information Theory* 1:3–16, 1989

M. Studeny. Conditional Independence Relations Have No Finite Complete Characterization. *Kybernetika* 25:72–79. Institute of Information Theory and Automation, Prague, Czech Republic, 1990

Chapter 24
Evidence Propagation

After having discussed efficient representations for expert and domain knowledge, we intent to exploit them to draw inferences when new information (evidence) becomes known. Using the Volkswagen example from the last chapter, an inference could be the update of the probabilities of certain car parts combinations when the customer has chosen, say, the engine type to be m^*. The objective is to propagate the evidence through the underlying network to reach all relevant attributes. Obviously, the graph structure will play an important role.

Given a graphical model (no matter whether a Markov or Bayes network) $G = (V, E)$ over an attribute set V with underlying distribution p_V and an observed value a_o of the attribute $A_o \in V$, propagating this evidence corresponds formally to the calculation of the following probabilities:

$$\forall A \in V \setminus \{A_o\} : \forall a \in \text{dom}(A) : \quad P(A = a \mid A_o = a_o).$$

Evidence propagation is thoroughly treated in the literature and a multitude of different algorithms exist. They can be distinguished by model type (Markov or Bayes network), by graph topology (tree, polytree, general graph or completely independent of any graph structure), or by type of computation (exact or approximative), just to name the most important criteria for comparison. A polytree is a directed tree where nodes can have more than one parent. The graph G_1 in Fig. 24.2 on page 457 is a polytree while G_2 is a simple tree.

It is not our intention to provide a wide coverage over those algorithms. We rather intent to discuss one particular evidence propagation algorithm in greater detail in order to stress the underlying ideas.

We already saw in the introduction Chap. 21 that we can exploit the graph structure to guide the evidence propagation: First, the (undirected) tree structure guaranteed unique paths between the attributes and second, the tree decomposes into multiple isolated subgraphs when any node is deleted. The latter coincides with the statement that (in an undirected tree) an instantiated attribute (where instantiated here means that for this attributes we have observed evidence) separates the attributes in the subtrees as depicted in Fig. 24.1. An alike separation can also be

R. Kruse et al., *Computational Intelligence*, Texts in Computer Science,
DOI 10.1007/978-1-4471-5013-8_24, © Springer-Verlag London 2013

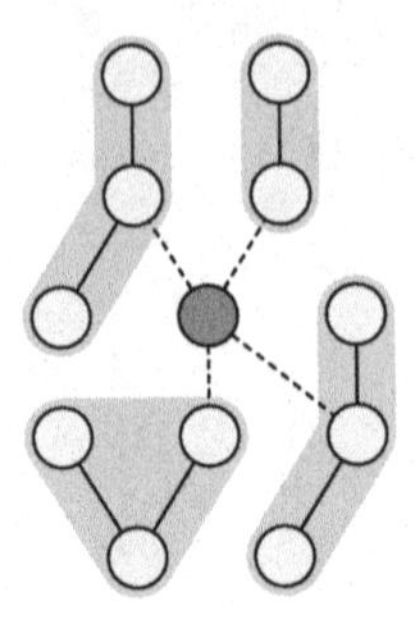

Fig. 24.1 The gray node separates the notes in all four subtrees. We will use that idea to simplify the evidence propagation

observed in directed trees: Given an attribute A, the attributes of the child trees (i.e., the subtrees having a child of A as the root node) become conditionally independent from each other. However, we have to take into account that in a polytree we might have to deal with multiple parent nodes. These form a converging connection with A which is activated when A is instantiated. The matter becomes even more complex when dealing with general graphs, that is, arbitrary undirected or directed acyclic graphs: Since there might be multiple paths between two nodes, we do not have a unique route for propagating evidence. There exist algorithms that deal with all of the mentioned problems. We will shortly discuss a solution that can deal with all of the above-mentioned network and graph types.

The evidence propagation itself will be carried out on a join tree. The tree structure guarantees the uniqueness of the evidence flow while the choice of a join tree is justified by the fact that every general (undirected or directed) graph can be transformed into a semantically equivalent join tree. Semantic equivalence means here that the resulting join tree is still a conditional independence map of the underlying distribution.

How any undirected graph can be transformed into a join tree was shown in Sect. 22.2.2. Let us therefore focus on the transformation of a directed acyclic graph into a join tree. The idea to just drop the edge directions in order to get an equivalent undirected graph does unfortunately not work as we can easily see. Consider the directed acyclic graph G_1 in Fig. 24.2. It encodes only the following conditional independence statements (although the symmetric counterparts do, of course, hold but we omit them here for the sake of brevity):

$$\mathcal{I}_{G_1} = \{A \perp\!\!\!\perp_{G_1} D \mid C,\ B \perp\!\!\!\perp_{G_1} D \mid C,\ A \perp\!\!\!\perp_{G_1} B \mid \emptyset\}.$$

By dropping the edge directions, we arrive at the undirected graph G_u as depicted in Fig. 24.2. Its set of encoded conditional independence statements is, alas, no subset of those of G_1:

$$\mathcal{I}_{G_u} = \{A \perp\!\!\!\perp_{G_u} D \mid C,\ B \perp\!\!\!\perp_{G_u} D \mid C,\ A \perp\!\!\!\perp_{G_u} B \mid C\} \not\subseteq \mathcal{I}_{G_1}.$$

There is one new conditional independence statement (here: $A \perp\!\!\!\perp B \mid C$) that is not valid in the original graph G_1. The fact that we lost a previously valid independence, namely $A \perp\!\!\!\perp_{G_1} B \mid \emptyset$, is unfortunate but does not invalidate the character of the in-

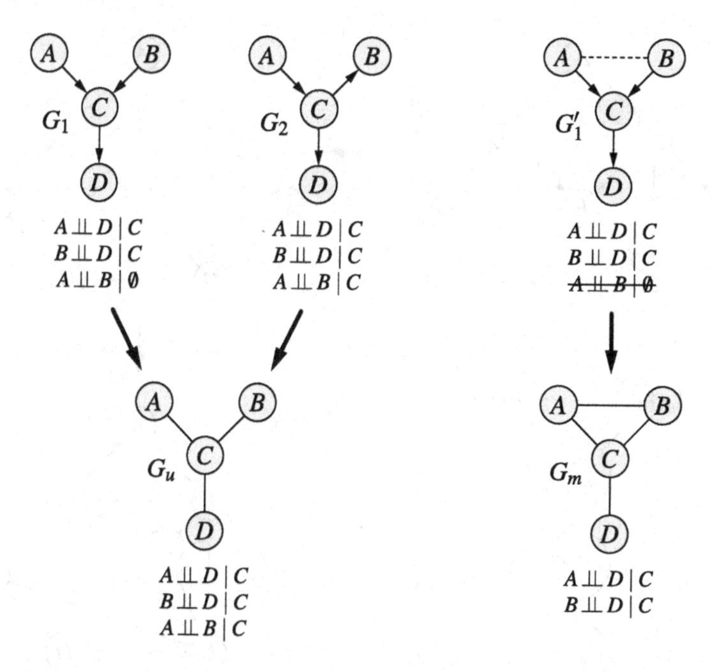

Fig. 24.2 In general, ignoring the edge directions in general does not result in an equivalent (w.r.t. the encoded conditional independence statements) undirected graph

dependence map. Hence, G_u is not an equivalent undirected graph for G_1! The goal must be to find a transformation that by no means introduces any new conditional independences—and be it at the loss of existing ones.

Note, that not every directed acyclic graph suffers from the illustrated phenomenon. Consider graph G_2 in Fig. 24.2. For its undirected pendant G_u, we indeed have

$$\mathcal{I}_{G_u} \subseteq \mathcal{I}_{G_2}.$$

(We even achieve equality here which is not necessarily always the case.)

Obviously, the illustrated problems arise due to the asymmetric definition of the d-separation which cause problems when dealing with converging nodes. It must be prevented that both parent nodes of a common child node become conditionally independent given the child node. This is achieved by connecting parent nodes that are not yet connected (the edge direction can be chosen arbitrarily as we drop it in the next step). This is depicted on the right of Fig. 24.2 and graph G'_1. By connecting the nodes A and B, we loose a (marginal) independence statement, but we prevent the introduction of a new one which leads to $\mathcal{I}_{G_u} \subseteq \mathcal{I}_{G'_1}$.

The transformation of connecting unconnected parent nodes we already know as moralization (see Definition 22.29 on page 428). With these prerequisites, we can transform any given graphical model into a common input format (namely a join tree): If the given graph is a Markov network, we might need to triangulate the graph before it is transformed into a join tree. If we are given a Bayes network,

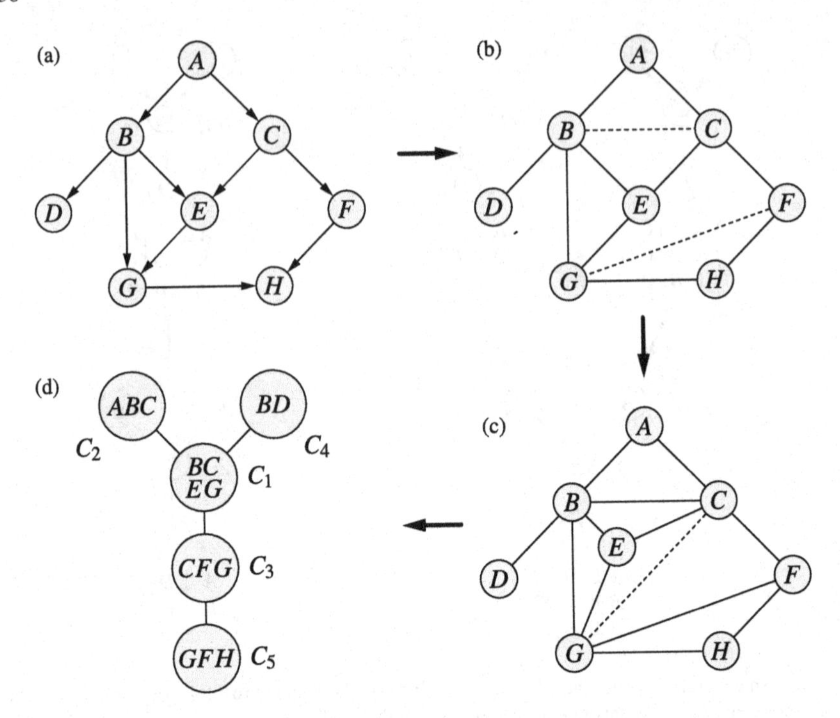

Fig. 24.3 Transformation of a Bayes network into a join tree. (**a**) Bayes network, (**b**) Moral graph, (**c**) Triangulated moral graph, (**d**) Join tree

we moralize the graph, triangulate it (if necessary) and transform it into a join tree afterwards. We will illustrate this with an example that we will reuse later when running the evidence propagation algorithm.

Example 24.1 (Join tree from a Bayes network) We will shortly consider an propagation of evidence inside the Bayes network depicted in Fig. 24.3(a). The result of the moralization step can be seen in Fig. 24.3(b). The circle E–C–F–G has no chord which leads to the insertion of one. We decide for chord C–G (E–F would have been possible as well) and arrive at the triangulated graph in Fig. 24.3(c). We know this graph already from Fig. 23.3 and Example 23.2 which allows us to directly reuse the cliques identified back then, resulting in the join tree in Fig. 24.3 (d). Note, that during triangulation as well as moralization only edges might be added (never deleted) and hence the character of a conditional independence map never changes.

The evidence propagation will consist of the following principal steps:

Initialization: Evidences, that is, the known attribute values will be incorporated into all relevant clique potentials.

Message passing: Neighboring cliques send a message to each other to announce the changes. Hence, there are exactly two messages sent across each edge. Having r cliques in the tree, $2(r-1)$ messages will be sent.

Update: After all messages have been sent, each clique can update the joint probability of its attributes.

Marginalization: Since we are interested in the marginal distribution of single attributes, we finally marginalize out each attribute from the smallest clique that contains it.

We now investigate these four steps more closely.

24.1 Initialization

We already know Algorithm 23.1 from page 448 to determine the clique potentials. If the join tree resulted from an initial Bayes network (as in our example case), we can easily determine the clique potentials of all cliques $C \in \mathcal{C}$ as the product of the conditional probabilities of all node families which are properly contained in C:

$$\phi_C(\mathbf{c}) = 1 \cdot \prod_{\text{fa}(A) \subseteq C} P\big(A = \text{proj}_{\{A\}}^{C}(\mathbf{c}) \mid \text{pa}(A) = \text{proj}_{\text{pa}(A)}^{C}(\mathbf{c})\big). \tag{24.1}$$

If the values of one or more attributes become known, it leads to the modification of those cliques that contain these attributes: All attribute value combinations that are incompatible with the evidence get assigned a value of zero by the potential functions. In all other cases, the potential functions remain unchanged:

$$\phi'_C(\mathbf{c}) = \begin{cases} 0, & \text{if } \text{proj}_{E \cap C}^{C}(\mathbf{c}) \neq \text{proj}_{E \cap C}^{E}(\mathbf{e}), \\ \phi_C(\mathbf{c}), & \text{otherwise.} \end{cases}$$

24.2 Message Passing

After initialization, neighboring cliques exchange messages to inform each other about (a possible) change in their potential functions. Each clique sends exactly one message to each neighbor. The message $M_{B \to C}$ from clique B to clique C is declared on the attributes of the separator set S_{BC}. The values of the message from B to C depends on the potential of the sending clique ϕ_B and the messages of all *other* neighbors of B. The values are multiplied and marginalized on the attributes in S_{BC}:

$$M_{B \to C}(\mathbf{s}_{BC}) = \sum_{\mathbf{b} \setminus \mathbf{s}_{BC}} \Big[\phi_B(\mathbf{b}) \cdot \prod_{D \in \text{adj}(B) \setminus \{C\}} M_{D \to B}(\mathbf{s}_{DB}) \Big]. \tag{24.2}$$

Since we need for the computation of the message from B to C all received messages from all other neighbors of B, it is clear that only cliques with no other neigh-

$P(A)$	
a_1	0.6
a_2	0.4

$P(B \mid A)$	a_1	a_2
b_1	0.2	0.1
b_2	0.8	0.9

$P(C \mid A)$	a_1	a_2
c_1	0.3	0.7
c_2	0.7	0.3

$P(D \mid B)$	b_1	b_2
d_1	0.4	0.7
d_2	0.6	0.3

$P(F \mid C)$	c_1	c_2
f_1	0.1	0.4
f_2	0.9	0.6

$P(E \mid B, C)$	b_1		b_2	
	c_1	c_2	c_1	c_2
e_1	0.2	0.4	0.3	0.1
e_2	0.8	0.6	0.7	0.9

$P(G \mid B, E)$	b_1		b_2	
	e_1	e_2	e_1	e_2
g_1	0.95	0.4	0.7	0.5
g_2	0.05	0.6	0.3	0.5

$P(H \mid G, F)$	g_1		g_2	
	f_1	f_2	f_1	f_2
h_1	0.2	0.4	0.5	0.7
h_2	0.8	0.6	0.5	0.3

Fig. 24.4 Parameters of the Bayes network in Fig. 24.3(a)

bors can send its messages first: namely the outer nodes of the join tree. We will later see how this leads to a message sending cascade across the join tree.

24.3 Update

After all messages have been sent, each clique C can compute is joint probability distribution $P(\mathbf{c})$ as the product of its potential function and all received messages from its neighbors:

$$P(\mathbf{c}) \propto \phi_C(\mathbf{c}) \cdot \prod_{B \in \text{adj}(C)} M_{B \to C}\big(\text{proj}_{S_{BC}}^{C}(\mathbf{c})\big). \tag{24.3}$$

The $\propto$-sign denotes that the distribution $P(\mathbf{c})$ needs to be normalized in case it does not add up to one (over all $\mathbf{c}$).

24.4 Marginalization

After each clique has updated its joint distribution, we look for each attribute A for the smallest clique C that contains it in order to minimize the marginalization effort (a marginalization from other cliques that contain A would be also possible, of course):

$$P(a) = \sum_{\mathbf{c} \setminus a} P(\mathbf{c}).$$

Let us now study a full run of the algorithm. The parameters of the Bayes network in Fig. 24.3(a) are shown in Fig. 24.4.

Example 24.2 (Initialization, propagating "zero" evidence) For the join tree in Fig. 24.3(d), we use the initial Bayes network from Fig. 24.3(a) and can compute the following potentials:

$$\phi_{C_1}(b, c, e, g) = P(e \mid b, c) \cdot P(g \mid e, b)$$
$$\phi_{C_2}(a, b, c) = P(b \mid a) \cdot P(c \mid a) \cdot P(a)$$

$$\phi_{C_3}(c, f, g) = P(f \mid c)$$
$$\phi_{C_4}(b, d) = P(d \mid b)$$
$$\phi_{C_5}(g, f, h) = P(h \mid g, f).$$

Since each conditional probability occurs in exactly on potential, it is immediately clear that the product of all potentials equals the network's decomposition formula. The clique potentials derived via Algorithm 23.1 based on separator and residual sets would lead to the following potentials:

$$\phi_{C_1}(b, c, e, g) = P(b, e \mid c, g)$$
$$\phi_{C_2}(a, b, c) = P(a \mid b, c)$$
$$\phi_{C_3}(c, f, g) = P(c \mid f, g)$$
$$\phi_{C_4}(b, d) = P(d \mid b)$$
$$\phi_{C_5}(g, f, h) = P(h, g, f).$$

In the remainder of this example, we will use potentials retrieved from the Bayes network. Since we want to compute the a-priori distribution of all attributes first, no clique potentials are modified in this run.

Example 24.3 (Message passing) Figure 24.5 shows the join tree from Fig. 24.3(d) together with all eight messages which are computed as follows:

$$\begin{aligned}
M_{21}(b, c) &= \textstyle\sum_a \quad \phi_2(a, b, c), \\
M_{41}(b) &= \textstyle\sum_d \quad \phi_4(b, d), \\
M_{53}(f, g) &= \textstyle\sum_h \quad \phi_5(f, g, h), \\
M_{13}(c, g) &= \textstyle\sum_{b,e} \ \phi_1(b, c, e, g)\, M_{21}(b, c) \quad M_{41}(b), \\
M_{31}(c, g) &= \textstyle\sum_f \quad \phi_3(c, f, g) \quad M_{53}(f, g), \\
M_{12}(b, c) &= \textstyle\sum_{e,g} \ \phi_2(b, c, e, g)\, M_{31}(c, g) \quad M_{41}(b), \\
M_{35}(f, g) &= \textstyle\sum_c \quad \phi_3(c, f, g) \quad M_{13}(c, g), \\
M_{14}(b) &= \textstyle\sum_{c,e,g} \phi_1(b, c, e, g)\, M_{21}(b, c) \quad M_{31}(c, g).
\end{aligned}$$

As can easily be seen, the messages M_{41}, M_{53} and M_{21} can be computed (in arbitrary order) first because they do not require other messages for their computation. This is immediately clear as they originate from leaf nodes which do not have any other neighbors except the recipient of the message. After these three messages are sent, we can compute (and send) M_{31} and M_{13}. Only then we are able to compute M_{12}, M_{14} and M_{35}. Figure 24.5 depicts the just mentioned dependencies as a directed graph with edges directed to the dependent messages. The computed messages are as follows:

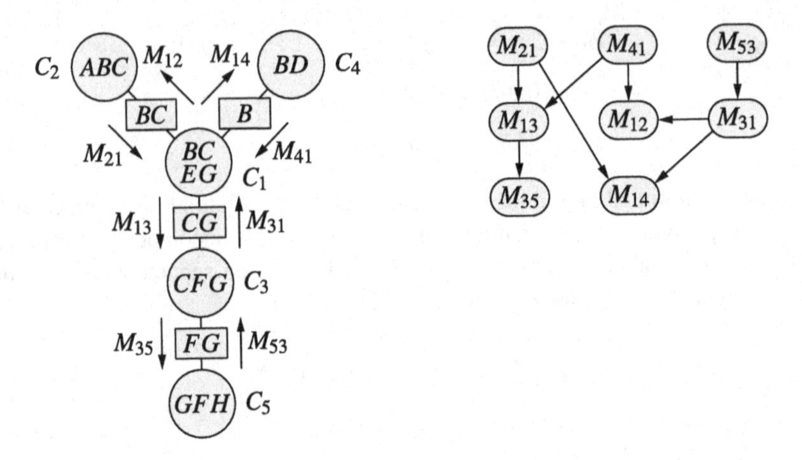

Fig. 24.5 Messages that need to be sent across the join tree together with a dependency graph of the messages

$$M_{21} = (\overset{b_1,c_1}{0.06}, \overset{b_1,c_2}{0.10}, \overset{b_2,c_1}{0.40}, \overset{b_2,c_2}{0.44}) \qquad M_{41} = (\overset{b_1}{1}, \overset{b_2}{1})$$

$$M_{13} = (\overset{c_1,g_1}{0.254}, \overset{c_1,g_2}{0.206}, \overset{c_2,g_1}{0.290}, \overset{c_2,g_2}{0.250}) \qquad M_{35} = (\overset{f_1,g_1}{0.14}, \overset{f_1,g_2}{0.12}, \overset{f_2,g_1}{0.40}, \overset{f_2,g_2}{0.33})$$

$$M_{53} = (\overset{f_1,g_1}{1}, \overset{f_1,g_2}{1}, \overset{f_2,g_1}{1}, \overset{f_2,g_2}{1}) \qquad M_{31} = (\overset{c_1,g_1}{1}, \overset{c_1,g_2}{1}, \overset{c_2,g_1}{1}, \overset{c_2,g_2}{1})$$

$$M_{12} = (\overset{b_1,c_1}{1}, \overset{b_1,c_2}{1}, \overset{b_2,c_1}{1}, \overset{b_2,c_2}{1}) \qquad M_{14} = (\overset{b_1}{0.16}, \overset{b_2}{0.84}).$$

Example 24.4 (Update) After all messages have been sent, the cliques can update their joint probability distribution. For the five cliques of our example, we get:

$$\begin{aligned}
P(c_1) &= P(b,c,e,g) = \phi_1(b,c,e,g) \cdot M_{21}(b,c) \cdot M_{31}(c,g) \cdot M_{41}(b)\\
P(c_2) &= P(a,b,c) \quad \propto \phi_2(a,b,c) \quad \cdot M_{12}(b,c)\\
P(c_3) &= P(c,f,g) \quad \propto \phi_3(c,f,g) \quad \cdot M_{13}(c,g) \cdot M_{53}(f,g)\\
P(c_4) &= P(b,d) \quad \propto \phi_4(b,d) \quad \cdot M_{14}(b)\\
P(c_5) &= P(f,g,h) \quad \propto \phi_5(f,g,h) \quad \cdot M_{35}(f,g).
\end{aligned}$$

The numbers can be found in the P-columns of the potential tables in Table 24.1. In this case, no normalization is necessary.

Example 24.5 (Marginalization) Finally, we determine the marginal probabilities for each attribute from the clique potentials. To keep the effort low, we choose the smallest clique that contains the attribute. We arrive at the following marginalizations:

$$\begin{aligned}
P(a) &= \textstyle\sum_{b,c} P(a,b,c), & P(b) &= \textstyle\sum_{d} P(b,d),\\
P(c) &= \textstyle\sum_{a,b} P(a,b,c), & P(d) &= \textstyle\sum_{b} P(b,d),\\
P(e) &= \textstyle\sum_{b,c,g} P(b,c,e,g), & P(f) &= \textstyle\sum_{c,g} P(c,f,g),\\
P(g) &= \textstyle\sum_{c,f} P(c,f,g), & P(h) &= \textstyle\sum_{g,f} P(g,f,h).
\end{aligned}$$

The actual numbers can be found in Table 24.2.

Table 24.1 The potential tables of the join tree from Fig. 24.3(d). The first column contains the potentials according to Eq. (24.1), the second column contains the updated joint probabilities of the cliques according Eq. (24.3) after zero evidence has been propagated. The third column contains the probabilities after the evidence $H = h_1$ has been propagated

ϕ_1					P	P'
b_1	c_1	e_1	g_1	0.19	0.0122	0.0095
			g_2	0.01	0.0006	0.0009
		e_2	g_1	0.32	0.0205	0.0161
			g_2	0.48	0.0307	0.0431
	c_2	e_1	g_1	0.38	0.0365	0.0241
			g_2	0.02	0.0019	0.0025
		e_2	g_1	0.24	0.0230	0.0152
			g_2	0.36	0.0346	0.0443
b_2	c_1	e_1	g_1	0.21	0.0832	0.0653
			g_2	0.09	0.0356	0.0501
		e_2	g_1	0.35	0.1386	0.1088
			g_2	0.35	0.1386	0.1947
	c_2	e_1	g_1	0.07	0.0311	0.0205
			g_2	0.03	0.0133	0.0171
		e_2	g_1	0.45	0.1998	0.1321
			g_2	0.45	0.1998	0.2559

ϕ_2				P	P'
a_1	b_1	c_1	0.036	0.0360	0.0392
		c_2	0.084	0.0840	0.0753
	b_2	c_1	0.144	0.1440	0.1523
		c_2	0.336	0.3360	0.3220
a_2	b_1	c_1	0.028	0.0280	0.0305
		c_2	0.012	0.0120	0.0108
	b_2	c_1	0.252	0.2520	0.2665
		c_2	0.108	0.1080	0.1035

ϕ_3				P	P'
c_1	f_1	g_1	0.1	0.0254	0.0105
		g_2	0.1	0.0206	0.0212
	f_2	g_1	0.9	0.2290	0.1892
		g_2	0.9	0.1850	0.2675
c_2	f_1	g_1	0.4	0.1162	0.0480
		g_2	0.4	0.0998	0.1031
	f_2	g_1	0.6	0.1742	0.1440
		g_2	0.6	0.1498	0.2165

ϕ_4			P	P'
b_1	d_1	0.4	0.0640	0.0623
	d_2	0.6	0.0960	0.0934
b_2	d_1	0.7	0.5880	0.5910
	d_2	0.3	0.2520	0.2533

ϕ_5				P	P'
f_1	g_1	h_1	0.2	0.0283	0.0585
		h_2	0.8	0.1133	0
	g_2	h_1	0.5	0.0602	0.1243
		h_2	0.5	0.0602	0
f_2	g_1	h_1	0.4	0.1613	0.3331
		h_2	0.6	0.2419	0
	g_2	h_1	0.7	0.2344	0.4841
		h_2	0.3	0.1004	0

Table 24.2 Marginal probabilities of the Bayes network (w.r.t. the corresponding join tree) in Fig. 24.3 without evidence present. They represent the a-priori distribution of the attributes and were computed by propagating "zero evidence."

$P(\cdot)$	A	B	C	D	E	F	G	H
$\cdot_1$	0.6000	0.1600	0.4600	0.6520	0.2144	0.2620	0.5448	0.4842
$\cdot_2$	0.4000	0.8400	0.4500	0.3480	0.7856	0.7380	0.4552	0.5158

Example 24.6 (Second propagation run, now with evidence) We will now do a second propagation run and propagate the evidence $H = h_1$ in order to arrive at the conditional probability distributions $P(A \mid H = h_1)$ to $P(G \mid H = h_1)$. As the steps are identical to the previous run, we will only tell the changed parameters and intermediate values.

For initialization we set all those clique potential entries to zero that contradict the evidence, i.e., for which $H \neq h_1$. Since H is only contained in clique C_5, we only need to change one potential table:

Table 24.3 Marginal probabilities of the Bayes network (w.r.t. the corresponding join tree) in Fig. 24.3 given evidence $H = h_1$. They represent the a-posteriori distributions of the attributes

$P(\cdot \mid h_1)$	A	B	C	D	E	F	G	H
$\cdot_1$	0.5888	0.1557	0.4884	0.6533	0.1899	0.1828	0.3916	1.0000
$\cdot_2$	0.4112	0.8443	0.5116	0.3467	0.8101	0.8172	0.6084	0.0000

ϕ'_5				P
f_1	g_1	h_1	0.2	
		h_2	**0**	
	g_2	h_1	0.5	
		h_2	**0**	
f_2	g_1	h_1	0.4	
		h_2	**0**	
	g_2	h_1	0.7	
		h_2	**0**	

The message computation and passing is equivalent to the previous run. The messages are now as follows:

$$M_{21} = (\overset{b_1,c_1}{0.06}, \overset{b_1,c_2}{0.10}, \overset{b_2,c_1}{0.40}, \overset{b_2,c_2}{0.44}) \qquad M_{41} = (\overset{b_1}{1}, \overset{b_2}{1})$$

$$M_{13} = (\overset{c_1,g_1}{0.254}, \overset{c_1,g_2}{0.206}, \overset{c_2,g_1}{0.290}, \overset{c_2,g_2}{0.250}) \qquad M_{35} = (\overset{f_1,g_1}{0.14}, \overset{f_1,g_2}{0.12}, \overset{f_2,g_1}{0.40}, \overset{f_2,g_2}{0.33})$$

$$M_{53} = (\overset{f_1,g_1}{0.2}, \overset{f_1,g_2}{0.5}, \overset{f_2,g_1}{0.4}, \overset{f_2,g_2}{0.7}) \qquad M_{31} = (\overset{c_1,g_1}{0.38}, \overset{c_1,g_2}{0.68}, \overset{c_2,g_1}{0.32}, \overset{c_2,g_2}{0.62})$$

$$M_{12} = (\overset{b_1,c_1}{0.527}, \overset{b_1,c_2}{0.434}, \overset{b_2,c_1}{0.512}, \overset{b_2,c_2}{0.464}) \qquad M_{14} = (\overset{b_1}{0.075}, \overset{b_2}{0.409})$$

After all messages have been sent, we can again update the joint probability distributions according to Eq. (24.3). These result in the values of the P'-column of Table 24.1. In this case a normalization is necessary. Finally, we marginalize over the clique joint distributions to get the updated marginal attribute distributions (given $H = h_1$). These are shown in Table 24.3.

Figure 24.6 contains notations that will help to illustrate the propagation algorithm (Castillo et al. 1997).

1. $\mathcal{C}_{CB}$ be the set of cliques of the join subtree containing C that results when edge C–B is removed. If the cliques differ only in the indices of its names, we will use these indices in the notation.
   Example: $\mathcal{C}_{C_1,C_3} = \mathcal{C}_{13} = \{C_1, C_2, C_4\}$ and $\mathcal{C}_{C_3,C_1} = \mathcal{C}_{31} = \{C_3, C_5\}$

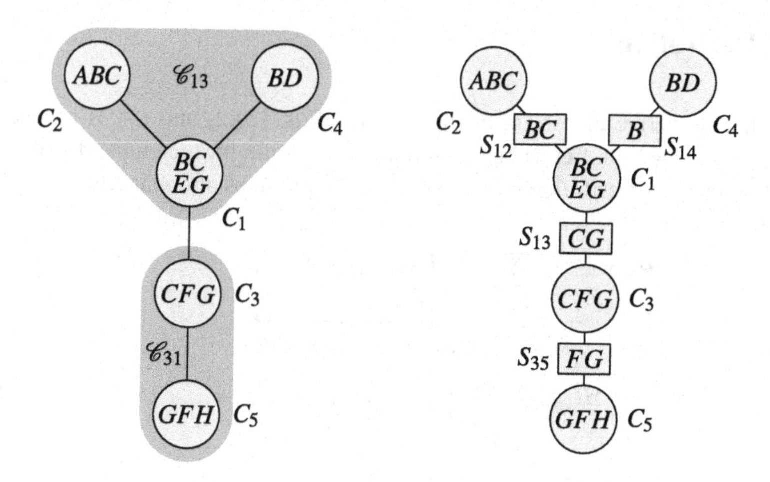

Fig. 24.6 Example for the notation used to derive the evidence propagation steps

2. The union of all attributes of all cliques of C_{CB} is denoted as X_{CB}.
   Example: $X_{C_1,C_3} = X_{13} = \{A, B, C, D, E, G\}$ and $X_{C_3,C_1} = X_{31} = \{C, F, G, H\}$
3. We obviously get $V = X_{CB} \cup X_{BC}$.
4. Separator and residual sets are already known. We will now index them with the cliques between which they are defined:

$$S_{CB} = S_{BC} = C \cap B$$

and

$$R_{CB} = X_{CB} \setminus S_{CB}.$$

   Example: $R_{C_1,C_3} = R_{13} = \{A, B, D, E\}$, $R_{C_3,C_1} = R_{31} = \{F, G\}$ and $S_{C_1,C_3} = S_{13} = \{C, G\}$
5. From 3. and 4. follows that the entire node set V partitions into the following three disjoint node sets given any edge $(C, B) \in E$:

$$V = R_{CB} \cup S_{CB} \cup R_{BC}.$$

Hence, the attributes R_{CB} are u-separated from the attributes R_{BC} by the attributes S_{BC}.

$$R_{BC} \perp\!\!\!\perp R_{CB} \mid S_{CB}.$$

24.5 Derivation

We will now derive the origin of the essential Eqs. (24.2) and (24.3) needed for evidence propagation. Let us begin with the update rule for the clique distribution. This can be written as follows (we discuss the single steps afterwards):

$$
\begin{aligned}
P(\mathbf{c}) &\overset{(1)}{=} \underbrace{\sum_{\mathbf{v}\setminus\mathbf{c}}}_{\text{Marginalization}} \underbrace{\prod_{D\in\mathcal{C}} \phi_D\left(\text{proj}_D^V(\mathbf{v})\right)}_{\text{Decomposition of } P(\mathbf{v})} \\
&\overset{(2)}{=} \phi_C(\mathbf{c}) \sum_{\mathbf{v}\setminus\mathbf{c}} \prod_{D\neq C} \phi_D\left(\text{proj}_D^V(\mathbf{v})\right) \\
&\overset{(3)}{=} \phi_C(\mathbf{c}) \sum_{\bigcup_{B\in\text{adj}(C)} \mathbf{r}_{BC}} \prod_{D\neq C} \phi_D\left(\text{proj}_D^V(\mathbf{v})\right) \\
&\overset{(4)}{=} \phi_C(\mathbf{c}) \prod_{B\in\text{adj}(C)} \underbrace{\left(\sum_{\mathbf{r}_{BC}} \prod_{D\in\mathcal{C}_{BC}} \phi_D\left(\text{proj}_D^V(\mathbf{v})\right)\right)}_{M_{B\to C}(\mathbf{s}_{BC})} \\
&= \phi_C(\mathbf{c}) \prod_{B\in\text{adj}(C)} M_{B\to C}(\mathbf{s}_{BC}).
\end{aligned}
$$

(1) The vector $\mathbf{v}$ is defined on all attributes. Since we are only interested in attributes in C, we marginalize all remaining attributes' value combinations $\mathbf{v}\setminus\mathbf{c}$. The product is the application of Definition 23.7 on page 446 about the decomposition of a distribution w.r.t. an undirected graph.
(2) The product in (1) ran over all cliques, including C. This factor is now pulled out of the product and since it is independent of the sum, also pulled out of the sum.
(3) We simplify the marginalization. Instead to run over all value combinations $\mathbf{v}\setminus\mathbf{c}$, we exploit the following relationship that we can derive from the separations induced by the running intersection property in Eq. (22.2) on page 440. Applied to the join tree, we get:

$$
V \setminus C = \left(\bigcup_{B\in\text{adj}(C)} X_{BC}\right) \setminus C = \bigcup_{B\in\text{adj}(C)} (X_{BC} \setminus C) = \bigcup_{B\in\text{adj}(C)} R_{BC}.
$$

The sum index $\bigcup_{B\in\text{adj}(C)} \mathbf{r}_{BC}$ means the iteration over all value combinations of all attributes in all R_{BC}.
(4) The sum is split up by assigning the residuals to the respective neighboring cliques. The resulting factors are the starting point for the messages.

The messages $M_{B\to C}(\mathbf{s}_{BC})$ declared on the separators can be further simplified to finally arrive at the Eq. (24.2):

$$M_{B\to C}(\mathbf{s}_{BC}) = \underbrace{\sum_{\mathbf{x}_{BC}\setminus\mathbf{s}_{BC}}}_{\mathbf{r}_{BC}} \prod_{D\in\mathcal{C}_{BC}} \phi_D(\mathbf{d})$$
$$= \sum_{\mathbf{b}\setminus\mathbf{s}_{BC}} \phi_B(\mathbf{b}) \prod_{D\in\mathrm{adj}(B)\setminus\{C\}} M_{D\to B}(\mathbf{s}_{DB}).$$

24.6 Other Propagation Algorithms

We will sketch two related evidence propagation algorithms at the end of this chapter. Information on other approaches can be found in for example, Castillo et al. (1997), Borgelt et al. (2009).

The algorithm described above (Jensen 1996, 2001; Jensen and Nielsen 2007) allows any directed (but acyclic) graph structure as input. If simpler structures are employed, one can use simpler algorithms, of course. One example are polytree propagation algorithms (Pearl 1988). Since polytrees are singly connected (i.e., they fall apart into two subgraphs if an edge is removed), the propagation path is already clear without the need to construct a secondary structure. Because of the edge directions, two message types are distinguished: λ-messages from child to parent nodes and π-messages from parent to child nodes. Of course, one can apply the join tree propagation algorithm on polytrees. This will lead to cliques consisting of nodes together with its parent nodes.

Another technique that not necessarily needs to operate on graph structures is the so-called bucket elimination (Dechter 1996; Zhang and Poole 1996). Given a factorization of a probability distribution, an attribute can be eliminated by summing over all factors containing the attribute. Successive summing allows to isolate target attributes. The efficiency of this approach depends heavily on the order of the summations. A conditional independence graph can be used to suggest an order by means of its edges.

References

C. Borgelt, M. Steinbrecher and R. Kruse. *Graphical Models—Representations for Learning, Reasoning and Data Mining*, 2nd ed. J. Wiley & Sons, Chichester, United Kingdom, 2009

E. Castillo, J.M. Gutiérrez and A.S. Hadi. *Expert Systems and Probabilistic Network Models*. Springer-Verlag, New York, NY, USA, 1997

R. Dechter. Bucket Elimination: A Unifying Framework for Probabilistic Inference. *Proc. 12th Conf. on Uncertainty in Artificial Intelligence (UAI'96, Portland, OR, USA)*, 211–219. Morgan Kaufmann, San Mateo, CA, USA, 1996

F.V. Jensen. *An Introduction to Bayesian Networks*. UCL Press, London, United Kingdom, 1996

F.V. Jensen. *Bayesian Networks and Decision Graphs*. Springer-Verlag, Berlin, Germany, 2001

F.V. Jensen and T.D. Nielsen. *Bayesian Networks and Decision Graphs*, 2nd ed. Springer-Verlag, London, United Kingdom, 2007

J. Pearl. *Probabilistic Reasoning in Intelligent Systems: Networks of Plausible Inference*. Morgan Kaufmann, San Mateo, CA, USA, 1988

N.L. Zhang and D. Poole. Exploiting Causal Independence in Bayesian Network Inference. *Journal of Artificial Intelligence Research* 5:301–328. Morgan Kaufmann, San Mateo, CA, USA, 1996

Chapter 25
Learning Graphical Models

We will now address the third question from Chap. 21, namely how graphical models can be learned from given data. Until now, we were given the graphical structure. Now, we will introduce heuristics that allow us to induce these structures.

In principle, we seek for a graph G that fits best to a given database D. Pretty much all learning algorithms for graphical models consist of the following two parts:

- A heuristic that efficiently traverses the search space and generates promising graph candidates.
- An evaluation measure that assigns to each candidate graph a goodness value (w.r.t. the database). This value is then used to guide the heuristic through the search space.

An exhaustive search through all possible graphs excludes itself as the size of the search space grows huge even for small numbers of nodes. For example, the set of directed acyclic graphs grows super-exponentially in the number of nodes (Robinson 1977): For 10 nodes, the graph set has already a cardinality of $4.18 \cdot 10^{18}$.

The evaluation function quantifies how "good" the candidate graph "explains" the database. One way to do this is to compute the probability with which the graph might have generated the database. We will see shortly, that the pure form of this idea does not work for a learning algorithm. However, adaptations of it are used in many learning algorithms, like the K2 algorithm that we will discuss later on. Generating a database based on a graph means that the probability distribution described by the graphical model was used to sample tuples that make up the database.

In the following, we will refer to the graph structure (no matter whether directed or undirected) of a graphical model as B_S. The set of all probability parameters (entries of the potential tables in the undirected case or the conditional probabilities in the directed case) be B_P. Let us consider the component B_P in case of a Bayes network. It contains the specific entries of the conditional probability distributions $P(A \mid \mathrm{pa}(A))$. Each node maintains a potential table which contains for each combination of the parent attributes values of a node A_i the probability of the attribute values of A_i. In general, the potential table of attribute A_i contains the columns $Q_{i1}, \dots, Q_{iq_i}$. Each of these q_i columns corresponds to a conditional

R. Kruse et al., *Computational Intelligence*, Texts in Computer Science,
DOI 10.1007/978-1-4471-5013-8_25, © Springer-Verlag London 2013

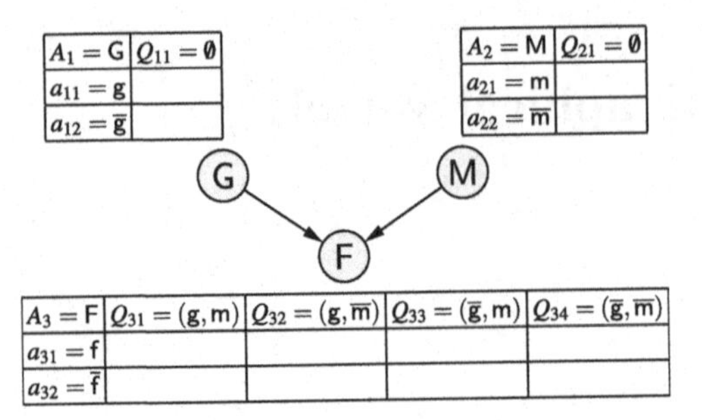

Fig. 25.1 Induction of the potential tables structure B_S of the depicted Bayes network

A_i	Q_{i1}	$\cdots$	Q_{ij}	$\cdots$	Q_{iq_i}
a_{i1}	θ_{i11}	$\cdots$	θ_{ij1}	$\cdots$	θ_{iq_i1}
$\vdots$	$\vdots$	$\ddots$	$\vdots$	$\ddots$	$\vdots$
a_{ik}	θ_{i1k}	$\cdots$	θ_{ijk}	$\cdots$	θ_{iq_ik}
$\vdots$	$\vdots$	$\ddots$	$\vdots$	$\ddots$	$\vdots$
a_{ir_i}	θ_{i1r_i}	$\cdots$	θ_{ijr_i}	$\cdots$	$\theta_{iq_ir_i}$

Fig. 25.2 General potential table of attribute A_i. Each column represents a probability distribution

probability distribution of the attribute values $a_{i1}, \ldots, a_{ir_i}$ given the attribute value combination associated with the column.

As an example, consider the simple Bayes network in Fig. 25.1. All attributes be binary and represent Grippe, Malaria and Fever. Note that we use the dated notion *grippe* here as the modern word *flu* would obviously interfere with the variable naming. The node F has two parent attributes (G and M) which are both binary. Hence, there exist four parent attributes value combinations: (g, m), $(\mathsf{g}, \overline{\mathsf{m}})$, $(\overline{\mathsf{g}}, \mathsf{m})$ and $(\overline{\mathsf{g}}, \overline{\mathsf{m}})$ which correspond to the four columns in the potential table for F in Fig. 25.1. If the node A_i is a root node, the potential table contains the marginal distribution $P(A_i)$. This artificial column is denoted by $\emptyset$.

The entries of these tables are denoted as θ_{ijk} and represent the probability of attribute A_i assuming value a_{ik} while its parent attributes $\mathrm{pa}(A_i)$ assume the jth value combination Q_{ij}. Figure 25.2 shows a general potential table whose indices we use in the remainder: the index i runs over the attributes ($i = 1, \ldots, n$), j runs over the different parent attributes' value combinations of attribute A_i ($j = 1, \ldots, q_i$) and k runs over all r_i values of A_i ($k = 1, \ldots, r_i$). The values $(\theta_{ij1}, \ldots, \theta_{ijr_i})$ represent a probability distribution: They correspond to the parameters of multinomial distribution of order r_i.

Let us know discuss two examples for computing the parameters B_P from given data D and given structure B_S. With D and B_S provided, we can easily determine the parameters B_P by counting. This is justified because the conditional distributions are multinomial distributions and relative frequencies constitute optimal estimators. Table 25.1 shows an example database with 100 cases.

Table 25.1 An example database with 100 cases

Gripppe	$\overline{\mathsf{g}}$	$\overline{\mathsf{g}}$	$\overline{\mathsf{g}}$	$\overline{\mathsf{g}}$	g	g	g	g
Malaria	$\overline{\mathsf{m}}$	$\overline{\mathsf{m}}$	m	m	$\overline{\mathsf{m}}$	$\overline{\mathsf{m}}$	m	m
Fever	$\bar{\mathsf{f}}$	f	$\bar{\mathsf{f}}$	f	$\bar{\mathsf{f}}$	f	$\bar{\mathsf{f}}$	f
#	34	6	2	8	16	24	0	10

In the first part, we assume the network structure B_S to be an edgeless graph, that is, we deal with three marginal distributions. The "decomposition" of the joint distribution then reads:

$$P(\mathsf{G}=g, \mathsf{M}=m, \mathsf{F}=f) = P(\mathsf{G}=g)\, P(\mathsf{M}=m)\, P(\mathsf{F}=f)$$

$$\text{with } g \in \{\mathsf{g}, \overline{\mathsf{g}}\},\ m \in \{\mathsf{m}, \overline{\mathsf{m}}\},\ f \in \{\mathsf{f}, \bar{\mathsf{f}}\}.$$

Consequently, the distributions $P(\mathsf{G})$, $P(\mathsf{M})$ and $P(\mathsf{F})$ have to be estimated from data.

$$P(\mathsf{G}=g) \approx \widehat{P}(\mathsf{G}=g) = \frac{\#(\mathsf{G}=g)}{|D|}$$

where the statements $\#(X=x)$ and $\#(x)$ denote the number of tuples in D having the value x for attribute X, that is, $\#(X=x) = \#(x) \stackrel{\text{Def}}{=} |\{t \in D \mid t(X)=x\}| = |\{X=x\}|$. Estimated from the example distribution in Table 25.1, we get:

$$\begin{aligned}
\widehat{P}(\mathsf{G}=\mathsf{g}) &= {}^{50}/_{100} = 0.50, & \widehat{P}(\mathsf{G}=\overline{\mathsf{g}}) &= 1 - \widehat{P}(\mathsf{G}=\mathsf{g}) = 0.50,\\
\widehat{P}(\mathsf{M}=\mathsf{m}) &= {}^{20}/_{100} = 0.20, & \widehat{P}(\mathsf{M}=\overline{\mathsf{m}}) &= 1 - \widehat{P}(\mathsf{M}=\mathsf{m}) = 0.80,\\
\widehat{P}(\mathsf{F}=\mathsf{f}) &= {}^{50}/_{100} = 0.48, & \widehat{P}(\mathsf{F}=\bar{\mathsf{f}}) &= 1 - \widehat{P}(\mathsf{F}=\mathsf{f}) = 0.52.
\end{aligned}$$

In the second part, we assume the network structure B_S to be as depicted in Fig. 25.1. The decomposition reads:

$$P(\mathsf{G}=g, \mathsf{M}=m, \mathsf{F}=f) = P(\mathsf{G}=g)\, P(\mathsf{M}=m)\, P(\mathsf{F}=f \mid \mathsf{G}=g, \mathsf{M}=m).$$

The estimations for $P(\mathsf{G})$ and $P(\mathsf{M})$ are computed as in the previous example while we use the following formula to estimate the conditional probability distribution $P(\mathsf{F} \mid \mathsf{G}, \mathsf{M})$:

$$\widehat{P}(f \mid g, m) = \frac{\widehat{P}(f, g, m)}{\widehat{P}(g, m)} = \frac{\frac{\#(g,m,f)}{|D|}}{\frac{\#(g,m)}{|D|}} = \frac{\#(g, m, f)}{\#(g, m)}.$$

We get the following numbers for the distribution $\hat{P}(\mathsf{F} \mid \mathsf{G}, \mathsf{M})$ from the example data:

$$\begin{aligned}
\widehat{P}(\mathsf{F}=\mathsf{f} \mid \mathsf{G}=\mathsf{g}, \mathsf{M}=\mathsf{m}) &= \frac{{}^{1}/_{100}}{{}^{1}/_{100}} = 1.00,\\
\widehat{P}(\mathsf{F}=\mathsf{f} \mid \mathsf{G}=\mathsf{g}, \mathsf{M}=\overline{\mathsf{m}}) &= \frac{{}^{24}/_{100}}{{}^{40}/_{100}} = 0.60,\\
\widehat{P}(\mathsf{F}=\mathsf{f} \mid \mathsf{G}=\overline{\mathsf{g}}, \mathsf{M}=\mathsf{m}) &= \frac{{}^{8}/_{100}}{{}^{10}/_{100}} = 0.80,\\
\widehat{P}(\mathsf{F}=\mathsf{f} \mid \mathsf{G}=\overline{\mathsf{g}}, \mathsf{M}=\overline{\mathsf{m}}) &= \frac{{}^{6}/_{100}}{{}^{40}/_{100}} = 0.15.
\end{aligned}$$

Given a database D, a potential network structure B_S and the estimated parameters B_P, we can compute $P(D \mid B_S, B_P)$ when accepting the following assumptions.

1. The data-generating process can be modeled by the Bayes network represented by (B_S, B_P).
2. The tuples in the database occur independently from each other.
3. All tuples are complete, that is, there are no missing values.

Assumption 1 legitimates the investigation for a Bayes network as the underlying model since in case of a violation of this assumption, the entire search would be pointless. Assumption 2 states that the occurrence of a tuple does not affect the probability of the other tuples. It is not to be confused with the statement of all tuples being equally probable! Assumption 3 finally allows us to use the above-mentioned counting as we do not need to take care of missing values.

The probability of database D (given the graph candidate) can now be computed as follows:

$$
\begin{aligned}
P(D \mid B_S, B_P) &\\
= \; & \prod_{h=1}^{100} P(c_h \mid B_S, B_P) \\
= \; & \underbrace{\overbrace{P(\mathsf{g},\mathsf{m},\mathsf{f})}^{\text{case 1}} \cdot \overbrace{P(\mathsf{g},\mathsf{m},\mathsf{f})}^{\text{case 10}}}_{\text{10 times}} \cdot \underbrace{\overbrace{P(\bar{\mathsf{g}},\mathsf{m},\mathsf{f})}^{\text{case 51}} \cdot \overbrace{P(\bar{\mathsf{g}},\mathsf{m},\mathsf{f})}^{\text{case 58}}}_{\text{8 times}} \cdot \underbrace{\overbrace{P(\bar{\mathsf{g}},\bar{\mathsf{m}},\bar{\mathsf{f}})}^{\text{case 67}} \cdot \overbrace{P(\bar{\mathsf{g}},\bar{\mathsf{m}},\bar{\mathsf{f}})}^{\text{case 100}}}_{\text{34 times}} \\
= \; & P(\mathsf{g},\mathsf{m},\mathsf{f})^{10} \cdot P(\bar{\mathsf{g}},\mathsf{m},\mathsf{f})^{8} \cdot P(\bar{\mathsf{g}},\bar{\mathsf{m}},\bar{\mathsf{f}})^{34} \\
= \; & P(\mathsf{f}|\mathsf{g},\mathsf{m})^{10} P(\mathsf{g})^{10} P(\mathsf{m})^{10} \cdot P(\mathsf{f}|\bar{\mathsf{g}},\mathsf{m})^{8} P(\bar{\mathsf{g}})^{8} P(\mathsf{m})^{8} \cdot P(\bar{\mathsf{f}}|\bar{\mathsf{g}},\bar{\mathsf{m}})^{34} P(\bar{\mathsf{g}})^{34} P(\bar{\mathsf{m}})^{34} \\
= \; & P(\mathsf{f} \mid \mathsf{g},\mathsf{m})^{10} P(\bar{\mathsf{f}} \mid \mathsf{g},\mathsf{m})^{0} P(\mathsf{f} \mid \mathsf{g},\bar{\mathsf{m}})^{24} P(\bar{\mathsf{f}} \mid \mathsf{g},\bar{\mathsf{m}})^{16} \\
& \cdot P(\mathsf{f} \mid \bar{\mathsf{g}},\mathsf{m})^{8} P(\bar{\mathsf{f}} \mid \bar{\mathsf{g}},\mathsf{m})^{2} P(\mathsf{f} \mid \bar{\mathsf{g}},\bar{\mathsf{m}})^{6} P(\bar{\mathsf{f}} \mid \bar{\mathsf{g}},\bar{\mathsf{m}})^{34} \\
& \cdot P(\mathsf{g})^{50} P(\bar{\mathsf{g}})^{50} P(\mathsf{m})^{20} P(\bar{\mathsf{m}})^{80}.
\end{aligned}
$$

The last equation shows the principle by which the factors were reordered: First, sort by attribute (in the example F, G then M). Within an attribute, we group by parent value combinations (in the example, for F we do (g,m), $(\mathsf{g},\bar{\mathsf{m}})$, $(\bar{\mathsf{g}},\mathsf{m})$ and $(\bar{\mathsf{g}},\bar{\mathsf{m}})$). Finally, we sort by equal attribute values (in the example, for attribute F: first f, then $\bar{\mathsf{f}}$).

The general computation of the probability of a database D is as follows:

$$
P(D \mid B_S, B_P) = \prod_{i=1}^{n} \prod_{j=1}^{q_i} \prod_{k=1}^{r_i} \theta_{ijk}^{\alpha_{ijk}}. \tag{25.1}
$$

We saw that with known structure B_S we could estimate the parameters B_P from the database. This could lead us to the idea to use a maximum likelihood approach to infer the structure from data:

$$\widehat{B_S} = \operatorname*{argmax}_{B_S \in \mathcal{B}_R} P(D \mid B_S, B_P).$$

This approach, alas, has the drawback that the probability increases with the number of parameters which would lead to a model with maximal parameter number, that is, a fully connected graph B_S.

We can easily tackle this problem using a maximum a-posteriori estimator. The approach hence would be as follows:

$$\begin{aligned}
\widehat{B_S} &= \operatorname*{argmax}_{B_S} P(B_S \mid D) = \operatorname*{argmax}_{B_S} \frac{P(D \mid B_S)\,P(B_S)}{P(D)} \\
&= \frac{P(D, B_S)\,P(B_S)}{P(D)\,P(B_S)} = \operatorname*{argmax}_{B_S} \frac{P(B_S, D)}{P(D)} \\
&= \operatorname*{argmax}_{B_S} P(B_S, D).
\end{aligned}$$

Consequently, we seek a computational expression for the term $P(B_S, D)$. The result of the following derivation will be the K2 metric. An elaborate treatment can be found in Cooper and Herskovits (1992).

First, we consider $P(B_S, D)$ to be the marginalization of $P(B_S, B_P, D)$ over all possible parameters B_P which is known as model averaging. The number of parameters is fixed (finite number of entries in the potential tables of the attributes), however, the values of these entries—the θ_{ijk}—are continuous values which requires us to integrate over all models. Indeed, we deal with a multiple integral over all θ_{ijk} which is later made explicit.

$$\begin{aligned}
P(B_S, D) &= \int_{B_P} P(B_S, B_P, D)\,dB_P && (25.2) \\
&= \int_{B_P} P(D \mid B_S, B_P)\,P(B_S, B_P)\,dB_P && (25.3) \\
&= \int_{B_P} P(D \mid B_S, B_P)\,f(B_P \mid B_S)P(B_S)\,dB_P && (25.4) \\
&= \underbrace{P(B_S)}_{\text{A-priori prob.}} \int_{B_P} \underbrace{P(D \mid B_S, B_P)}_{\text{Prob. of data}}\ \underbrace{f(B_P \mid B_S)}_{\text{Parameter densities}}\ dB_P. && (25.5)
\end{aligned}$$

The a-priori distribution can be used to put a bias on certain classes of graph structures (e.g., by assigning a low probability to overly complex structures). Assuming that there is an underlying Bayes network, that the database cases are independent of each other and that the tuples are complete, we can apply Eq. (25.1):

$$P(B_S, D) = P(B_S) \int_{B_P} \left[\prod_{i=1}^{n} \prod_{j=1}^{q_i} \prod_{k=1}^{r_i} \theta_{ijk}^{\alpha_{ijk}} \right] f(B_P \mid B_S)\,dB_P.$$

The parameter densities $f(B_P \mid B_S)$ make a statement how probable the respective parameters B_P of a given network structure are. Hence, they are second-order densities as they represent densities about probability distributions. A vector $(\theta_{ij1}, \ldots, \theta_{ijr_i})$ is a probability distribution for fixed i and j (the jth column of the ith potential table, see Fig. 25.2). Assuming the densities of all columns of all attributes are mutually independent, we can simplify the expression for $f(B_P \mid B_S)$ to:

$$f(B_P \mid B_S) = \prod_{i=1}^{n} \prod_{j=1}^{q_i} f(\theta_{ij1}, \ldots, \theta_{ijr_i}).$$

Plugging this term into $P(B_S, D)$, we arrive at the following expression:

$$\begin{aligned}
&P(B_S, D)\\
&= P(B_S) \int \cdots \int_{\theta_{ijk}} \left[\prod_{i=1}^{n} \prod_{j=1}^{q_i} \prod_{k=1}^{r_i} \theta_{ijk}^{\alpha_{ijk}} \right] \cdot \left[\prod_{i=1}^{n} \prod_{j=1}^{q_i} f(\theta_{ij1}, \ldots, \theta_{ijr_i}) \right] d\theta_{111} \cdots d\theta_{nq_nr_n}\\
&= P(B_S) \prod_{i=1}^{n} \prod_{j=1}^{q_i} \int \cdots \int_{\theta_{ijk}} \left[\prod_{k=1}^{r_i} \theta_{ijk}^{\alpha_{ijk}} \right] \cdot f(\theta_{ij1}, \ldots, \theta_{ijr_i})\, d\theta_{ij1} \cdots d\theta_{ijr_i}.
\end{aligned}$$

The last simplifying assumption, again, regards the parameter densities. For fixed i and j, the density $f(\theta_{ij1}, \ldots, \theta_{ijr_i})$ be uniform. Hence, we get:

$$f(\theta_{ij1}, \ldots, \theta_{ijr_i}) = (r_i - 1)!$$

$$\begin{aligned}
P(B_S, D) &= P(B_S) \prod_{i=1}^{n} \prod_{j=1}^{q_i} \int \cdots \int_{\theta_{ijk}} \left[\prod_{k=1}^{r_i} \theta_{ijk}^{\alpha_{ijk}} \right] \cdot (r_i - 1)!\, d\theta_{ij1} \cdots d\theta_{ijr_i}\\
&= P(B_S) \prod_{i=1}^{n} \prod_{j=1}^{q_i} (r_i - 1)! \underbrace{\int \cdots \int_{\theta_{ijk}} \prod_{k=1}^{r_i} \theta_{ijk}^{\alpha_{ijk}}\, d\theta_{ij1} \cdots d\theta_{ijr_i}}_{\text{Dirichlet integral} = \frac{\prod_{k=1}^{r_i} \alpha_{ijk}!}{(\sum_{k=1}^{r_i} \alpha_{ijk} + r_i - 1)!}}.
\end{aligned}$$

Finally, we get the equation for $P(B_S, D)$ which is referred to as the K2 metric of the network structure B_S given the data D:

$$P(B_S, D) = \mathrm{K2}(B_S \mid D) = P(B_S) \prod_{i=1}^{n} \prod_{j=1}^{q_i} \left[\frac{(r_i - 1)!}{(N_{ij} + r_i - 1)!} \prod_{k=1}^{r_i} \alpha_{ijk}! \right]$$

$$\text{with } N_{ij} = \sum_{k=1}^{r_i} \alpha_{ijk}. \tag{25.6}$$

The parameter independences are two important properties of the K2 metric. They can be distinguished into *global* and *local* properties (see Heckerman et al. 1994, pp. 13f):

- *Global*—This property manifests itself in the outer product of the K2 formula: The product runs over all K2 values of the families of the attributes. It originates from the likelihood equation (25.1).
- *Local*—The Eq. (25.1) assumes the independence of the child attribute values given the parent attributes values. This is manifested in the product over all q_i different parent attributes' value combinations of the attribute A_i. The equation for the K2 metric also contains this product.

We exploit the global parameter independence to represent the K2 metric as follows:

$$\mathrm{K2}(B_S \mid D) = P(B_S) \prod_{i=1}^{n} \mathrm{K2}_{\mathrm{local}}(A_i \mid D) \quad \text{with}$$

$$\mathrm{K2}_{\mathrm{local}}(A_i \mid D) = \prod_{j=1}^{q_i} \left[\frac{(r_i - 1)!}{(N_{ij} + r_i - 1)!} \prod_{k=1}^{r_i} \alpha_{ijk}! \right].$$

The presented K2 metric computes for a given database D the quality of a network candidate B_S. We saw that the set of candidate graphs $\mathcal{B}_R$ for the given attribute set R becomes way too large to allow to consider each individual graph. Hence we need a heuristic how to check only a computationally feasible subset of $\mathcal{B}_R$.

The following algorithm uses the K2 metric as its evaluation metric and a greedy search as search heuristic. It further requires a topological ordering (see Definition 22.33 on page 429) on the attributes (nodes).

The search starts with an isolated graph, that is, with n isolated nodes. That is, the parent sets $\mathrm{pa}(A_i)$ are empty at the beginning. The indices $1 \le i \le n$ shall represent the topological ordering. The function q_i is defined as the local K2 value of node A_i given the parent set M:

$$q_i(M) = \mathrm{K2}_{\mathrm{local}}(A_i \mid D) \quad \text{with } \mathrm{pa}(A_i) = M.$$

The parent sets $\mathrm{pa}(A_i)$ for all nodes A_i are determined incrementally using the following heuristic:

1. Determine for a parentless node A_i the quality measure $q_i(\emptyset)$.
2. Then, all predecessors $\{A_1, \ldots, A_{i-1}\}$ are tested individually as potential parent nodes and the quality measure is recomputed. Let Y be the node that leads to the best quality:

$$Y = \operatorname*{argmax}_{1 \le l \le i-1} q_i\big(\{A_l\}\big)$$

   This best quality be $g = q_i(\{Y\})$.
3. If g is better than $q_i(\emptyset)$, the node Y is permanently added as a parent of A_i: $\mathrm{pa}(A_i) = \{Y\}$.
4. Steps 2 and 3 are repeated to augment the parent node set until no potential attributes are left, the quality cannot be increased or a certain maximal number of parent nodes is reached.

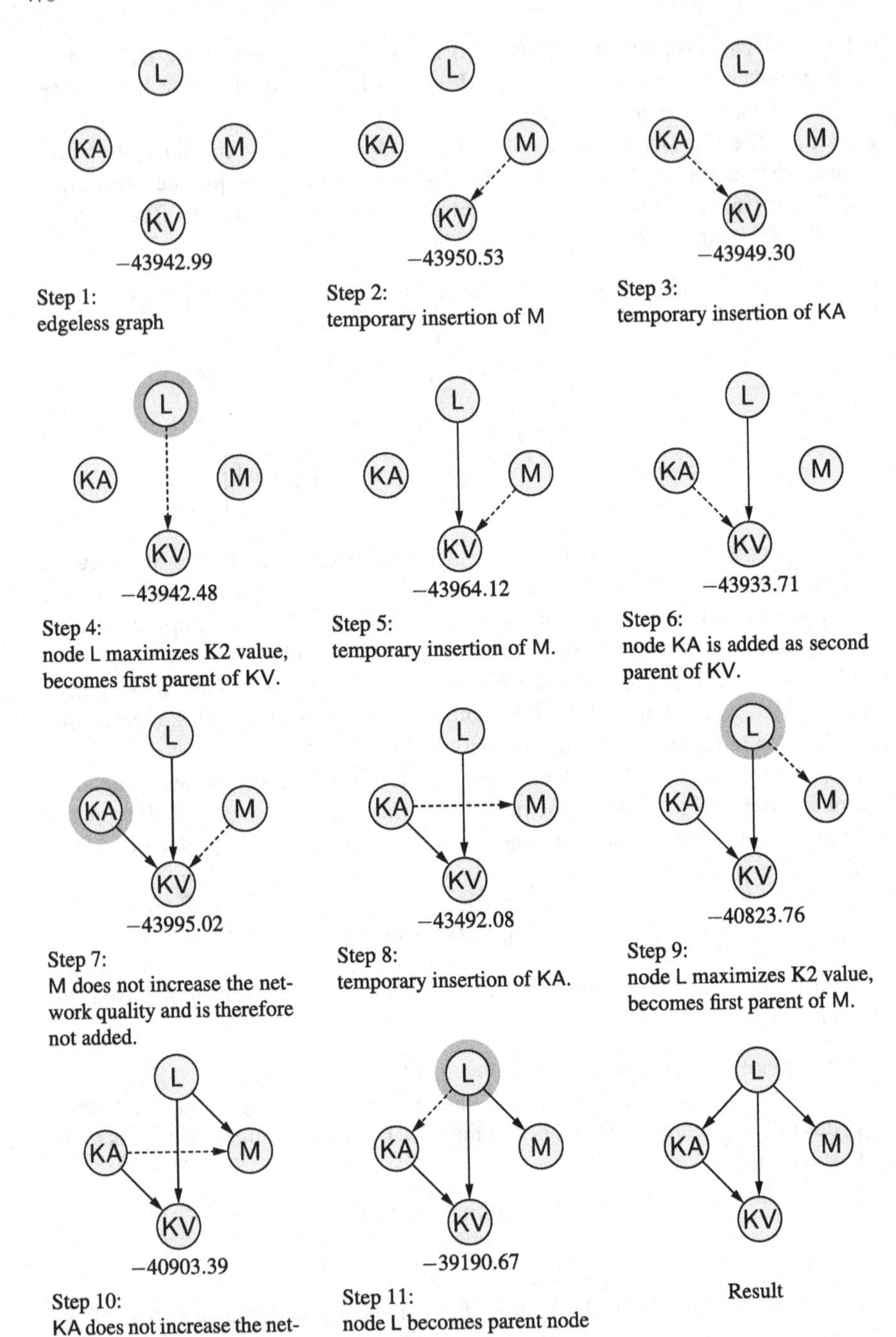

Fig. 25.3 An example run of the K2 algorithm. The topological ordering is L ≺ KA ≺ M ≺ KV

Algorithm 25.1 shows the K2 algorithm in pseudocode. Figure 25.3 shows a specific example.

Algorithm 25.1 (K2 Algorithm)
procedure K2;
begin
  for $i \leftarrow 1 \ldots n$ **do** (* initialization *)
    $\text{pa}(A_i) \leftarrow \emptyset$;
  for $i \leftarrow n \ldots 1$ **do begin** (* iteration *)
    repeat
      Choose $Y \in \{A_1, \ldots, A_{i-1}\} \backslash \text{pa}(A_i)$ which maximizes $g = q_i(\text{pa}(A_i) \cup \{Y\})$;
      $\delta \leftarrow g - q_i(\text{pa}(A_i))$;
      if $\delta > 0$ **then** $\text{pa}(A_i) \leftarrow \text{pa}(A_i) \cup \{Y\}$; **end**
    until $\delta \leq 0$ **or** $\text{pa}(A_i) = \{A_1, \ldots, A_{i-1}\}$ **or** $|\text{pa}(A_i)| = n_{\max}$;
  end
end

References

G.F. Cooper and E. Herskovits. A Bayesian Method for the Induction of Probabilistic Networks from Data. *Machine Learning* 9:309–347. Kluwer, Dordrecht, Netherlands, 1992

D. Heckerman, D. Geiger, and D.M. Chickering. *Learning Bayesian Networks: The Combination of Knowledge and Statistical Data*, MSR-TR-94-09. Microsoft Research, Advanced Technology Division, Redmond, WA, USA, 1994

R.W. Robinson. Counting Unlabeled Acyclic Digraphs. In: C.H.C. Little (ed.) *Combinatorial Mathematics V*. LNMA 622:28–43. Springer-Verlag, Heidelberg, Germany, 1977

Index

R. Kruse et al., *Computational Intelligence*, Texts in Computer Science,
DOI 10.1007/978-1-4471-5013-8, © Springer-Verlag London 2013